then be easily retrieved via consultation of Chemical Abstracts. A desired side-effect is to familiarize the student with author's names and their fields of endeavor. The many coworkers, who actually did the work, may forgive us that only the name of the respective boss is given.

Among our own coworkers who helped to bring this English Edition to completion, the native speakers Pamela Alean (Great Britain, now a resident of Zürich, Switzerland) and James Hurley (USA; resident of Marburg, Germany) stand out. They went a long way to eliminate our worst excesses of "Gerglish". The bulk of the structural formulae was drawn by one of the authors (A.S.) thereby keeping things in the right perspective and making the book easy to use. Monika Scheld, Marburg, helped with the preparation of the indexes and by checking the cross references. We are grateful to the editor Dr. Michael Weller and the production manager Bernd Riedel (both of VCH Publishers) for a pleasant form of cooperation and their toleration of several last-minute changes. Finally, the authors mutually acknowledge their unflagging support during the various stages of the enterprise.

Ch. Elschenbroich
Marburg
Germany

March
1989

A. Salzer
Zürich
Switzerland

Ch. Elschenbroich, A. Salzer

Organometallics

Distribution:
VCH, P.O. Box 101161, D-6940 Weinheim (Federal Republic of Germany)

Switzerland: VCH, P.O. Box, CH-4020 Basel (Switzerland)

United Kingdom and Ireland: VCH (UK) Ltd., 8 Wellington Court, Cambridge CB1 1HZ (England)

USA and Canada: VCH, Suite 909, 220 East 23rd Street, New York, NY 10010-4606 (USA)

Softcover ISBN 3-527-28164-9 (VCH, Weinheim) ISBN 0-89573-983-6 (VCH, New York)
Hardcover ISBN 3-527-28165-7 (VCH, Weinheim) ISBN 0-89573-984-4 (VCH, New York)

Christoph Elschenbroich
Albrecht Salzer

Organometallics

A Concise Introduction

Second, Revised Edition

Weinheim · New York · Basel · Cambridge

Publisher of the German edition:
B. G. Teubner, Stuttgart, 1991: Elschenbroich/Salzer: Organometallchemie, 3. Aufl.

Prof. Dr. Christoph Elschenbroich
Fachbereich Chemie
Philipps-Universität Marburg
Hans-Meerwein-Str.
D-3550 Marburg

Priv.-Doz. Dr. Albrecht Salzer
Anorgan.-chemisches Institut
Universität Zürich
Winterthurer Str. 190
CH-8057 Zürich

2nd edition 1992

Published jointly by
VCH Verlagsgesellschaft mbH, Weinheim (Federal Republic of Germany)
VCH Publishers Inc., New York, NY (USA)

Editorial Director: Karin von der Saal
Production Manager: Dipl.-Wirt.-Ing. (FH) Bernd Riedel
Cover design: TWI, H. Weisbrod, D-6943 Birkenau

Library of Congress Card No. applied for

A catalogue record for this book is available from the British Library

Die Deutsche Bibliothek – CIP-Einheitsaufnahme

Elschenbroich, Christoph:
Organometallics : a concise introduction / Christoph Elschenbroich ; Albrecht Salzer. – 2., rev. ed. – Weinheim ; New York ; Basel ; Cambridge : VCH, 1992
Einheitssacht.: Organometallchemie ⟨engl.⟩
ISBN 3-527-28164-9 (Weinheim ...)
ISBN 0-89573-983-6 (New York)
Ne: Salzer, Albrecht:

Printed on acid-free and low-chlorine paper.

Composition: Graphischer Betrieb Konrad Triltsch, D-8700 Würzburg
Printing: Graphischer Betrieb Konrad Triltsch, D-8700 Würzburg
Bookbinding: Graphischer Betrieb Konrad Triltsch, D-8700 Würzburg

Printed in the Federal Republic of Germany

Preface to the Second Edition

To revise, according to the Oxford American Dictionary, means to reexamine and alter or correct. That is what we have done in this Second Edition which is characterized by elimination of errors, substitution of old examples with new ones, which illustrate certain statements more effectively, and addition of recent findings which are apt to become landmarks. From the last category, the following entries should be mentioned: the fundamental new main-group element clusters R_4B_4, $R_{12}Al_{12}^{2-}$, and $Cp_4^* Al'_4$, the heterocycle tellurepin; the first stable carbonyl complex of silver, Ag(CO) $[B(OTeF_5)_4]$; the coordination of buckminsterfullerene to platinum; the unsupported Co=Co double bond in Cp*Co=CoCp*; and the stabilization of the disilylene Me_4Si_2 through coordination to tungsten (the cover molecule). Further important additions include mention of homogeneous Ziegler-Natta type catalysts based on Cp_2ZrMe^+ derivatives, the prescription of how to arrive at the "magic numbers" which are supposed to govern cluster geometries, an extension of the section on C–H activation, and a very brief treatment of organolanthanoid chemistry, which concludes the final chapter of the book.
All in all, we have made more than a hundred small changes, overt and covert, which may escape immediate detection since we have kept the format of the book virtually unchanged (A Concise Introduction). However, in order to lure the reader into looking up further details of interesting results, we now give in the author index the complete literature citations of work referred to in the text.
Once again, we thank our publisher for splendid cooperation, in particular Karin von der Saal, who in her gentle but persistent way helped to complete the job in time.
Apart from ourselves, no one read the manuscript; so nobody else is to blame for the remaining errors and omissions.

Ch. Elschenbroich
Marburg
Germany

November
1991

A. Salzer
Zürich
Switzerland

Preface to the First Edition

The present volume is the translation of the Second Edition (1988) of our text "Organometallchemie – Eine kurze Einführung"; corrections and a few results of very recent origin were included but otherwise the body was left unchanged.

Can a 500 page treatise on a branch of chemistry still be called "concise"? On the other hand, a section of only 20 pages covering transition-metal olefin complexes certainly must be regarded as short. This contrast illustrates the dilemma encountered if one sets out to portray the whole of organometallic chemistry in a single volume of tolerable size.

The book developed from an introductory course (one semester, about 30 lectures) on organometallic chemistry for students confronted with the field for the first time. The material covered is a mixture of indispensible basic facts and selected results of most recent vintage. Attempts to systematize organometallic chemistry by relating molecular structures to the number and nature of the valence electrons are presented as are applications of organometallics in organic synthesis and in industrial processes based on homogeneous catalysis.

An apparent omission is the absence of a chapter specifically dealing with organometallic reaction mechanisms. It is our contention, however, that mechanistic organometallic chemistry has not yet reached the stage which would warrant a short overview from which useful generalizations could be drawn by the beginner. Note, for example, that even reactions as fundamental as metal carbonyl substitution are currently under active investigation, the intermediacy of 17 or 19 valence electron species opening up new possible pathways. Interspersed within the text, however, the reader finds several comments and mechanistic proposals ranging from well established kinetic studies to catalysis loops which at times have more the character of mnemotechnic devices than of kinetic schemes based on experimental evidence. Detailed mechanistic considerations should be deferred to the second act of the study of organometallic chemistry and several textbooks, mainly concentrating on organotransition metal compounds, offer a wealth of material with which to pursue this goal.

We have structured the text in the traditional way – following the periodic table for main-group element organometallics and according to the nature of the ligand for transition-metal complexes – which we find most suitable for an introduction. Apart from the Chapters 16 and 17 (Metal-metal bonds, clusters, catalysis) somewhat more specialized material is presented in sections called "Excursions". Rigorous scientific referencing would be inappropriate in a text of the present scope. At the end, a literature survey (300 odd entries) is given which leads the reader to important review articles and key papers, including several classics in the field. Furthermore, in the running text authors names are linked to the facts described whereby the form (Author, year) designates the year of the discovery, usually in a short communication, and the form (Author, year R) the appearance of the respective full paper or review. The complete citation can

Contents

Preface

ORGANOMETALLIC COMPOUNDS OF THE TRANSITION ELEMENTS

Introduction

1 Historical Development and Current Trends in Organometallic Chemistry

1760 The cradle of organometallic chemistry is a Paris military pharmacy. It is there that Cadet works on sympathetic inks based on cobalt salts. For their preparation, he uses cobalt minerals which contain arsenic.

$$As_2O_3 + 4\,CH_3COOK \longrightarrow \text{“Cadet's fuming liquid”}$$

contains cacodyloxide $[(CH_3)_2As]_2O$
($\kappa\alpha\kappa\omega\delta\eta\varsigma$ = malodorous)
first organometallic compound

1827 Zeise's salt $Na[PtCl_3C_2H_4]$, **first olefin complex**

1840 R. W. Bunsen continues the study of cacodyl compounds which he names "alkarsines". The weakness of the As–As bond in molecules of the type $R_2As{-}AsR_2$ led to a profusion of derivatives like $(CH_3)_2AsCN$ whose taste (!) is checked by Bunsen.

1849 E. Frankland, student of Bunsen's at Marburg, attempts the preparation of an "ethyl radical" (cacodyl as well was taken to be a radical).

$$3\,C_2H_5I + 3\,Zn \not\longrightarrow ZnI_2 + 2\,C_2H_5$$

$$3\,C_2H_5I + 3\,Zn \longrightarrow (C_2H_5)_2Zn \text{ (a pyrophoric liquid)} + C_2H_5ZnI \text{ (a solid)} + ZnI_2$$

Frankland possesses admirable skill in the manipulation of air-sensitive compounds. As a protective atmosphere he uses hydrogen gas!

1852 Frankland prepares the important alkylmercury compounds:

$$2\,CH_3X + 2\,Na/Hg \longrightarrow (CH_3)_2Hg + 2\,NaX$$

additionally: $(C_2H_5)_4Sn$, $(CH_3)_3B$ (1860).
In the following years, **alkyl transfer reactions** using R_2Hg and R_2Zn serve in the synthesis of numerous main-group organometallics.
Frankland also introduced the concept of valency ("combining power") and the term organometallic.

1852 C. J. Löwig and M. E. Schweizer in Zürich first prepare $(C_2H_5)_4Pb$ from ethyliodide and Na/Pb alloy. In a similar manner, they also obtain $(C_2H_5)_3Sb$ and $(C_2H_5)_3Bi$.

1859 W. Hallwachs and A. Schafarik generate alkylaluminum iodides:

$$2\,Al + 3\,RI \longrightarrow R_2AlI + RAlI_2$$

1863 C. Friedel and J. M. Crafts prepare **organochlorosilanes:**

$$SiCl_4 + m/2\ ZnR_2 \longrightarrow R_mSiCl_{4\text{-}m} + m/2\ ZnCl_2$$

1866 J. A. Wanklyn develops a method for the synthesis of halide-free magnesium alkyls:

$$(C_2H_5)_2Hg + Mg \longrightarrow (C_2H_5)_2Mg + Hg$$

1868 M. P. Schützenberger obtains $[Pt(CO)Cl_2]_2$, **first metal carbonyl complex.**

1871 D. I. Mendeleev uses organometallic compounds as test cases for his periodic table. Example:

Known:	**predicted:**	**found:**	
$Si(C_2H_5)_4$	Eka-$Si(C_2H_5)_4$	$Ge(C_2H_5)_4$	(C. Winkler, 1887)
	$d = 0.96$	$d = 0.99$	
$Sn(C_2H_5)_4$	bp: 160 °C	bp: 163.5 °C	

1890 L. Mond: $Ni(CO)_4$, **first binary metal carbonyl,** used in a commercial process for refining nickel. Mond is the founder of the English company ICI (Imperial Chemical Industries) as well as a renowned collector and patron of the arts.

1899 P. Barbier replaces Mg for Zn in reactions with alkyl iodides:

$$(H_3C)_2C{=}CH{-}CH_2{-}C({=}O)CH_3 \xrightarrow[2.\ H_2O]{1.\ CH_3I,\ Mg} (H_3C)_2C{=}CH{-}CH_2{-}C(OH)(CH_3){-}CH_3$$

explored in more detail by Barbier's student V. Grignard (Nobel prize 1912 shared with P. Sabatier). Although less sensitive than ZnR_2, RMgX is a more potent alkyl group transfer reagent.

1901 L. F. S. Kipping prepares $(C_6H_5)_2SiO$, suspects its high molecularity, and calls the material **diphenylsilicone**.

1909 W. J. Pope: formation of $(CH_3)_3PtI$, **first σ-organotransition-metal compound.**

1909 P. Ehrlich (inventor of chemotherapy, Nobel prize 1908) introduces Salvarsan for the treatment of syphilis.

1917 W. Schlenk: Lithium alkyls via transalkylation.

$$2\,Li + R_2Hg \longrightarrow 2\,LiR + Hg$$
$$2\ EtLi + Me_2Hg \longrightarrow 2\,MeLi + Et_2Hg$$

1919 F. Hein from $CrCl_3$ and PhMgBr synthesizes polyphenylchromium compounds, now known to be sandwich complexes.

1922 T. Midgley and T. A. Boyd introduce $Pb(C_2H_5)_4$ as an antiknock additive in gasoline.

1928 W. Hieber inaugurates his systematic study of metal carbonyls:

$$Fe(CO)_5 + H_2NCH_2CH_2NH_2 \longrightarrow (H_2NCH_2CH_2NH_2)Fe(CO)_3 + 2\,CO$$
$$Fe(CO)_5 + X_2 \longrightarrow Fe(CO)_4X_2 + CO$$

1929 F. A. Paneth generates alkyl radicals through PbR_4 pyrolysis, radical identification by means of their ability to cause the transport of a metallic mirror. Paneth thus reaches a goal set by Frankland in 1849.

1930 K. Ziegler encourages more extensive use of organolithium compounds in synthesis by developing a simpler way of preparation.

$PhCH_2OMe + 2\,Li \longrightarrow PhCH_2Li + MeOLi$ (ether cleavage)
H. Gilman: $RX + 2\,Li \longrightarrow RLi + LiX$ (procedure used today)

1931 W. Hieber prepares $Fe(CO)_4H_2$, **first transition-metal hydride complex.**

1938 O. Roelen discovers **hydroformylation** (the oxo process).

1939 W. Reppe starts work on the transition-metal catalyzed reactions of acetylenes.

1943 E. G. Rochow: $2\,CH_3Cl + Si \xrightarrow{\text{Cu-cat., 300 °C}} (CH_3)_2SiCl_2 + \ldots$
This "direct synthesis" triggers large scale production and use of **silicones**. Preliminary work by R. Müller (Radebeul near Dresden) was interrupted by the Second World War.

1951 P. Pauson (GB) and S. A. Miller (USA) obtain ferrocene $(C_5H_5)_2Fe$, **first sandwich complex.**

1953 G. Wittig discovers the reaction bearing his name.

1955 E. O. Fischer: rational synthesis of bis(benzene)chromium $(C_6H_6)_2Cr$.

1955 K. Ziegler, G. Natta: **polyolefins** from ethylene and propylene, respectively, in a **low pressure process** employing mixed metal catalysts (transition-metal halide/AlR_3).

1956 H. C. Brown: **hydroboration**.

1959 J. Smidt, W. Hafner: preparation of $[(C_3H_5)PdCl]_2$, installation of the field of π-allyl transition-metal complexes.

1959 R. Criegee: stabilization of cyclobutadiene by complexation in $[(C_4Me_4)NiCl_2]_2$ verifying a prediction by H. C. Longuet-Higgins and L. Orgel (1956).

1960 M. F. Hawthorne: **carboranes**.

1961 L. Vaska: $(PPh_3)_2Ir(CO)Cl$ reversibly binds O_2.

1963 Nobel prize to K. Ziegler and G. Natta.

1964 E. O. Fischer: $(CO)_5WC(OMe)Me$, first **carbene complex**.

1965 G. Wilkinson, R. S. Coffey: $(PPh_3)_3RhCl$ acts as a homogeneous catalyst in the hydrogenation of alkenes.

1968 A. Streitwieser: preparation of uranocene, $(C_8H_8)_2U$.

1969 P. L. Timms: synthesis of organotransition-metal complexes by means of metal-atom ligand-vapor cocondensation.

1970 G. Wilkinson: kinetically inert transition-metal alkyls through blockage of β-elimination.

1972 H. Werner: $[(C_5H_5)_3Ni_2]^+$, first **triple-decker sandwich complex**.

1973 E. O. Fischer: $I(CO)_4Cr(CR)$, first **carbyne complex**.

1973 Nobel prize to E. O. Fischer and G. Wilkinson.

1976 Nobel prize to W. N. Lipscomb: theoretical and experimental clarification of structure and bonding in boranes.

1979 Nobel prize to H. C. Brown and G. Wittig: applications of organoboranes and methylenephosphoranes, respectively, in organic synthesis.

1981 R. West: $(1,3,5\text{-}Me_3C_6H_2)_4Si_2$, first stable compound with a $>Si{=}Si<$ **double bond**.

1981 Nobel prize to R. Hoffmann and K. Fukui: semiempirical MO-concepts in a unified discussion of structure and reactivity of inorganic, organic and organometallic molecules, **isolobal analogies**.

1983 R. G. Bergman, W. A. G. Graham: intermolecular reactions of organotransition-metal compounds with alkanes **(C–H activation)**.

Current Trends

① Participation of higher main-group elements in $p_\pi - p_\pi$ multiple bonds

Examples:

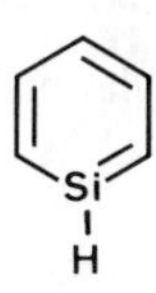

Siline, accessible in the gasphase and in matrix isolation

Arsaalkyne $R = t$-Bu

Distibene $Ar = 1,3,5\text{-}(t\text{-}Bu)_3C_6H_2$

② Stabilization of highly reactive species by metal coordination

Examples: cyclobutadiene, aryne, carbyne, thiocarbonyl, selenoformaldehyde.

$(Me_3CO)_3W \equiv C-Me$

③ Metallacycles

Examples: $(AgCl)(CR)Os(CO)(PR_3)_2Cl$ $Cp_2W-CH_2-CH_2-CH_2$

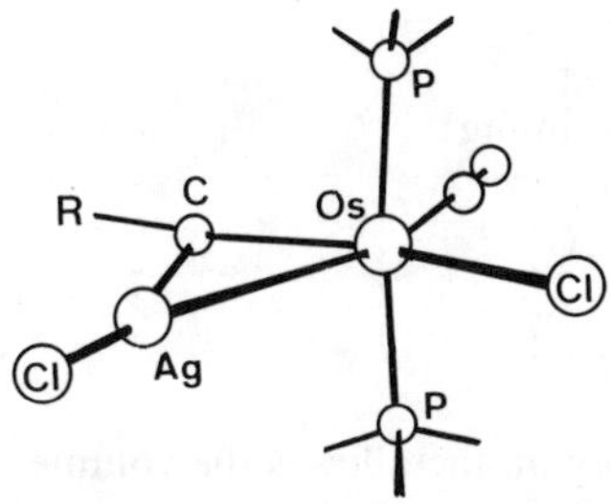

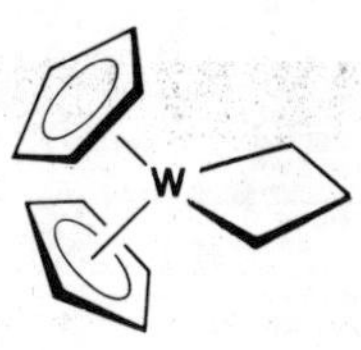

④ Study of organometallic clusters

Example: $[Me_3NCH_2Ph]_5[Ni_{38}Pt_6(CO)_{48}H]$

Metal skeleton of the cluster anion (X-ray diffraction)

Electron micrograph of the supported material

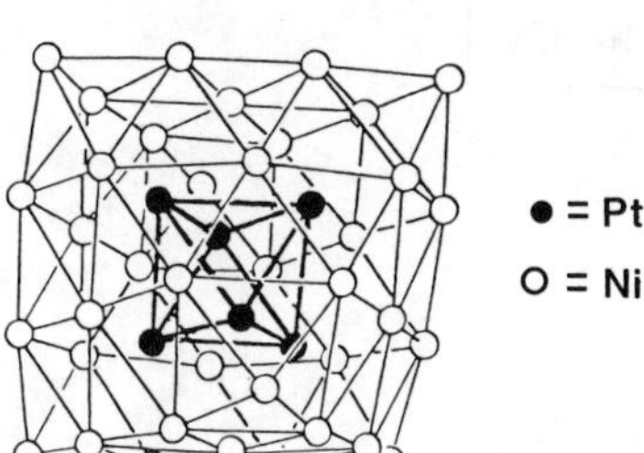

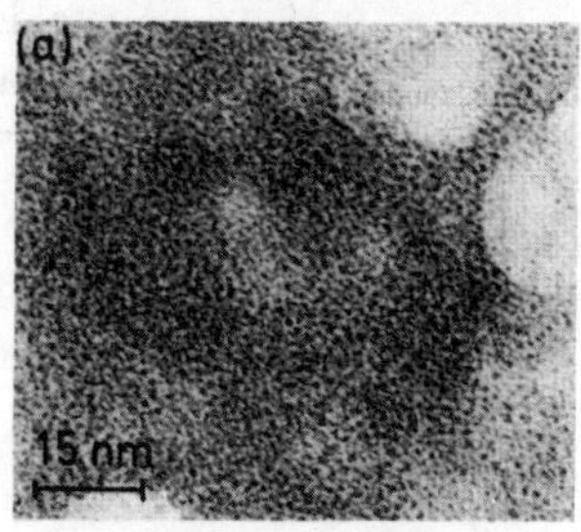

⑤ Catalysis via organometallic intermediates

Examples of *industrial processes*: technical improvements in the hydroformylation reaction (oxo process), developments in the areas of Fischer-Tropsch chemistry, alkene metathesis and C–H activation.

Example of an *organic synthesis*: PdL_4-catalyzed enantioselective allylic alkylation (transfer of chirality).

95% diastereomeric purity:

⑤ profits from fundamental work carried out on the topics ②–④.

Production figures of organometallic compounds:

1. Silicones	700 000 t/a
2. Pb-alkyls	600 000 t/a (declining)
3. Organo-Al	50 000 t/a
4. Organo-Sn	35 000 t/a
5. Organo-Li	900 t/a

Even more impressive than the figures for main-group organometallics is the volume of organic intermediates and of polymers which are produced through the use of organo-transition-metal catalysis:

1. Polypropylene	7.7×10^6 t/a
2. "Oxo products"	5.0×10^6 t/a
3. Acetaldehyde	2.2×10^6 t/a
4. Acetic acid	1.0×10^6 t/a

2 Demarcation and Classification of Organometallic Compounds

Organometallic compounds (metal organyls, organometallics) are defined as materials which possess direct, more or less polar bonds $M^{\delta+} - C^{\delta-}$ between metal and carbon atoms. In many respects, the organic chemistry of the elements B, Si, P and As resembles the chemistry of the respective metallic homologues. Therefore, the term organoelement compounds is used occasionally in order to include for consideration the aforementioned non- and semi-metals. A convenient classification of organometallics is based on the bond type:

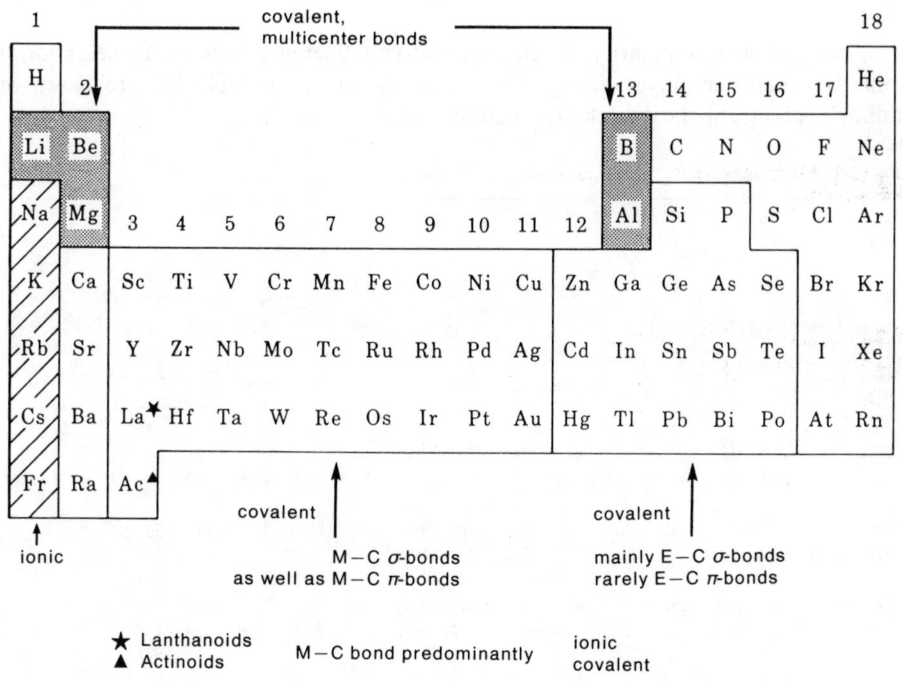

The designation σ-, π-, δ-bond is defined as follows:

Overlap	Number of nodal Planes including the Bond Axis	Bond Type	Example
	0	σ	$>B-CH_3$
	1	π	$(CO)_5Cr{=}CR_2$
	2	δ	$[R_4Re{\equiv}ReR_4]^{2-}$

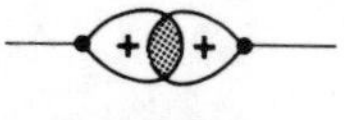

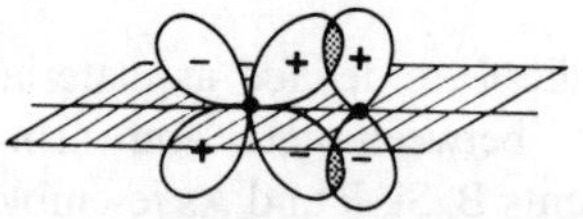

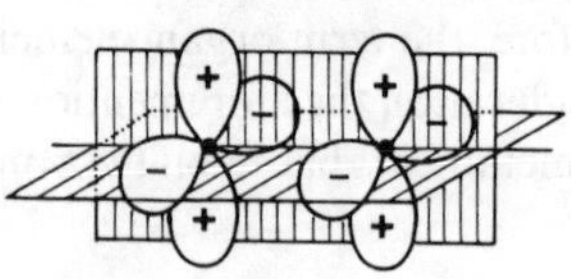

In evaluations of bond polarity, the electronegativity difference between the neighboring atoms is usually employed. The electronegativity values in table 2-1 are based on Pauling's thermochemical method of determination:

Table 2-1: Electronegativity values according to Pauling

H 2.2																
Li 1.0	Be 1.6											B 2.0	C 2.5	N 3.0	O 3.4	F 4.0
Na 0.9	Mg 1.3											Al 1.6	Si 1.9	P 2.2	S 2.6	Cl 3.1
K 0.8	Ca 1.0	Sc 1.3	Ti 1.5	V 1.6	Cr 1.6	Mn 1.6	Fe 1.8	Co 1.9	Ni 1.9	Cu 1.9	Zn 1.7	Ga 1.8	Ge 2.0	As 2.2	Se 2.6	Br 2.9
Rb 0.8	Sr 1.0	Y 1.2	Zr 1.3	Nb 1.6	Mo 2.1	Tc 1.9	Ru 2.2	Rh 2.3	Pd 2.2	Ag 1.9	Cd 1.7	In 1.8	Sn 1.8	Sb 2.0	Te 2.1	I 2.6
Cs 0.8	Ba 0.9	La 1.1	Hf 1.3	Ta 1.5	W 2.3	Re 1.9	Os 2.2	Ir 2.2	Pt 2.3	Au 2.5	Hg 2.0	Tl 1.6	Pb 1.9	Bi 2.0	Po 2.0	At 2.2
		Lanthanoids:			1.1–1.3											
		Actinoids:			1.1–1.3											

Source: L. Pauling, *The Nature of the Chemical Bond, 3. Ed.*, Ithaca (1960); A. L. Allred, *J. Inorg. Nucl. Chem.* *17* (1961) 215.

In accordance with the similar electronegativities of carbon and hydrogen, the division ionic/covalent of organoelement compounds bears a strong resemblance to the classification of element hydrides.

Contrary to the element hydrides, however, the dependence of the electronegativity of carbon EN (C) on the hybridization ratio has to be taken into account in organoelement compounds. Since s electrons are exposed to a stronger effective nuclear charge than p electrons of the same principal quantum number, EN (C) increases with increasing s character in the hybrid: the value $EN(C_{sp^3}) = 2.5$ *applies to* sp^3 *hybridized carbon, whereas for cases of higher s character the values* $EN(C_{sp^2}) = 2.75$ *and* $EN(C_{sp}) = 3.29$ *have been proposed (Bent, 1960). This gradation also reflects the increase in CH acidity according to* $C_2H_6 < C_2H_4 \ll C_2H_2$ *and suggests that the M–C bond is considerably more polar in alkynylmetal complexes (p. 208) than in metal alkyls. Furthermore, it should not be overlooked that in compounds of the type* L_nM-CR_3*, the electronegativity of the metal EN (M) will be modified by the ligands L and the electronegativity of carbon EN (C) by the groups R. A lucid discussion of the concept of electronegativity and its applications is found in J. E. Huheey, Inorganic Chemistry – Principles of Structure and Reactivity, New York: Harper and Row,* 1983.

By way of generalization, it may be stated that the chemistry of main-group organometallics is governed by the group the metal belongs to, whereas for organotransition-metal compounds, the nature of the ligand dominates. Consequently, the material in chapters 4–11 is arranged in conformity with the periodic table, while that of chapters 12–15 is presented according to the types of ligand.

3 Energy, Polarity and Reactivity of the M–C Bond

In discussions of the properties of organometallics it is important to distinguish between **thermodynamic (stable, unstable)** and **kinetic (inert, labile)** factors.
M–C single bonds are encountered throughout the periodic table (examples: $MgMe_2$, PMe_3, MeBr, $[LaMe_6]^{3-}$, WMe_6). For organotransition-metal compounds special rules apply which are derived from the large number of valence orbitals and the higher tendency of transition-metal atoms to engage in multiple bonding (see chapter 16).

***Typical* M–C *bond lengths d in pm* and calculated *covalent radii r* for main-group elements,** $r = d - r_{carbon} = d - 77$.

Group											
2, 12			13			14			15		
M	*d*	*r*	*M*	*d*	*r*	*M*	*d*	*r*	*M*	*d*	*r*
Be	179	102	B	156	79	C	154	77	N	147	70
Mg	219	142	Al	197	120	Si	188	111	P	187	110
Zn	196	119	Ga	198	121	Ge	195	118	As	196	119
Cd	211	134	In	223	146	Sn	217	140	Sb	212	135
Hg	210	133	Tl	225	148	Pb	224	147	Bi	226	149

Source: *Comprehensive Organometallic Chemistry* (*COMC*) ***1*** (1982) 10

3.1 Stability of Main-group Organometallics

Compared with the strengths of M–N, M–O and M–Hal bonds, **M–C bonds** must be deemed **weak**. This bond weakness is reflected in the uses that organometallics find in synthesis. Since standard entropies are seldom known for organometallic compounds, enthalpies of formation ΔH_f^0 are often used in place of free enthalpies of formation ΔG_f^0, when evaluating thermodynamic stabilities. A decisive factor in the low ΔH_f^0 values of organometallics is the high bond energy of the constituent elements (M, C, H) in their respective standard states (see table 3-1).
A limitation in the use of mean bond enthalpies $\bar{E}$(M–C) when assessing the reactivity of organometallics is the fact that stepwise bond dissociation energies $D_{1,2\ldots n}$ may deviate strongly from the mean value $\bar{E} = 1/n \sum_i^n D_i$. As an example dimethylmercury may be considered:

$$(CH_3)_2Hg \longrightarrow CH_3Hg + CH_3 \qquad D_1(CH_3Hg-CH_3) = 214\ \text{kJ/mol}$$
$$CH_3Hg \longrightarrow Hg + CH_3 \qquad D_2(CH_3-Hg) = 29\ \text{kJ/mol}$$

Table 3-1: Comparison of standard enthalpies ΔH_f^0 in kJ/mol and mean bond enthalpies $\bar{E}$(M−C) in kJ/mol of methyl derivatives in the gas phase with values $\bar{E}$(M−X), X = Cl, O

Group											
12 MMe_2			13 MMe_3			14 MMe_4			15 MMe_3		
M	ΔH_f^0	$\bar{E}$	*M*	ΔH_f^0	$\bar{E}$	*M*	ΔH_f^0	*E*	*M*	ΔH_f^0	$\bar{E}$
			B	−123	365	C	−167	358	N	−24	314
			Al	−81	274	Si	−245	311	P	−101	276
Zn	50	177	Ga	−42	247	Ge	−71	249	As	13	229
Cd	106	139	In	173	160	Sn	−19	217	Sb	32	214
Hg	94	121	Tl	—	—	Pb	136	152	Bi	194	141
c.f.			B−O		526	Si−O		452	As−O		301
			B−Cl		456	Si−Cl		381	Bi−Cl		274
			Al−O		500	Si−F		565			
			Al−Cl		420	Sn−Cl		323			

Data for M−C: *Comprehensive Organometallic Chemistry, 1* (1982) 5
Data for M−X: J. E. Huheey, *Inorganic Chemistry*, 3. Ed., A-32

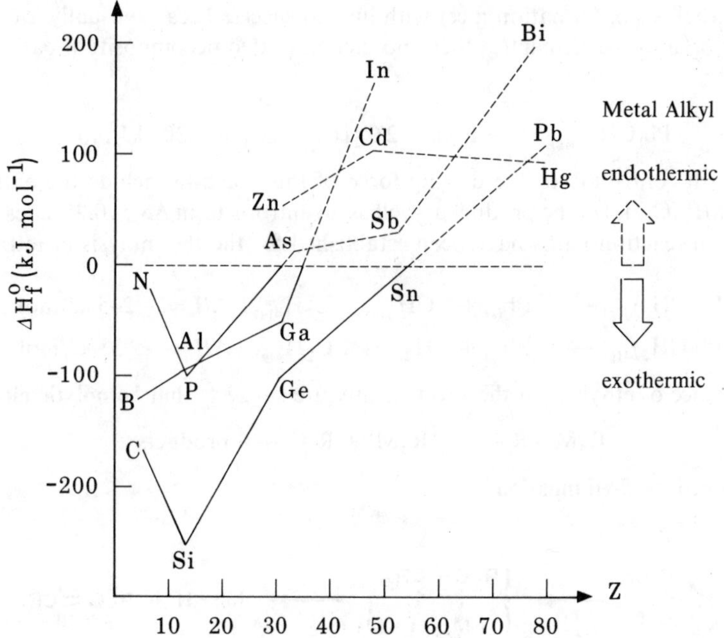

Generalizations:

- **M – C bond energies cover a wide range**

Compound	$(CH_3)_3B$	$(CH_3)_3As$	$(CH_3)_3Bi$
$\bar{E}$(M – C) kJ/mol	365	229	141
bond	relatively strong	medium	weak

- The **mean bond energy** $\bar{E}$(M – C) within a main-group decreases with increasing atomic number. This trend also applies to the bonds of M to other elements of the second row. A rationale for this effect is the increasing disparity in the radial extension and concomitant unfavorable overlap of the atomic orbitals contributing to the M – C bond.
- **Ionic bonds** are encountered if M is particularly electropositive and/or the carbanion is especially stable.
  Examples:
  $$Na^+[C_5H_5]^-,\ K^+[CPh_3]^-,\ Na^+[C{\equiv}CR]^-.$$
- **Multicenter bonding ("electron deficient bond")** arises if the valence shell of M is less than half filled and the cation M^{n+} is strongly polarizing (possesses a large charge/radius ratio).
  Examples:
  $$[LiCH_3]_4,\ [Be(CH_3)_2]_n,\ [Al(CH_3)_3]_2\ (\text{but } K^+[C_nH_{2n+1}]^-).$$

3.2 Lability of Main-group Organometallics

Predictions of the thermal behavior of organometallics which are based on the respective standard enthalpies of formation meet with limited success because usually, rather than decomposition into the elements, other, more complicated decomposition pathways are observed.

Examples:

$$Pb(CH_3)_{4(g)} \longrightarrow Pb_{(s)} + 2\,C_2H_{6(g)} \quad \Delta H = -307\ \text{kJ/mol} \tag{1}$$

Factors which contribute to the driving force of this reaction include the enthalpy of formation $\Delta H_f^0(C_2H_6)$ of the product as well as an entropy term $\Delta S > 0$. Besides reaction (1), additional reaction paths have been established for the thermolysis of tetramethyllead:

$$Pb(CH_3)_{4(g)} \longrightarrow Pb_{(s)} + 2\,CH_{4(g)} + C_2H_{4(g)} \quad \Delta H = -235\ \text{kJ/mol} \tag{2}$$

$$Pb(CH_3)_{4(g)} \longrightarrow Pb_{(s)} + 2\,H_{2(g)} + 2\,C_2H_{4(g)} \quad \Delta H = -33\ \text{kJ/mol} \tag{3}$$

The appearance of ethylene in the product mixture suggests that **homolytic cleavage**

$$R_3M{-}R \longrightarrow \{R_3M^{\cdot} + R^{\cdot}\} \longrightarrow \text{products} \tag{4}$$

is accompanied by ***β*-elimination**

$$\overset{\alpha\quad\beta}{\begin{matrix} R_2C - CR_2 \\ | \qquad | \\ M \qquad H \end{matrix}} \longrightarrow \left\{\begin{matrix} R_2C - CR_2 \\ \vdots \qquad \vdots \\ M \cdots H \end{matrix}\right\} \longrightarrow M{-}H + R_2C{=}CR_2 \tag{5}$$

The concerted nature of decomposition pathway (5) entails a lowering of the activation energy. However, this path is limited to molecules with hydrogen atoms in β-position. It is plausible, therefore, that the temperature of decomposition is higher for $Pb(CH_3)_4$ than for $Pb(C_2H_5)_4$ (p. 139).
A further condition for β-elimination to occur is the availability of an empty valence orbital on M to interact with the electron pair of the $C_\beta-H$ bond. It is for this reason that the β-elimination mechanism plays a more important role for the organometallics of groups 1, 2 and 13 (valence configurations s^1, s^2 and $s^2 p^1$) than for those of groups 14, 15 and 16 ($s^2 p^2$, $s^2 p^3$ and $s^2 p^4$). If a binary organometallic species has an empty coordination site at its disposal, β-elimination can be blocked and thermal stability increased through formation of a Lewis base adduct [Example: (bipy)$Be(C_2H_5)_2$, bipy = 2,2′-bipyridyl]. β-Elimination assumes a position of prime importance in the chemistry of organotransition-metal compounds (chapter 13).
As with organic compounds, all organometallic materials are thermodynamically unstable with respect to oxidation to MO_n, H_2O and CO_2. Nevertheless, large differences in the ease of handling of organometallics are encountered which may be traced back to differences in **kinetic intertness.** Example:

	Heat of combustion	Thermo-dynamics	Property	Kinetics
$Zn(C_2H_5)_2$	−1920 kJ/mol	unstable	pyrophoric	labile
$Sn(CH_3)_4$	−3590 kJ/mol	unstable	airstable	inert

Particularly labile against O_2 and H_2O are organometallic molecules which possess free electron pairs, low lying empty orbitals and/or high polarity of the M−C bonds. Compare:

	In air:	In water:	Relevant factors:
Me_3In	pyrophoric	hydrolysed	electron gap at In, high bond polarity.
Me_4Sn	inert	inert	Sn shielded well, low bond polarity.
Me_3Sb	pyrophoric	inert	free electron pair on Sb.
Me_3B	pyrophoric	inert	electron gap at B is closed by means of hyperconjugation, low bond polarity.
Me_3Al	pyrophoric	hydrolysed	electron gap at Al in the monomer, nucleophilic attack via Al(3*d*) in the dimer, high bond polarity.
SiH_4	pyrophoric	hydrolysed	Si shielded ineffectively, attack of O_2 and of nucleophiles via Si(3*d*).
$SiCl_4$	inert	hydrolysed	Si relatively electron-poor, high polarity of Si−Cl bonds, nucleophilic attack via Si(3*d*).
$SiMe_4$	inert	inert	Si shielded effectively, low polarity of Si−C bonds.

This brief survey is only intended to furnish a few qualitative arguments which have to be weighed with respect to their specific applications.

MAIN-GROUP ORGANOMETALLICS

4 Methods of Preparation in Perspective

Procedures for the formation of bonds between main-group metals and carbon may roughly be divided into the following reaction types:

oxidative addition [1], exchange [2] – [7], insertion [8] – [10] and elimination [11] – [12].

Metal + Organic Halide **Direct Synthesis [1]**

$$2\,M + n\,RX \longrightarrow R_nM + MX_n \text{ (or } R_nMX_n)$$

Examples :

$$2\,Li + C_4H_9Br \longrightarrow C_4H_9Li + LiBr$$
$$Mg + C_6H_5Br \longrightarrow C_6H_5MgBr$$

The high enthalpy of formation of the salt MX_n generally renders this reaction exothermic. This is, however, not true for elements of high atomic number (M = Tl, Pb, Bi, Hg) which form weak M–C bonds. Here $\Delta H_f^0(R_nM) > 0$ is not compensated for by $\Delta H_f^0(MX_n) < 0$ and an additional contribution to the driving force must be provided. One possibility is the use of an alloy which, in addition to the metal to be alkylated, contains a strongly electropositive element.

Mixed Metal Synthesis

$$2\,Na + Hg + 2\,CH_3Br \longrightarrow (CH_3)_2Hg + 2\,NaBr \quad \Delta H = -530 \text{ kJ/mol}$$
$$4\,NaPb + 4\,C_2H_5Cl \longrightarrow (C_2H_5)_4Pb + 3\,Pb + 4\,NaCl$$

[ΔH_f^0(NaX) boosts the driving force]

By their very nature direct syntheses are **oxidative additions** of RX to M^0 whereby $RM^{II}X$ is produced. The generation of new M–C bonds by means of the addition of RX to a low-valent metal compound is closely related to direct synthesis. Example:

$$Pb^{II}I_2 + CH_3I \longrightarrow CH_3Pb^{IV}I_3$$

Metal + Organometallic **Transmetallation [2]**

$$M + RM' \longrightarrow RM + M'$$
$$Zn + (CH_3)_2Hg \longrightarrow (CH_3)_2Zn + Hg \quad \Delta H = -35 \text{ kJ/mol}$$

This general method may be applied to M = Li–Cs, Be–Ba, Al, Ga, Sn, Pb, Bi, Se, Te, Zn, Cd. RM′ should be of weakly exothermic or, preferentially, of endothermic nature [e.g. $(CH_3)_2Hg$, $\Delta H = +94$ kJ/mol]. The decisive factor in its feasibility really is the difference in the free enthalpies of formation $\Delta(\Delta G_f^0)$ RM, RM′.

Organometallic + Organometallic **Metal Exchange** [3]

$$RM + R'M' \longrightarrow R'M + RM'$$
$$4\,PhLi + (CH_2{=}CH)_4Sn \longrightarrow 4\,(CH_2{=}CH)Li + Ph_4Sn$$

Precipitation of Ph_4Sn shifts this equilibrium to the right and good yields of vinyllithium are obtained. Vinyllithium is accessible by other methods only with difficulty.

Organometallic + Metal Halide **Metathesis** [4]

$$RM + M'X \longrightarrow RM' + MX$$
$$3\,CH_3Li + SbCl_3 \longrightarrow (CH_3)_3Sb + 3\,LiCl$$

The equilibrium lies on the side of the products if M is more electropositive than M′. For RM = alkali metal organyl, this procedure has wide applicability since the formation of MX makes a large contribution to the driving force.

Organometallic + Aryl Halide **Metal Halogen Exchange** [5]

$$RM + R'X \longrightarrow RX + R'M \qquad M = Li$$
$$n\text{-}BuLi + PhX \longrightarrow n\text{-}BuX + PhLi$$

The equilibrium is shifted to the right if R′ is superior to R in stabilizing a negative charge. Therefore, this reaction is practicable for aryl halides only (X = I, Br, rarely Cl, never F). Competing reactions are alkylation (Wurtz coupling) and the metallation of R′X. Metal halogen exchange is, however, a comparatively fast reaction and is favored at low temperatures (**kinetic control**).
Current mechanistic ideas include the intermediate formation of radicals (Reutov, 1976 R). Accordingly, metal halogen exchange is initiated by a single electron transfer (**SET**) step:

$$RLi + R'X \longrightarrow \underset{\text{(solvent cage)}}{\left\{\begin{matrix} R^{\cdot} & Li^+ \\ R'^{\cdot} & X^- \end{matrix}\right\}} \longrightarrow RX + R'Li$$

F in C_6H_5F is not exchanged for Li. Instead, the sequence: orthometallation, LiF elimination to yield an aryne and addition of RLi to the $-C{\equiv}C-$ triple bond affords the coupling product $R-R'$.

Organometallic + C−H Acid **Metallation** [6]

$$RM + R'H \rightleftharpoons RH + R'M \qquad M = \text{alkali metal}$$
$$PhNa + PhCH_3 \rightleftharpoons PhH + PhCH_2Na$$

Metallations (**replacement of H by M**) are acid/base equilibria $R^- + R'H \rightleftharpoons RH + R'^-$ which, with increasing acidity of R′H, are shifted to the right side. The practical success of a metallation is intimately linked to the **kinetic CH-acidity** (p. 28).
Substrates with an exceptionally high CH-acidity (acetylenes, cyclopentadienes) may also be metallated by alkali metals in a redox reaction:

$$C_5H_6 + Na \xrightarrow{THF} C_5H_5Na + \tfrac{1}{2}\,H_2$$

Mercuric Salt + C–H Acid **Mercuration 7**

By their very nature, mercurations are also metallations. In the case of **nonaromatic substrates,** they are confined to molecules of high CH-acidity (alkynes, carbonyl-, nitro-, halogeno-, cyano compounds etc.).

$$Hg[N(SiMe_3)_2]_2 + 2\,CH_3COCH_3 \longrightarrow (CH_3COCH_2)_2Hg + 2\,HN(SiMe_3)_2$$

If $Hg(CH_3COO)_2$ is used as a mercurating agent, the second step usually requires forcing conditions. A reaction of very wide scope is the **mercuration of aromatic compounds:**

$$Hg(CH_3COO)_2 + ArH \xrightarrow[\text{cat}\cdot HClO_4]{MeOH} ArHg(CH_3COO) + CH_3COOH$$

From a mechanistic point of view, this reaction is an electrophilic aromatic substitution.

Metal Hydride + Alkene (Alkyne) **Hydrometallation 8**

$$M{-}H + \;\rangle C{=}C\langle \longrightarrow M{-}\overset{|}{\underset{|}{C}}{-}\overset{|}{\underset{|}{C}}{-}H$$

M = B, Al; Si, Ge, Sn, Pb; Zr

$$(C_2H_5)_2AlH + C_2H_4 \longrightarrow (C_2H_5)_3Al \quad \text{Hydroalumination}$$

Propensity to addition: Si–H < Ge–H < Sn–H < Pb–H

Organometallic + Alkene (Alkyne) **Carbometallation 9**

$$M{-}R + \;\rangle C{=}C\langle \longrightarrow M{-}\overset{|}{\underset{|}{C}}{-}\overset{|}{\underset{|}{C}}{-}R$$

$$n\text{-BuLi} + Ph{-}C{\equiv}C{-}Ph \xrightarrow[2.\ H^+]{1.\ Et_2O} (Ph)(n\text{-Bu})C{=}C(Ph)(H) \quad \textit{cis addition}$$

In contrast to M–H, insertions into M–C bonds only proceed if M is very electropositive (M = alkali metal, Al).

Organometallic + Carbene **Carbene Insertion 10**

$$PhSiH_3 + CH_2N_2 \xrightarrow{h\nu} PhSi(CH_3)H_2 + N_2$$
$$Me_2SnCl_2 + CH_2N_2 \longrightarrow Me_2Sn(CH_2Cl)Cl + N_2$$
$$Ph_3GeH + PhHgCBr_3 \longrightarrow Ph_3GeCBr_2H + PhHgBr$$
$$RHgCl + R'_2CN_2 \longrightarrow RHgCR'_2Cl + N_2$$

Note that insertions of carbenes into M–C bonds are avoided, the insertion into an M–H or an M–X bond being strongly favored.

Pyrolysis of Carboxylates **Decarboxylation** 11

$$HgCl_2 + 2\,NaOOCR \longrightarrow Hg(OOCR)_2 \xrightarrow{\Delta T} R_2Hg + 2\,CO_2$$

The group R should contain electron withdrawing substituents (R = C_6F_5, CF_3, CCl_3 etc.). Decarboxylation of organometallic formiates leads to hydrides:

$$(n\text{-Bu})_3SnOOCH \xrightarrow[\text{reduced pressure}]{170\,°C} (n\text{-Bu})_3SnH + CO_2$$

Metal Halide (Hydroxide) + Aryldiazonium Salt 12

In synthetic organometallic chemistry, this is a method of limited importance.

$$ArN_2^+Cl^- + HgCl_2 \longrightarrow ArN_2^+HgCl_3^- \xrightarrow[-N_2]{} ArHgCl,\ Ar_2Hg$$

(depending on the catalyst)

$$ArN_2^+X^- + As(OH)_3 \longrightarrow ArAsO(OH)_2 + N_2 + HX \text{ (Bart reaction)}$$

5 Alkali Organometallics

5.1 Organolithium Compounds

PREPARATION: With methods [1], [2], [3], [5], [6], [9]

Methods **1** (starting from Li metal) and **6** are of prime importance (*n*-BuLi is commercially available).

$$CH_3Br + 2\,Li \xrightarrow[20\,°C]{Et_2O} CH_3Li + LiBr \qquad [1]$$

$$C_5Me_5H + n\text{-}BuLi \xrightarrow[-78\,°C]{THF} C_5Me_5Li + n\text{-}BuH \qquad [6]$$

Perlithiated hydrocarbons are formed in the cocondensation of Li vapor with chlorohydrocarbons (Example: CLi_4 from CCl_4 and Li). The air sensitivity of organoalkali compounds requires them to be manipulated under the **protective atmosphere** of an inert gas (**N_2, Ar**). Whereas, for reasons of solubility, Grignard reagents must be prepared in ethers, the preparation of organolithium compounds may be carried out more economically in inert hydrocarbons such as hexane.

For the **assay of RLi** solutions, volumetric methods are available. A simple acid/base titration following hydrolysis

$$RLi + H_2O \longrightarrow RH + LiOH$$
$$LiOH + HX \longrightarrow LiX + H_2O$$

is inapplicable since alkoxides, stemming from the reaction of RLi with O_2 or from ether cleavage, would suggest too high an RLi content. Therefore, a double titration method has been developed (Gilman, 1964). The RLi concentration amounts to the difference $(m + n) - n$:

$$1.\ m\,RLi + n\,ROLi + \boldsymbol{(m + n)}\,HX \longrightarrow m\,RH + n\,ROH + (m + n)\,LiX$$
$$2.\ m\,RLi + n\,ROLi + m\,BrCH_2CH_2Br \longrightarrow m\,RBr + m\,LiBr + m\,C_2H_4 + n\,ROLi$$
$$n\,ROLi + \boldsymbol{n}\,HX \longrightarrow n\,ROH + n\,LiX$$

More recently, methods of determination which include self-indicating reagents like 4-(hydroxymethyl)biphenyl have been employed. They permit direct titrations (Juaristi, 1983):

$$\text{Ph–C}_6\text{H}_4\text{–CH}_2\text{OH} \xrightarrow[-\,RH]{RLi} \text{Ph–C}_6\text{H}_4\text{–CH}_2\text{–O}^{\ominus}\,Li^{\oplus} \xrightarrow[-\,RH]{RLi} \text{Ph–C}_6\text{H}_4\text{–C}^{\ominus}\text{H–O}^{\ominus}\,2\,Li^{\oplus}$$

colorless — end point — orange red

Solutions of concentrations [RLi] > 0.1 M can be conveniently checked by means of ^{1}H-NMR spectroscopy using an internal concentration standard ($-1.0 > \delta\, RCH_2Li > -1.3$ ppm).

STRUCTURE AND BONDING

A conspicuous feature of organolithium compounds is their tendency to form **oligomeric units** in solution as well as in the solid state. A classic example is the structure of solid **methyllithium** which is best described as cubic body-centered packing of $(LiCH_3)_4$ units, the latter consisting of Li_4-tetrahedra with methyl groups capping the triangular faces (E. Weiss, 1964).

(*a*) *Unit cell of* $(LiCH_3)_4$(s)

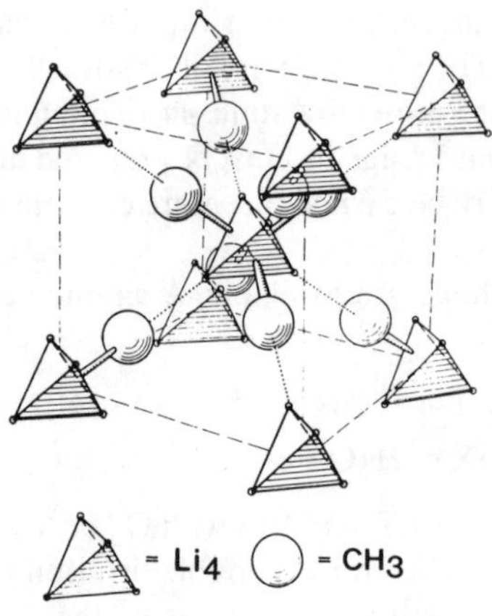

(*b*) *Schematic drawing of the unit* $(LiCH_3)_4$

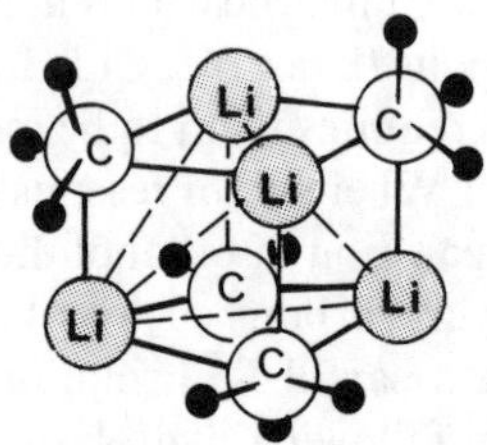

	$d(Li-C) = 231$ pm	$(LiCH_3)_4$
	$d(Li\cdots C) = 236$ pm	$(LiCH_3)_4$
	$d(Li-Li) = 268$ pm	$(LiCH_3)_4$
compare:	$d(Li-Li) = 267$ pm	Li_2(g)
	$d(Li-Li) = 304$ pm	Li(m)

The building blocks of the lattice are distorted cubes, with alternate occupation of the corners by C and Li atoms (**b**). This type of **heterocubane** arrangement is encountered frequently for species of constitution $\mathbf{(AB)_4}$.
A closer inspection of the Li–C distances reveals that the methyl groups of one $(LiCH_3)_4$ unit interact with the Li atoms of a neighboring Li_4-tetrahedron. These intermolecular forces are responsible for the low volatility and the insolubility of $LiCH_3$ in non-solvating media.
The structure of ***t*-butyllithium** is very similar to that of methyllithium, the intermolecular forces are however, weaker. As opposed to MeLi, *t*-BuLi is soluble in hydrocarbons and sublimes at 70 °C/1 mbar.
The **degree of association** of organolithium compounds is strongly dependent on the nature of the solvent:

LiR	Solvent	Aggregation
$LiCH_3$	THF, Et_2O	tetramer (Li_4 tetrahedron)
	$Me_2NCH_2CH_2NMe_2$(TMEDA)	monomer, dimer
Li-n-C_4H_9	cyclohexane	hexamer
	Et_2O	tetramer
Li-t-C_4H_9	hydrocarbon	tetramer
	THF	monomer
LiC_6H_5	THF, Et_2O	dimer
$LiCH_2C_6H_5$ (benzyl)	THF, Et_2O	monomer
LiC_3H_5 (allyl)	Et_2O	columnar structure (p. 23)
	THF	dimer

The presence of oligomers $(LiR)_n$ in solution is substantiated by osmometric measurements, by Li-NMR spectroscopy and by ESR experiments (p. 35). The mass-spectrometric observation of the fragment ion $[Li_4(t\text{-}Bu)_3]^+$ shows that the association is maintained in the gas phase.

Thorough NMR studies by T. L. Brown (1970, *R*) *and by G. Fraenkel* (1984, *R*) *have established that, much like Grignard reagents* (*p.* 43), *solutions of organolithium compounds represent complicated equilibrium mixtures. The dynamic processes in these solutions comprise* ***intramolecular bond fluctuation*****:**

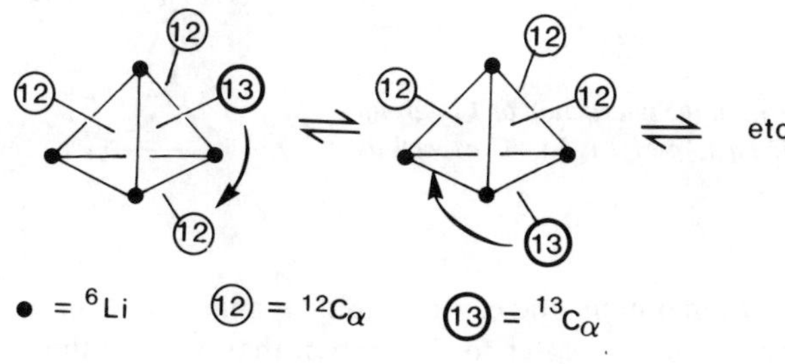

In the case of **[(*t*-Bu)Li]$_4$** *at* $-22\,°C$ $^{13}C\{^1H\}$NMR *features a septet caused by coupling to the three nearest neighbour* ^{6}Li *nuclei,* $^1J(^{13}C, {}^6Li) = 5.4$ Hz, *coupling to the remote* ^{6}Li *nuclei not being resolved. At* $-5\,°C$ *in the fast fluxional limit a nonet is observed,* $J(^{13}C, {}^6Li) = 4.1$ Hz, *which represents the weighted average of the three adjacent and the vanishingly small remote couplings* (*Thomas*, 1986).

as well as ***intermolecular exchange***:

$$\left.\begin{array}{c} R_4Li_4 \rightleftharpoons 2\,R_2Li_2 \\ + \qquad\qquad + \\ R'_4Li_4 \rightleftharpoons 2\,R'_2Li_2 \end{array}\right\} \rightleftharpoons 2\,R_2R'_2Li_4 \rightleftharpoons 4\,RR'Li_2 \rightleftharpoons \left\{\begin{array}{c} R_3R'Li_4 \\ + \\ RR'_3Li_4 \end{array}\right.$$

It has been demonstrated by means of mass spectrometry that these scrambling reactions proceed via cleavage of the Li_4 *units rather than by ligand transfer, whereby the integrity of the* Li_4 *units would be maintained* (*T. L. Brown*, 1970):

$$t\text{-}Bu_4\,{}^6Li_4 + t\text{-}Bu_4\,{}^7Li_4 \xrightarrow{\text{cyclopentane}} t\text{-}Bu_4\,{}^6Li_3\,{}^7Li + t\text{-}Bu_4\,{}^6Li_2\,{}^7Li_2 + \ldots .$$

The kinetic parameters of these processes are strongly influenced by the nature of the medium and the group R.

The tendency of organolithium compounds to associate in the solid state as well as in solution is due to the fact that in a single molecule LiR, the number of valence electrons is too low to use all the available Li valence orbitals for two-electron two-center ($2e\,2c$) bonding. In the aggregates $(LiR)_n$ this "electron deficiency" is compensated for by the formation of multicenter bonds. This is illustrated for the tetrahedral species $(LiCH_3)_4$:

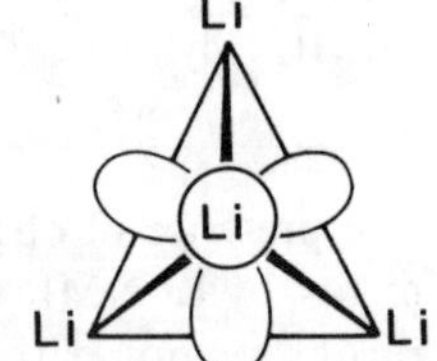

Li_4 *skeleton with* 4 Li (sp^3) *hybrid orbitals per* Li *atom. Directional properties of the* Li *valence orbitals*:
1 × ***axial***, *identical with one of the threefold axes of the tetrahedron,*
3 × ***tangential***, *pointing towards the normals of the triangular faces.*

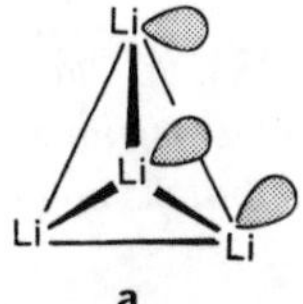

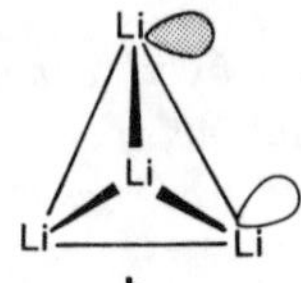

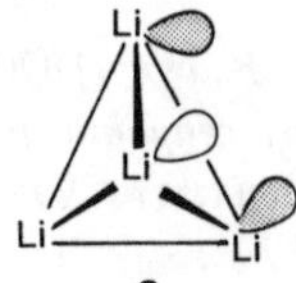

Group orbitals *formed from three tangential* Li (sp^3) *hybrid orbitals originating at the corners of the* Li_3 *triangle.*

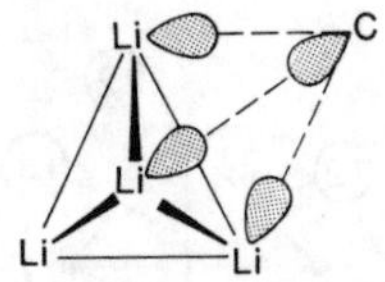

Four-center bonding molecular orbital *from the interaction of* Li_3 *group orbital* ***a*** *with a* C (sp^3) *hybrid orbital. This* 4*c*-*MO is* Li–C *as well as* Li–Li *bonding.*

The electronegativity difference between lithium and carbon should manifest itself in the $2e\,4c$ bonding pair of electrons being located closer to the carbon than to the lithium atoms. The **bond polarity** $Li^{\delta+}-C^{\delta-}$ can be demonstrated experimentally, e.g. by means of NMR spectroscopy. $LiCH_3$ molecules in matrix isolation were reported to possess a dipole moment of about 6 Debye (Andrews, 1967), a value of 9.5 Debye being expected for the case of full charge separation (ionic limit). The degree of covalent and ionic character of the Li–C bond is still an open question; sizeable covalent contributions being favored by some authors (Lipscomb, 1980; Ahlrichs, 1986) and essentially ionic character by others (Streitwieser, 1976; v. Ragué Schleyer, 1988 R).

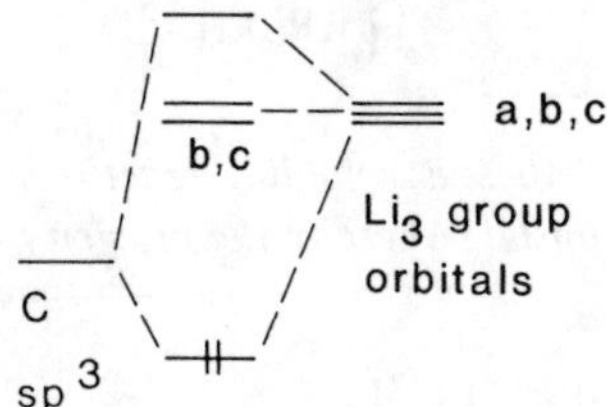

MO *diagram* *for one of the four* ***2e 4c bonds*** *in* R_4Li_4

The axial Li(sp^3) hybrid orbitals, which are unoccupied in an isolated molecule of $(LiCH_3)_4$, are used in the crystal for interaction with methyl groups of neighboring $(LiCH_3)_4$ units and in solution for the coordination of σ-donors (Lewis bases, solvent molecules). An example is the tetrameric etherate of **phenyllithium**, $(\mu^3\text{-}C_6H_5Li \cdot OEt_2)_4$ (Power, 1983). In the presence of the chelating ligand N,N,N',N'-tetramethylethylenediamine (TMEDA, N⌒N), however, phenyllithium crystallizes in a dimeric structure which is reminiscent of triphenylaluminum (Al_2Ph_6, p. 79) (E. Weiss, 1978):

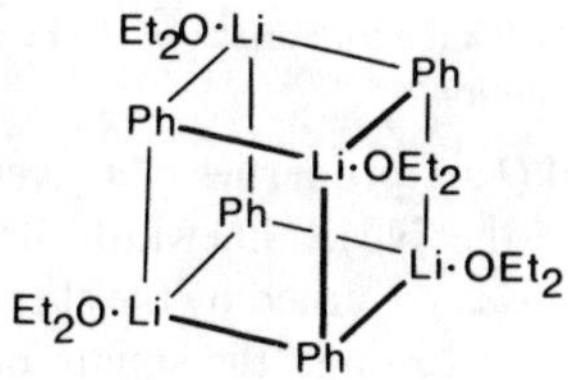

Et$_2$O·Li, Ph, Li·OEt$_2$, Ph, Li·OEt$_2$, Et$_2$O·Li, Ph, Ph

67°, 221, 221, N, Li, Li, N, N, 218, 228, N, 106°

d(Li–Li) = 250–270 pm

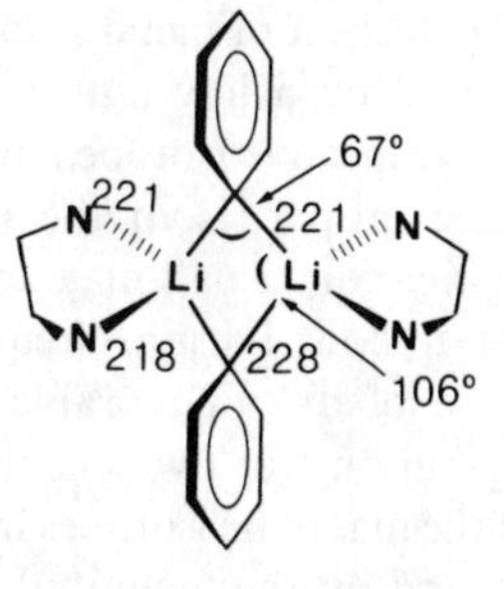

d(Li–Li) = 249 pm

A totally different type of association is encountered for organolithium compounds in which the carbanion represents a delocalized π-system. Instead of reducing the electron deficiency by means of clustering to Li_n ($n = 2, 4, 6$), interactions of partially solvated Li^+ ions with the π-electron system of the carbanion arise. In the case of **1,3-(*diphenyl*)*allyllithium***, a columnar structure with bridging $\mu(\eta^3:\eta^3)$allyl units is realized which resembles $(C_5H_5)In$ (p. 89) or a polydecker sandwich complex (p. 383). In contrast to the latter, however, ionic bonding appears to prevail in the η^3-allyl lithium compound (Boche, Massa, 1986):

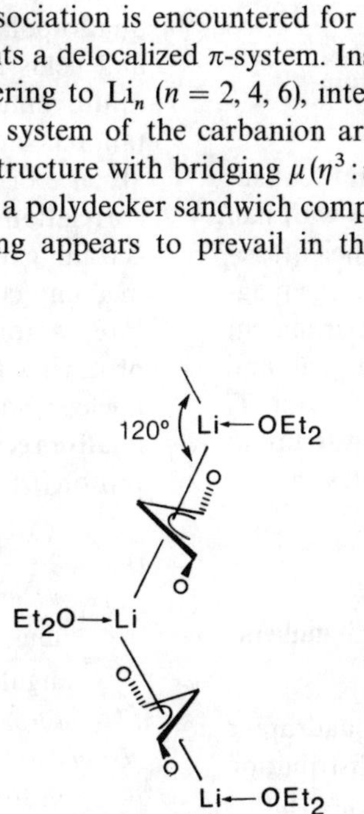

120°, Li←OEt$_2$, Et$_2$O→Li, Li←OEt$_2$, m, t, —o = —Ph

$d(Li–C_m)$ = 231 pm
$d(Li–C_t)$ = 248 pm
(mean values)

Excursion

^{6}Li- and ^{7}Li NMR of Organolithium Compounds

In organometallic research, NMR spectroscopy of the less common nuclei is often required. Experimental problems during the detection of metal resonances may be caused by a low natural abundance of the respective isotope and/or by the magnetic properties of the nucleus under investigation. Small magnetic moments lead to low Larmor frequencies and, because of an unfavorable Boltzmann-distribution, to low sensitivity. A division of the magnetic isotopes into two classes is based on the magnitude of the nuclear spin quantum number I.

a) Nuclei with the Spin Quantum Number $I = 1/2$

For small molecules, $I = 1/2$ nuclei usually yield sharp resonance lines with half widths $W_{1/2}$ (linewidth at half height) between 1 and 10 Hz. If, however, the magnetic interactions with the environment are weak, very long longitudinal and transverse relaxation times T_1 and T_2 may result [example: $T_1(^{109}Ag)$ up to 10^3 s]. This severely complicates the detection of these resonances.

b) Nuclei with Spin Quantum Numbers $I \geqq 1$

These nuclei possess electric quadrupole moments (deviations of the distribution of nuclear charge from spherical symmetry) which can cause extremely short nuclear relaxation times and concomitant large half widths $W_{1/2}$ (up to several ten thousand Hz).

$$W_{1/2} \sim \frac{(2I+3)\,Q^2\,q_{zz}^2\,\tau_c}{I^2(2I-1)}$$

Q = quadrupole moment, q_{zz} = electric field gradient, τ_c = correlation time of molecular reorientation (characteristic of the extent of tumbling motion).

I and Q are properties of a given nucleus. Thus, the NMR linewidth $W_{1/2}$ of this nucleus is governed by the chemical environment through the square of the electric field gradient q_{zz}^2 and through the correlation time τ_c. Relatively narrow lines are obtained for molecules of low molecular weight (τ_c is small) if the quadrupolar nuclei are embedded in ligand fields of cubic (tetrahedral, octahedral) symmetry. In these cases, an important relaxation path is blocked due to the absence of an electric field gradient. The correlation time τ_c may be controlled to a certain extent by the viscosity of the medium (choice of solvent and temperature). A rough measure of the chance of observing the magnetic resonance of a nucleus X is given by its receptivity. The **relative receptivity**, related to that of the proton ($D_x^p = 1.000$), is defined as follows:

$$D_x^p = \frac{\gamma_x^3 N_x I_x(I_x+1)}{\gamma_p^3 N_p I_p(I_p+1)}$$

$$\gamma = \frac{\text{magnetic moment}}{\text{angular momentum}} = \frac{\mu_x}{I \cdot h/2\pi}\ \frac{\text{rad}}{\text{s T}}$$

I = nuclear spin
N = natural abundance (%)
μ_x = magnetic moment [J T^{-1}]
γ = magnetogyric ratio [rad s^{-1} T^{-1}]

In table 5-1 the properties of "unconventional" nuclei are contrasted with the routine cases ^{1}H, ^{11}B, ^{13}C, ^{19}F and ^{31}P:

Table 5-1: Magnetic properties of some "unconventional" nuclei in contrast with "routine" cases ^{1}H, ^{11}B, ^{13}C, ^{19}F *and* ^{31}P

Nucleus	N in %	I	μ_x pos. + neg. −	Q in 10^{-28} m^2 (barn)	NMR frequency in MHz at 2.35 T	Standard	Relative Receptivity D_x^p
^{1}H	99.9	1/2	+		100	Me_4Si	1.000
^{2}H	0.015	1	+	$2.73 \cdot 10^{-3}$	15.4		$1.45 \cdot 10^{-6}$
^{6}Li	7.4	1	+	$-8.0 \cdot 10^{-4}$	14.7	Li^+ (aq)	$6.31 \cdot 10^{-4}$
^{7}Li	92.6	3/2	+	$-4.5 \cdot 10^{-2}$	38.9	Li^+ (aq)	0.27
^{11}B	80.4	3/2	+	$3.55 \cdot 10^{-2}$	32.1	$BF_3 \cdot OEt_2$	0.13
^{13}C	1.1	1/2	+		25.1	Me_4Si	$1.76 \cdot 10^{-4}$
^{15}N	0.37	1/2	−		10.14	$MeNO_2$, NO_3^-	$3.85 \cdot 10^{-6}$
^{19}F	100	1/2	+		94.1	CCl_3F	0.83
^{23}Na	100	3/2	+	0.12	26.5	Na^+ (aq)	$9.25 \cdot 10^{-2}$
^{25}Mg	10.1	5/2	−	0.22	6.1	Mg^{2+} (aq)	$2.71 \cdot 10^{-4}$
^{27}Al	100	3/2	+	0.15	26.1	$Al(acac)_3$	0.21
^{29}Si	4.7	1/2	−		19.9	Me_4Si	$3.7 \cdot 10^{-4}$
^{31}P	100	1/2	+		40.5	H_3PO_4	$6.63 \cdot 10^{-2}$
^{51}V	99.8	7/2	+	0.3	26.3	$VOCl_3$	0.38
^{57}Fe	2.19	1/2	+		3.2	$Fe(CO)_5$	$7.39 \cdot 10^{-7}$
^{59}Co	100	7/2	+	0.4	23.6	$[Co(CN)_6]^{3-}$	0.28
^{71}Ga	39.6	3/2	+	0.11	30.5	Ga^{3+} (aq)	$5.62 \cdot 10^{-2}$
^{77}Se	7.6	1/2	+		19.1	Me_2Se	$5.26 \cdot 10^{-4}$
^{103}Rh	100	1/2	−		3.2	$Rh(acac)_3$	$3.12 \cdot 10^{-5}$
^{119}Sn	8.6	1/2	−		37.3	Me_4Sn	$4.44 \cdot 10^{-3}$
^{125}Te	7.0	1/2	−		31.5	Me_2Te	$2.2 \cdot 10^{-3}$
^{183}W	14.4	1/2	+		4.2	WF_6	$1.04 \cdot 10^{-5}$
^{195}Pt	33.8	1/2	+		21.4	$[Pt(CN)_6]^{2-}$	$3.36 \cdot 10^{-3}$

Source: R. K. Harris, B. E. Mann, NMR and the Periodic Table, Academic Press, New York, 1978.

The extensive use of organolithium compounds in organic synthesis justifies a brief discussion of Li NMR spectroscopic characteristics. The choice between the two isotopes ^{6}Li ($I = 1$) and ^{7}Li ($I = 3/2$) must take the nature of the specific problem into account.

^{7}Li displays higher receptivity; due to the larger quadrupole moment, the linewidths are large, however. ^{6}Li presents reduced linewidths, albeit at the cost of lower receptivity (Wehrli, 1978). Therefore, **^{7}Li NMR** offers **higher sensitivity** and **^{6}Li NMR superior resolution** of coupling patterns (^{6}Li carries the smallest of all known nuclear quadrupole moments).

Among the alkali metal organometallics, organolithium compounds are not only the most widely used in the laboratory, they also boast the greatest variety of structures and bonding situations. The study of lithium organometallics in solution has greatly benefited from the application of Li NMR. The NMR spectra of the organometallics of the heavier alkali metals Na – Cs reflect the nature of the $M(solv)^+$ complex, since ionic bonding dominates and the organic counterion contributes to the shielding of M^+ only insignificantly. Li NMR spectra show a larger diversity because the bonding modes range from mainly covalent, in the case of lithium alkyls, to ion pairing as encountered for lithium tetraorganometallates and for compounds in which the organic group is effectively stabilized by resonance (triphenylmethyl, cyclopentadienyl). As would be expected, solvent effects play an important role in Li NMR, in that the solvating power affects the polarity of the Li–C bond and governs the degree of association (p. 21). The NMR shifts $\delta(^7Li)$ cover a small range of only 10 ppm, which shrinks to about 2 ppm if only those organolithium compounds are considered in which the bonding is thought to be essentially covalent. This sets a limit to the value of Li NMR in routine analysis.

- The method of choice for investigations of structural dynamics in organolithium chemistry is the application of ^6Li NMR to ^{13}C-enriched materials.

Examples:

Structure of Ion Pairs $Li^+C_mH_n^-$

If the group $C_mH_n^-$ of an organolithium compound is endowed with significant resonance stabilization, ionic bonding will dominate and the role of the solvent will be to favor one of two main kinds of structures known as **contact ion pairs** and **solvent separated ion pairs**. Since the en-

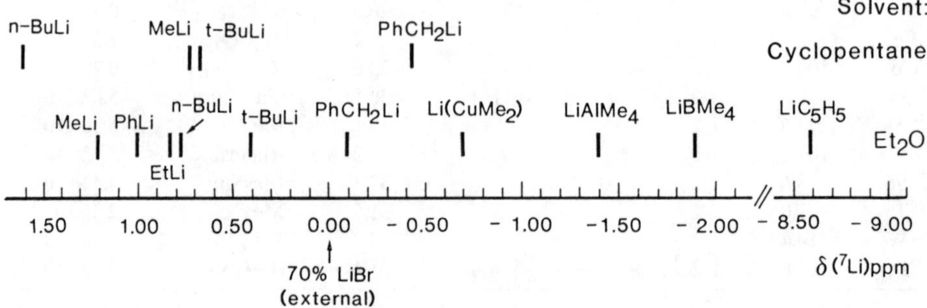

However, the following generalizations may be made:

- ^7Li NMR signals for the more covalent organolithium compounds appear at low magnetic field, those for species with a large ionic contribution to the bond are shifted to higher field.
- Solvent shifts are pronounced, although it is difficult to predict their direction.
- The observation of scalar couplings $^1J(^{13}C, ^6Li)$ or $^1J(^{13}C, ^7Li)$ may be taken as evidence for a covalent contribution to the Li–C bond. Surprisingly, couplings $J(^6Li, ^7Li)$ have not as yet been discovered.

vironment of the Li^+ cation strongly differs in these situations, characteristic ^7Li NMR shifts result. Thus, in addition to the diagnosis of the prevailing bond type, ^7Li NMR furnishes information on the disposition of the cation Li^+ relative to the organic anion. This will be illustrated for cyclopentadienyllithium and for triphenylmethyllithium (Cox, 1974). In the case of $Li^+C_5H_5^-$, the strong high-field shift of the ^7Li NMR signal points to a structure in which the Li^+ cation resides above the plane of the cyclic carbanion, that is in the shielding region of the aromatic ring current.

$\delta(^7Li)$ (External standard: 1.0 M LiCl in H_2O)

Organo-Li compound	Solvent			
	Et_2O	THF	DME	HMPA
$Li^+C(C_6H_5)_3^-$		−1.11	−2.41	−0.88
$Li^+C_5H_5^-$	−8.60	−8.37	−8.67	−0.88

⇧ Contact ion pair

⇧ Solvent separated ion pair

Whereas in the weakly solvating media tetrahydrofuran (THF) and dimethoxyethane (DME), $Li^+C(C_6H_5)_3^-$ and $Li^+C_5H_5^-$ form different types of ion pairs, in the strongly solvating medium hexamethylphosphoric trisamide (HMPA) both compounds exist as solvent-separated ion pairs, as inferred from the identical chemical shifts $\delta(^7Li)$.

The Tetrameric Structure of *t*-Butyllithium in Solution

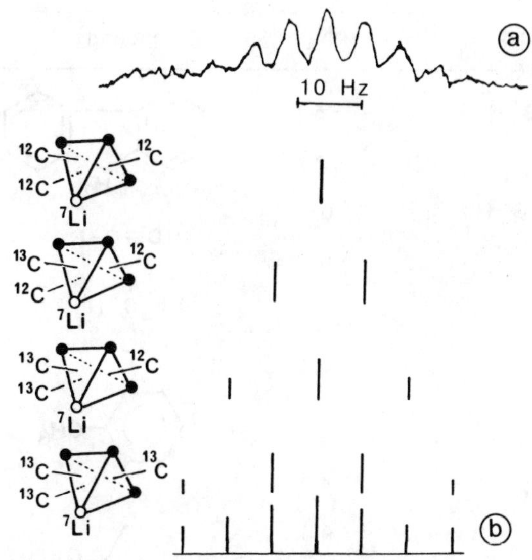

The reconstruction ⓑ *of the experimental spectrum* ⓐ [^{7}**Li NMR** *of* *t*-BuLi (0.1 M) *in cyclohexane, RT,* 57% ^{13}*C-enriched at the α-positions*] *is a superposition of isotopomeric species in which the observed nucleus* ^{7}Li *is coupled to* 0, 1, 2 *and* 3 *neighboring* ^{13}C *nuclei,* $^1J(^{13}C, ^7Li) = 11$ Hz. *The agreement between* ⓐ *and* ⓑ *demonstrates that the association of* (t-BuLi)$_4$ *units, which structural analysis revealed for the solid state, is maintained in solution. This also applies to methyllithium in the solid state and in THF at* −70 °C (*McKeever,* 1969).

REACTIONS OF ORGANOLITHIUM COMPOUNDS

In their chemical behavior, organolithium compounds resemble Grignard reagents; they are, however, more reactive.

a) Metallation and Subsequent Reactions

$$R-Li + R'-H \rightleftharpoons R-H + R'-Li$$

A prominent feature of this important method of introducing lithium is the discrepancy between equilibrium position and acceptable rate of reaction. If organometallic compounds R – M are regarded as salts of the corresponding CH-acids R – H, with increasing **CH-acidity** of R' – H the metallation equilibrium is shifted to the right. Table 5-2 compares the CH-acidity of organic CH-acids to inorganic acids.
Consequently, the stronger CH-acid, benzene, should be amenable to high yield metallation by *n*-butyllithium, a salt of the weaker CH-acid, butane:

$$C_6H_6 + n\text{-}C_4H_9Li \rightleftharpoons C_6H_5Li + n\text{-}C_4H_{10}$$

Table 5-2: CH acid exponents for organic CH-acids in nonaqueous media and values of inorganic acids for comparison:*

Compound	pK$_a$	Compound	pK$_a$
$(CN)_3C-H$	-5	fluorene (CH_2)	21
H_2SO_4	-2	$HC\equiv C-H$	24
$(NO_2)_3C-H$	0	Ph_3C-H	30
$HClO_3$	0	toluene ($C_6H_5-CH_3$)	~35
cyclopentane-1,3-dione (CH_2)	4.5	benzene (C_6H_5-H)	~37
CH_3COOH	4.7	$C_3H_7CH_2-H$ (alkanes)	~44
HCN	9.4		
O_2N-CH_3	10		
cyclopentadiene (CH_2)	15		
H_2O	15.7		

* The pK_a values are linked to the conventional scale in aqueous media by means of a conversion factor.

This reaction is unmeasurably slow. Rapid Li/H exchange occurs, however, upon addition of a strong σ-donor like tetramethylethylenediamine (TMEDA) or *t*-butoxide (t-BuO$^-$):

$$(n\text{-}C_4H_9Li)_6 + \xrightarrow[\text{6 TMEDA}]{C_6H_{12}} 6\ (\text{TMEDA})Li^+C_4H_9^- \xrightarrow[\text{fast},\ -C_4H_{10}]{C_6H_6} (\text{TMEDA})Li^+ + C_6H_5^-$$

TMEDA effects both the cleavage of *n*-BuLi oligomers and, through complexation of the Li^+ cation, a polarization of the Li–C bond. In this way, the carbanionic character and therefore the reactivity of the butyl group is increased. The fact that monomerization is essential for the metallation reaction to proceed smoothly is indicated by the observation that the rates of metallation by $PhCH_2Li$ (monomeric in THF) exceed those of CH_3Li (tetrameric in THF) by a factor of 10^4, although CH_3^- is the stronger base as compared with $PhCH_2^-$.

Example:

$$\text{Fe}(C_5H_5)_2 \xrightarrow[\text{TMEDA, 70}^\circ,\ \text{cyclohexane}]{n\text{-BuLi}} \text{Fe}(C_5H_4Li)_2 \xrightarrow{Ph_2PCl} \text{Fe}(C_5H_4PPh_2)_2$$

Metallations of geminal dichlorides are followed by LiCl elimination, resulting in **chlorocarbene** formation:

$$CH_2Cl_2 + n\text{-BuLi} \xrightarrow[-C_4H_{10}]{} LiCHCl_2 \xrightarrow[-LiCl]{} :CHCl \quad \text{follow-up reactions}$$

Often, there is no experimental evidence for the intermediacy of free carbenes in these reaction sequences. The α-haloalkyllithium compounds are then more generally referred to as ***carbenoids.***

b) Deprotonation of Organophosphonium Ions

$$Ph_3PCH_3^+ + RLi \longrightarrow RH + \left\{ \underset{\text{Ylid}}{Ph_3\overset{\oplus}{P}-\overset{\ominus}{\underline{C}}H_2} \longleftrightarrow \underset{\text{Ylene}}{Ph_3P{=}CH_2} \right\} + Li^+$$

$$\xrightarrow[\text{Wittig reaction}]{R_2CO} Ph_3P{=}O + R_2C{=}CH_2$$

This reaction is employed in the synthesis of terminal olefins.

c) Addition to Multiple Bonds (Carbolithiation)

In their tendency to add to **C–C multiple bonds**, organolithium compounds lie between Grignard reagents and organoboron- or organoaluminum compounds, respectively: RMgX < **RLi** < R_3B, $(R_3Al)_2$. Under mild conditions, organolithium compounds only add to conjugated dienes and to styrene derivatives. As in the case of metallation, the presence of strong σ-donors like TMEDA exerts an activating effect on RLi additions (example: *n*-BuLi/TMEDA initiates the polymerization of ethylene). The alkyllithium-initiated polymerization of isoprene produces a synthetic rubber which mimics the natural product in many respects (Hsieh, 1957). This discovery led to the first large scale application of organolithium compounds.

Isoprene $CH_2{=}C(CH_3){-}CH{=}CH_2$ $\xrightarrow[\text{pentane}]{\text{EtLi}}$ $[{-}CH_2{-}C(CH_3){=}CH{-}CH_2{-}]_n$ cis-1,4-Polyisoprene Synthetic Rubber

If Et_2O or THF is used as a solvent, the undesired trans-1,4, -3,4, and -1,2 additions are favored.

Carbolithiations of C–C multiple bonds may also proceed intramolecularly:

1-Br-8-(PhC≡C)-naphthalene $\xrightarrow[-\ n\text{-BuBr}]{n\text{-BuLi}}$ [1-Li-8-(PhC≡C)-naphthalene] ⟶ 1-Ph-2-Li-acenaphthylene

Among the additions of RLi to **C–N multiple bonds**, the reaction with nitriles deserves mention since it constitutes a versatile method for the preparation of ketones,

$$R{-}C{\equiv}N \xrightarrow{R'Li} \underset{R'}{\overset{R}{>}}C{=}NLi \xrightarrow{\text{hydrolysis}} \underset{R'}{\overset{R}{>}}C{=}NH \xrightarrow{\text{hydrolysis}} \underset{R'}{\overset{R}{>}}C{=}O$$

as well as the reaction with pyridine in which, similar to the Chichibabin reaction, 2-substituted pyridine derivatives are obtained:

Additions of organolithium compounds to **C−O multiple bonds** closely resemble the reactions of Grignard reagents, however the tendency for side reactions to occur is lower with RLi. As an example, the addition of RLi to *N*,*N*-dimethylformamide, which affords aldehydes, may be cited:

$$RLi + H{-}C(=O){-}NR'_2 \longrightarrow R{-}CH(OLi){-}NR'_2 \xrightarrow{H_3O^+} R{-}C(=O){-}H + H_2NR'^{+}_2$$

Whereas reactions of organolithium compounds with free carbon monoxide are nonspecific and have found little use, additions of RLi to CO, bound to transition metals, are of great importance from a principal as well as from a practical point of view. Note that this reaction led to the discovery of transition-metal carbene complexes (p. 210).

d) Reactions with Main-Group- and Transition-Metal Halides

Conversions of the type RLi + MX → MR + LiX have already been introduced as a synthetic procedure [4] for **main-group organometallics** (p. 16); due to their wide applicability, these may well be the most frequently conducted reactions in organometallic chemistry. For the halides of high valent elements, the reaction proceeds in several steps which do not always lend themselves to efficient control, necessitating separations of product mixtures. In some cases, addition of an excess of RLi leads to the formation of **"ate-complexes"**.

		Examples:
$n\,RLi + MX_n$	$\longrightarrow R_nM + n\,LiX$	Me_3Sb, Ph_4Sn
$RLi + MX_n$	$\longrightarrow RMX_{n-1} + LiX$	$RMgCl$, $MeSnCl_3$
$RLi + R'MX_{n-1}$	$\longrightarrow RR'MX_{n-2} + LiX$	$Ph(Me)SnCl_2$
$RLi + R_nM$	$\longrightarrow (R_{n+1}M)^- + Li^+$	Ph_4B^- "ate-complex"

In the case of **transition metals** which, compared to main-group elements, display a larger variety of bonding types to organic groups, coupling reactions are often followed by ligand displacement (p. 245, 281):

$$(\eta^5\text{-}C_5H_5)Mo(CO)_3Cl + LiC_3H_5 \longrightarrow (\eta^5\text{-}C_5H_5)Mo(CO)_3(\eta^1\text{-}C_3H_5) + LiCl$$

$$\downarrow \sigma/\pi\text{-rearrangement} \quad -CO$$

$$(\eta^5\text{-}C_5H_5)Mo(CO)_2(\eta^3\text{-}C_3H_5)$$

The intermediate alkylation of titanium is the basis of a procedure which achieves high ***chemo- and stereoselectivity of carbanion addition*** *to carbonyl groups* (*Reetz, Seebach,* 1980). *In contrast to* RLi *or* RMgX, *the organotitanium reagent* $RTi(O\text{-}iso\text{-}Pr)_3$ *attacks aldehyde functions in a highly chemoselective way*:

$$ClTi(OCHMe_2)_3 \xrightarrow[-\,LiCl]{CH_3Li\ (CH_3MgX)}$$

$$PhC(O)H + PhC(O)CH_3 \xrightarrow[-20°]{1)\ CH_3Ti(OCHMe_2)_3 \quad 2)\ H^+} PhCH(OH)CH_3 + PhC(OH)(CH_3)CH_3$$

1 : 1 → >99 : <1

In comparison, the analogous reaction of CH_3Li *produces a* 50 : 50 *mixture of the secondary and tertiary alcohols. Under the mild reaction conditions, other functional groups like* CN, NO_2 *and* Br *remain unaffected.*

It is assumed that the Ti–C *bond is significantly less polar than the* Li–C *and* XMg–C *bonds, resulting in a lower rate of reaction and higher selectivity for the former. An additional aspect is the higher steric demand of the* $Ti(O\text{-}iso\text{-}Pr)_3$ *group, which leads to particularly effective discrimination between different transition states.*

As an example for stereoselectivity, the reaction of CH_3TiCl_3 *with* α*-alkoxyaldehydes will be mentioned*:

$$TiCl_4 \xrightarrow[-\,LiCl]{CH_3Li}$$

$$H_3C\text{–}CH(OCH_2Ph)\text{–}CHO \xrightarrow[CH_2Cl_2/-78°]{CH_3TiCl_3} [\text{chelate complex with } Ti(CH_3)Cl_3] \longrightarrow$$

92 : 8

As opposed to the titanium alkoxides, CH_3TiCl_3 *still acts as a Lewis acid. Presumably, during the course of this reaction, an octahedral chelate complex is formed as an intermediate in which intra- or intermolecular transfer of the methyl group to the less hindered side of the coordinated aldehyde is favored* (*Reetz*, 1987).

5.2 Organyls of the Heavier Alkali Metals

Compared to the eminent importance of organolithium compounds to almost every branch of organometallic chemistry, the organyls of the heavier alkali metals, with the exception of C_5H_5Na (NaCp), play only a limited role.

PREPARATION:

$$2\,Na_{(dispersion)} + n\text{-}C_5H_{11}Cl \longrightarrow n\text{-}C_5H_{11}Na + NaCl \quad \boxed{1}$$

$$2\,K_{(mirror)} + (CH_2{=}CHCH_2)_2Hg \longrightarrow 2\,CH_2{=}CHCH_2K + Hg \quad \boxed{2}$$

$$Na + C_5H_6 \longrightarrow \underset{(NaCp)}{C_5H_5Na} + 1/2\,H_2 \quad \boxed{6}$$

STRUCTURE AND PROPERTIES:

In the group of organoalkali metal compounds, the ionic character of the M–C bond increases from Li to Cs. Whereas $NaCH_3$ has the predominantly covalent structure of $LiCH_3$, **KCH_3** is best described as an ionic lattice of the **NiAs type** (CH_3^- in a trigonal prismatic environment of $6\,K^+$, K^+ surrounded by an octahedron of $6\,CH_3^-$). The heavier alkali metal organyls are extremely reactive compounds which slowly metallate even saturated hydrocarbons (in which they are insoluble). Solubility in other media is usually accompanied by attack of the solvent. The high reactivity of these alkali metal organyls is caused by the pronounced carbanionic character. Ethers are slowly metallated at the α-position, an elimination of alkoxide follows (**ether cleavage**):

$$CH_3CH_2OCH_2CH_3 + KC_4H_9 \longrightarrow C_4H_{10} + K^{\oplus}[\overset{\ominus}{C}H(CH_3)-O-C_2H_5]$$

$$\downarrow$$

$$KOC_2H_5 + H_2C{=}CH_2$$

Ether cleavage is particularly rapid with cyclic ethers like THF. Additional modes of reaction which limit the stability are self-metallation:

$$2\,C_2H_5Na \longrightarrow C_2H_4Na_2 + C_2H_6 \longrightarrow \cdots$$

and β-hydride elimination:

$$C_2H_5Na \longrightarrow NaH + C_2H_4$$

More stable products are obtained if the respective carbanion is stabilized by resonance, i.e. if the negative charge is delocalized effectively as in C_5H_5Na or in Ph_3CK.

ADDITION COMPOUNDS OF THE ALKALI METALS

Besides alkali metal organyls, the generation of which involves the cleavage of a bond in the organic educt, there is a second class of compounds which are formed by an electron transfer from the alkali metal to the organic partner, with no bonds being split:

$$M + ArH \rightleftharpoons M^+ + ArH^{\bullet -}$$
$$2\,M + ArH \rightleftharpoons 2\,M^+ + ArH^{2-}$$

In this way **radical anions** $ArH^{\bullet -}$ and dianions ArH^{2-} are obtained, the latter being diamagnetic in the majority of cases. They are of theoretical as well as of practical significance. Sodium naphthalenide $Na^+C_{10}H_8^{\bullet -}$ forms moss-green solutions in ethers like DME or THF and serves as a **self-indicating reducing agent** in the synthesis of metal complexes in low oxidation states. An attractive feature here is the availability of a very strong reducing agent which operates in homogeneous phase $[E_{1/2}(C_{10}H_8^{0/-1}) = -2.5$ V versus saturated calomel electrode]. Example:

$$(n\text{-BuO})_4Ti + 2\,Na^+C_{10}H_8^{\bullet -} \xrightarrow{THF} (n\text{-BuO})_2Ti(THF)_2 + 2\,C_{10}H_8 + 2\,n\text{-BuONa}$$

Alkali metal-arene addition compounds have also been obtained in crystalline form. The compound $\mathbf{[Li(TMEDA)]_2C_{10}H_8}$ *may be regarded as an arene complex of a main-group element – no decision as to the prevailing bond type being implied by this designation!*

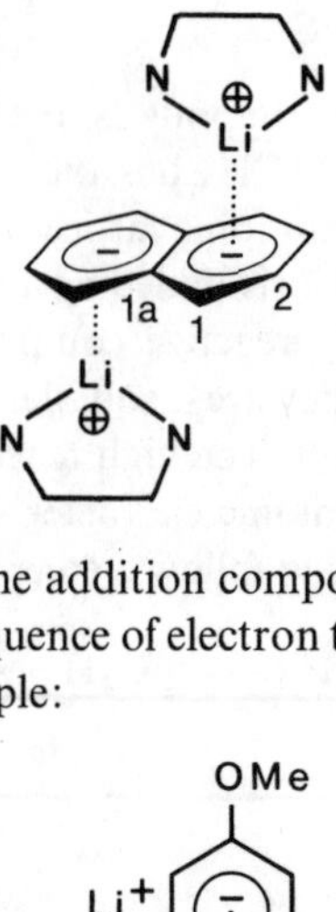

$d(Li-C_{1a}) = 266$ pm
$d(Li-C_1) = 233$ pm
$d(Li-C_2) = 226$ pm
(Stucky, 1972)

In the presence of proton donors the addition compounds undergo follow-up reactions. The **Birch reduction** consists of a sequence of electron transfer (ET)- and protonation steps which are used in synthesis. Example:

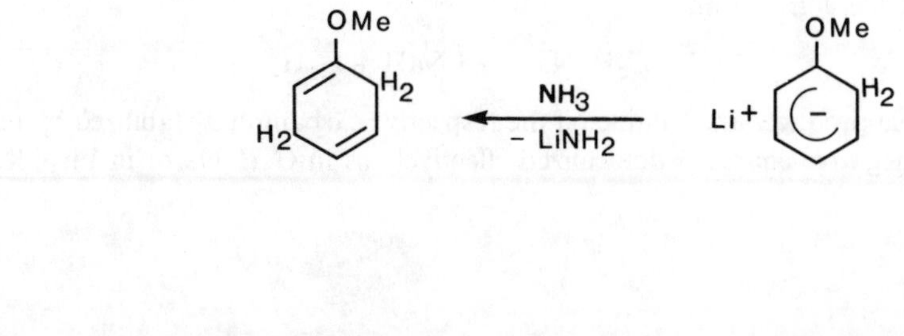

Electron transfer to cyclic conjugated systems often has **structural consequences:**

$$2\,C_8H_8 \xrightarrow[-K^+]{K} 2\,C_8H_8^{\cdot-} \rightleftharpoons C_8H_8 + C_8H_8^{2-}$$

nonplanar
$4n$ π-electrons
antiaromatic

planar
$(4n+1)\,\pi$-electrons

planar
$(4n+2)\,\pi$-electrons
aromatic

Whereas the radical anions of cyclic conjugated systems usually remain monomeric, radical anions of acyclic substrates tend to dimerize or to initiate polymerization. In the case of diphenylacetylene this tendency is exploited in a versatile **synthesis of five-membered heterocycles** containing B, Si, Sn, As, and Sb as heteroatoms (see p. 63, 110, 162).

$$\text{Ph}{-}\text{C}{\equiv}\text{C}{-}\text{Ph} \xrightarrow[-\,\text{Li}^+]{\text{Li}} \text{Ph}{-}\text{C}{\equiv}\text{C}{-}\text{Ph}^{\cdot-} \xrightarrow{\times 2} \text{Ph}_4\text{C}_4^{2-}$$

subsequent reaction

Excursion

ESR Spectroscopy of Organoalkali Metal Compounds

a) Radical Anions of Aromatic π-Systems

In alkali-metal addition compounds $M^+ArH^{\cdot-}$, an unpaired electron resides in the lowest unoccupied molecular orbital (LUMO) of the neutral molecule ArH. Correspondingly, in radical cations $ArH^{\cdot+}$, the highest occupied molecular orbital (HOMO) contains an unpaired electron. The ESR spectra of $ArH^{\cdot+}$ and of $ArH^{\cdot-}$ furnish **information about the composition of the frontier orbitals HOMO and LUMO** from $C(p_\pi)$atomic orbitals of the molecular frame. In view of the fact that regioselectivity of chemical reactions can be a result of frontier orbital control (Fukui), knowledge of the shape of these molecular orbitals is of practical significance.

The form of the singly occupied MO ψ_K may be derived from an analysis of the ESR spectrum **(McConnell relation):**

$$a(^1H_\mu) = Q \cdot c^2_{K\mu}$$

$c_{K\mu}$ = *coefficient of the* π-AO *at the atom* C_μ *in the singly occupied* MO ψ_K

$a(^1H)$ = *isotropic hyperfine coupling constant (fluid solution)*

$Q = -2.3$ mT

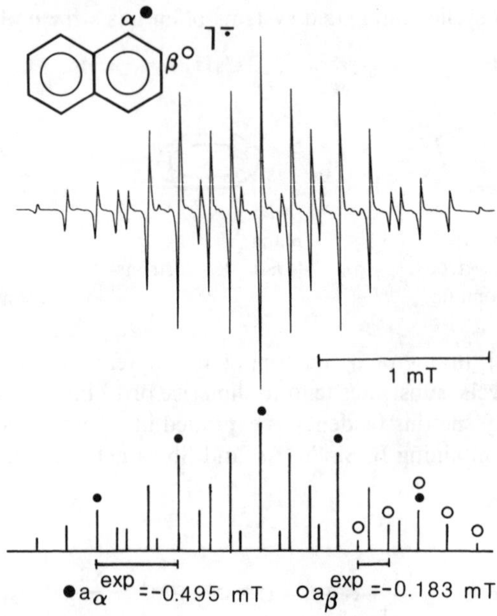

In the naphthalene radical anion $C_{10}H_8^{\cdot-}$ the LUMO ψ_6 is singly occupied. Since the probability of finding the unpaired electron varies for the α- and for the β-position, differing electron-proton hyperfine coupling constants $a(^1H_\alpha)$ and $a(^1H_\beta)$ are encountered (Intensities not to scale).

The parameter Q decribes the efficiency of π-σ spin polarization. This is the mechanism, which in π-radicals is responsible for the occurrence of spin density at the proton and thus for the observation of isotropic hyperfine coupling.

The McConnell relation (and its refined version) has been applied to a large number of π-radicals and has confirmed quantum chemical calculations of molecular electronic structure.

It frequently happens that, in addition to hyperfine coupling to magnetic nuclei of the radical anion, splitting caused by the counterion (e.g. Na^+, nuclear spin I = 3/2) is detected. In these cases, ESR spectroscopy yields unambiguous information about the structure of the ion pair $M^+ArH^{\cdot-}$.

Like NMR, ESR spectroscopy lends itself to the **study of dynamic processes.** The ion pair M^+ pyrazine$^{\cdot-}$ in THF may serve as an example (Atherton, 1966):

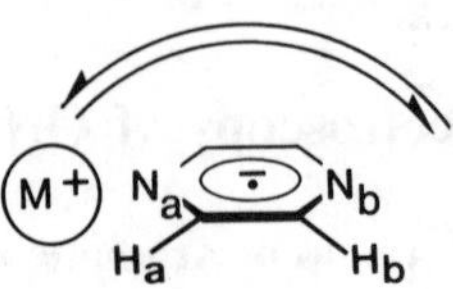

Li⁺ *is firmly coordinated to one of the two N atoms. This follows from the hyperfine pattern which is governed by unequal coupling constants $a(^1H_a) \neq a(^1H_b)$ and $a(^{14}N_a) \neq a(^{14}N_b)$.*

Na⁺, *however, undergoes an exchange between the two N sites (activation energy 30 kJ/mol) which at −65 °C is slow and at +23 °C is fast on the ESR time scale (10^{-6}–10^{-8} s). In the fast exchange case a hyperfine structure with the coupling constants* $a(\mathbf{4}\,^1H) = 0.27$ mT $a(\mathbf{2}\,^{14}N) = 0.714$ mT *and* $a(^{23}Na) = 0.055$ mT *is observed.*

b) Preservation of the Tetrameric Structure of $(LiCH_3)_4$ in Solution

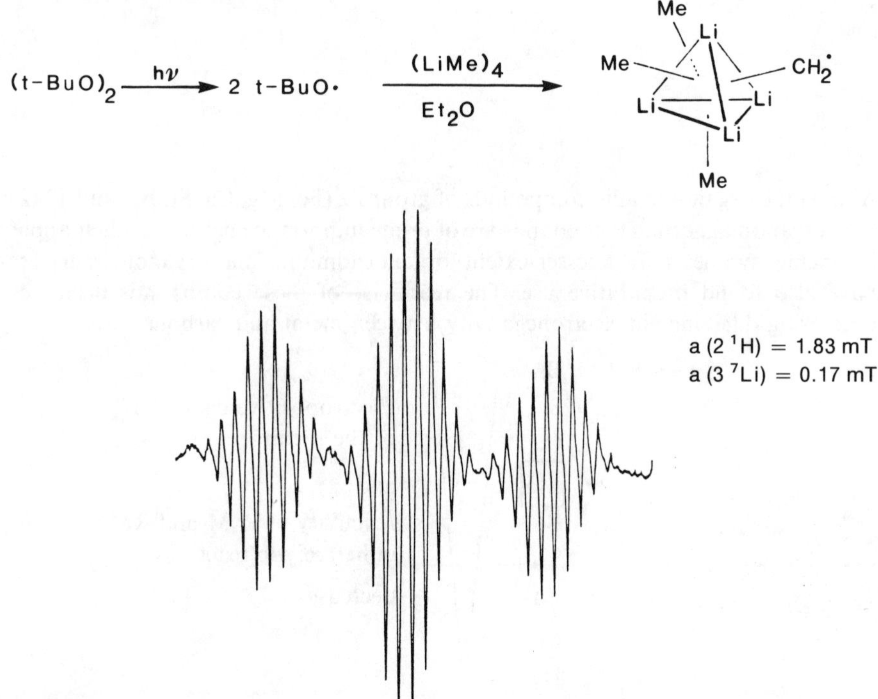

Photolytically generated radicals t-BuO· abstract an H atom from the $(LiCH_3)_4$ *unit. The identity of the species* $(CH_3)_3Li_4CH_2\cdot$ *is revealed by the ESR spectrum which consists of a triplet* ($2\ ^1H$, $I = 1/2$) *of decets* ($3\ ^7Li$, $I = 3/2$) *(Kochi, 1973).*

In addition to structural information, the observation of hyperfine coupling in fluid solution also provides hints as to the degree of covalency of the Li–C *bond.*

6 Organometallics of Groups 2 and 12

Among the organometallic compounds of groups 2 (Be, Mg, Ca, Sr, Ba) and 12 (Zn, Cd, Hg), organomagnesium compounds are of prime importance because of their application in organic synthesis. To a lesser extent, organocadmium- and organomercury reagents have also found preparative use. The reactivity of these compounds decreases with decreasing difference in electronegativity between metal and carbon:

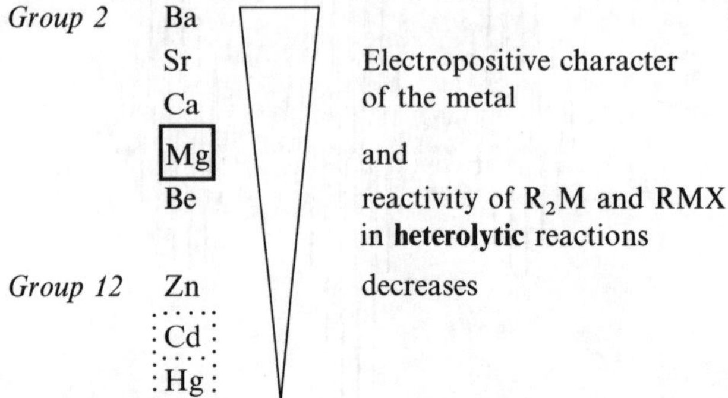

Organomagnesium compounds combine in a unique way high reactivity and ease of access. The high reactivity of R_2Hg in transmetallations may be traced to ready **homolytic** cleavage of the Hg–C bond (p. 51).

6.1 Organometallics of the Alkaline Earths (Group 2)

6.1.1 Organoberyllium Compounds

This group comprises highly toxic as well as air- and moisture-sensitive materials which display a number of structural peculiarities.

PREPARATION AND STRUCTURES

$$Be + R_2Hg \longrightarrow R_2Be \longrightarrow Hg \qquad \boxed{2}$$

$$BeCl_2 + 2\,RLi\,(RMgX) \longrightarrow R_2Be + 2\,LiCl \quad (MgXCl) \qquad \boxed{4}$$

As a Lewis acid, BeR_2 forms stable etherates $(Et_2O)_2BeR_2$; they are obtained free of solvent only with difficulty.

Like $BeCl_2$, *the compound* $\mathbf{Be(CH_3)_2}$ *is a polymeric solid (structural type* SiS_2*). The bonding situation in the chain resembles that in the aluminum alkyls (2e 3c bonds, compare* $Al_2(CH_3)_6$ *p. 78) thereby reflecting the "diagonal relationship"* Be–Al.

$$\left(-Be(\mu\text{-}CH_3)_2Be(\mu\text{-}CH_3)_2Be- \right)_n$$

d(Be–Be) = 210 pm, *angle* Be–C–Be = 66° (*Rundle*, 1951).

In the gas phase, BeR_2 is monomeric and linear and therefore is to be described by Be(*sp*) hybridization. For steric reasons, Be(*t*-Bu)$_2$ is monomeric even in the solid state. The thermolysis of Be(*t*-Bu)$_2$ yields pure, unsolvated beryllium hydride:

$$Be(t\text{-}Bu)_2 \xrightarrow{T>100^{\circ}} H_2C{=}C(CH_3)_2 + BeH_2 \qquad \textit{β-Elimination}$$

$Be(CH_3)_2$, which is void of β-H atoms, only decomposes at $T > 200\,°C$. Similar to magnesium, organoberyllium compounds equilibrate with beryllium halides:

$$BeR_2 + BeX_2 \rightleftharpoons 2\,RBeX\,.$$

Organoberyllium hydrides, which can be prepared from RBeX and metal hydrides, according to their ^{1}H-NMR spectra feature hydride bridges and terminal methyl groups in *cis*- and *trans* configuration.

$$2\,MeBeBr + 2\,LiH \xrightarrow{Et_2O} [(Et_2O)(Me)Be(\mu\text{-}H)_2Be(Me)(OEt_2)] + 2\,LiBr$$

Thus, the formation of a **hydride bridge is favored over an alkyl bridge**, and isomerizations via intermediate cleavage of hydride bridges are slow on the NMR time scale.

Beryllocene, accessible according to (E. O. Fischer, 1959)

$$BeCl_2 + 2\,NaC_5H_5 \longrightarrow (C_5H_5)_2Be$$

still poses structural problems since investigations in different states of aggregation lead to different molecular geometries:

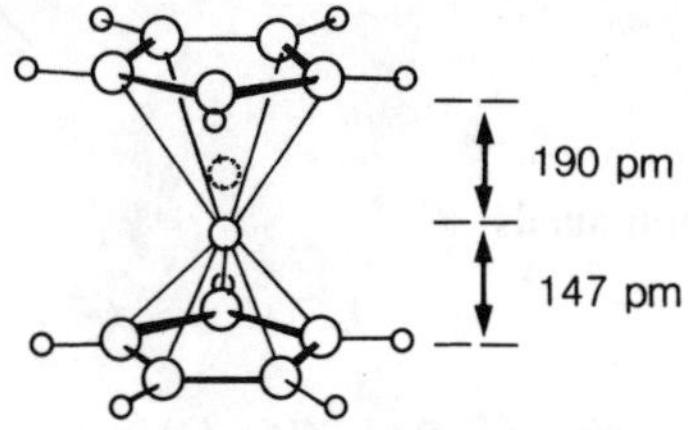

Structure of $\mathbf{(C_5H_5)_2Be}$ *in the gas phase (electron diffraction): symmetry* C_{5v}. *The energy minima of the* Be *atomic positions lie on the fivefold axis. In both positions, the* Be *atom has different distances to the two ring centers.*

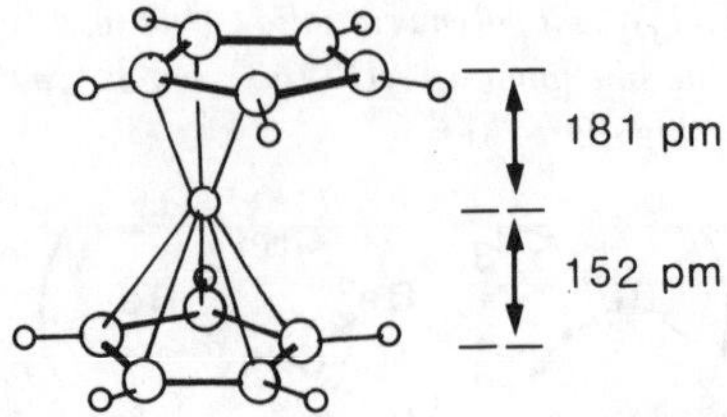

Structure of $\mathbf{(C_5H_5)_2Be}$ *in the crystal (X-ray diffraction,* $-120\,°C$*): symmetry* C_s*,* η^3*-,* η^5*-structure, "slipped sandwich". A re-interpretation of the electron diffraction data led to the conclusion that this structure may also pertain to beryllocene in the gas phase* (*Haaland*, 1979).

On the one hand, $(C_5H_5)_2Be$ possesses a dipole moment ($\mu = 2.24$ Debye in cyclohexane). On the other hand, judging from the ^1H-NMR spectrum, beryllocene displays two equivalent C_5H_5 rings. Evidently, an exchange of the Be atom between two positions on the fivefold axis takes place, which is fast on the ^1H-NMR time scale (**fluxional structure**). The bonding in beryllocene is essentially ionic. The structural peculiarities may arise from the fact that the optimal bond distance between Be^{2+} and $C_5H_5^-$ is smaller than half the van der Waals distance between two $C_5H_5^-$ ions.
From beryllocene and dimethylberyllium a half-sandwich compound is obtained which, in the spirit of Wade's rules (p. 68), may be referred to as a nido cluster:

$$(C_5H_5)_2Be + Be(CH_3)_2 \longrightarrow 2\ (C_5H_5)BeCH_3$$

CH3 — Be: 171 pm; Be — C_5H_5: 150 pm

The compound $(C_5H_5)BeCl$ *belongs to the same category,* $d(Be-Cl) = 187$ pm, $d(Be-C_5H_{5\,center}) = 145$ pm. *The structure of cyclopentadienylberyllium boranate demonstrates that* Be *and* B *share the propensity to engage in 2e 3c bridge bonding:*

$$(C_5H_5)BeCl + LiBH_4 \xrightarrow[- LiCl]{} (C_5H_5)Be(\mu\text{-}H)_2BH_2$$

6.1.2 Organomagnesium Compounds

PREPARATION

$$Mg + RX \xrightarrow{Et_2O} RMgX(Et_2O)_n \quad X = Br, I \qquad \boxed{1}$$

The addition of I_2 activates the Mg surface; MgI_2 thus formed, binds the last traces of water in the reaction mixture. In a more economical technical variation, hydrocarbon/

ether mixtures are used as solvents:

$$Mg + PhCl \xrightarrow{\text{petrolether/THF}} PhMgCl(THF)_n \quad THF:Mg > 1$$

Unsolvated **Grignard reagents** can be prepared by means of cocondensation (CC) of magnesium vapor with the vapor of the alkyl halide on a cooled surface (Klabunde, 1974):

$$Mg(g) + RX(g) \xrightarrow[\text{2. } RT]{\text{1. CC, } -196\,°C} RMgX$$

An extremely reactive form of magnesium is obtained according to Rieke (1977), if anhydrous $MgCl_2$ is reduced with potassium:

$$MgCl_2 \xrightarrow[-2\,KCl]{\text{K, THF}} Mg_{active} \left(\xrightarrow[25°\,C,\ 3\,h]{C_8H_{17}F} C_8H_{17}MgF \quad 89\%! \right)$$

The preparation of **binary organomagnesium compounds** is accomplished by transmetallation

$$Mg + R_2Hg \longrightarrow R_2Mg + Hg$$ 2

as well as by solvent-induced shifts in the equilibrium position:

$$\underset{\text{(solution)}}{2\,RMgX} + 2\,\text{dioxane} \rightleftharpoons \underset{\text{(solution)}}{R_2Mg} + \underset{\text{(precipitate)}}{MgX_2(\text{dioxane})_2}$$

The latter procedure may lead to **magnesacycles:**

$$2\,BrMg(CH_2)_nMgBr \xrightarrow[n=4,\,5,\,6]{\text{dioxane}} \overline{(CH_2)_n\,Mg} + MgBr_2(\text{dioxane})_2$$

Besides organic halides, a number of compounds with conjugated C=C double bonds react with magnesium:

(Yasuda, 1976) (Ramsden, 1967)

Mg-butadiene, [(2-butene-1,4-diyl)magnesium] a white, polymeric, sparingly soluble material of unknown structure acts as a source for butadiene dianions $C_4H_6^{2-}$. It is the reagent of choice for the introduction of butadiene as a ligand in transition-metal complexes (p. 255).

Mg-anthracene, orange-yellow crystals, reacts with anhydrous transition-metal halides to form catalysts which serve in the hydrogenation of magnesium under mild conditions (Bogdanović, 1980):

$$Mg + H_2\,(1-80\text{ bar}) \xrightarrow[20-65\,°C]{\text{Mg-anthracene}\,[MX_n]\ \text{THF}} MgH_2$$

$$MX_n = TiCl_4,\ CrCl_3,\ FeCl_2$$

MgH_2 adds to 1-alkenes (**hydromagnesation**) and, upon cleavage above 300 °C, affords pyrophoric Mg free of solvent and halide for synthetic use. Furthermore, MgH_2 is a high-temperature storage medium for molecular hydrogen (Bogdanović, 1990 R).

MECHANISM OF FORMATION AND CONSTITUTION OF ORGANOMAGNESIUM COMPOUNDS

Notwithstanding the extensive use of Grignard reagents in organic synthesis, details concerning their modes of formation, aggregation in solution, and mechanisms of consecutive reactions are still topics of current research (Ashby and others).
Results by Walborsky (1990 R) suggest that the formation of "RMgX" is initiated by an electron transfer step (*ET*). A hint as to the intermediacy of radicals R· is furnished by the reaction with the trapping agent 2,2,6,6-tetramethylpiperidine nitroxyl, TMPO (Whitesides, 1980):

$$RX + Mg(s) \xrightarrow{ET} RX^{\bar{\cdot}} + Mg(s)^{+} \longrightarrow R^{\cdot} + XMg^{\cdot}(s)$$

R· + TMPO → TMPO–R (N–OR); R· + XMg·(s) + solvent → RMgX(solv)

TMPO

The possibility that *ET*-reactions are involved, and the experience, that the latter are catalyzed by transition-metal ions dictates the use of ultrapure magnesium in mechanistic studies and invalidates many previous investigations in this field.
$Mg(C_2H_5)_2$, like $Be(C_2H_5)_2$, possesses a **polymeric chain structure** with Mg$\diagup$Et$\diagdown$Mg($2e3c$) bridge bonds. From ethers, Grignard reagents crystallize as solvates, e.g. $RMgX(Et_2O)_2$. Exceedingly bulky groups R cause MgR_2 to crystallize in unsolvated, monomeric form: $Mg[C(SiMe_3)_3]_2$ appears to be the first example of a two-coordinate magnesium compound in the solid state (Eaborn, 1989). The readily accessible **magnesocene,** $Mg(C_5H_5)_2$ (Fischer, Wilkinson, 1954), is a useful reagent for the introduction of C_5H_5 groups:

$$C_2H_5MgBr \xrightarrow[-C_2H_6]{C_5H_6,\ Et_2O} C_5H_5MgBr \xrightarrow[-\,MgBr_2]{x2,\ 220\,°C,\ 10^{-4}\ mbar} Mg(C_5H_5)_2 \xrightarrow[-\,MgCl_2]{MCl_2} M(C_5H_5)_2$$

$$Mg + 2\,C_5H_6 \xrightarrow[-H_2]{500\,°C} Mg(C_5H_5)_2$$

$Mg(C_5H_5)_2$ *forms white, pyrophoric crystals which start to sublime at* 50 °C/10^{-3} mbar, *dissolve in non-polar as well as in polar aprotic media and undergo vigorous hydrolysis. The structural parameters apply to the crystalline state (X-ray diffraction, E. Weiss, 1975). In the gas phase, the bond distances are slightly elongated and the eclipsed conformation is favored (electron diffraction, Haaland, 1975).*

139
Mg
198
230

The driving force of cyclopentadienylations of transition-metal halides by means of $Mg(C_5H_5)_2$ is largely provided by the formation of MgX_2. The bonding situation in magnesocene, especially the question of the dominance of covalent or ionic bonding, is still a matter of debate. The structural similarities to ferrocene and the absence of a dipole moment do not constitute unequivocal evidence for covalent bonding since in the case of purely ionic bonding, an axially symmetric sandwich structure would also be the (electrostatically) favored configuration. Furthermore, the realization of a molecular crystal lattice does not necessarily imply covalent bonding within the $Mg(C_5H_5)_2$ units because, due to the disparate sizes of Mg^{2+} and $C_5H_5^-$, an arrangement in which the lattice points are occupied by triple-ions $Mg^{2+}(C_5H_5^-)_2$ is more favorable than a genuine ionic lattice. Observations which ascribe a high degree of polarity to the bonding in $Mg(C_5H_5)_2$ are the electric conductivity of solutions in NH_3 or in THF, brisk hydrolysis to $Mg(OH)_2$ and C_5H_6 and the similarity of the ^{13}C-NMR shifts to those of alkali metal cyclopentadienyls:

Dominant bond type:	Ionic			Covalent
	$Li(C_5H_5)$	$Na(C_5H_5)$	$\mathbf{Mg(C_5H_5)_2}$	$Fe(C_5H_5)_2$
^{13}C-NMR δ/ppm	103.6	103.4	**108.0**	68.2

^{25}Mg-NMR evidence which implies extensive charge neutralization between magnesium and cyclopentadienyl has added new momentum to the discussion of bonding in magnesocene (Benn, 1986 R).

GRIGNARD REAGENTS IN SOLUTION

A concise description of the situation prevailing in solution is given by the **Schlenk equilibrium** (1929):

$$L(R)Mg(\mu\text{-}X)_2Mg(R)L \rightleftharpoons 2\ RMgX \overset{K}{\rightleftharpoons} R_2Mg + MgX_2 \rightleftharpoons R_2Mg(\mu\text{-}X)_2MgL_2$$

L = solvent molecule with donor properties, usually ether
K = 0.2 for EtMgBr (equilibrium constant)

An equimolar solution of MgX_2 *and* R_2Mg *in ether behaves identically to a conventional Grignard reagent* RMgX. *Radioactive* ^{28}Mg (β^- *emission,* $\tau_{1/2} = 21.2$ h), *added to the solution as* $^{28}MgBr_2$, *is rapidly scrambled among the species* MgX_2, RMgX *and* R_2Mg. *The dynamic character of the Schlenk equilibrium also manifests itself in the* ^{1}H-*NMR spectrum of* CH_3MgBr *in solution. At room temperature, only one signal is observed for the methyl protons, attesting to rapid exchange of* CH_3 *between* CH_3MgBr *and* $(CH_3)_2Mg$. *Separate signals appear at* $T < -100\,°C$ *(slow exchange). The most direct approach in the study of the Schlenk equilibrium is provided by* ***^{25}Mg-NMR spectroscopy*** *through which the various species involved can be characterized separately (Benn, 1986 R).*
Example (THF, 37 °C): Et_2Mg $\delta = 99.2$, EtMgBr $\delta = 56.2$, $MgBr_2$ $\delta = 13.9$ ppm. *At* 67 °C, *coalescence to a single signal,* $\delta = 54$ ppm, *occurs. These data reflect the kinetics of exchange as well as the temperature dependence of the Schlenk equilibrium.*

The detailed study of the influence of solvent, concentration and the nature of R has revealed that the reality must be more complicated than the description given by the Schlenk equilibrium. In THF, RMgX exists as the monomeric species $RMgX(THF)_2$ over a wide range of concentrations. In Et_2O, however, RMgX is monomeric only in dilute solution (< 0.1M) whereas at higher concentrations oligomers (chains and rings) are formed:

$$(Et_2O)_2Mg(Ph)Br \cdots \left[(Et_2O)Mg(Ph)Br \right]_n \cdots (Et_2O)Mg(Ph)Br$$

In general, **halide bridges are preferred over alkyl (2*e* 3*c*) bridges.** In the case of *t*-BuMgX, only monomeric and dimeric units are encountered.

Solutions of RMgX in Et_2O show electrical conductivity, although the extent of dissociation is small:

$$2\,RMgX \rightleftharpoons RMg^+ + RMgX_2^-$$

$$RMg^+ \xrightarrow{+e^-} R\cdot + Mg \quad \text{(Cathode)}$$

$$RMgX_2^- \xrightarrow[\text{electrolysis}]{-e^-} R\cdot + MgX_2 \quad \text{(Anode)}$$

During electrolysis, radicals R· are generated at both electrodes. If these radicals are relatively long lived, dimerization can occur:

$$2\,PhCH_2MgBr \xrightarrow[\text{electrolytic coupling}]{} PhCH_2CH_2Ph + Mg + MgBr_2$$

Alternatively, the radicals can react with the electrode material as in a commercial process for the production of tetraethyllead (p. 139):

$$4\,C_2H_5MgCl \xrightarrow[\text{Pb electrodes}]{\text{electrolysis}} Pb(C_2H_5)_4 + 2\,Mg + 2\,MgCl_2$$

REACTIONS OF ORGANOMAGNESIUM COMPOUNDS

From the breadth of applications of Grignard reagents in organic synthesis, only a small selection will be presented here (products after hydrolysis):

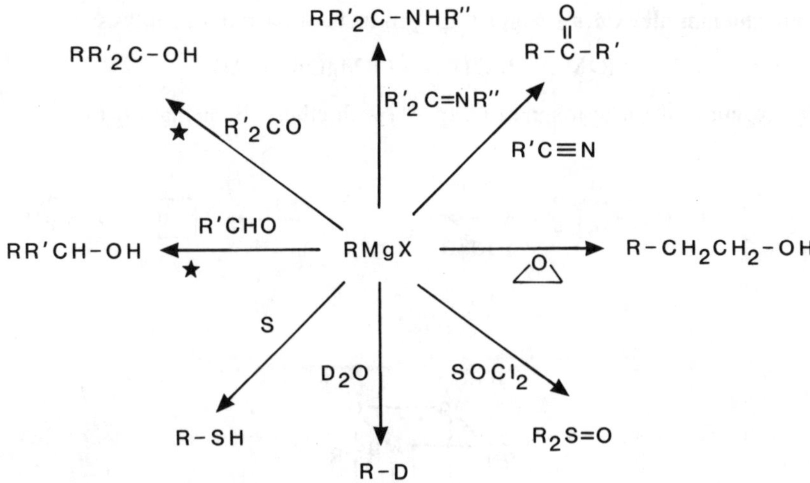

* The intermediate conversion of RMgX into $RTi(OCHMe_2)_3$ renders the addition to carbonyl functions highly chemo- and stereoselective (p. 32).

Like organolithium compounds, Grignard reagents serve as alkylating and arylating agents for main-group and transition-metal halides:

$$SbCl_3 + 3\,CH_3MgX \xrightarrow[-\,3MgXCl]{} (CH_3)_3Sb \overset{CH_3MgX}{\not\longrightarrow}$$

RMgX is less reactive than RLi, however. In contrast to the latter, no "ate complexes" like $[(C_6H_5)_6Sb]^-$ are obtained from reactions with Grignard reagents.
The following example illustrates the synthesis of a metallacycle:

$$(\eta^5\text{-}C_5H_5)_2MCl_2 + C_6H_4(CH_2MgBr)_2 \xrightarrow[-\,2MgBrCl]{} (\eta^5\text{-}C_5H_5)_2M(CH_2)_2C_6H_4$$

Metallacycle

M=Ti, Zr, Hf, Nb

Organomagnesium hydrides RMgH are prepared according to:

$$R_2Mg + MgH_2 \xrightarrow{THF} [R(THF)Mg(\mu\text{-}H)_2Mg(THF)R]$$

Note the preference for (2*e* 3*c*)hydride- over (2*e* 3*c*)alkyl bridging.

Organomagnesium alkoxides RMgOR are products of partial alcoholysis:

$$R_2Mg + R'OH \longrightarrow RMgOR' + RH$$

or of the reaction of Rieke magnesium (p. 41) with ethers (Bickelhaupt, 1977):

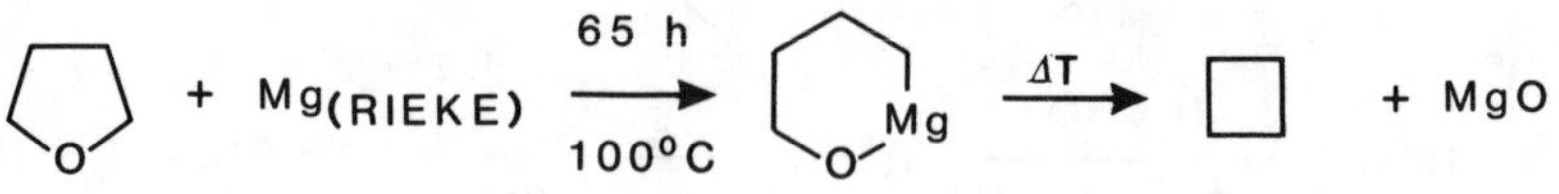

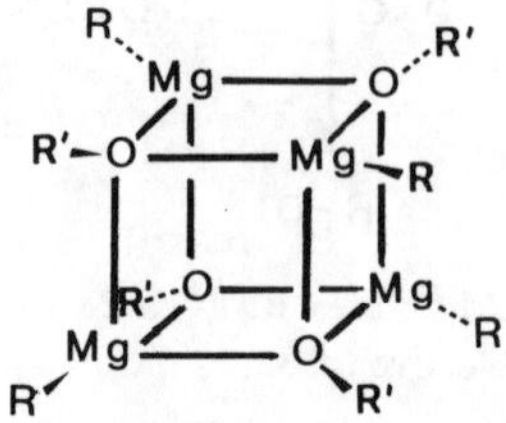

$$\text{THF} + \text{Mg(RIEKE)} \xrightarrow[100^\circ C]{65\ h} \text{(oxamagnesacyclohexane)} \xrightarrow{\Delta T} \square + \text{MgO}$$

The structure of the aggregates $(RMgOR')_n$ *is governed by the ability of alkoxide ions to function as triply bridging ligands. The tetramer* $(RMgOR')_4$ *is an example of a heterocubane* $(AB)_4$.

Magnesium-"ate-complexes" $M_xMg_yR_z$ (M = group 1, 2 or 13 metal) were first described by Wittig (1951). The less electropositive metal usually appears in the "ate-complex"-anion.

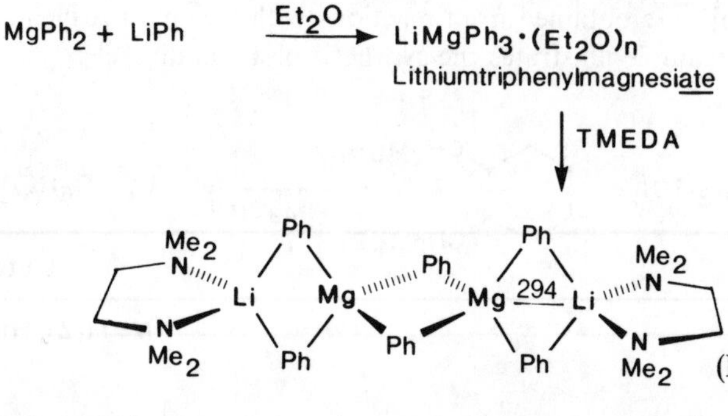

$$MgPh_2 + LiPh \xrightarrow{Et_2O} LiMgPh_3 \cdot (Et_2O)_n$$

Lithiumtriphenylmagnesi<u>ate</u> solution

↓ TMEDA

crystal (E. Weiss, 1978)

The central unit $[Mg_2Ph_6]^{2-}$ is isoelectronic and isostructural to $[Al_2Ph_6]$ (p. 79). In the case of two metals of comparable electropositive character, an essentially covalent structure with ($2e\,3c$) alkyl bridges is realized (Stucky, 1969):

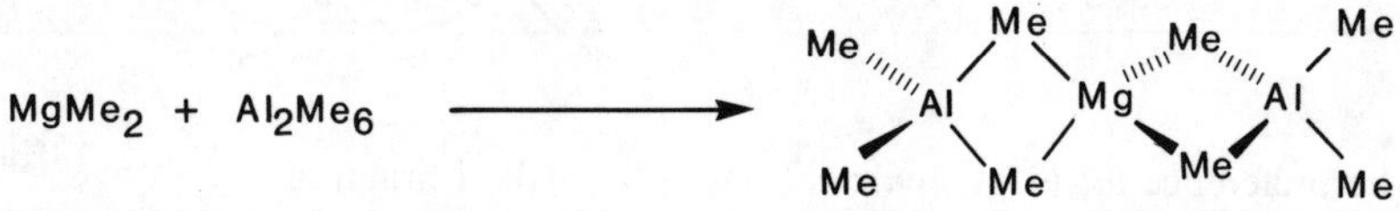

$$MgMe_2 + Al_2Me_6 \longrightarrow Me_2Al(\mu\text{-}Me)_2Mg(\mu\text{-}Me)_2AlMe_2$$

6.1.3 Organocalcium-, Strontium- and Barium Compounds

The preparation of the heavier alkaline earth organometallics is difficult, requiring meticulous control of the reaction conditions. Since the products do not offer any advantages over Grignard reagents, their chemistry has not been studied extensively. Organobarium compounds have found limited application as initiators for polymerizations. Like the organyls of the heavier alkali metals, the Grignard homologues RMX (M = Ca, Sr, Ba) are best described by ionic bonding; they also attack ethers at the α-position. This property, as well as the lack of solubility in nonpolar solvents complicates the study of RMX in homogeneous solution.
The preparation of the di(cyclopentadienyl) compounds of Ca, Sr and Ba should not be dismissed as an exercise in "bean counting", since a number of thought-provoking structural features has emerged from their study. $Ca(C_5H_5)_2$ in the crystal forms polymeric chains (Stucky, 1974). Structural information concerning the isolated molecules is available only for the permethylated derivatives $M(C_5Me_5)_2$.

Special features of the ligand ***pentamethylcyclopentadienyl*** $\mathbf{C_5Me_5^-}$ **(Cp*)**, *compared to* $C_5H_5^-$ (Cp) (Jutzi, 1987 R):

- *stronger π-donor, weaker π-acceptor properties* (*p.* 319)
- *increased covalent character of the cyclopentadienyl-metal bond*
- *increased thermal stability of the metal complexes*
- *kinetic stabilization effected by steric shielding of the central metal*
- *attenuation of intermolecular interactions, decreased tendency towards polymeric structures, increased vapor pressure and solubility.*

The latter aspect allowed the structural analyses of $M(C_5Me_5)_{2(g)}$ (M = Ca, Sr, Ba) by means of electron diffraction (Blom, 1987) which have recently been supplemented by X-ray crystallography (Hanusa, 1990):

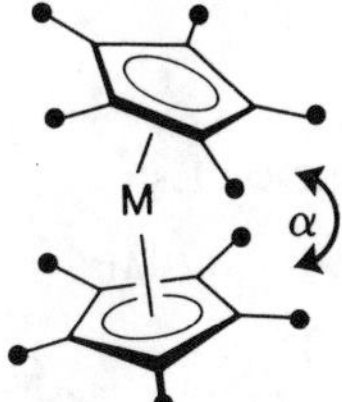

	α	
M	$M(C_5Me_5)_2$	MI_2
Mg	180	180
Ca	154	148
Sr	149	144
Ba	148	138

It came as a surprise to find that the di(cyclopentadienyl)metal complexes of the heavier alkaline earths possess ***bent sandwich*** *structures. A comparable bending is, however, also encountered within the corresponding metal halides in the gas phase* (*Kasparov*, 1979) – *the reasons for the lowering of symmetry in these two classes of compounds should be related. Whereas in the case of the compounds* $M(C_5H_5)_2$, M = Ge, Sn, Pb, *the existence of a lone pair of electrons at the central metal atom may enter into explanations of the bent structure* (*p.* 134), *its absence in complexes with* M = Ca, Sr, Ba *calls for a more subtle interpretation. Ironically, both covalent* (Coulson, Hayes, 1973) *and ionic bonding models* (*Guido*, 1976) *have been used in this context* (*cf. p.* 443).

6.2 Organometallics of Zn, Cd, Hg (Group 12)

The elements of group 12 possess completely filled *d* shells of low energy, void of donor or acceptor properties. Therefore, a discussion of the organometallic chemistry of Zn, Cd, and Hg in connection with that of the group 2 elements is warranted.

6.2.1 Organozinc Compounds

PREPARATION

$$C_2H_5I + Zn(Cu) \longrightarrow \text{“}C_2H_5ZnI\text{”} \xrightarrow{\Delta T} (C_2H_5)_2Zn + ZnI_2$$ 1

$$Zn + R_2Hg \longrightarrow R_2Zn + Hg$$ 2

$$ZnCl_2 + 2\,RLi\ (RMgX) \longrightarrow R_2Zn + 2\,LiCl\ (MgXCl)$$ 4

$$3\,Zn(OAc)_2 + 2\,R_3Al \longrightarrow 3\,R_2Zn + 2\,Al(OAc)_3$$

STRUCTURE AND PROPERTIES

In contrast to BeR_2 and MgR_2, binary organozinc compounds ZnR_2 (R = alkyl or aryl) are monomeric. The molecules are linear and have low melting- and boiling points [e.g. $Zn(C_2H_5)_2$: mp −28 °C, bp 118 °C, pyrophoric]. Whereas self-association via Zn–R–Zn (2*e* 3*c*) bridges is apparently disfavored, Zn–H–Zn (2*e* 3*c*) bridges are formed readily:

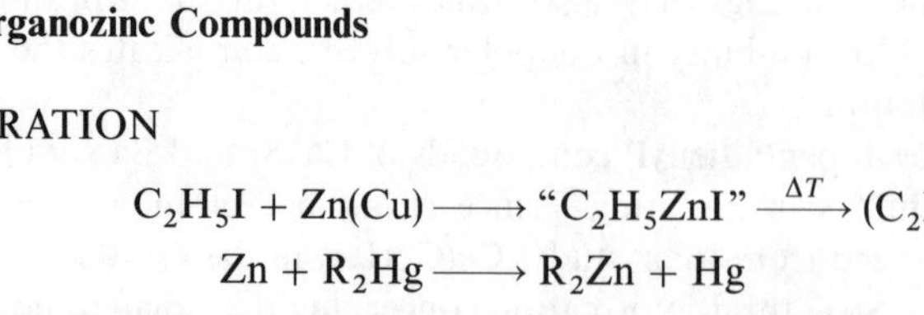

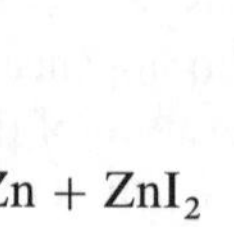

$$Et_2Zn + ZnH_2 \xrightarrow[\text{THF}]{\text{pyridine}} [(Py)(Et)Zn(\mu\text{-}H)]_3$$

Zn–H–Zn
2e3c

σ-Donor ligands also coordinate to binary organozinc molecules:

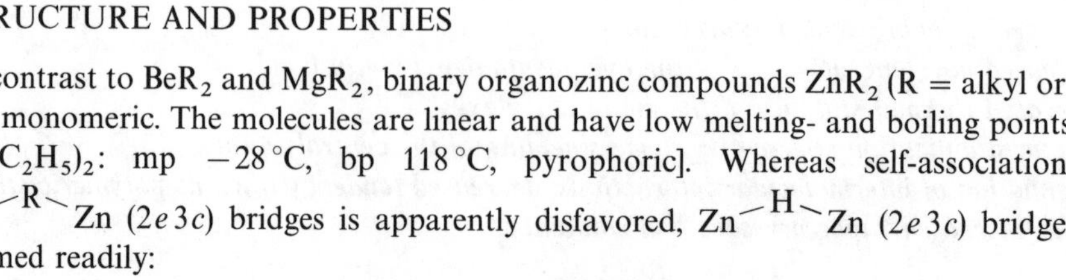

Me–Zn–Me $Me_2Zn–OMe_2$ $Bu_2Zn(TMEDA)$ $[ZnMe_4]^{2-}$

They effect the association of molecules of the type RZnX:

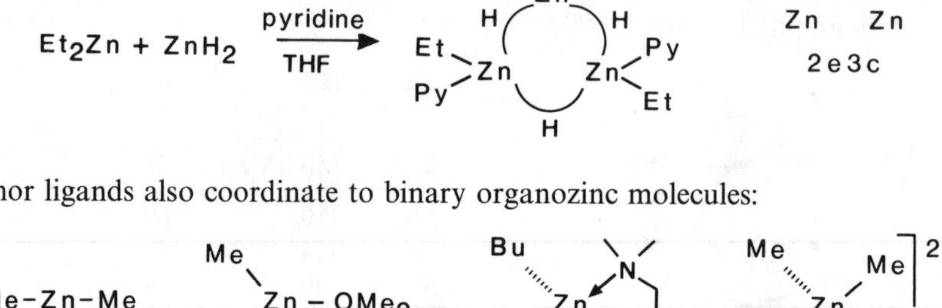

$[RZnX]_2$ $[RZnX]_3$ $[RZnX]_4$

X = halide, t-BuO^- a heterocubane

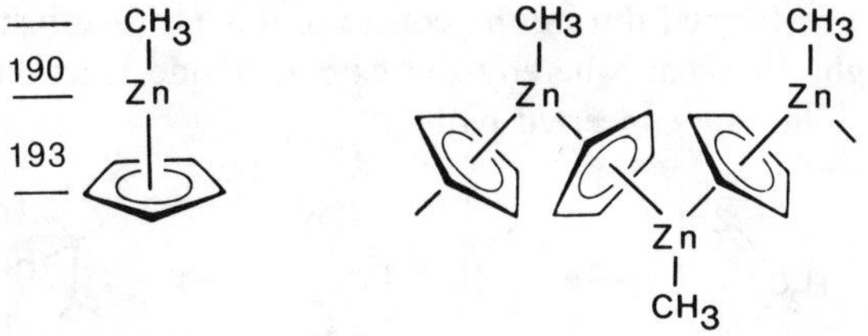

(Cyclopentadienyl) (methyl)zinc, *which is monomeric in the gas phase, forms chains with bridging* Cp *ligands in the crystal.*

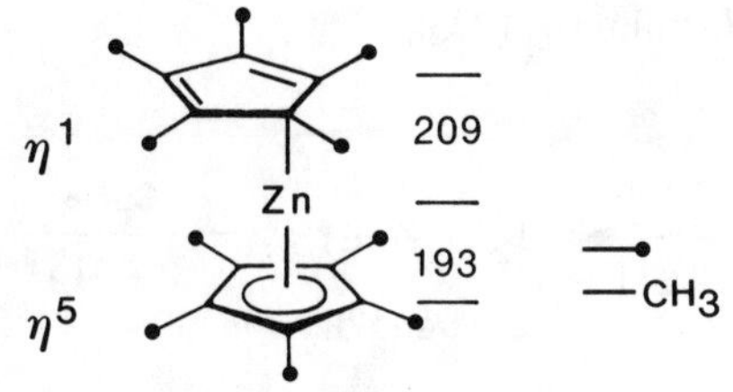

The structure of **bis(pentamethylcyclopentadienyl)zinc** *in the gas phase features* η^1-, η^5*-coordination. In this way, as in the case of* CpZnMe, *an* 18 *valence-electron shell is created for the* Zn *atom.* Cp^*_2Zn *in solution gives rise to one* ^{1}H-*NMR signal only, which indicates a rapid haptotropic change between* η^1*- and* η^5*-coordination* (*Haaland*, 1985).

Cp_2Zn, like CpZnMe, is polymeric in the solid state. In their chemical behavior, organozinc compounds resemble organomagnesium- and organolithium compounds, their reactivity in addition reactions being reduced, however. In organometallic synthesis, ZnR_2 replaces LiR and RMgX if relatively mild, non-basic conditions are called for:

$$Me_2Zn + NbCl_5 \longrightarrow Me_2NbCl_3 + ZnCl_2$$

The well known **Reformatsky reaction** proceeds via an organozinc intermediate:

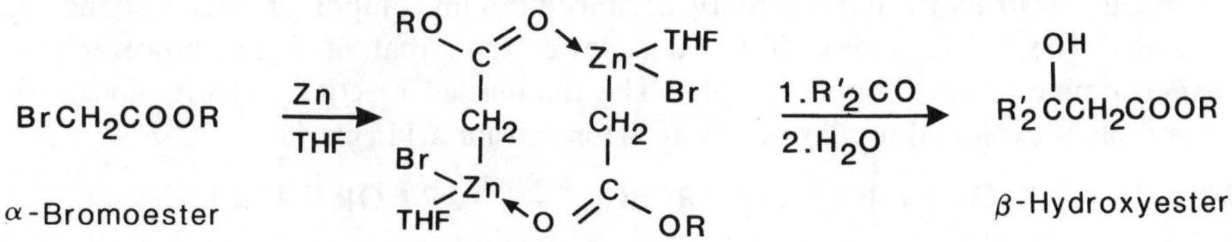

The dimeric structure of the Reformatsky reagent is maintained in solution (van der Kerk, 1984).

Organozinc carbenoids play an important role in organic synthesis. An example is the **Simmons-Smith reaction** (1973 R), by which cyclopropanations are effected:

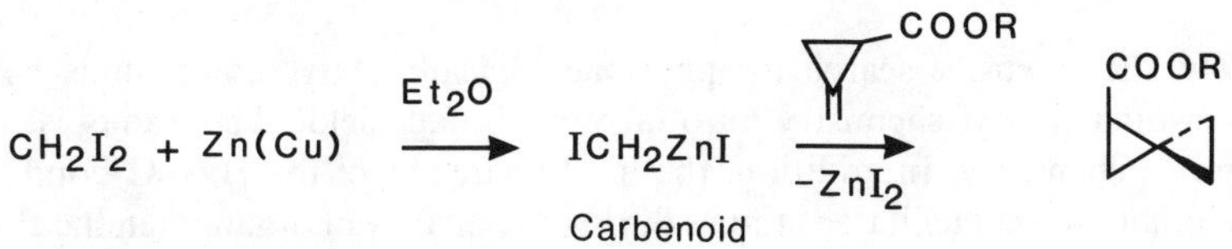

Free carbenes are not formed during the course of this process; instead, attack by the carbenoid is thought to occur whereby the extent of side reactions is reduced. The addition probably follows a concerted path:

$$\text{>C=C<} + H_2C(ZnI)(I) \longrightarrow [\text{transition state}: CH_2 \cdots ZnI \cdots I] \longrightarrow \text{cyclopropane-}CH_2 + ZnI_2$$

Organozinc carbenoids also lend themselves to ring expansions of arenes to yield cycloheptatriene derivatives (Hashimoto, 1973):

$$CHI_3 + Et_2Zn \xrightarrow[-EtI]{} \underset{\text{Carbenoid}}{I_2CHZnEt} \xrightarrow[-ZnI_2]{C_6H_6} \text{7-H-7-Et-cycloheptatriene}$$

6.2.2 Organocadmium Compounds

The method of choice for the preparation of binary organocadmium compounds is metathesis

$$CdCl_2 + 2\,LiR\,(RMgX) \longrightarrow R_2Cd + 2\,LiCl\,(MgXCl)$$ 4

Organocadmium halides equilibrate with their binary counterparts:

$$Me_2Cd + CdI_2 \overset{THF}{\rightleftharpoons} 2\,MeCdI \quad K \approx 100$$

Organocadmium alkoxides are obtained by partial alcoholysis:

$$R_2Cd + R'OH \longrightarrow RCdOR' + RH$$

With regard to structure and reactivity, organocadmium compounds resemble their zinc congeners. The Lewis acidity of R_2Cd is lower than that of R_2Zn, however (tetraorganocadmiates CdR_4^{2-} are unstable). The diminished reactivity of organocadmium compounds is exploited in a **synthesis of ketones from acid chlorides:**

$$R_2Cd + CdCl_2 \rightleftharpoons 2\,RCdCl \xrightarrow{2\,R'COCl} 2\,R'COR + CdCl_2$$

This procedure is practicable in the presence of the functionalities $\rangle C{=}O$, $-COOR$, and $-C{\equiv}N$ to which organocadmium reagents, as opposed to Grignard reagents, do not add.

6.2.3 Organomercury Compounds

Historically, the extensive search for pharmacologically active compounds has transformed organomercury chemistry into a very broad field. The same is true for organoarsenic chemistry. In addition, the inert character of the Hg−C bond towards attack by air and water facilitated studies during the infancy of organometallic chemistry.

Traditionally, organomercury alkoxides PhHgOR′ have found application as **fungicides, bactericides and antiseptics,** but their use is currently on the decline. Today, a few organomercury compounds are still utilized in organic synthesis.

PREPARATION

$$RI + Hg \xrightarrow{\text{sunlight}} RHgI \qquad \text{(historical)} \; \boxed{1}$$

$$ArN_2Cl + Hg \xrightarrow{0\,°C} ArHgCl + N_2$$

$$HgCl_2 \xrightarrow{RLi} RHgCl \xrightarrow{RLi} R_2Hg \qquad \boxed{4}$$

$$Hg(OAc)_2 + R'_2BR \longrightarrow RHgOAc + R'_2BOAc$$

(R = primary alkyl, alkenyl, R′ = cyclohexyl)

$$PhH + Hg(OAc)_2 \xrightarrow{MeOH} PhHgOAc + HOAc \quad \text{(mercuration)} \; \boxed{7}$$

$$(CF_3COO)_2Hg \xrightarrow{K_2CO_3,\ 200\,°C} (CF_3)_2Hg + 2\,CO_2 \qquad \boxed{11}$$

$$Na(MeAlCl_3) \xrightarrow[\text{Hg anode}]{\text{salt melt}} Me_2Hg \qquad \text{(electroalkylation)}$$

STRUCTURE AND PROPERTIES

In accordance with the similar electronegativities of mercury and carbon, the Hg–C bond is essentially covalent. The organometallic chemistry of mercury is almost exclusively that of the oxidation state Hg^{II}. Species of the kind $R-Hg^{I}-Hg^{I}-R$ have occasionally been proposed, but never unequivocally identified. Thermochemically, the instability of organomercury (I) compounds is reflected in the strongly differing dissociation energies for the first and the second methyl group in $(CH_3)_2Hg$, $D(MeHg-Me) = 214\,kJ/mol$, $D(Hg-Me) = 29\,kJ/mol$. Therefore:

$$RHg^+ \xrightarrow{e^-} \{RHg\cdot\} \longrightarrow R\cdot + Hg$$

R_2Hg species readily undergo homolytic cleavage if energy is supplied thermally or photochemically. This reaction represents a convenient **source of radicals** and can be applied to homolytic aromatic substitution:

$$R_2Hg \xrightarrow[-Hg]{h\nu} 2\,R\cdot \xrightarrow{ArH} ArR$$

The utility of organomercury compounds as transmetallating agents (p. 15) is based on this inherent weakness of the Hg–C bond. Conversely, organomercury compounds are practically inert against air and moisture. This may be traced to the weak acceptor character of molecules R_2Hg and RHgX. The formation of adducts with concomitant increase of the coordination number is observed for R_2Hg only if R bears an exceptionally high electron affinity. Example: $(CF_3)_2Hg(R_2PCH_2CH_2PR_2)$.

If donor ligands are proffered to organomercury halides, a redistribution often occurs wherein the organometallic component maintains its coordination number 2:

$$2\,RHgX + 2\,PR'_3 \longrightarrow R_2Hg + HgX_2(PR'_3)_2$$

The molecules of RHgX and R_2Hg are **linear** (*sp* or $d_{z^2}s$ hybridization of Hg). Attempts to force bent C−Hg−C moieties by means of "tricks" are evaded by oligomerization (D. S. Brown, 1978):

$$C_6H_4Br_2 + Na/Hg \xrightarrow{CH_3COOEt} [o\text{-}C_6H_4Hg]_3$$

o-Phenylenemercury
$d(Hg-C) = 210$ pm
$d(Hg\cdot\cdot Hg) = 358$ pm
compare: $d(Hg-Hg) = 302$ pm (*metal*)

Di(cyclopentadienyl)mercury **$(C_5H_5)_2Hg$,** can be prepared in aqueous media

$$2\,C_5H_5Tl + HgCl_2 \xrightarrow{H_2O} (C_5H_5)_2Hg + 2\,TlCl$$

Although $(C_5H_5)_2Hg$ forms a Diels-Alder adduct with maleic anhydride and displays a $Hg \overset{\sigma}{-} C$ stretching vibration in the IR spectrum, only one signal is observed in the ^{1}H-NMR spectrum. These findings are in accord with **structural fluxionality** for $(\eta^1\text{-}C_5H_5)_2Hg$, such that a sequence of **haptotropic shifts** renders all the ring protons equivalent on the ^{1}H-NMR time scale (Wilkinson, 1956):

$$\eta^1\text{-}C_5H_5Hg\text{–}C_5H_5 \rightleftharpoons \eta^1\text{-}C_5H_5Hg^{\oplus}\cdots[C_5H_5]^{\ominus} \rightleftharpoons \eta^1\text{-}C_5H_5Hg\text{–}C_5H_5$$

The pentamethylcyclopentadienyl ligand Cp* replaces only one chloride ion at $HgCl_2$ leading to **Cp*HgCl,** a compound which in the solid state displays an interesting folded ladder structure (Lorberth, Massa, 1988):

The most important reactions with synthetic appeal, involving organomercury compounds as intermediates, are mercuration, solvomercuration/demercuration and carbene transfer.

- **Mercurations** are metallation reactions (H/M exchange) in which mercury(II)-acetate is used. They are applicable to arenes and to nonaromatic compounds of sufficient CH acidity (alkynes, nitro compounds, 1,3-diketones etc.):

$$\mathrm{PhH + Hg(OAc)_2 \longrightarrow PhHgOAc + HOAc}$$

Mechanistically, this is an electrophilic substitution (S_E). Mercurations are catalyzed by strong, non-coordinating acids which generate the attacking electrophile $HgOAc^+$:

$$\mathrm{Hg(OAc)_2 + HClO_4 \longrightarrow HgOAc^+ + ClO_4^- + HOAc}$$

- **Solvomercuration/demercuration** constitutes an addition of HgX^+ and Y^- to an alkene with subsequent cleavage of the Hg–C bond:

$$\mathrm{RCH{=}CH_2 \xrightarrow[HY]{HgX_2\ (X=NO_3,\ OAc)} R\overset{\displaystyle Y}{\overset{|}{C}}HCH_2HgX \xrightarrow{H_2,NaBH_4} R\overset{\displaystyle Y}{\overset{|}{C}}HCH_3}$$

(HY = *solvent or component of the medium*)
(Y = OH, OR, OAc, O_2R, NR_2 *etc.*)

Example:

$$\text{(cyclohexylidene)}{=}CH_2 \xrightarrow[H_2O/THF]{Hg(OAc)_2} \text{1-(}CH_2HgOAc\text{)cyclohexan-1-ol} \xrightarrow[NaOH]{NaBH_4} \text{1-}CH_3\text{-cyclohexan-1-ol}$$

This procedure has been widely applied; it operates under particularly mild conditions, tolerates many functional groups and is rarely accompanied by rearrangements. The first step, called **oxymercuration** in this case, follows Markovnikov direction, and thus complements hydroborations (p. 60 f), which proceed in an anti-Markovnikov fashion. The utility of the mercuration products stems from the possibility of replacing mercury by hydrogen, halogen or other functional groups.

- **Carbene Transfer from Phenyl(α-halomethyl)mercury (Seyferth-Reagent)**
Compounds of the type $PhHgCX_3$ readily extrude dihalocarbenes CX_2 which can be used in consecutive reactions. From precursors $PhHgCX_2X'$ elimination occurs in such a way that PhHgX′ retains the heavier halide. Examples:

$$PhHgCl + CHCl_2Br + t\text{-}BuOK \xrightarrow[-25°]{THF} PhHgCCl_2Br + KCl + t\text{-}BuOH$$

$$PhHgCCl_2Br \xrightarrow[-\,PhHgBr]{C_6H_6,\ 80°,\ 2h} CCl_2$$

$$CCl_2 + \text{cyclohexene} \longrightarrow \text{7,7-dichlorobicyclo[4.1.0]heptane } (C_6H_{10}CCl_2)$$

$$PhHgCX_2Br \xrightarrow[-\,PhHgBr]{\Delta T} CX_2 \xrightarrow{Ar{-}C{\equiv}C{-}R} \text{1,1-dihalo-2-Ar-3-R-cyclopropene} \xrightarrow{H_2O} \text{2-Ar-3-R-cyclopropenone}$$

X = Cl, Br

These **cyclopropanations** proceed under rather mild conditions and, most notably, in the absence of basic reagents; the intermediate $PhHgCX_2X'$ has to be isolated, however. Functional groups like $-COOH$, $-OH$, and $-NR_2$ create problems, since they react with $PhHgCX_2X'$. In contrast to the Simmons-Smith reaction, the mechanism of the Seyferth dihalocyclopropanation is thought to involve free carbene intermediates.

Excursion

Organomercury compounds in vivo

As a result of the Minimata catastrophy in Japan (1953–1960) and of accidental mass poisonings by cereal seed in Iraq (1970/71), the reputation of the element mercury has been considerably maligned. In the first case, mercury from industrial waste waters found its way into marine organisms and sediments. In the second case, wheat, which had been treated with the seed disinfectant ethylmercury *p*-toluenesulfonic anilide, was consumed. The threat posed by mercury in vivo is closely affiliated with organomercury chemistry:

- Biological methylation converts inorganic Hg^{II}-compounds into the **extremely toxic ion CH_3Hg^+** (Jernelöv,

1969). Daily oral intake of 0.3 mg of CH_3HgX by man leads to symptoms of poisoning. The generation of RHgX through oxidative alkylation of Hg^0 in the environment has not yet been established.

- Several compounds of the type CH_3HgX are water soluble. Thus, they are rapidly distributed in aqueous ecosystems and accumulate in certain organisms (particularly fish). Example: Water (Lake Powell, Arizona, 0.01 ppb) → food chain → trout (84 ppb), carp (250 ppb).

ple in chloralkali electrolysis, amounts to > 90%.

The **biological methylation** of Hg^{2+} ions to yield CH_3Hg^+, and to a limited extent also $(CH_3)_2Hg$, is effected by microorganisms, which utilize methylcobalamin $CH_3[Co]$, a derivative of vitamin B_{12} coenzyme. $CH_3[Co]$ is the only natural product which is able to transfer the methyl group as a carbanion, as required in the reaction with Hg^{2+} (p. 203):

$$CH_3[Co] + Hg^{2+}_{aq} \xrightarrow{H_2O} (H_2O)[Co]^+ + CH_3Hg^+_{aq}$$

In this context, two **model reactions** are of interest:

$$[CH_3Co^{III}(CN)_5]^{3-} + Hg^{2+}_{aq} + H_2O \longrightarrow CH_3Hg^+_{aq} + [(H_2O)Co^{III}(CN)_5]^{2-}$$

(Halpern, 1964)

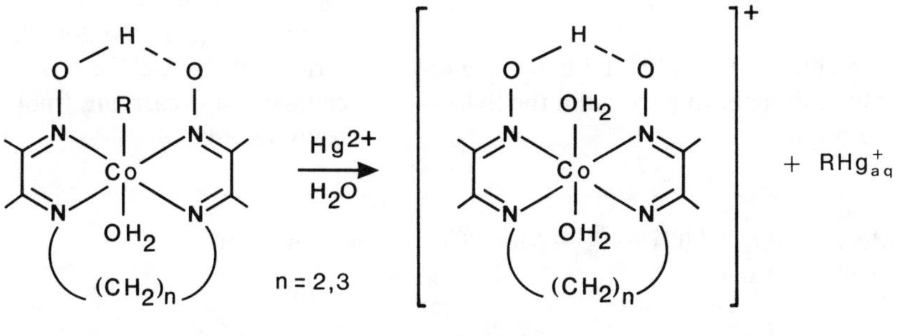

n = 2,3

With regard to the **origin** of mercury in natural waters, anthropogenic sources of modern times (retrieval of Hg from ores, fossil fuels, chloralkali electrolysis) are globally comparable to the natural sources (weathering, vulcanism). Because of the higher local concentrations, the former are more of an ecological threat, however. Admittedly though, the last decade has witnessed a remarkable reduction of Hg emission which, for exam-

Other metals which are subject to biological methylation include tin, lead and arsenic. Alkylated metal ions are usually more poisonous than the purely inorganic forms, maximal toxicity being reached for the permethylated monocations $MeHg^+$, Me_3Sn^+ and Me_2As^+.

The **distribution** of the CH_3Hg^+ cation between aqueous systems and living organisms is controlled by its coordination chemistry. Since organometallic Hg^{II} occurs almost exclusively with the coordination number 2, the following equilibria only have to be considered:

$$[MeHgOH_2]^+ + OH^- \rightleftharpoons MeHgOH + H_2O \qquad \text{acid/base}$$

$$[MeHgOH_2]^+ + MeHgOH \rightleftharpoons [(MeHg)_2OH]^+ + H_2O \qquad \text{condensation}$$

$$[MeHgOH_2]^+ + X^- \rightleftharpoons MeHgX + H_2O \qquad \text{complex formation}$$

Formation constants: MeHgF $\ll$ MeHgCl $<$ MeHgBr $<$ MeHgI $\ll$ MeHgSMe

The ion **$MeHg^+$** therefore should be classified as a **soft Lewis acid.** The preference of $MeHg^+$ for soft Lewis bases governs the solubility behavior of compounds MeHgX:

X	MeHgX
Hal^-, CN^-, SCN^-, SR^- (soft bases)	covalent bond MeHg-X, soluble in organic media
NO_3^-, SO_4^{2-} (hard bases)	ionic bond $MeHg^+X^-$ soluble in water as $MeHg^+_{aq}$

The uptake of $MeHg^+$ by living organisms is thought to proceed in the following way:

In addition to the blocking of thiol groups of enzymes by $MeHg^+$, bonding of the methylmercury cation to the pyrimidine bases uracil and thymine has also been suggested. The discovery that $MeHg^+$ causes chromosomal aberrations, i.e. is mutagenic, is thus explained. **MeHgSR** complexes are **thermodynamically stable,** but **kinetically labile.** This lability makes the ion $MeHg^+$ immediately available upon changes of the medium. It also forms the basis for chemotherapy which consists of the conversion of $MeHg^+$ into a chelate complex, which then is excreted. For this purpose, penicillamine (2-amino-3-methyl-3-thiobutyric acid) as well as synthetic ion-exchange resins carrying thiol groups, are being tested.

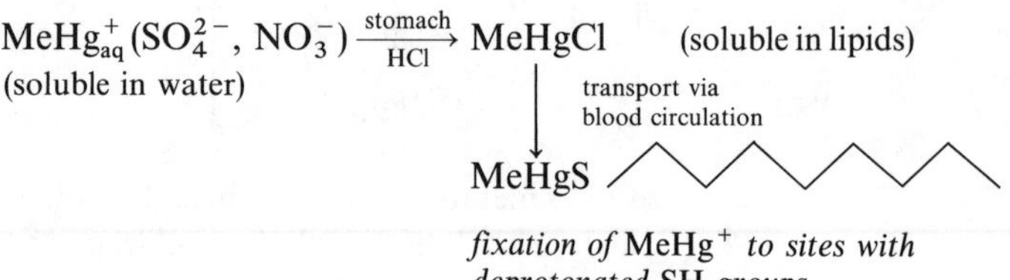

fixation of $MeHg^+$ *to sites with deprotonated* SH *groups*

7 Organometallics of the Boron Group (Group 13)

Among the elements B, Al, Ga, In and Tl the organometallic chemistry of boron and aluminum clearly predominates. Organoboron chemistry is closely related to the chemistry of organoboranes; therefore boranes will also be touched upon in this chapter. Boranes and organoboranes are intriguing from a structural (borane clusters), bond theoretical (multicenter bonding) as well as from a practical point of view (hydroboration, carbaboration). At one time, boranes and carbaboranes were of interest as potential rocket propellants since their specific combustion value exceeds that of hydrocarbons by about 40%. However, complete combustion to form B_2O_3 and H_2O was never achieved in practice. Instead of B_2O_3, suboxides like BO are always formed to a certain extent, negating their advantage over hydrocarbons. Certain organoboron compounds have entered nuclear medicine as neutron absorbers. The practical value of aluminum organyls is based on their use in a number of technical processes and, in the laboratory, as economical sources of carbanions.

7.1 Organoboron Compounds

7.1.1 Organoboranes

PREPARATION

Organic derivatives of BH_3 are accessible in a multitude of ways. Direct syntheses are of no importance since pure elemental boron is both expensive and rather unreactive.

Binary organoboranes:

$$Et_2OBF_3 + 3\,RMgX \longrightarrow R_3B + 3\,MgXF + Et_2O \quad (R = \text{alkyl, aryl})$$ 4

In the presence of an excess of RMgX, tetraorganoborates BR_4^- are formed.

$$B(OEt)_3 + 1/2\,Al_2R_6 \longrightarrow R_3B + Al(OEt)_3$$

Contrary to lithium- and magnesium organyls, organoaluminum compounds do not effect quaternation at boron.

$$HB\langle + RCH{=}CH_2 \longrightarrow RCH_2{-}CH_2B\langle$$ **Hydroboration** 8

Organoboron halides:

$$BCl_3 + SnPh_4 \longrightarrow PhBCl_2 + Ph_3SnCl \quad \text{or} \quad Ph_2BCl + Ph_2SnCl_2$$ 4

$$BCl_3 + ArH \xrightarrow{AlCl_3} ArBCl_2 + HCl$$

$$BR_3 + I_2 \longrightarrow R_2BI + RI$$

$$BCl_3 + HC{\equiv}CH \longrightarrow Cl_2BCH{=}CHCl \qquad \text{Haloboration}$$

$$B_2Cl_4 + CH_2{=}CH_2 \longrightarrow Cl_2BCH_2{-}CH_2BCl_2 \qquad \text{Diboration}$$

Organoboron hydrides $(R_2BH)_2$ and $(RBH_2)_2$ may be prepared from the respective organoboron halides R_nBX_{3-n} by means of substitution of H for X, using LiH or $LiAlH_4$. Organoboron halides also serve as intermediates in the synthesis of numerous other derivatives:

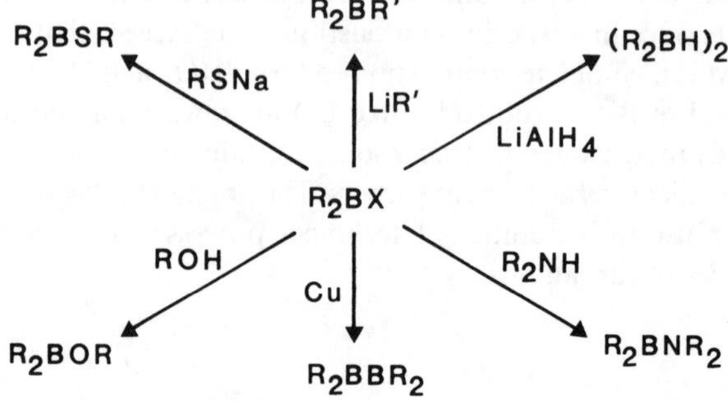

Tetraorganoborates BR_4^- are obtained from tetrahaloborates:

$$NaBF_4 + 4\,PhMgBr \longrightarrow NaBPh_4 + 4\,MgBrF$$

$NaBPh_4$ is sold in Germany under the name Kalignost® ("recognizing potassium"). The ion BPh_4^- forms precipitates with K^+, Rb^+, Cs^+, Tl^+, Cp_2Co^+ etc. and as the ammonium salt, can be used in small scale preparations of BPh_3:

$$NH_4BPh_4 \longrightarrow BPh_3 + NH_3 + PhH$$

The **radical anion** $Ph_3B^{\cdot-}$, which is isoelectronic with $Ph_3C\cdot$, is in equilibrium with its dimer (J. L. Mills, 1976):

$$2\,Ph_3B \xrightarrow{K,\,DME} 2\,Ph_3B^{\cdot-} \rightleftharpoons [Ph_3B{-}BPh_3]^{2-}$$

No tendency towards dimerization is exhibited by the anion $(mesityl)_3B^{\cdot-}$ (Kaim, 1989). **Organoborinium ions** R_2B^+ only exist as Lewis base adducts (Nöth, 1985 R):

$$Ph_2BCl + AgClO_4 \xrightarrow[-AgCl]{MeNO_2,\,bipy} [Ph_2B(bipy)]^+ + ClO_4^-$$

Borinium ion (C.N. 4)

PROPERTIES

Because of the low polarity of the B–C bond, binary boranes R_3B are stable in water; they are easily oxidized, however. The more volatile boron trisalkyls are pyrophoric.

Contrary to BH_3, BR_3 is monomeric. This may be traced to hyperconjugation with the alkyl substituents whereby the electron deficiency at boron is alleviated. Electron rich groups like vinyl or phenyl provide the B – C bond with partial double bond character. Substituents E with lone pairs of electrons act similarly:

$$\left\{ >B-CH=CH_2 \longleftrightarrow >\overset{\ominus}{B}=CH-\overset{\oplus}{C}H_2 \right\} \qquad \left\{ >B-\bar{E} \longleftrightarrow >\overset{\ominus}{B}=\overset{\oplus}{E} \right\}$$

The strength of the boron-heteroatom π-bonds in R_2BE increases in the order $Cl < S < O < F < N$. This is derived from the following observations:

E:	Cl SMe	OMe F NR_2
Formation of an adduct $R_2BE \cdot NR_3$:	observed	not observed

Organoboron hydrides R_2BH and RBH_2 form dimers which always display **hydride bridges** rather than alkyl bridges:

The structural elucidation of these derivatives of diborane by IR spectroscopy is based on the strongly differing B – H stretching frequencies for bridging (H^b) and terminal (H^t) positions:

	Stretching frequency	Intensity
$\nu(B{-}H^b{-}B)$ symm.	$1500-1600\ cm^{-1}$	strong
$\nu(B{-}H^b{-}B)$ asymm.	1850	medium
$\nu(B-H^t)$	2500 – 2600	

The lower stretching frequency of hydride bridges is a consequence of the diminished bond order (0.5) which can be deduced from the interaction diagram of a **B–H–B (2*e*3*c*) bond**.

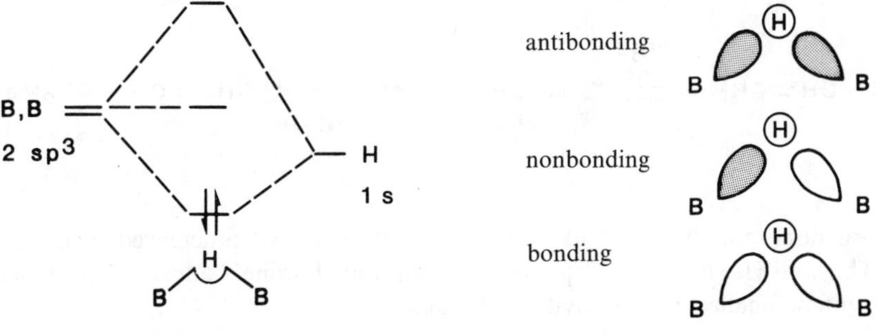

REACTIONS

Controlled oxidation of organoboron compounds R_3B affords the **alkoxyboranes** R_2BOR (borinic esters), $RB(OR)_2$ (boronic esters) and $B(OR)_3$ (organoborates) which are also accessible through alcoholysis of the respective organoboron halides R_nBX_{3-n}. If an amine oxide is used as a source of oxygen, the liberated amine may be titrated; this is a method for the **determination of the number of B–C bonds** initially present:

$$R_3B + 3\,Me_3NO \longrightarrow B(OR)_3 + 3\,Me_3N$$

One of the most versatile methods in organic synthesis is **hydroboration** (H. C. Brown, 1956 f.). Herein, rather than the resulting organoboranes themselves, the products of their subsequent reactions are important:

$$RCH{=}CH_2 + H{-}B\langle \;\underset{150°}{\overset{Et_2O,\ 25°}{\rightleftharpoons}}\; RCH(H){-}CH_2(B\langle)$$

$$RCH(H){-}CH_2B\langle \xrightarrow{H^+} RCH_2{-}CH_3$$

$$RCH(H){-}CH_2B\langle \xrightarrow{H_2O,\ OH^-} RCH_2{-}CH_2OH$$

$$RCH(H){-}CH_2B\langle \xrightarrow{HCrO_4^-} RCH_2{-}COOH$$

$$RCH(H){-}CH_2B\langle \xrightarrow{Ag_2O} (RCH_2CH_2)_2$$

$$RCH(H){-}CH_2B\langle \xrightarrow{Et_2NCl} RCH_2CH_2Cl$$

$$RCH(H){-}CH_2B\langle \xrightarrow{I_2,\ OH^-} RCH_2CH_2I$$

Hydroboration is **regioselective (*anti*-Markovnikov direction)**, electropositive boron being attached to the carbon atom bearing the larger number of hydrogen atoms, and **stereoselective (*cis* addition)**. Hydroboration is **reversible** which can be exploited for counter-thermodynamic isomerizations of olefins:

$$MeCH{=}CHMe \;\overset{>B-H}{\rightleftharpoons}\; MeCH(B\langle){-}CH_2Me \;\underset{>B-H}{\overset{180°}{\rightleftharpoons}}\; CH_2{=}CH{-}CH_2Me$$

With organoboranes bearing bulky substituents, the selectivities achieved are particularly high. Thus, bis-(1,2-dimethylpropyl)borane ("disiamylborane") adds to 1-pentene, leaving 2-pentene unaffected (selectivity > 99%).

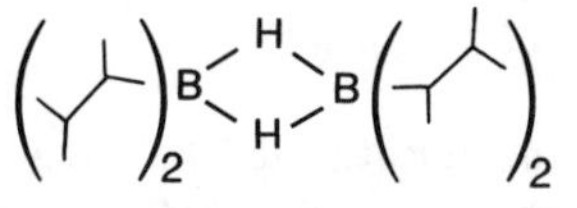

disiamylborane

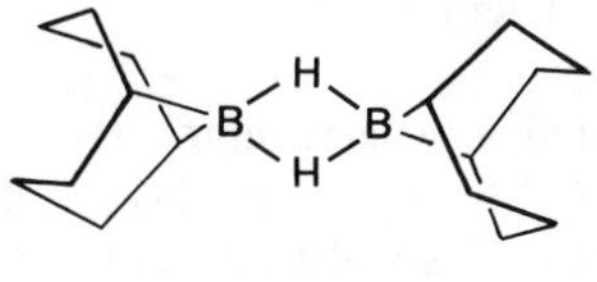

9-BBN

Bis-(9-borabicyclo[3.3.1]nonane) ("9-BBN"), which can be prepared from diborane and 1,5-cyclohexadiene, has the additional advantage of being easy to handle (solid, stable up to 200 °C, precautions required are comparable to those recommended for $LiAlH_4$). 9-BBN excels in the **chemoselective reduction** of functional groups as in the conversion of α, β-unsaturated aldehydes and ketones to the corresponding allyl alcohols:

$$Ph-CH=CH-C(=O)H \xrightarrow[2.\ H_2NCH_2CH_2OH]{1.\ 9\text{-BBN},\ 25\,°C} Ph-CH=CH-CH_2OH \qquad 99\,\%$$

Enantioselective syntheses of chiral alcohols from prochiral alkenes can be effected by means of chiral boranes (H. C. Brown, 1961).
Exceptionally high enantiomeric excesses (*ee*) are obtained in hydroborations with the reagent *trans*-2,5-dimethylborolane (Masamune, 1985):

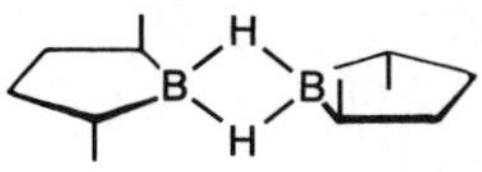

R,R

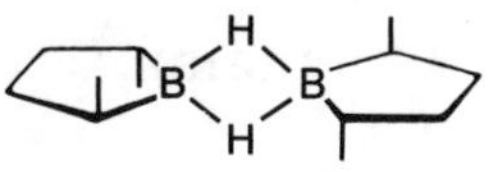

S,S

Early observations that allylboronic esters readily add across aldehyde functions (Gaudemar, 1966) have added impetus to the use of **carbaborations** in **organic synthesis**. They effect **diastereogenic C–C coupling** of high enantioselectivity. Example (R. W. Hoffmann, 1982 R):

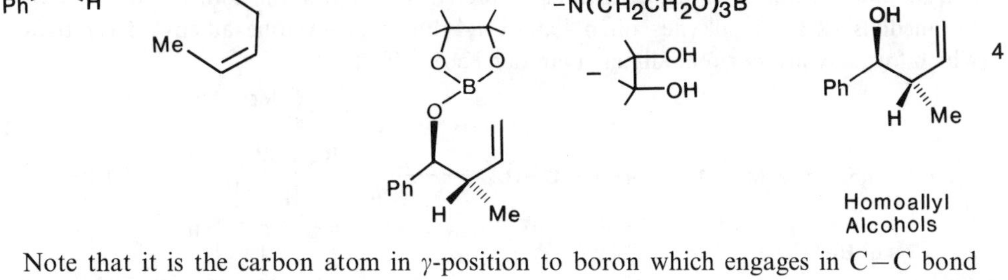

$N(CH_2CH_2OH)_3$

$-N(CH_2CH_2O)_3B$

96 : 4

Homoallyl Alcohols

Note that it is the carbon atom in γ-position to boron which engages in C–C bond formation.

7.1.2 Boron Heterocycles

Boron is found in a large number of saturated and unsaturated heterocycles in such segments as C–B–C, C–B–N, N–B–N, O–B–O, C–B–O, C–B–S etc. Here, only a few examples will be given in which organometallic aspects dominate.

Boracycloalkanes (boracyclanes) are products of metathesis reactions or of cyclohydroborations:

Parent Compound:

$Li(CH_2)_4Li$ + $PhBF_2$ $\xrightarrow{-2LiF}$ (cyclic B–Ph)

Borolane

2 (diene) + 2 $BH_3 \cdot THF$ $\xrightarrow[2.\ 1h\ \Delta]{1.\ 0^o}$ (H-bridged dimer)

Borinane

(diene) + 1/2 $(EtBH_2)_2$ ⟶ (cyclic B–Et)

Borepane

12 (t-Bu)BF_2 + 8 Na/K $\xrightarrow{-\ 8\ Na/K[RBF_3]}$

(Paetzold, 1991)

Boratetrahedrane

Boracycloalkenes and **boraarenes** are of interest in the context of π-electron delocalization across B(sp^2) centers. Therefore, the synthesis of **borirene** (boracyclopropene $\mathbf{C_2BH_3}$), which is isoelectronic with the smallest Hückel system $C_3H_3^+$, has been the subject of considerable synthetic effort. Attempts to prepare borirenes via addition of organoboranediyls (RB:) to alkynes often led to 1,4-dibora-2,5-cyclohexadienyl derivatives which formally are borirene dimers (van der Kerk, 1983):

4 C_8K + 2 $MeBBr_2$ + R–C≡C–R $\xrightarrow[C_6H_6]{\Delta}$ (1,4-dimethyl-2,3,5,6-tetra-R-1,4-dibora-2,5-cyclohexadiene) + 4 KBr

(R = n-Bu)

Monomeric borirenes can be obtained photochemically if protection by sterically demanding substituents R is provided (Eisch, 1987):

$R_2B{-}C{\equiv}C{-}R \xrightarrow[\text{benzene, pyridine}]{h\nu,\ 300\ \text{nm}}$ (borirene) R = mesityl (2,4,6-trimethylphenyl)

Structural and spectroscopic data point to extensive π-electron delocalization:

The isoelectronic series can be extended:

$C_3R_3^+$, C_2BR_3, $CB_2R_3^-$ (Berndt, 1985).

An intriguing aspect of the **dihydrodiboretes** (diboracyclobutenes) is the structural disparity of the two isomers (Siebert, 1985 R):

R = N(i-Pr)$_2$; 2 K, −2 KCl; Δ

	planar	folded ("butterfly", p. 394)
π-electrons:	localized	delocalized

The higher thermodynamic stability as well as the folded structure of the 1,3-isomer had been predicted from theoretical considerations (v. R. Schleyer, 1981).
The antiaromatic **borole** C_4H_4BH (4 π-electrons) is labile even in perarylated form (Eisch, 1969):

2 Li$^+$ (2−); Me$_2$SnCl$_2$, −2 LiCl; PhBCl$_2$, −Me$_2$SnCl$_2$; 2 Li; —○ = —Ph

Pentaphenylborole

Cocondensation of BF vapor with alkynes leads to 1,4-diboracyclohexa-2,5-dienes (Timms, 1968):

$$BF_3\,(g) \xrightarrow[1800^\circ]{B} \{BF\} \xrightarrow{RC\equiv CR}$$

F B R R R R B F

Borabenzene (Borin) C_5H_5B as yet has only been obtained as Lewis base adduct or as an aromatic, anionic η^6-ligand (p. 380) in transition-metal complexes (Ashe, 1971):

n-Bu_2SnH_2 → Sn (n-Bu)$_2$ → $PhBBr_2$, −(n-Bu)$_2SnBr_2$ → B Ph → $LiCPh_3$, $-H^+$ → B Ph

Hydrostannation

Phenyl-boratabenzene

B OMe → $LiN(C_6H_{11})_2$, $-HN(C_6H_{11})_2$ → Li^+ B OMe → Me_3SiCl, −LiCl → B OMe SiMe$_3$ H → $-Me_3SiOMe$, Py → B—N

*The prefix "**bora**" designates the replacement of a* CH *fragment by* B, *"**borata**" that of* CH *by* BH^-.

The pyridine adduct of borabenzene (G. Maier, 1985) is isoelectronic with biphenyl. As gauged from spectroscopic data (^{1}H-NMR in particular), boratabenzene $C_5H_5BR^-$ displays cyclic π-conjugation. Due to the difference in electron affinity, replacement of carbon by boron, however, acts as a perturbation.
The tropylium ion $C_7H_7^+$ has found its isoelectronic counterpart in derivatives of **borepin** C_6BH_7, the least substituted, known to date, being the Me−B derivative (Sakurai, 1987):

(n-Bu)$_2SnH_2$ → Sn(n-Bu)$_2$ → $MeBBr_2$, −20°, −(n-Bu)$_2SnBr_2$ → B—Me

Stannepin

Borepin

Boron heterocycles which only contain heteroatoms include the boroxins (condensation products of boronic acids), the cycloaminoboranes, and the borazines.

$$3\ RB(OH)_2 \longrightarrow (RBO)_3 + 3\ H_2O$$

Boroxine

Boroxines are planar; yet, compared to the borazines, the free electron pairs in the former remain localized to a large extent. **Aminoboranes** as monomers are also planar. They equilibrate with puckered cyclic dimers and trimers, the position of these equilibria being governed by the size of the substituents R. In the monomeric as well as in the oligomeric forms, boron gains an octet configuration:

$$R_2BCl + R_2NLi \xrightarrow{-\ LiCl} \left\{ R_2B-\bar{N}R_2 \longleftrightarrow R_2\overset{\ominus}{B}=\overset{\oplus}{N}R_2 \right\} \text{Aminoborane}$$

$$\rightleftarrows (R_2BNR_2)_2 + (R_2BNR_2)_3$$

The isoelectronic nature of the units $>C=C<$ and $>B=N<$ suggests replacements of the former by the latter. In the case of benzene this leads into the class of **borazines** ($B_3N_3H_6$, cyclotrisborazine), also termed "borazoles" or "inorganic benzenes" since, in many regards, they mimic the properties of arenes:

$$3\ MeNH_3Cl + 3\ BCl_3 \xrightarrow{-9HCl} (ClBNMe)_3 \xrightarrow{MeMgBr} (MeBNMe)_3$$

Hexamethyl-cyclotrisborazine

If $>C=C<$ units are only partially replaced by $>B=N<$, the analogy to benzene is lost, however:

$$\text{C}_4\text{H}_8\text{BH}{=}\text{NMe} \xrightarrow[-2H_2]{Pd/C} \text{1,2-Azaborine} \longrightarrow \text{rapid polymerisation}$$

1,2-Azaborine

Since the electronegativities of B and N are very different, 1,2-azaborine acts like a strongly polarized butadiene.

7.1.3 Polyhedral Boranes, Carbaboranes and Heterocarbaboranes

"Boranes: Rule breakers become pattern makers" K. Wade (1974)

In the context of this section, a cursory treatment of the binary boranes is indispensable. Preparative results, in part dating back to A. Stock (1876–1946) revealed that, as for hydrocarbons, borane chemistry may also be organized in terms of homologous series:

B_nH_{n+6}	e.g. B_4H_{10}, B_9H_{15}
B_nH_{n+4}	B_2H_6, B_5H_9
B_nH_{n+2} [only as $(B_nH_n)^{2-}$, $n = 5–12$]	$B_5H_5^{2-}$, $B_6H_6^{2-}$

As the structures of these boranes were unraveled, the desire grew to understand the bonding situation. In both of these areas, W. N. Lipscomb has earned merits.
Lipscomb devised a system to organize ***borane topology*** *and he described the electronic structure even of the more complicated boranes in terms of the* ***valence bond (VB) model.*** *In boranes, the number of valence electron pairs is always less than the number of interatomic connecting lines which depict the molecular framework. Therefore, a certain number of 2e3c bonds must be included. On p. 67, for a few boranes, molecular structures (left) are juxtaposed with the respective valence bond formulae (right). Whereas in the structural representations the number of connecting lines does* ***not*** *correspond to the number of electron pairs available for bonding, in the valence formula this* ***is*** *the case.*
The symbols have the following meaning:*

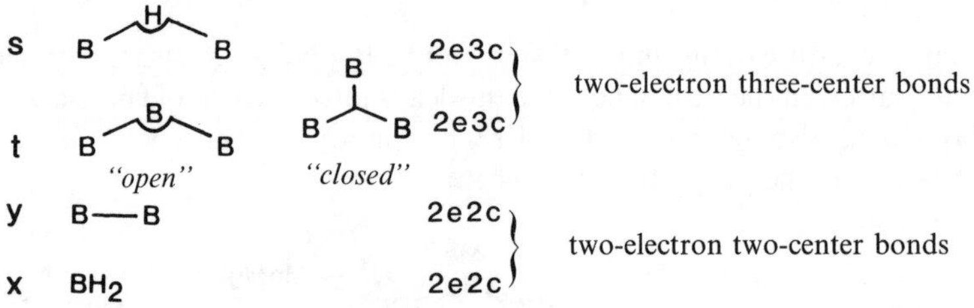

s B–H–B 2e3c
t B–B–B "open" / B(B)(B) "closed" 2e3c — two-electron three-center bonds
y B—B 2e2c
x BH_2 2e2c — two-electron two-center bonds

"Open" and "closed" three center bonds differ in the nature of the constituent atomic orbitals.
According to the numbers s, t, y, x which tally the different bonding fragments present, a particular borane is assigned its specific ***styx code****. Example: styx(B_5H_9) = 4120. For a borane of known composition, usually one styx code only exists, which is compatible with the total number of valence electrons and the number of* B- *and* H *atoms. In this way, structural predictions become possible, which then have to be confirmed by spectroscopic techniques or by diffraction methods.*

* In the literature as well as at various places in this text, $2e3c$ bonds are also represented by two lines. Examples:

B–H–B Al–CH_3–Al

B_2H_6

DIBORANE (6) 2002

B_4H_{10}

TETRABORANE (10) 4012

B_5H_9

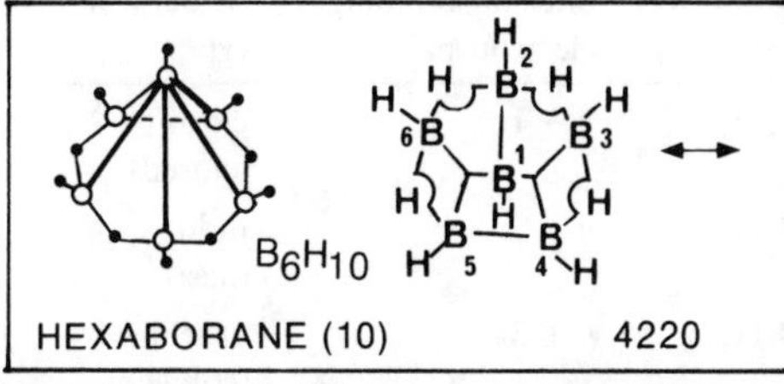

PENTABORANE (9) 4120

B_5H_{11}

PENTABORANE (11) 3203

B_6H_{10}

HEXABORANE (10) 4220

B_8H_{12}

OCTABORANE (12) 4420

B_9H_{15}

NONABORANE (15) 5421

$B_{10}H_{14}$

DECABORANE (14) 4620

Frequently, it is impossible to do justice to the symmetry of a borane molecule by writing a single valence-electron formula employing the symbols discussed on p. 66. In these cases, several canonical forms must be given which together represent a resonance hybrid. Example: B_5H_9, *symmetry* C_{4v}. *For decaborane(14), 24 canonical forms have to be considered. For the highly symmetrical borane anions* $B_nH_n^{2-}$, *valence bond representations become unwieldy because of the large number of canonical forms.*

Contemporary descriptions of polyhedral boranes are based on **MO methods** introduced in the now classical work of Hoffmann and Lipscomb (1962) which constitutes an early example of the application of the Extended-Hückel method. Thus, given a certain **number *n* of skeletal atoms** (B, C or heteroatom), the number of electrons available for skeletal bonding fixes a certain structure. The following system has become known as **Wade's rules;** it is particularly successful in discussing the heteroboranes.

- *Each of the n units* :B–H *furnishes* 2 *skeletal bonding electrons*
- *Each additional* ·H *furnishes* 1 *skeletal bonding electron*
- *Ionic charges must be included in the electron count*

Borane	Skeletal bonding electron pairs	Structural type	Structure	Example
B_nH_{n+2}*	$n+1$	closo (closed)	Polyhedron with n vertices 0 vertices unoccupied	$B_{12}H_{12}^{2-}$
B_nH_{n+4}	$n+2$	nido (nest)	Polyhedron with $n+1$ vertices 1 vertex unoccupied	B_5H_9
B_nH_{n+6}	$n+3$	arachno (cobweb)	Polyhedron with $n+2$ vertices 2 vertices unoccupied	B_4H_{10}

* For binary boranes, the closo structure is only realized in the anions $B_nH_n^{2-}$

Therefore, the structures of the boranes $B_6H_6^{2-}$, B_5H_9, and B_4H_{10} are derived from an octahedron with 0, 1 or 2 vertices, respectively, remaining unoccupied:

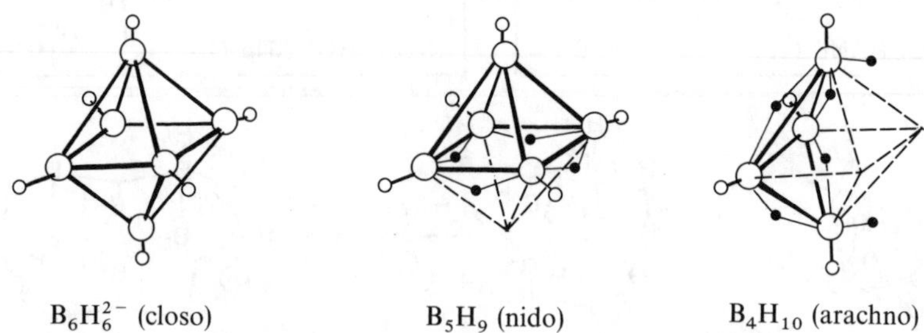

$B_6H_6^{2-}$ (closo) B_5H_9 (nido) B_4H_{10} (arachno)

Increasing hydrogen content in the boranes leads to successively more "open" structures since the required number of BBB ($2e\,3c$) bonds decreases.

Idealized Structures of closo-, nido- and arachno-Boranes

Only the $(BH)_n$ *skeleton is shown.* BH_2 *groups and* B–H–B *bridges occur at the open face of the respective incomplete polyhedron. In the heteroboranes, one or more* BH-*units are replaced by other building blocks.*

STRUCTURAL TYPE		closo	nido	arachno
SKELETAL BONDING ELECTRON PAIRS		n + 1	n + 2	n + 3
TRIANGULAR FACES	VERTICES			
6	n = 5			
8	6			
10	7			
12	8			
14	9			
16	10			
18	11			
20	12			

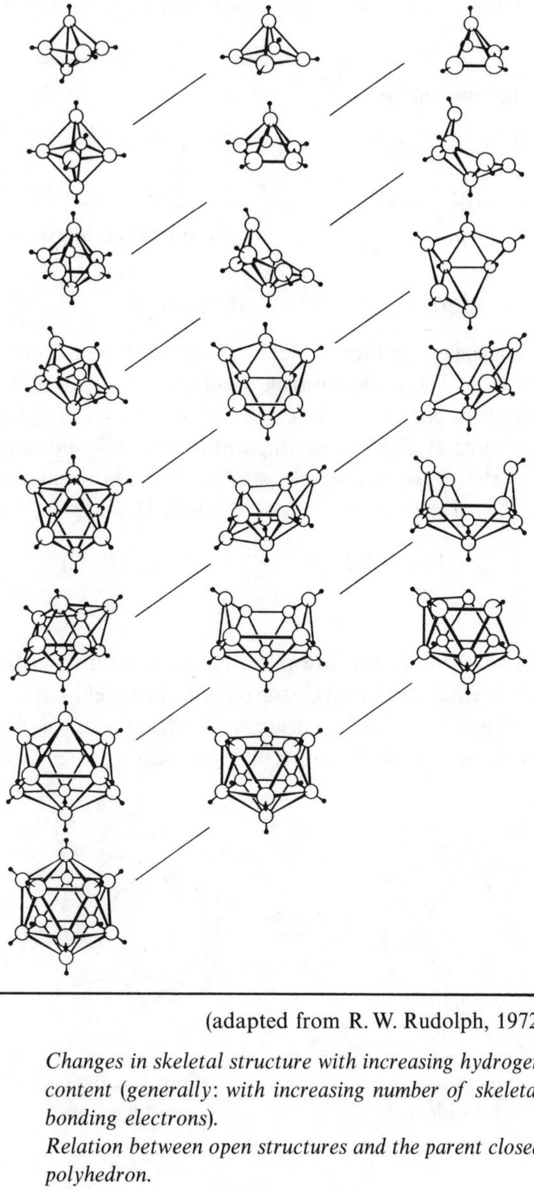

(adapted from R. W. Rudolph, 1972)

Horizontal: *Changes in skeletal structure with increasing hydrogen content (generally: with increasing number of skeletal bonding electrons).*

Diagonal: *Relation between open structures and the parent closed polyhedron.*

PREPARATION

Nido- and arachno-boranes:

$$Mg_3B_2 + H_3PO_{4\,aq} \longrightarrow B_4H_{10}, B_5H_9, B_6H_{10} \quad \text{(A. Stock, 1912 f)}$$

$$4\,NaH + B(OMe)_3 \longrightarrow NaBH_4 + 3\,NaOMe$$

$$2\,NaBH_4 + H_3PO_4 \longrightarrow B_2H_6 \xrightarrow[-H_2]{\Delta T} B_4H_{10}, B_5H_9, B_5H_{11} \quad \text{etc.}$$

Borane pyrolysis

In addition, several special methods for the synthesis of the more exotic boranes have been devised. They rely on the meticulous control of temperature and pressure.

Closo-borane anions:

$$B_2H_6 + 2\,Et_3N \longrightarrow 2\,Et_3NBH_3 \xrightarrow[-H_2]{B_4H_{10},\ 190\,°C} 2\,[Et_3NH]^+ + [B_{12}H_{12}]^{2-}$$

$$[Et_4N]BH_4 \xrightarrow[-H_2]{185\,°C,\ 16\,h} 2\,[Et_4N]^+ + [B_{10}H_{10}]^{2-}$$

Tetrahydroborate pyrolysis

HETEROBORANES, CARBABORANES

Heteroboranes contain skeletal atoms different from boron (C, Sn, Pb, Al, transition metals etc.). The **carbaboranes**, pioneered by Hawthorne and Grimes, may be formally derived from boranes by isoelectronic replacement of BH^- by CH. Therefore, the neutral carbaborane $B_4C_2H_6$, like the dianion $B_6H_6^{2-}$, possesses a closo structure.
One of the most extensively studied carbaboranes is dicarba-closo-dodecaborane(12), $C_2B_{10}H_{12}$. Like the isoelectronic anion $B_{12}H_{12}^{2-}$ it forms an icosahedron.

$$B_{10}H_{14} + 2\,Et_2S \xrightarrow[-H_2]{(n\text{-}Pr)_2O} B_{10}H_{12}(Et_2S)_2 \xrightarrow[-H_2,\ -Et_2S]{HC\equiv CH} C_2B_{10}H_{12}$$

"ortho-carbaborane"

The two carbon atoms are initially introduced in vicinal position (ortho-carbaborane). Under thermal conditions, isomerization to meta-carbaborane ($T = 600\,°C$) and to para-carbaborane ($T \approx 700\,°C$) occurs. This migration of carbon atoms on the surface of the icosahedron is thought to involve transition states of cuboctahedral symmetry:

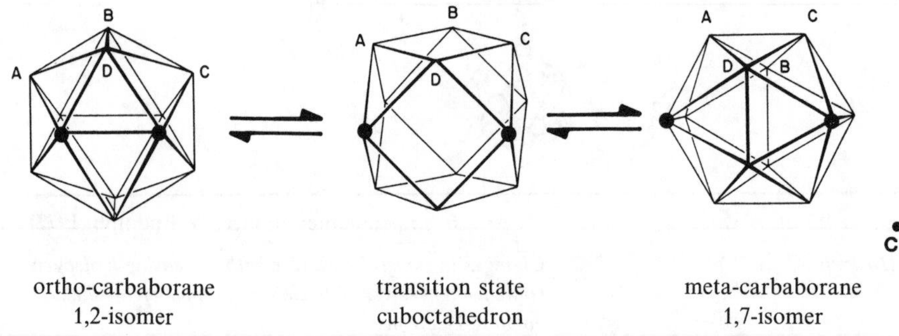

ortho-carbaborane 1,2-isomer — transition state cuboctahedron — meta-carbaborane 1,7-isomer

Carbaboranes of the nido- and arachno type have also been prepared. 1,2,3,4-Tetraethyl-1,2,3,4-tetracarbadodecaborane(12), which according to Wade's rules has a nido structure, may serve as an example. This carbaborane, like $B_{12}H_{12}^{4-}$, possesses $n + 2$ skeletal electron pairs.

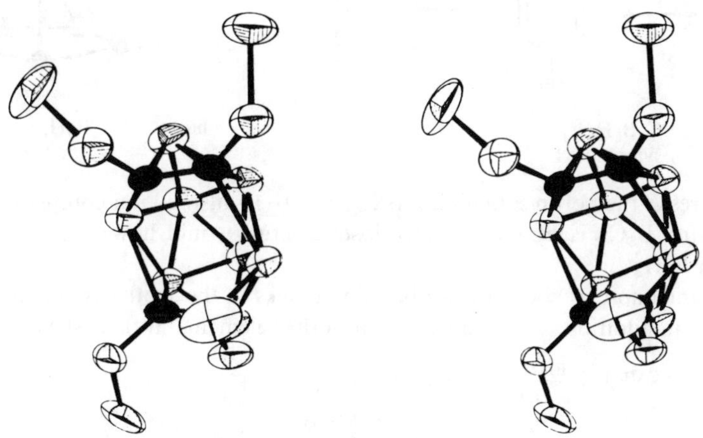

Stereoview of the structure of $\mathbf{Et_4C_4B_8H_8}$ (*Grimes*, 1984)

Guide to three-dimensional viewing: *Place two fingers between the eyes and the plane of the paper so that the left eye sees only the left-side image, similarly, the right eye should only see the right-side image. The eyes, thus fixed along parallel axes, are then relaxed and adjusted to an "infinity setting". In a short time (and with a little practice!), the three-dimensional image of the molecule should appear, which remains even after removing the guide fingers. A prerequisite is, that the distance between the two stereo pictures approximates the separation between the eyes.*

Nido-carbaboranes of the type $[C_5BH_6]^+$ bear resemblance to half-sandwich complexes (p. 309); a highly substituted derivative of this cation has been prepared by Jutzi (1977):

$$(CH_3)_5C_5Ge(CH_3)_3 \xrightarrow[-(CH_3)_3GeX]{BX_3} (CH_3)_5C_5BX_2 \xrightarrow{BX_3} [(CH_3)_5C_5BX]^+ \; BX_4^-$$

The nido structure of this carbaborane may be traced back to the isoelectronic nature of $C_5BH_6^+$ and $B_6H_6^{4-}$ ($n + 2$ skeletal electron pairs). The carbaborane anion $C_5BH_6^-$ ($n + 3$ skeletal electron pairs), in accordance with Wade's rules, features an arachno structure (compare the planar configuration of boratabenzene $C_5H_5BR^-$, p. 64). A limitation to Wade's structural scheme is imposed by substituents like halogens or amino groups, which can satisfy the electronic demand of boron by means of π back-bonding (p. 64). Thus, 1,4-difluoro-1,4-diboracyclohexadiene $C_4H_4B_2F_2$ has a planar rather than a nido structure, the latter being predicted by Wade's rules for the parent carbaborane $C_4B_2H_6$ ($n + 2$ skeletal electron pairs):

$C_4B_2H_4F_2$
symmetry D_{2h}

but $C_4B_2H_6$
symmetry C_s

The structures of the boron subhalides $(BX)_n$, $n = 4-10$, also do not comply with Wade's rules. Example: B_4Cl_4 is tetrahedral; this closo structure would, however, be predicted for the dianion $B_4H_4^{2-}$.

Nido- and arachno-carbaboranes are building blocks in the synthesis of **heterocarbaboranes** which, in addition to B and C, contain other elements in their skeleton.

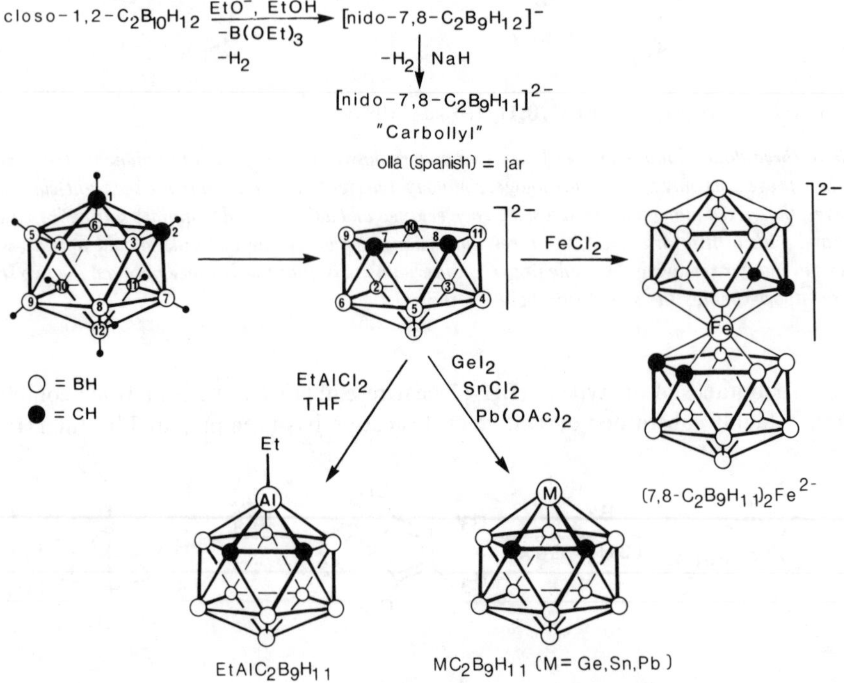

This participation of Ge, Sn, Pb and Fe in icosahedral frameworks is unique in the structural chemistry of these elements. Very recently, from (i-Bu)$_2$AlCl and K the anion $[(i\text{-Bu})_{12}Al_{12}]^{2-}$ has been prepared (Uhl, 1991). This is the first example of a homonuclear icosahedral core built from atoms other than boron.

The anion $[(7,8\text{-}C_2B_9H_{11})_2Fe]^{2-}$ is just one of numerous cases where the five-membered ring of a partially degraded icosahedral carbaborane assumes the role of a cyclopentadienyl ligand in metallocenes. The relation of di(carbollyl)metal complexes to the metallocenes is based on the isolobal (p. 396) nature of the ligands carbollyl $7,8\text{-}C_2B_9H_{11}^{2-}$ and cyclopentadienyl $C_5H_5^-$.

Excursion

^{11}B-NMR of Organoboron Compounds

Of the two boron isotopes, ^{10}B (20%, $I = 3$) and ^{11}B (80%, $I = 3/2$), the latter possesses superior NMR properties (cf. p. 24). Therefore, ^{11}B-NMR is performed almost exclusively.

The chemical shifts $\delta(^{11}B)$ cover a range of about 250 ppm; they are ruled by the charge, the coordination number, and the substituents at boron. For trisorganoboranes, the chemical shifts $\delta(^{11}B)$ correlate with $\delta(^{13}C)$ values of isoelectronic carbenium ions.

Chemical Shifts $\delta(^{11}B)$ of Selected Organoboranes (external standard $Et_2O \cdot BF_3$):

venes the electronegativities of F and Cl, has been related to stronger π-bonding $\rangle B^{\ominus}=F^{\oplus}$ as compared with $\rangle B^{\ominus}=Cl^{\oplus}$, a consequence of the shorter length of the B–F σ-bond. An even stronger π-donor capacity is exhibited by the substituent NR_2.

The ability of α,β-unsaturated organic groups to engage in π-conjugation with three-coordinate boron can be assessed from an inspection of $\delta(^{11}B)$- and $\delta(^{13}C)$ values of **vinyl boranes** (Odom, 1975).

In this context, $\delta(^{13}C_\beta)$ rather than $\delta(^{13}C_\alpha)$ data are examined, since the shielding of the nuclei C_α, which are directly bonded to boron, is governed by mesomeric (π) **and** inductive (σ) effects.

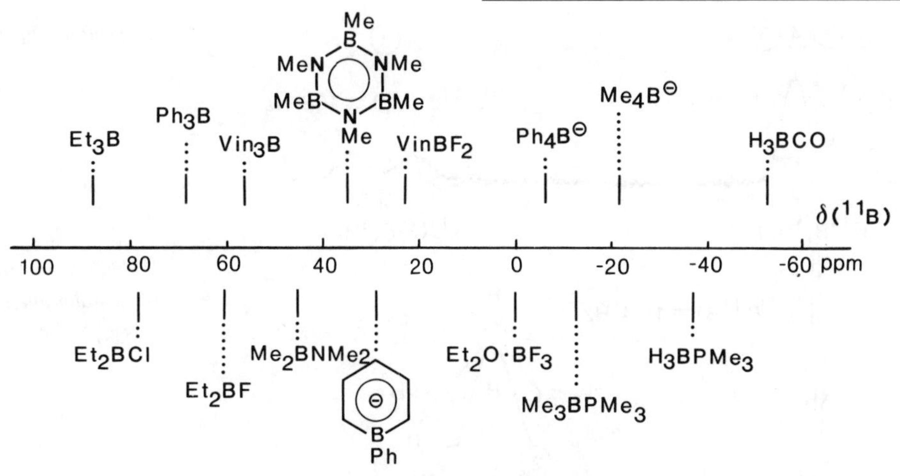

Substituent effects on $\delta(^{11}B)$ values are influenced by σ- as well as by π-bonding contributions. Thus, the nucleus ^{11}B is shielded more effectively in Et_2BF than in Et_2BCl. This gradation, which contra-

Furthermore, due to their vicinity to ^{11}B, a nucleus with a quadrupole moment, C_α nuclei often give rise to NMR signals which are broadened beyond detection (p. 295f).

^{11}B- and ^{13}C-NMR Data

$B(CH_2CH_3)_3$		$B(CH{=}CH_2)_3$		$CH_2{=}CH_2$	
$\delta^{11}B$	86.8	$\delta^{11}B$	56.4		
$\delta^{13}C$ (CH_2)	19.8	$\delta^{13}C$ (CH) α	141.7		
$\delta^{13}C$ (CH_3)	8.5	$\delta^{13}C$ (CH_2) β	138.0	$\delta^{13}C$ (CH_2)	122.8

$$\left\{ \;{>}B{-}CH{=}CH_2 \;(\text{a},\ \alpha,\ \beta) \longleftrightarrow {>}\overset{\ominus}{B}{=}CH{-}\overset{\oplus}{C}H_2 \;(\text{b}) \longleftrightarrow \right\}$$

The contribution of canonical structure ⓑ in $B(CH{=}CH_2)_3$ manifests itself in the shielding of ^{11}B [compared to $B(C_2H_5)_3$] and in the deshielding of $^{13}C(CH_2)$ [compared to $CH_2{=}CH_2$]. Therefore, boron in α-alkenylboranes acts as a π-acceptor.

As expected, considerably stronger shielding is experienced by ^{11}B in **Lewis base adducts** [solvent effects on $\delta(^{11}B)$!] and in tetraorganoborate anions. A striking effect is the very strong shielding of ^{11}B in borane carbonyl H_3BCO. Note, however, that in borane carbonyl, as opposed to transition-metal carbonyls, CO can only function as a σ-donor. Another factor which may contribute to ^{11}B shielding in borane carbonyl is the diamagnetic susceptibility of the $C{\equiv}O$ triple bond [compare $\delta(^{11}B)$ for H_3BPMe_3 and for H_3BCO].

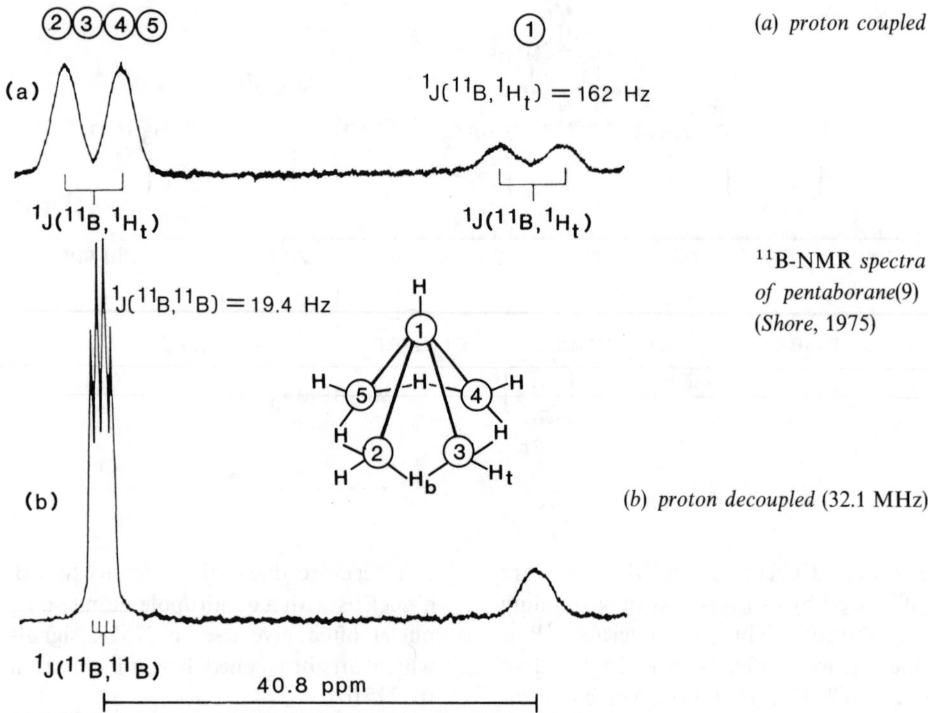

(a) proton coupled

^{11}B-NMR *spectra of pentaborane*(9) (*Shore*, 1975)

(b) proton decoupled (32.1 MHz)

^{11}B-NMR signals for **carbaboranes** appear in a large shift range between + 53 and − 60 ppm. Usually, nuclei with higher connectivity to other skeletal atoms absorb at higher field. The large line-width of these signals is caused, i.a., by unresolved multiplet structure stemming from extensive B, B- and B, H coupling.

Spin-Spin coupling constants $^1J(^{11}B, ^1H_t)$ for terminal B−H units lie between 100 and 200 Hz, $^1J(^{11}B, ^1H_b)$ values for $B\overset{H}{\frown}B$ ($2e3c$) bridges between 30 and 60 Hz (Example: B_2H_6, 135 and 46 Hz resp.). The splitting patterns assist in the elucidation of borane structures.

An instructive, albeit deceptively simple example is given in the ^{11}B-NMR spectrum of pentaborane (9) (p. 74). These spectra reflect the point group C_{4v} of B_5H_9. This rather high symmetry is also responsible for the fact that, despite quadrupole relaxation of the nucleus ^{11}B, line widths of a few Hz only are observed.

7.2 Organoaluminum Compounds

Although known as a highly reactive species for more than a century it is only since 1950 that organoaluminum compounds have gained in interest. This development was triggered by the pioneering work of K. Ziegler. The relatively belated appreciation of aluminum organyls may be traced to the longstanding confinement to ethereal solvents, which attenuate the reactivity of monomeric R_3Al through the formation of solvates $R_3Al \cdot OEt_2$. Compared with the organometallics of groups 1 and 2, aluminum organyls excel in the ease of their addition to alkenes and alkynes. The regio- and stereoselectivity of these carbaluminations, as well as of the related hydroaluminations using R_2AlH, are an additional asset. Aluminum organyls, which are derivatives of the cheapest of the active metals, may gradually substitute lithium- and magnesium organyls as more economical reducing and alkylating agents.

PREPARATION

In accordance with the large practical importance of aluminum alkyls, a number of **industrial processes** have been developed for their production:

$$4\,Al + 6\,MeCl \longrightarrow \underset{\text{Sesquichloride}}{2\,Me_3Al_2Cl_3} \rightleftharpoons Me_4Al_2Cl_2 + Me_2Al_2Cl_4 \quad \boxed{1}$$

$$\downarrow 2\,NaCl$$

$$Me_4Al_2Cl_2(l) + 2\,Na[MeAlCl_3](s)$$

$$\times 3 \downarrow 6\,Na$$

$$2\,Me_6Al_2 + 2\,Al + 6\,NaCl$$

Aluminum is activated by grinding with Et_3Al in order to remove surface oxides. This process, patented by **Chemische Werke Hüls**, is only used to produce trimethyl- and triethylaluminum.
In the **Ziegler direct process** two observations (in the presence of aluminum alkyls, Al reacts with H_2; $\gt Al-H$ adds to alkenes) are exploited:

$$Al + 1.5\,H_2 + 2\,Et_3Al \xrightarrow[100-200\text{ bar}]{80-160\,^\circ C} 3\,Et_2AlH \quad \text{"Vermehrung" (increase)}$$
$$3\,Et_2AlH + 3\,C_2H_4 \xrightarrow[1-10\text{ bar}]{80-110\,^\circ C} 3\,Et_3Al \quad \text{"Anlagerung" (attachment)}$$
$$\text{Sum}\quad Al + 1.5\,H_2 + 3\,C_2H_4 \longrightarrow Et_3Al$$

The reactivity of Al is enhanced by alloying it with 0.01–2% of Ti. If aluminum and the alkene are used in a molar ratio of 1:2, dialkylaluminum hydride is obtained.
Since hydroalumination is reversible and the affinity of $\gt Al-H$ for alkenes increases in the order $CH_2{=}CR_2 < CH_2{=}CHR < CH_2{=}CH_2$, several other aluminum organyls can be prepared starting from *tris*(*iso*-butyl)aluminum:

$$Al + 1{,}5\,H_2 + 3\,CH_2{=}CMe_2 \xrightarrow[200\text{ bar}]{100^\circ} (iso\text{-}Bu)_3Al$$
$$(iso\text{-}Bu)_3Al \xrightarrow[-3\,CH_2=CMe_2]{+3\,CH_2=CHMe} (n\text{-}Pr)_3Al \xrightarrow[-3\,CH_2=CHMe]{+3\,CH_2=CH_2} Et_3Al$$
"Verdrängung" (displacement)
$$(iso\text{-}Bu)_3Al \underset{140^\circ,\ 20\text{ mbar}}{\rightleftharpoons} \text{via } (iso\text{-}Bu)_2AlH + CH_2{=}CMe_2$$
$$(n\text{-}Pr)_3Al \rightleftharpoons (n\text{-}Pr)_2AlH + CH_2{=}CHMe$$

The less volatile, more heavily substituted alkene is condensed from the gas stream and the equilibrium shifted in favor of the desired, thermodynamically more stable aluminum alkyl. This is the method of choice for the generation of higher aluminum trisalkyls like *tris*(*n*-octyl)aluminum from *tris*(*iso*-butyl)aluminum.
Due to the commercial availability of numerous aluminum alkyls, the task of preparing them in the laboratory will seldom arise. Nevertheless, a few approaches will be mentioned:

$$3\,Ph_2Hg + 2\,Al \xrightarrow{\Delta T} 2\,Ph_3Al + 3\,Hg \qquad \boxed{2}$$
$$3\,RLi + AlCl_3 \xrightarrow{\text{Heptane}} R_3Al + 3\,LiCl \qquad \boxed{4}$$
$$3\,RCH{=}CH_2 + AlH_3\cdot OEt_2 \longrightarrow (RCH_2CH_2)_3Al\cdot OEt_2 \qquad \boxed{9}$$

TECHNICAL APPLICATIONS OF TRIS(ALKYL)ALUMINUM COMPOUNDS

a) The multiple insertion of ethylene into the Al–C bond, discovered by K. Ziegler, has become known as the **Aufbaureaktion (growth reaction).** It is used to produce 1-alkenes and unbranched primary alcohols.

$$Et_3Al \xrightarrow[\text{110°/100 bar, growth}]{CH_2{=}CH_2} Al[(C_2H_4)_mEt][(C_2H_4)_nEt][(C_2H_4)_oEt] \xrightarrow{O_2} Al[O(C_2H_4)_mEt][O(C_2H_4)_nEt][O(C_2H_4)_oEt]$$

displacement: $CH_2{=}CH_2$, 200–300°

$$Et_3Al + 3\ CH_2{=}CH{-}(CH_2CH_2)_{m,n,o}H$$

hydrolysis: H_2O

$$3\ Et(CH_2CH_2)_{m,n,o}OH + Al(OH)_3$$

The sequence of insertions can proceed up to chain lengths of about C_{200}. A limiting factor is the competition between the growth and displacement reactions. Among the products of this process, unbranched even 1-alkanols of chain lengths $C_{12}-C_{16}$ (**"detergent alcohols"**) are of prime importance, since their sulfates are produced in bulk quantities to serve as biodegradable surfactants $ROSO_3H$.

b) If, instead of ethylene, propene or another 1-alkene is used, only a single insertion into the Al–C bond takes place. An example is the **catalytic dimerization of propene** which forms the basis of an important technical process for the production of isoprene:

① carbalumination: $(n\text{-}Pr)_3Al + CH_2{=}CH{-}CH_3 \rightarrow (n\text{-}Pr)_2Al{-}CH_2{-}CH(Me)(n\text{-}Pr)$

② dehydroalumination: $(n\text{-}Pr)_2Al{-}CH_2{-}CH(Me)(n\text{-}Pr) \rightarrow (n\text{-}Pr)_2AlH + CH_2{=}C(Me)(n\text{-}Pr)$

③ hydroalumination: $(n\text{-}Pr)_2AlH + CH_2{=}CH{-}CH_3 \rightarrow (n\text{-}Pr)_3Al$

$CH_2{=}C(Me)(n\text{-}Pr) \xrightarrow{-CH_4\ \text{cracking}}$ Isoprene $\xrightarrow{\text{polymerisation}}$ Synthetic Rubber (1,4-cis-Polyisoprene)

c) Olefin polymerization employing mixed catalysts like $Et_3Al/TiCl_4$ in heptane (**Ziegler-Natta low-pressure process**) will be discussed in section 17.6.

PROPERTIES

The binary aluminum alkyls R_3Al are colorless, mobile liquids which violently react with air and water: Aluminum alkyls with short chain lengths are pyrophoric, with water they react explosively and their manipulation calls for utmost care and the **protective atmosphere of an inert gas** (N_2 or Ar). With the exception of alkanes and aromatic hydrocarbons, all solvents are rapidly attacked by aluminum alkyls. Thermal cleavage to form R_2AlH and alkene for aluminum alkyls with β-branched alkyl groups becomes appreciable at 80 °C, for *tris*(n-alkyl)aluminum at about 120 °C. The reactivity of organoaluminum halides R_nAlX_{3-n} and -alkoxides $R_nAl(OR')_{3-n}$ is considerably lessened.

STRUCTURE AND BONDING OF BINARY ALUMINUM ORGANYLS

Aluminum organyls have a pronounced tendency towards the formation of **dimeric units Al_2R_6**. This association is opposed by bulky ligands R:

	Solid state	Solution (in hydrocarbons)	Gas phase
$AlMe_3$	dimeric	dimeric	dimeric $\rightleftharpoons$ monomeric
$AlEt_3$, $Al(n\text{-}Pr)_3$	dimeric	dimeric	monomeric
$Al(iso\text{-}Bu)_3$	dimeric	monomeric	monomeric
$AlPh_3$	dimeric	dimeric $\rightleftharpoons$ monomeric	monomeric

Among the organometallics MR_3 of group 13, dimerization is only observed for the aluminum organyls.

Structural data for $Al_2(CH_3)_6$

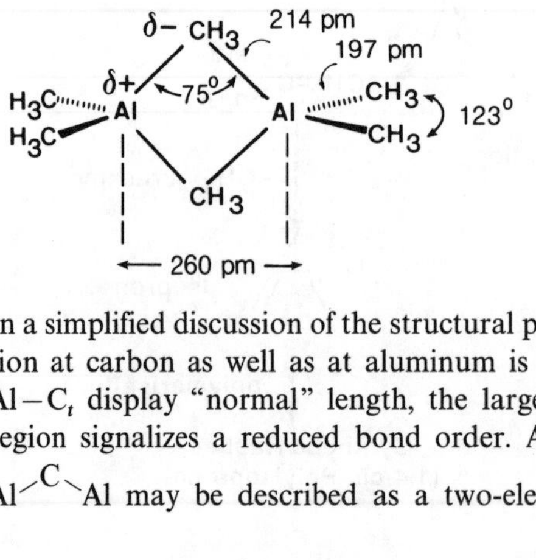

$Al\overset{C}{\frown}Al$ *(2e 3c) bonds in* $Al_2(CH_3)_6$

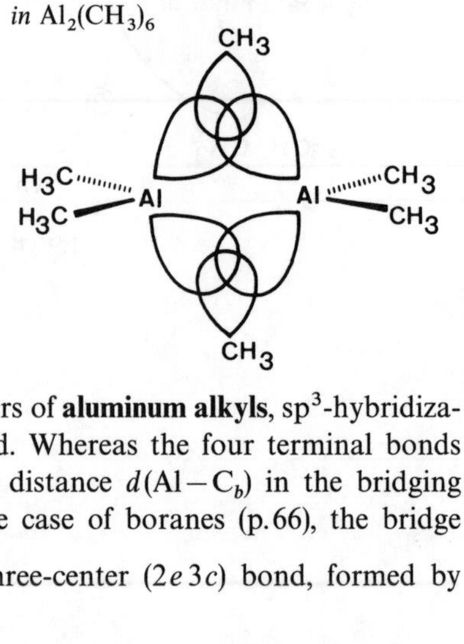

In a simplified discussion of the structural parameters of **aluminum alkyls**, sp^3-hybridization at carbon as well as at aluminum is assumed. Whereas the four terminal bonds $Al-C_t$ display "normal" length, the larger bond distance $d(Al-C_b)$ in the bridging region signalizes a reduced bond order. As in the case of boranes (p.66), the bridge $Al\overset{C}{\frown}Al$ may be described as a two-electron three-center ($2e\,3c$) bond, formed by the interaction of one C(sp^3) orbital with two Al(sp^3) orbitals. Since the fragments 2 $(CH_3)_2Al\cdot$ and 2 $CH_3\cdot$ furnish a total of 4 electrons for bonding, 2 electrons per Al$^{\diagup C \diagdown}$Al-bridge are available.

The bond angles and the Al—Al distance prompt a more refined interpretation, however. In particular, the distance d(Al—Al) = 260 pm in Al_2Me_6 is significantly shorter than the corresponding distance in dimeric aluminum halides which possess Al$^{\diagup X \diagdown}$Al bridges, consisting of two ($2e\,2c$) bonds [example: Al_2Cl_6, d(Al—Al) = 340 pm]. The short distance d(Al—Al) in Al_2Me_6 is indicative of a direct Al—Al interaction, for this length only marginally surpasses the sum of the covalent radii of two Al atoms (2 × 126 pm).

In an extreme representation, which presupposes an Al—Al σ-bond, Al(sp^2) ⓑ rather than Al(sp^3)-hybridization ⓐ is assumed.

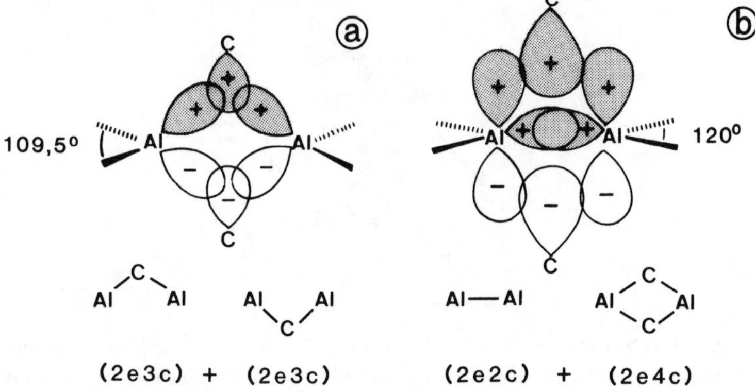

Description ⓑ concords with the considerable opening of the angle C_tAlC_t beyond 109.5°. The smaller bond angle AlC_bAl reflects a compromise between maximal orbital overlap and tolerable $Al^{\delta+}/Al^{\delta+}$ repulsion. The actual situation probably resides between the alternatives ⓐ and ⓑ.

The structure of **triphenylaluminum** Al_2Ph_6, which is dimeric in the solid state, also points at a bonding situation which lies between two limiting cases. Two pertinent aspects, the small angle C$^{\diagup C_b \diagdown}$C at the bridging carbon atom and the disposition of the plane of the phenyl ring, perpendicular to the Al—Al axis, must be mentioned.

Structural data for $Al_2(C_6H_5)_6$

Alternative hybridizations at the bridging C atom

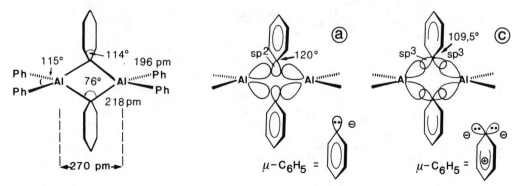

In model ⓐ the phenyl carbanion contributes a $C_b(sp^2)$ orbital and one electron pair to the $(2e\,3c)$ bond in the bridge $Al\overset{C_b}{\diagup\diagdown}Al$. Aromatic conjugation would remain unaffected in this mode. In description ⓒ, the ion $C_6H_5^-$ acts as a 4e ligand, supplying two $C_b(sp^3)$ orbitals to the bridge bond $Al\overset{C_b}{\diagup\diagdown}Al$. In this variant, the bridge would be formed by two $(2e\,2c)$ bonds, and the electron deficiency in the bridging region abated at the expense of aromatic conjugation in the ligand. The angle CC_bC of 114° again implies a bonding situation between the two borderline cases.

Aluminum organyls offer yet another bridging mode: **alkynyl groups** form particularly strong bridges which are maintained even in the gas phase (Haaland, 1978). They are characterized by disparate bonding relations of the bridging atom C_α to the Al atoms.

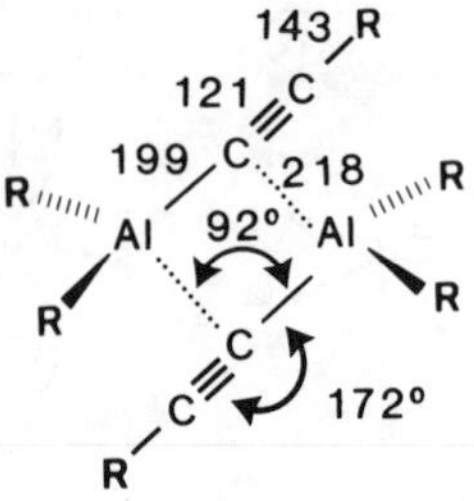

143 R
121 C
199 C 218
R Al 92° Al R
R R
C
C 172°
R

R=Ph

Structure of **$[Ph_2Al-C{\equiv}CPh]_2$** *in the crystal* (*Stucky*, 1974): C_α *forms a* σ *bond to one and a* π*-donor bond to the other* Al *atom. This mode is reminiscent of* σ/π *bridges, which occur for transition-metal carbonyls (**p. 225**). The alkynyl ligand may be regarded as a donor of three electrons* ($1\,\sigma$*-e*, $2\,\pi$*-e*) *which renders alkynyl bridges less electron deficient than alkyl bridges, thereby accounting for their robust nature.*

The association of certain aluminum organyls **in solution** is confirmed by spectroscopic methods, especially by ^1H- and ^{13}C-NMR. Thus, trimethylaluminum at − 50 °C shows two ^1H-NMR signals of relative intensities 1:2 [$\delta(CH_3$, bridging) 0.50 ppm; $\delta(CH_3$, terminal) − 0.65 ppm] which coalesce at − 25 °C and appear as one sharp line [$\delta(CH_3)$ − 0.30 ppm] at + 20 °C. This result is sufficiently accounted for by intramolecular exchange:

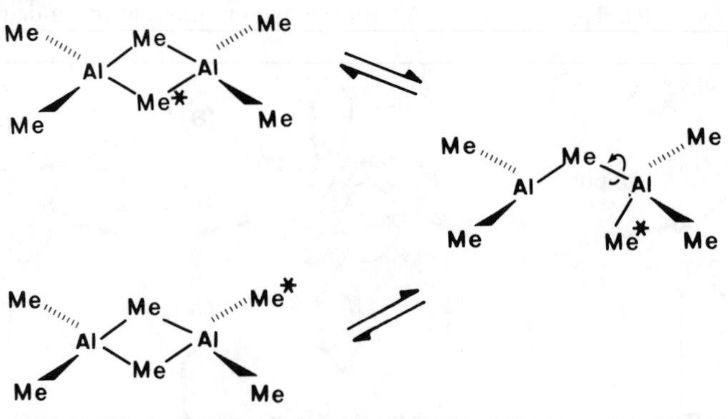

Me Me Me
Al Al
Me*
Me Me
⇌
Me Me Me
Al Al
Me Me* Me

Me Me Me*
Al Al
Me
Me Me
⇌

Additionally, intermolecular exchange also takes place. Therefore, pure compounds Al-RR′R″, which bear different alkyl groups at aluminum, cannot be isolated:

$$R_2Al(\mu\text{-}R')_2AlR_2 \rightleftharpoons 2\,R_2R'Al \rightleftharpoons RR'Al(\mu\text{-}R)(\mu\text{-}R')AlRR' \rightleftharpoons \ldots$$

etc. $\rightleftharpoons$ R_3Al + RR'_2Al

Values for the enthalpy change ΔH, affiliated with the monomer/dimer equilibrium, are accessible by ^{1}H-NMR:

$$Al_2(CH_3)_6 \rightleftharpoons 2\,Al(CH_3)_3 \qquad \Delta H = 84\ \text{kJ mol}^{-1}$$

Furthermore, NMR studies of species with a mixed ligand sphere allow an evaluation of the **relative ability of substituents to form bridges:**

$R_2N > RO > Cl > Br$	>	$Ph-C\equiv C- > Ph$	>	Me > Et > *iso*-Pr > *t*-Bu
Two ($2e\,2c$) bonds, trend follows the Lewis basicity		unsaturated bridges, σ- and π-interactions		one ($2e\,3c$) bond, steric demand is decisive

Bridges of exceptional strength are formed by the hydride ion in compounds $(R_2AlH)_n$:

solution	neat, liquid	neat, gas phase
$n = 3$	$n > 3$	$n = 2$

$(Me_2AlH)_3$: cyclic trimer with three Al–H–Al bridges.

$(Me_2AlH)_2$: Al–H–Al 103°, H–Al–H 167 (as labelled), Me–Al–Me 119°, Al···Al 252.

*In this context, a brief **discussion of the energetics** is instructive. The enthalpy changes $\Delta H_{react.}$, pertaining to the cleavage of $[R_2AlX]_2$ into monomeric units, decrease in the following order:*

$$[R_2AlX]_2 \longrightarrow 2\,R_2AlX \qquad \Delta H_{react.}$$

X:	**H**	>	Cl	>	Br	>	I	>	CH_3
$\Delta H_{react.}$:	150		124		121		102		84 kJ/mol

*This is, however, **not** the sequence of decreasing bond energy $D(Al-X)$ in the bridges $Al-X-Al$ because the listed values for $\Delta H_{react.}$ relate to a process which, apart from bond* *cleavage, includes additional changes. First of all, it must be borne in mind that in the dimer, the aluminum atom is in a tetrahedral environment whereas in the monomer it finds itself in a trigonal planar coordination.*

$$R_2Al(\mu\text{-}X)_2AlR_2 \xrightarrow{\Delta H_{Al-X}} R_2AlX \text{ (pyramidal)} + R_2AlX \text{ (pyramidal)} \xrightarrow{\Delta H_{reorg.}} 2\ R_2Al{-}X \text{ (planar)}$$

bond cleavage + reorganization

$\Delta H_{react.} = \Delta H_{Al-X}$ + $\Delta H_{reorg.}$

measure of the bond energy $D(\text{Al}-\text{X})$ *in an* Al–X–Al *bridge*

(1) *intramolecular ligand repulsion decreases: pyramidal > planar*
(2) $\text{Al}(p_\pi)-\text{X}(p_\pi)$ *bonding is stronger in the planar than in the pyramidal configuration.*
(3) $\text{Al}\overset{\sigma}{-}\text{C}$ *bond energy decreases:* $sp^3 > sp^2$

After correction for $\Delta H_{reorg.}$, *the enthalpy changes accompanying* Al–X *cleavage in* Al–X–Al *bridges now vary as:*

$$\Delta H_{Al-X}\text{: Cl} > \text{Br} > \text{I} > \mathbf{H} > \text{Me}$$

The inescapable conclusion is that hydride bridges appear to be so strong, because the enthalpy of reorganization $\Delta H_{reorg.}$ *is small for* R_2AlH. *This must be related to the minor role, relief of interligand repulsion plays for the small hydride ligand on going from pyramidal to trigonal planar geometry, and to the absence of a stabilizing π-bonding contribution in the trigonal planar form.*

REACTIONS OF ALUMINUM ORGANYLS

The exceptional reactivity of organoaluminum compounds AlR_3 has led to a multitude of synthetic applications which can only be treated in a cursory fashion here. The reactivity of $RAlX_2$ is very much reduced as compared to that of AlR_3. Hence, in many cases only one of the three Al–C bonds can be utilized in synthesis.
Reactions of the higher halohydrocarbons with aluminum alkyls must be performed with circumspection: the **contact of AlR_3 with $CHCl_3$ or CCl_4 has led to explosions**. For synthetic applications, it is of significance that in their chemoselectivity, lithium organyls, Grignard reagents and aluminum organyls complement each other.
Whereas LiR and RMgX typically add to polar multiple bonds like C=O and –C≡N, addition to >C=C< double bonds usually being confined to conjugated systems (p. 30), AlR_3 also adds to isolated multiple bonds >C=C< and –C≡C– (**carbalumination**). As in the case of hydroboration, it is the products of subsequent reactions of the initial

(HOBERG, 1980) RAl(Ph)C=C(Ph)AlR ring with Ph, Ph

$$AlR_3 \xrightarrow{h\nu,\ PhC\equiv CPh} \text{(HOBERG, 1980) diphenyl-substituted } RAl\text{–}AlR \text{ ring}$$

$$AlR_3 \xrightarrow{R'NH_2} (RAlNR')_4$$

$$AlR_3 \xrightarrow{ROH} R_2Al(OR'),\ RAl(OR')_2,\ Al(OR')_3$$

$$AlR_3 \xrightarrow{O_2} Al(OR)_3$$

$$AlR_3 \xrightarrow{SnCl_4} SnR_4$$

$$AlR_3 \xrightarrow{AlCl_3} RAlCl_2 + R_2AlCl$$

$$AlR_3 \xrightarrow{LiR} Li[AlR_4]$$

$$AlR_3 \xrightarrow{H_2,\ 300\ bar,\ 150^\circ} R_2AlH$$

$$AlR_3 \xrightarrow{R'Cl} R_2AlCl + RR'$$

$$AlR_3 \xrightarrow{CH_2=CHR'} R_2AlCH_2CHRR'$$

organoaluminum adducts which are of practical interest. The rate law for carbalumination is compatible with a monomer/dimer pre-equilibrium:

$$1/2\ (Et_3Al)_2 \rightleftharpoons Et_3Al \xrightarrow[\text{rate-determ.}]{CH_2=CHR} \left[\begin{matrix} CH_2-\overset{\oplus}{C}HR \\ \overset{\ominus}{Et_2Al}-Et \end{matrix}\right] \longrightarrow \begin{matrix} CH_2-CHR \\ Et_2Al \quad\quad Et \end{matrix}$$

$$\text{Rate} = k[(Et_3Al)_2]^{1/2} \cdot [\text{Alkene}]$$

Carbaluminations invariably proceed as ***cis*-additions**; alkynes react faster than alkenes and terminal alkenes more rapidly than internal alkenes. The R_2Al fragment usually attaches itself to the least substituted (= terminal) carbon atom of the alkene.
Closely related to carbalumination is **hydroalumination** (cf. "attachment" in the Ziegler direct process, p. 76):

$$R_2AlH + \rangle C=C\langle \longrightarrow -\underset{R_2Al}{\overset{|}{\underset{|}{C}}}-\underset{H}{\overset{|}{\underset{|}{C}}}-$$

The readiness to undergo hydroalumination increases in the order:

$$RCH=CHR < R_2C=CH_2 < RCH=CH_2 < CH_2=CH_2$$

An inverse trend is observed for dehydroaluminations where alkenes are liberated (p. 76).

Hydroaluminations are highly **stereoselective (*cis*)**; the degree of **regioselectivity (*anti*-Markovnikov)** varies, however:

$RCH{=}CH_2$ / $H{-}AlR'_2$	$RCH{=}CH_2$ / $R'_2Al{-}H$	;	$PhCH{=}CH_2$ / $H{-}AlR'_2$	$PhCH{=}CH_2$ / $R'_2Al{-}H$
97%	3%		75%	25%

Example:

$$\xrightarrow[\substack{100-135° \\ -\ 2\ \text{iso-butene}}]{(\text{iso-Bu})_2AlH} \left[\ \ldots CH_2-\right]_3 Al \xrightarrow[2.\ H_2O]{1.\ O_2} \ldots CH_2OH$$

β-Citronellol

An advantage of hydroalumination over hydroboration is the ease of Al–C cleavage which does not require the use of peroxides. As a disadvantage, the lower chemoselectivity must be mentioned: besides C=C multiple bonds, functional groups are also attacked and, in the case of terminal alkynes, metallation at the CH-acidic position $R-C{\equiv}C-H$ occurs.

LEWIS-BASE ADDUCTS OF ALUMINUM ORGANYLS

Aluminum organyls are stronger Lewis acids than organoboron compounds. They number among the **hard acids** (in the Pearson sense) because Lewis basicity towards $AlMe_3$ decreases as follows:

$$Me_3N > Me_3P > Me_3As > Me_2O > Me_2S > Me_2Se > Me_2Te$$

hard bases — soft bases

With Me_2S and Me_2Se, complicated equilibrium mixtures are formed since the strength of the bridge $Al\overset{CH_3}{\diagup\ \diagdown}Al$ is comparable to that of the Lewis-acid/base interaction $Me_3Al \cdot EMe_2$. The finding that in group 13 only the aluminum organyls form dimers is another expression of the high Lewis acidity of AlR_3.

The determination of the coordination number (C.N.) *at* Al *for adducts present in solution often poses difficulty – especially in cases where multidentate Lewis bases are involved. For compounds of the type* $(R_nAlX_{3-n})_m$ *(m* = 1, 2, 3) *and* $[R_2\overline{AlO(CH_2)_2Y}]_2$ (Y = OR′, NR′$_2$) *chemical shifts in the* ^{27}**Al-*NMR spectra*** *can provide insights (Benn, 1987 R)*:

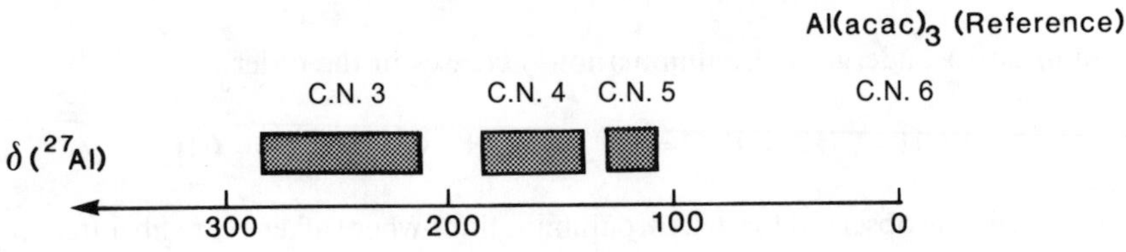

Reactions of R_3Al with **tertiary amines** usually terminate with adduct formation; in the cases of **secondary and primary amines**, consecutive reactions follow:

$$(Me_3Al)_2 \xrightarrow{Me_3N} Me_3Al\text{-}NMe_3$$

$$(Me_3Al)_2 \xrightarrow{Me_2NH} 2\ Me_3Al\text{-}NHMe_2 \xrightarrow[-\,2\ CH_4]{110°} (Me_2AlNMe_2)_2$$

$Me_2Al(\mu\text{-}NMe_2)_2AlMe_2$

$$(Me_3Al)_2 \xrightarrow[\Delta T,\ -8\ CH_4]{MeNH_2} (MeAlNMe)_4$$

heterocubane type

H_3C–$\overset{\ominus}{Al}$–$\overset{\oplus}{N}$–H

The Lewis-acid/base interaction results in an increase of carbanion character of the CH_3 *group as well as an increase in* NH *acidity. The elimination of* CH_4 *in the reactions shown above is thereby facilitated.*

Adducts of aluminum alkyls with unsaturated Lewis bases can rearrange via alkyl migration such that carbalumination of the multiple bond is effected. In this way, nitriles are transformed into dimeric bis(organo)aluminum ketimides:

$$(Me_3Al)_2 \xrightarrow{t\text{-}BuCN} 2\ t\text{-}Bu\overset{\oplus}{C}{\equiv}N\text{-}\overset{\ominus}{Al}Me_3 \xrightarrow{\Delta T} [(Me)(t\text{-}Bu)C{=}N\text{-}AlMe_2]_2$$

Carbanions add to AlR_3 to form tetraorganoalanates:

$$2\,LiC_2H_5 + Al_2(C_2H_5)_6 \longrightarrow 2\,LiAl(C_2H_5)_4$$

(soluble in hydrocarbons)

These **"ate complexes"** also arise from reductions with alkali metals:

$$3\,Na + 2\,Al_2(C_2H_5)_6 \longrightarrow 3\,NaAl(C_2H_5)_4 + Al$$

$LiAlEt_4$, like $BeMe_2$, has a polymeric structure which can be described as a chain of edge-sharing tetrahedra, in which alternating Li and Al atoms are connected by M–C–M ($2e\,3c$) bridges:

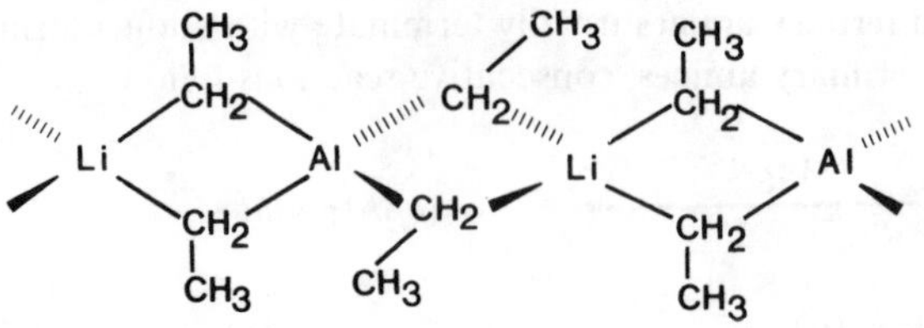

$[LiAlEt_4]_n$

Similar to R_2B^+ (p. 58), the extremely Lewis acidic **bis(organo)aluminum cation R_2Al^+** can only be isolated as its Lewis base adduct. As expected, aluminum adopts a higher coordination number than boron.

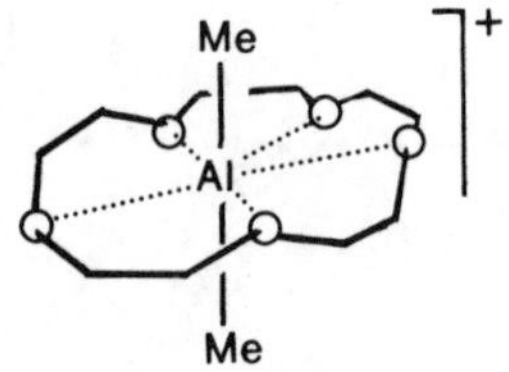

$\{Me_2Al[15]crown\text{-}5\}^+$: *the linear cation* Me_2Al^+ *assumes a central position here. In the interior of the larger macrocycle* [18]*crown*-6, Me_2Al^+ *is located excentrically* (*Atwood*, 1987).

An unusual Al^I compound which could be fully characterized, is the cluster $[(\eta^5\text{-}C_5Me_5)Al]_4$, prepared from $AlCl(Et_2O)_x$ and $(C_5Me_5)_2Mg$ (Schnöckel, 1991).

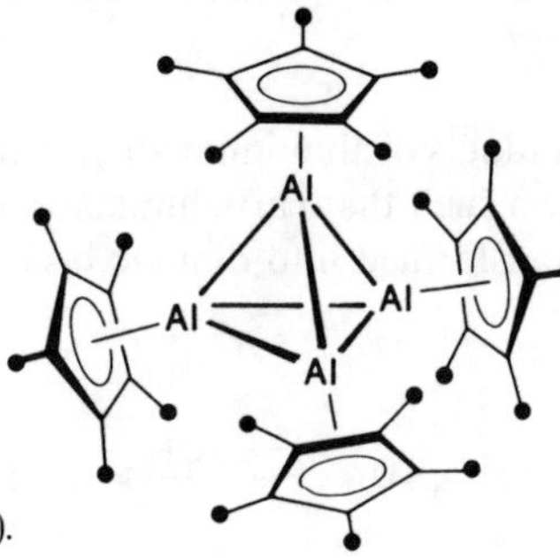

$[(\eta^5\text{-}Cp^*)Al]_4$ d(Al-*ring center*) = 201 pm
d(Al−Al) = 277 pm
compare: $d(\text{Al}-\text{Al})_{\text{metal}}$ = 286 pm
Note that the corresponding indium complex is a hexamer (p. 90).

7.3 Gallium-, Indium-, and Thallium Organyls

The organometallic compounds of gallium, indium, and thallium are far less important than those of boron and aluminum. Ga- and In-organyls serve as doping agents in the manufacture of semiconductors. Thus, via thermal decomposition of gaseous mixtures of trimethylgallium and arsane, layers of galliumarsenide can be deposited (MOCVD, metal organic chemical vapor deposition):

$$(CH_3)_3Ga(g) + AsH_3(g) \xrightarrow{700-900\,^\circ C} GaAs(s) + 3\,CH_4(g)$$

The organometallic chemistry of gallium in many respects still resembles that of aluminum; it is confined almost exclusively to the oxidation state Ga^{III}. For indium, besides In^{III}, a few organyls of In^{I} have been reported [Example: $(C_5H_5)In$]. In the case of thallium, Tl^{III} as well as Tl^{I} is encountered in organometallic compounds. Despite their high toxicity, organothallium compounds have found use in organic synthesis.

7.3.1 σ-Organyls of Ga, In, Tl and their Lewis-Base Adducts

The organyls R_3M (M = Ga, In, Tl) have virtually no tendency to form dimers. Furthermore, as dissociation enthalpies for the adducts $R_3M \cdot NMe_3$ indicate, the Lewis acidities of species R_3M decrease for the heavier homologues of aluminum ($B < \mathbf{Al} > Ga > In > Tl$).
The preparation of R_3Ga, R_3In and of the mixed ligand species R_nMX_{3-n} proceeds according to standard methods of organometallic chemistry:

$$2\,M + 3\,Me_2Hg \longrightarrow 2\,Me_3M + 3\,Hg \quad (M = Ga, In, \nparallel Tl)$$ 2

$$GaBr_3 + 3\,MeMgBr \longrightarrow Me_3Ga \cdot OEt_2 + 3\,MgBr_2$$ 4

$$InCl_3 + 2\,MeLi \longrightarrow Me_2InCl$$

$$Me_3M + HCN \longrightarrow Me_2MCN + CH_4 \quad (M = Ga, In)$$ 6

$$Et_2GaH + CH_2{=}CH{-}CH_3 \longrightarrow Et_2(n\text{-}Pr)Ga$$ 8

The binary organyls $\mathbf{R_3Ga}$ and $\mathbf{R_3In}$ are **monomeric** in solution and in the gas phase; they are highly air-sensitive. Alkyl-group scrambling among the mixed trisalkyls RR′R″M only takes place at elevated temperatures.
The electron deficiency at Ga can be relieved by means of **complex formation**:

$$Me_2GaCl \xrightarrow{NH_3} [Me_2Ga(NH_3)_2]^+Cl^-$$

$$Me_3Ga \xrightarrow[Et_2O]{H_2O} [Me_2GaOH]_4 \begin{cases} \xrightarrow{H^+_{aq}} [Me_2Ga(H_2O)_2]^+ \\ \xrightarrow{OH^-_{aq}} [Me_2Ga(OH)_2]^- \end{cases}$$

In view of the pyrophoric nature of the majority of binary organyls R_3Ga, the resistance of the Ga−C bonds towards hydrolysis and oxidation in the species $[R_2GaL_2]^{+,-}$ is remarkable.
Rather than by complex formation, an octet configuration at Ga and In can also be attained by **self-association**:

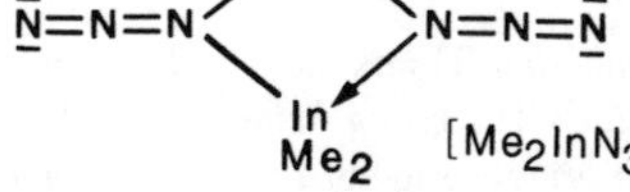

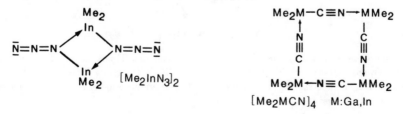

$[Me_2InN_3]_2$

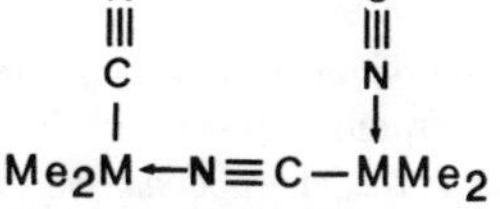

$[Me_2MCN]_4$ M:Ga,In

$[Me_2GaC_2Ph]_2$ $[Me_2InCl]_x$

The dimeric nature of $Me_2GaC{\equiv}CPh$, *which contrasts with most other compounds* R_3Ga, *is another example of the aforementioned strong inclination for bridge formation exhibited by alkynyl groups (p. 80) (Oliver, 1981).*

In the organometallic chemistry of thallium – as opposed to that of Ga and In – changes in oxidation state M^{III}/M^{I} play an important role. Thus, the yields of thallium trisalkyls in syntheses starting from inorganic Tl^{III} compounds and Li alkyls or Grignard reagents, are impaired by the oxidizing action of Tl^{III}. Conversely, *tris*(organo)thallium compounds may be formed via oxidative addition of RI to Tl^{I} compounds:

$$TlI + 2\,MeLi + MeI \longrightarrow Me_3Tl + 2\,LiI$$

$$MeTl + MeI \longrightarrow Me_2TlI$$

The air- and water-sensitive binary compounds R_3Tl are subject to ready, occasionally even explosive decomposition. Another indication of the weakness of the Tl–C bond is the alkylation of mercury by means of R_3Tl:

$$2\,R_3Tl + 3\,Hg \longrightarrow 3\,R_2Hg + 2\,Tl$$

Trimethylthallium, which is monomeric in solution and in the gas phase, forms a three-dimensional network in the crystalline state, whereby each thallium atom possesses three short and two long bonds to neighboring carbon atoms, with mean values of 228 and 324 pm respectively (Sheldrick, 1970):

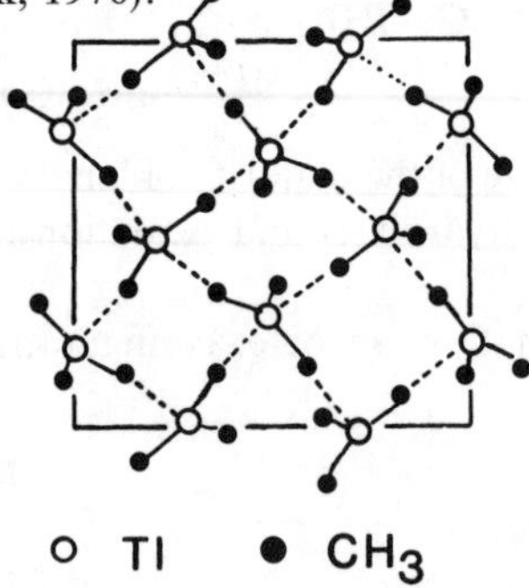

o Tl ● CH_3

A projection of the structure of $\mathbf{Me_3Tl}$ *in the crystal reveals that* Tl *has the coordination number* 5, *four* CH_3 *groups assuming bridging positions and the fifth* CH_3 *group being terminal. The large differences in the* Tl–C *bond lengths within the* Tl–C–Tl *bridges suggest that three center bonding (2e 3c) is only weakly present.*

Tl–C bond cleavage in R_3Tl by means of a dihalogen leads to bis(organo)thallium halides which in polar media form very stable cations:

$$\left.\begin{array}{l} R_3Tl + X_2 \xrightarrow{-RX} \\ TlCl_3 + 2\,RMgX \xrightarrow[-MgCl_2]{} \end{array}\right\} \longrightarrow R_2TlX \xrightarrow{H_2O} R_2Tl^+_{aq} + X^- \qquad \boxed{4}$$

Like the isoelectronic species R_2Hg and R_2Sn^{2+}, $\mathbf{R_2Tl^+}$ is **linear**. R_2TlOH in aqueous solution acts as a strong base:

$$R_2TlOH + H_2O \longrightarrow R_2Tl^+_{aq} + OH^-$$

With methylenephosphoranes, organometal halides of Ga^{III}, In^{III} and Tl^{III} form oligomeric ylid complexes via transylidation (p. 180):

$$Me_2MX + Me_3P{=}CH_2 \longrightarrow Me_2M^{\ominus}(X)(CH_2{-}\overset{\oplus}{P}Me_3)$$

$$\downarrow Me_3P{=}CH_2, \; -[Me_4P]X$$

$$Me_2M{-}CH_2{-}\overset{\oplus}{P}Me_2({-}\overset{\ominus}{:}CH_2) \xrightarrow{\times 2} [Me_2\overset{\ominus}{M}(CH_2)_2\overset{\oplus}{P}Me_2]_2 \text{ (cyclic dimer)}$$

M = Ga, In, Tl X = Cl, Br

7.3.2 π-Complexes of Ga, In, and Tl

Cyclopentadienyl compounds of indium are known for the oxidation states In^{III} (σ-organyls) and In^{I} (π-complexes) (E. O. Fischer, 1957):

$$InCl_3 + 3\,NaC_5H_5 \xrightarrow{THF} (\eta^1\text{-}C_5H_5)_3In$$

$$\downarrow 150\,°C,\ 1\ mbar,\ -C_{10}H_{10}$$

$$InCl + LiC_5H_5 \xrightarrow[60\,°C]{benzene} (\eta^5\text{-}C_5H_5)In$$

$(C_5H_5)_3In$ displays one ^1H-NMR signal only even at −90 °C; it is thought to be fluxional. $(C_5H_5)In$, a sublimable, water-stable but air-sensitive material, is the only easily accessible compound of In^I and serves as a starting material for the preparation of other In^I compounds:

$$(C_5H_5)In + HX \longrightarrow InX + C_6H_6$$

(X: oxinate, acetylacetonate etc.)

An interesting aspect of the structure of $(C_5H_5)In$ is the strongly differing metal-ring distance in the crystal and in the gas phase. This points at higher covalency of the In–C bond in the isolated molecule (p. 90).

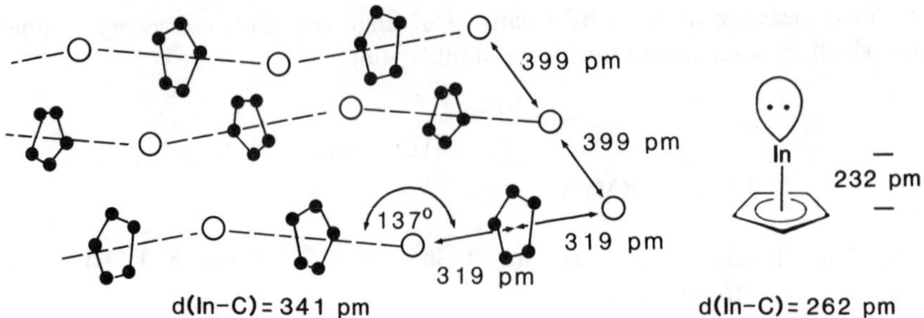

(C_5H_5)In *in the crystal: polymer chains* $[(C_5H_5)In]_x$ (*Panattoni*, 1963)

(C_5H_5)In *in the gas phase: monomeric half-sandwich complex* (*L. S. Bartell*, 1964)

(C_5H_5)In possesses a dipole moment of 2.2 Debye, the indium atom and its lone pair forming the negative end of the molecule (Tuck, 1982). The structure of (C_5Me_5)In in the solid state differs fundamentally from that of parent (C_5H_5)In: whereas the latter forms zig-zag polymeric chains, the former is packed as octahedral $(C_5Me_5)_6In_6$ units which bear resemblance to the cluster $(C_5H_5)_6Ni_6$ (p. 409). The high volatility of (pentamethylcyclopentadienyl)indium indicates, however, that bonding between the (C_5Me_5)In monomers in the hexamer is weak (Beachley, 1986).
The best known organometallic compound of Tl^I, (C_5H_5)Tl, can even be prepared in an aqueous medium:

$$Tl_2SO_{4(aq)} + 2\,C_5H_6 + 2\,NaOH \longrightarrow 2(C_5H_5)Tl + Na_2SO_4 + 2\,H_2O \qquad \boxed{4}$$

(C_5H_5)Tl is sublimable, sparingly soluble in organic solvents and can be handled in air; it is a convenient cyclopentadienyl-transfer agent for use with transition metal ions. The structure of (C_5H_5)Tl largely resembles that of (C_5H_5)In.

(C_5H_5)In *may serve as a model for discussing the bonding situation in* ***main-group element cyclopentadienyls*** (C_5H_5)M (*Canadell*, 1984). *The rather small difference in the electronegativities of* In *and* C *leads one to expect that the metal-ligand bond in isolated molecules of* (C_5H_5)In *should be predominantly covalent. If the molecule is mentally constructed from the fragments* :In^+ (*sp-hybrid*) *and* $C_5H_5^-$ ($6\,\pi$-*e*), *the following metal-ligand interactions are conceivable*:

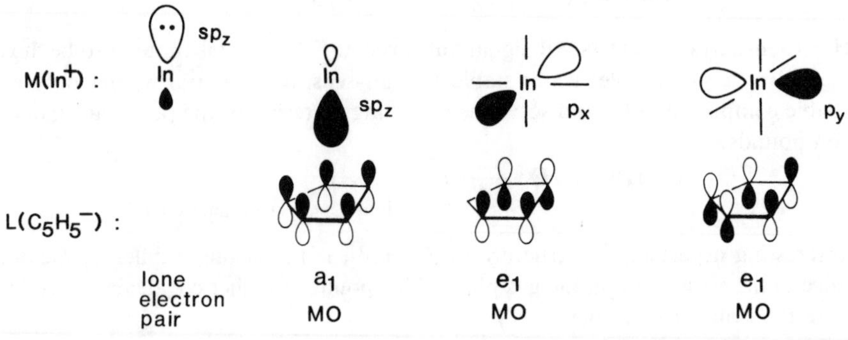

All three interactions generate a charge transfer L → M. *The extent of this charge transfer and the ensuing covalent nature of the bond are governed by the energies of the basis orbitals which in individual cases have to be established by means of quantum chemical calculations. Similar considerations apply to the isolobal* (*p.* 396) *fragments* $R-Be^+$, $R-Mg^+$, $R-B^{2+}$, $R-Al^{2+}$, *as well as* Ge^{2+}, Sn^{2+} (p. 136) *and* Pb^{2+}.

Besides variation of the metal component, varying the organic ligand is also conceivable. This leads to the question whether the ligand C_6H_6, which is iso-π-electronic with $C_5H_5^-$, can form **main-group element arene π-complexes**:

While the surprisingly high solubility of $Ga^I[Ga^{III}Cl_4]$ in benzene has been known for some time, implying some kind of Ga^I-arene interaction, the isolation and structural characterization of these adducts was achieved only recently (Schmidbaur, 1985 R). The compound $[(\eta^6\text{-}C_6H_6)_2Ga]GaCl_4 \cdot 3\,C_6H_6$ features tilted bis(arene)gallium^I cations, which are linked through $GaCl_4^-$-tetrahedra to form dimeric units.

Structure of $\mathbf{[(\eta^6\text{-}C_6H_6)_2Ga]GaCl_4 \cdot 3\,C_6H_6}$ *in the crystal. Three benzene molecules of the empirical formula are outside the coordination sphere of gallium. The distances* $d(Ga-Cl)$ *differ, as do the distances* $d(Ga-C)$ (*mean values* 308 *and* 323 pm *for the two* $Ga-C_6H_6$ *segments*).

In a related mesitylene π-complex of In^I, bridging by $InBr_4^-$-tetrahedra leads to a polymeric structure:

Unit from the chain structure of $\mathbf{\{[(\eta^6\text{-mesitylene})_2In]InBr_4\}_n}$ *in a schematic representation. The tilting angle amounts to* 133°, *the two arenes are almost equidistant from the central metal;* d(In-*ring center*) = 283 *and* 289 pm.

A property common to these materials is the ease with which the arene ligands are cleaved off thermally. Herein, complex stability decreases in the order $Ga^I > In^I > Tl^I$; it increases with increasing degree of alkylation of the arene. Furthermore, the tendency to maintain the favorable coordination to halide ions is evident.

Contrary to (arene)transition-metal π-complexes [$(C_6H_6)_2Cr$ melts without decomposition at 284 °C!], bonding in (arene)main-group π-complexes falls into the domain of **weak interactions**. Without delving into a detailed discussion here, only one plausible reason for these disparate properties will be mentioned: the stability of the transition metal-ligand bond is derived from the σ-donor/π-acceptor synergism (p. 185); d-electrons for backbonding are, however, not available for main-group elements, since electrons in the subultimate d-shell are essentially core-electrons, being strongly bonded to the nucleus (p. 190).

Hence, it is all the more remarkable that a mode of coordination of an arene to Ga^I has been discovered, which is without precedent in the area of transition metal π-complexes (Schmidbaur, 1987):

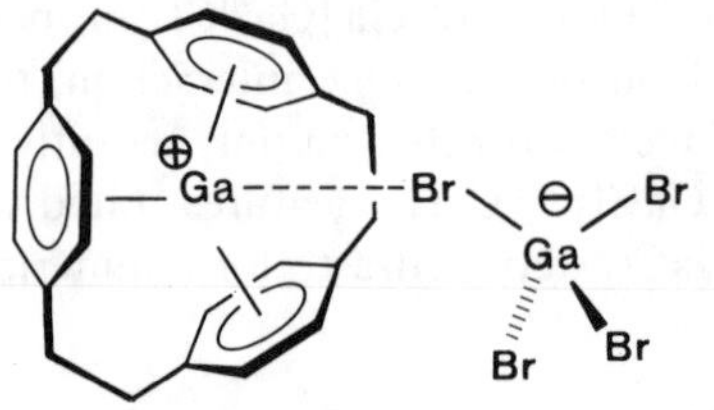

In the ion pair **{(η^{18}-[2.2.2]paracyclophane)Ga^I}$GaBr_4$** *the central atom* Ga^I *exhibits nearly equal distances to the three ring centers; it is, however, displaced away from the ligand center along the three-fold axis by* 43 pm *in the direction of the counter ion* $GaBr_4^-$. *Unquestionably, the chelate effect exerts a stabilizing function here.*

7.3.3 Applications of Thallium Organyls in Organic Synthesis

$Thallium^{III}$ salts attack alkenes, whereby a solvent anion enters vicinally to the Tl–C bond formed **(oxythalliation)**. These organothallium intermediates decompose more rapidly than the primary products of the otherwise similar oxymercuration (cf. p. 53). The metal leaves as Tl^I here:

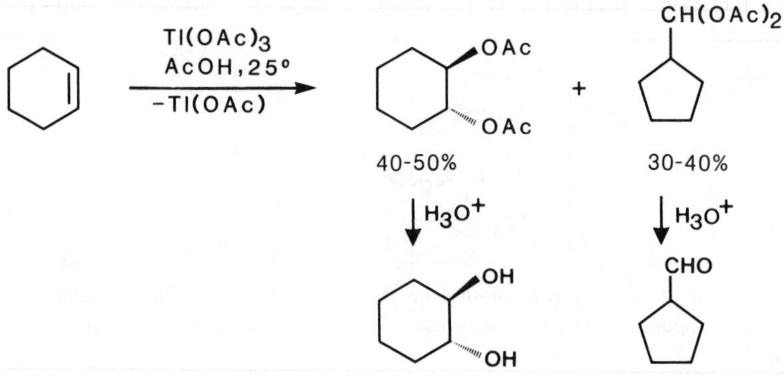

The product distribution in this olefin oxidation strongly depends on the nature of the medium and the counterions of Tl^{III}. Thus, oxythalliation of cyclohexene with $Tl(NO_3)_3$ in MeOH almost exclusively yields the cyclopentanecarbaldehyde. Mechanistically, this ring contraction may proceed as follows:

A process of high regioselectivity is the **electrophilic aromatic thalliation** by means of thallium trifluoroacetate, $Tl(TFA)_3$, in trifluoroacetic acid, TFAH. In subsequent reactions, which are accompanied by a change in oxidation state $Tl^{III} \rightarrow Tl^{I}$, the new substituent enters at the position of the former Tl−C bond. In this way, iodination of benzene can be effected under particularly mild conditions. Photolysis of the intermediary arylthallium trifluoroacetates in aromatic solvents lends itself to the preparation of unsymmetrical biphenyls:

8 Organoelement Compounds of the Carbon Group (Group 14)

Compounds in which carbon is bonded to one of its homologues – silicon, germanium, tin or lead – form that branch of organometallic chemistry which as yet has found the most extensive technical application (cf. p. 100, 137, 139):

- *Silicones are materials with unique properties*
- *Organotin compounds are widely applied as stabilizers for plastics and as fungicides in crop protection.*
- *Lead alkyls, because of their use as antiknock agents, were for a long time the top organometallic commodity produced by industry. Recently, production figures are declining.*

Accordingly, the literature abounds with reports dealing with their structures and reactions.
Two characteristics distinguishing the organoelement compounds of the carbon group from the organyls of the boron group are the **lower polarity of the E ◂ C bond and,** in compounds ER_4, the **presence of an octet configuration.** Consequently, molecules ER_4 show no tendency to associate via ($2e\,3c$) alkyl- or aryl bridges. The reactivity towards nucleophiles is diminished and, as opposed to the corresponding compounds of the neighboring groups like $AlMe_3$ or PMe_3, the species ER_4 usually are water-stable and often even air-stable (examples: $SiMe_4$, $SnPh_4$, **"soft organometallics"**). It is instructive to examine a few group trends concerning the E–C bond:

E	Thermal stability	Bond energy *E*(E–C) in kJ/mol	Bond length *d*(E–C) in pm	Bond polarity $E^{\delta+}$ ◂ $C^{\delta-}$	Electronegativity *EN*
C	▽ (decreasing)	358	154	△ (increasing)	2.5
Si		311	188		1.9
Ge		249	195		2.0
Sn		217	217		1.8
Pb		152	224		1.9

These trends shape the reactivities in **homolytic reactions:**

$$Et_4C \xrightarrow{Cl_2} Et_3C-C_2Cl_nH_{5-n}$$
$$Et_4Si \longrightarrow Et_3Si-C_2Cl_nH_{5-n}$$

Chlorination of the organic part, C–C *and* C–Si *bonds survive.*

but

$$Et_4Ge \xrightarrow{Cl_2} Et_3GeCl + EtCl$$
$$Et_4Sn \longrightarrow Et_nSnCl_{4-n} + 4\text{-}n\,EtCl$$
$$Et_4Pb \longrightarrow PbCl_4 + 4\,EtCl$$

Chlorinating cleavage of the Ge–C, Sn–C, *and* Pb–C *bonds.*

as well as in heterolytic reactions:

E = Si,Ge,Sn,Pb — $-E^{\delta+} \leftarrow C^{\delta-}-$; Nu (nucleophilic attack), El (electrophilic attack)

–C—C–

The availability of empty nd orbitals at E *renders* ***associative mechanisms*** **(*A or I_A*)** *of substitution* ***possible,*** *coordination number* 5 *being adopted by the intermediate. Example: hydrolysis of* R_3SiCl.

d-orbitals of appropriate energy are unavailable. Associative mechanisms of nucleophilic substitution at saturated carbon are unfavorable.

Successive replacement of R by more electronegative groups X in R_nEX_{4-n} increases the affinity of E for attacking nucleophiles. Thus, $SnMe_4$ is inert towards hydrolysis and $SnMe_6^{2-}$ is unknown, whereas $SnCl_4$ is readily hydrolysed and $SnCl_6^{2-}$ can be prepared. Characteristic differences are also found among the **modes of catenation:**

C,Si,(Ge)	(Ge),Sn,Pb	
$(-E-E-)_n$	$(-E-X-)_n$	e.g. $(Me_3SnCN)_n$
$(-E-C-)_n$	$(E(X)_2)_n$	e.g. $(Me_2PbCl_2)_n$
$(-E-O-)_n$		
E maintains its coordination number 4	*E increases its coordination number to* 5 *or* 6	

8.1 Organosilicon Compounds

8.1.1 Silicon Organyls of Coordination Number 4

PREPARATION

Because of their great technical importance, many of the simple organosilanes can be obtained commercially so that their preparation in the laboratory is seldom necessary.

In the technical **direct process** (Rochow-Müller) for the production of methylchlorosilanes, Si/Cu alloys are used:

$$2\,RCl + Si/Cu \xrightarrow{\Delta T} R_2SiCl_2 + \ldots \quad \text{(cf. p. 100)} \qquad \boxed{1}$$

(R = alkyl, aryl)

Special organosilanes are prepared by means of **metathesis reactions:**

$$SiCl_4 + 4\,RLi \longrightarrow R_4Si + 4\,LiCl \qquad \boxed{4}$$

$$R_3SiCl + R'MgX \longrightarrow R_3R'Si + MgXCl$$

$$2\,R_2SiCl_2 + LiAlH_4 \longrightarrow 2\,R_2SiH_2 + LiCl + AlCl_3$$

or via **hydrosilation (Speier process):**

$$HSiCl_3 + R{-}CH{=}CH_2 \longrightarrow RCH_2CH_2SiCl_3 \qquad \boxed{8}$$

Hydrosilations, which may also be initiated photochemically and which are catalyzed by radical forming reagents, transition-metal complexes, or Lewis bases, proceed in an ***anti-Markovnikov direction.*** *Polyenes are hydrosilated in a* ***regio- and stereospecific*** *fashion. If chiral metal complexes are employed as catalysts, enantioselective hydrosilation of prochiral alkenes becomes possible.*

Arylsilanes are also accessible by means of the **Vollhardt cyclization** (p. 277) of silylalkynes.

PROPERTIES AND REACTIONS

Reactions involving Si – C Bond Cleavage

C – C and Si – C bonds are energetically very similar

$$[D(C{-}C) = 334,\ D(Si{-}C) = 318\ \text{kJ/mol}].$$

For organosilanes, high thermal stability is therefore expected. Thus, **homolytic cleavage** of tetramethylsilane only sets in at temperatures exceeding 700 °C and tetraphenylsilane can be distilled in air at 430 °C without decomposition! Due to the low polarity of the Si – C bond, **heterolytic cleavage** also does not occur readily; this type of degradation requires harsh reaction conditions which have to be well adapted to the specific case. Since the organosilyl group formally leaves as R_3Si^+, the propensity for Si – C cleavage in R_3SiR' correlates with the CH-acidity of the parent hydrocarbon R'H (Example: Silylalkynes $R_3Si{-}C{\equiv}CR$ are easily desilylated).

Heterolytic Si – C cleavage can in principle be initiated in four different ways which differ in the nature of the attacking agent (electrophile El or nucleophile Nu) and in the site of attack (Si or C). The establishment of a generally applicable order of reactivities is complicated by the fact that substituents at Si or C may exert a considerable influence. Bearing in mind these complications, the following gradation of the **readiness to undergo Si – C bond cleavage** summarizes the experimental evidence:

Type:	***I***	>	***II***	>	***III***	>	***IV***
	Si – C(aryl)		Si – C(aryl)		Si – C(alkyl)		Si – C(alkyl)
	↑		↑		↑		↑
attacking species:	El		Nu		El		Nu

I. Si–C cleavage occurs most readily in **arylsilanes** and related compounds via **electrophilic attack at carbon.** Example:

$$\text{PhSiR}_3 + H^+ \underset{}{\overset{\text{HX, slow}}{\rightleftharpoons}} [\text{R}_3\text{Si(H)C}_6\text{H}_5]^+ \xrightarrow[\text{H}_2\text{O}]{\text{fast}} \text{C}_6\text{H}_6 + R_3SiOH_2^+$$

$HX = CF_3COOH$, R_FSO_3H

Mechanistically this **protodesilylation** resembles electrophilic aromatic substitution (Eaborn, 1975 R); it proceeds 10^4 times faster than proton exchange in the corresponding Si-free arene. Further examples for Si–C cleavages effected by electrophiles:

$$\text{PhSiMe}_3 \xrightarrow{\overset{\delta^+\,\delta^-}{ICl}} \text{PhI} + Me_3SiCl$$

$$\text{PhSiMe}_3 \xrightarrow[AlCl_3]{RCOCl} \text{PhCOR} + Me_3SiCl$$

$$\text{CH}_2{=}\text{CHCH}_2\text{SiMe}_3 \xrightarrow[RT]{I_2} \text{CH}_2{=}\text{CHCH}_2\text{I} + Me_3SiI$$

II. Si–C cleavage of **arylsilanes,** initiated by **nucleophilic attack at Si,** proceeds less readily. Yet, this variant has gained some practical importance since it constitutes a way of generating carbenes and carbanions which circumvents strongly basic or reducing conditions. The nucleophile of choice is the fluoride ion which possesses an exceptionally high affinity for Si [E(Si–F) = 565 kJ/mol, strongest of all single bonds]. The fluoride ion is usually introduced as a component of the salt $(n\text{-Bu})_4N^+F^-$ (Kuwajima, 1976). Example (R. W. Hoffmann, 1978):

$$\text{Me}_3\text{Si-C}_7\text{H}_6^+ \xrightarrow[-\,Me_3SiF]{(n\text{-Bu})_4N^+F^-,\ THF} \{\text{C}_7\text{H}_6^-\} \longrightarrow \text{Heptafulvalene}$$

Heptafulvalene

Si–C cleavage of **allylsilanes** is exploited in the **Sakurai reaction** (1982 R) which consists in the regiospecific addition of allylsilanes to carbonyl compounds. This reaction can be initiated by nucleophilic attack of F^- at Si thereby generating the allyl anion as an intermediate:

$$Me_3Si\text{-allyl} + RR'C{=}O \xrightarrow[\text{THF}]{(n\text{-Bu})_4N^+F^-} R'\text{-}C(R)(OSiMe_3)\text{-allyl} \xrightarrow[\text{MeOH}]{\text{HCl}} R'\text{-}C(R)(OH)\text{-allyl}$$

The fact that only catalytic amounts of $(n\text{-Bu})_4N^+F^-$ suffice may be traced to the trans-silylation step ③:

OSiMe₃, HCl, MeOH, OH, homoallyl alcohol, F⁻, ③, Me₃Si, ①, Me₃SiF, O⁻, ②

In the more common variant of the Sakurai reaction, the carbonyl compound is activated by a Lewis acid:

$$\alpha,\beta\text{-enone} + Me_3Si\text{-allyl} \xrightarrow[CH_2Cl_2]{TiCl_4} \xrightarrow{H_2O} \delta,\epsilon\text{-enone}$$

III. Si–C cleavages in **alkylsilanes,** which are initiated by **electrophilic attack at carbon,** require the presence of strong Lewis acids as catalysts:

$$Me_4Si + HCl \xrightarrow[C_6H_6]{AlCl_3} Me_3SiCl + CH_4$$

Slow cleavage is also effected by concentrated sulfuric acid, though:

$$2\,Me_4Si \xrightarrow{H_2SO_4} (Me_3Si)_2O + 2\,CH_4$$

IV. Si–C cleavages in **alkylsilanes** by means of **nucleophilic attack at Si** are generally slow; they are only observed for very strong nucleophiles in polar, aprotic media:

$$Me_3SiCR_3 + OR^- \xrightarrow[\text{HMPA}]{\text{slow}} Me_3SiOR + CR_3^- \xrightarrow[H^+]{\text{fast}} HCR_3$$

"from the medium"

HMPA = *hexamethylphosphoric trisamide*

The origin of the proton in the above reaction sequence is not always clear.
These desilylations proceed much more rapidly if they are accompanied by a relief of ring strain:

KOH, EtOH, 25° → $SiMe_2OH$

or if good leaving groups are present (**β-effect**):

$$Et_3SiCH_2CH_2Cl + OH^- \longrightarrow Et_3SiOH + C_2H_4 + Cl^-$$

In this way, the plant growth substance Alsol® $(PhCH_2O)_2MeSiCH_2CH_2Cl$ *releases ethylene, which accelerates the ripening of bananas.*

A sequence of Si–C bond cleavages and bond formations is also responsible for the dynamic behavior of trimethylsilylcyclopentadiene which manifests itself in the **^{1}H-NMR spectrum** (H. P. Fritz, 1965):

$$Me_3SiCl + NaC_5H_5 \longrightarrow Me_3Si(\eta^1\text{-}C_5H_5) + NaCl$$

$\delta(^1H)$	Si–CH_3	H_1	$H_{2,5}$	$H_{3,4}$	
−30 °C	0.2	3.31	6.44	6.55	ppm
+120 °C	0.2		5.75		ppm

These data are rationalized by a series of **metallotropic 1,2-shifts** ($k_{30°} = 10^3\,s^{-1}$, E_a = 55 kJ/mol), the molecule is **fluxional:**

H SiMe3 1 5 H 2 H 4 3 H H ⇌ 1 H 5 2 SiMe3 4 3 ⇌ H SiMe3 ⇌ etc.

In addition, considerably slower **prototropic 1,2-shifts** occur (relative rate 10^{-6}) which lead to an equilibrium mixture. The composition of this mixture can be determined by analyzing the Diels-Alder adducts (Ashe, 1970):

H SiMe3 ⇌ (K_1 = 0.15, 60°) SiMe3 H H ⇌ (K_2 = 0.35, 60°) SiMe3 H H ⇌ etc.

RC≡CR

H SiMe3 R R + R R SiMe3 + R R SiMe3

Reactions not involving Si – C Bond Cleavage

Due to the relatively inert character of the Si – C bond, R_3Si- and $R_2Si\langle$ units in many cases remain unaffected, playing a spectator role. Therefore, in the chemistry of organosilyl halides, -hydroxides, -alkoxides and -amides, organometallic aspects in the narrower sense recede. With the exception of the industrially important **organochlorosilane hydrolysis,** reactions of organosilyl derivatives containing silicon – heteroatom bonds will be assigned to the inorganic chemistry of the main-group elements, and will be treated here in a cursory way only. Another important aspect of organosilicon chemistry, the use of **R_3Si as a protecting group** in organic syntheses, will also be neglected here.

Organosilanols and Silicones

Organochlorosilanes R_nSiCl_{4-n} are subject to swift hydrolysis to form the corresponding silanols $R_nSi(OH)_{4-n}$ which, like silicic acid $Si(OH)_4$, readily undergo condensation reactions. In the simplest case, that of trimethylchlorosilane, hexamethyldisiloxane is obtained:

$$2\,Me_3SiCl \xrightarrow[-\,HCl]{H_2O} 2\,Me_3SiOH \xrightarrow[-\,H_2O]{} Me_3Si-O-SiMe_3$$

The driving force of this hydrolysis is largely governed by the high Si – O bond energy [cf.: $\bar{E}$(Si – Cl) = 381 kJ/mol, $\bar{E}$(Si – O) = 452 kJ/mol] and the solvation enthalpies of H^+ and Cl^-. The hydrolysis of bifunctional organosilanes Me_2SiCl_2 leads to higher condensation products:

$$n\,Me_2SiCl_2 \xrightarrow[-HCl]{H_2O} n\,Me_2Si(OH)_2 \xrightarrow[-\,H_2O]{} (Me_2SiO)_n$$

The formation of polysiloxane chains and rings $(R_2SiO)_n$ (poly**silico**ket**ones**), which contrasts with the monomeric nature of ketones R_2CO, reflects the weakness of the $Si=O\,(p_\pi - p_\pi)$ bond.

Silicone oils were first obtained by Ladenburg in 1872; pioneering work concerning silicones proper was carried out by Kipping, starting in 1901. However, large-scale production of silicones was only triggered by the demand for new materials with special properties, the elucidation of the principles of polymerization (Staudinger, Nobel prize in 1953) and by the development of a rational method to prepare the monomers (Rochow and Müller).

Industrial Production of the Monomers:

$$MeCl + Si/Cu \xrightarrow{300\,°C} Me_nSiCl_{4-n}$$

ca. 9 : 1 **direct process** (Rochow, Müller 1945)

Rather than being a nuisance, the formation of a mixture of various methylchlorosilanes is highly desirable since, after distillative separation, these intermediates fulfill different functions. Apart from Cu (5 – 10%, added as Cu_2O), 0.1 – 1% of electropositive metals like Ca, Mg, Zn or Al are added to Si. In this way, the product distribution of the species Me_nSiCl_{4-n} can be steered to a certain extent. Addition of small amounts (0.001 – 0.005%) of As, Sb or Bi as "promoters" increases the rate of reaction.

In a crude picture of the ***mechanism of the Müller-Rochow direct process,*** *a negative polarization of silicon (towards copper silicide) is assumed. This would increase the ease of electrophilic attack at* Si.

$$\begin{array}{c} \overset{\delta^-}{Cl}-\overset{\delta^+}{Me} \quad \overset{\delta^+}{Me}-\overset{\delta^-}{Cl} \\ -\overset{\delta^-}{Si}-\overset{\delta^+}{Cu}-\overset{\delta^-}{Si}-\overset{\delta^+}{Cu}-\overset{\delta^-}{Si} \end{array} \longrightarrow Si-Cu^+ \; ^-Cl \rightarrow SiMe_2 \leftarrow Cl^- \; ^+Cu-Si-$$

Recent studies have shown, that the reaction of CH_3Cl *occurs at the phase* η-Cu_3Si (*Falconer*, 1985).

An alternative view starts from the premise that the Si *surface is oxidized by* CuCl *which is formed from* Cu *and* CH_3Cl. *Finally, it has also been proposed that* CH_3Cu *is formed as an intermediate which decomposes at the surface into* Cu *and* $CH_3\cdot$ *radicals, the latter attacking silicon. The heterogeneous nature of the reaction and the large number of components which participate render the mechanistic study of the Müller-Rochow process exceedingly difficult.*

The production of bulk silicones involves organochlorosilane hydrolysis and subsequent thermal treatment in the presence of catalytic amounts of H_2SO_4, often after addition of certain cross-linking agents.

Composition of the crude organochlorosilane mixture from the direct process and applications of the individual components:

Me_2SiCl_2	$MeHSiCl_2$	$MeSiCl_3$	$SiCl_4$	Me_3SiCl	Disilane
ca. 80%	3%	8%	1%	3%	5%
$-O-SiMe_2-O-$		$-O-SiMe(O-)-O-$	$-O-Si(O-)_2-O-$	$-O-SiMe_3$	
chain formation		branching	cross-linking	chain termination	

The thermal treatment under acidic conditions serves to generate specific chain lengths:

$$Me_2SiCl_2 \xrightarrow[\text{condensation}]{\text{hydrolysis}} \underset{\text{rings}}{(Me_2SiO)_m} + \underset{\text{chains}}{HO-SiMe_2-(OSiMe_2)_n-OH}$$

$$\xrightarrow{H_2SO_4,\ \Delta T} HO-SiMe_2-(OSiMe_2)_{m+n}-OH$$

$$Me_3SiCl \xrightarrow[\text{condensation}]{\text{hydrolysis}} Me_3Si-O-SiMe_3$$

$$\longrightarrow Me_3SiO-SiMe_2-(OSiMe_2)_{m+n}-OSiMe_3$$

Final cross-linking is achieved in a number of ways:

DBPO = Dibenzoylperoxide

Depending on the structure of the siloxane skeleton, silicone oils, -elastomers or -resins are obtained. Thanks to their outstanding properties, silicones are almost ubiquitous in modern technology. Their advantages include high thermal stability and corrosion resistance, small temperature coefficients of viscosity, favorable dielectric properties, foam suppressing and water repelling action combined with their physiologically innocuous nature (applications in plastic surgery, use as inflatulence drugs).

The extraordinary properties of silicone materials may be traced back to the idiosyncrasies of the **Si – O – Si (siloxane) bond.** The high **flexibility** of $(-Me_2SiO-)_n$ chains suggests **low barriers of conformational changes.** This can be demonstrated by an examination of the rotational barriers in $(CH_3)_4E$:

E:	C	Si	Ge	Sn	Pb	
Rot.Bar.:	18	7	1.5	0		kJ/mol
d(E – C):	154	188	194	216	230 pm	

Flexibility is also ensured by the low energy of the **Si – O – Si bending vibration:**

$Me_3Si-O-SiMe_3$ (163 pm; $\alpha = 148°$) → $Me_3Si-O-SiMe_3$ ($\alpha = 180°$) → $Me_3Si-O-SiMe_3$

The potential energy graph of the bending vibration displays a flat curvature in the range $140° < \alpha < 220°$, the linear transition state of inversion ($\alpha = 180°$) being situated only 1 kJ/mol above the bent ground state. This may be a consequence of the fact that in the linear disposition Si–O–Si, O=Si($p_\pi - d_\pi$) bonding is maximal. In this context, it is worth mentioning that hexaphenyldisiloxane possesses a linear structure Si–O–Si even in the ground state.

The **low temperature coefficient of viscosity** which recommends silicone oils as lubricants for extreme temperature ranges, is probably caused by two counteracting effects: Siloxane chains tend to form helices which are stabilized by intramolecular interactions of the polar Si–O segments. At elevated temperatures, the helices open and the conventional temperature dependence of viscosity (η decreases with increasing T) is compensated for by increasing intermolecular interactions of the unfolded polysiloxane chains. Interfacial effects of silicones, which have led to important technical applications, are based on the polarity of the Si–O–Si bridge and the hydrophobic nature of the alkyl groups bonded to silicon. As examples, the impregnation of textile fibres and the use as mold-releasing agents in the manufacture of tires may be mentioned.

Other Compounds Containing the Units $-R_2Si-E-$ (E=S, N)

Apart from the siloxanes, a large number of compounds with chain and ring structures are known which contain other main-group elements in addition to organosilyl units $-SiR_2-$. Since the properties of these compounds are not dominated by the organoelement component, only a few representative examples will be mentioned here.

$Me_3SiCl \xrightarrow[-\,LiCl]{LiSH} Me_3SiSH \xrightarrow[-\,H_2S]{\times 2} Me_3Si{-}S{-}SiMe_3$

Trimethylsilanethiol; Hexamethyldisilathiane

$Me_3Si{-}S{-}SiMe_3 \xrightarrow[350°]{\times 2} 3\,Me_4Si + SiS_2$

$3\,Me_2SiCl_2 + 3\,H_2S + 6\,C_5H_5N \xrightarrow{-\,6\,pyHCl} (Me_2SiS)_3$ Hexamethylcyclotrisilathiane

$(Me_2SiS)_3 \xrightarrow{200°} (Me_2SiS)_2$ planar four-membered ring (S—SiMe$_2$ / Me$_2$Si—S)

The Si–S bond in these molecules is thermally quite stable; as opposed to Si–O bonds, however, Si–S bonds are easily hydrolyzed.

A particularly extensive field is that of **organosilicon-nitrogen chemistry.** Compared to $-R_2Si-O-$ units, characteristic aspects of $-R_2Si-NR-$ segments are the preference of cyclic over open-chain oligomers, and the ready hydrolytic cleavage of the Si–N bond.

$$Me_3SiCl \xrightarrow[-NH_4Cl]{NH_3} Me_3SiNH_2 \xrightarrow[-NH_3]{100°,\ \times 2} Me_3Si{-}NH{-}SiMe_3 \ \text{(Hexamethyldisilazane)}$$

$$Me_3Si{-}NH{-}SiMe_3 \xrightarrow[-H_2]{NaH} (Me_3Si)_2NNa \xrightarrow{MX_2} [(Me_3Si)_2N]_2M$$

$$Me_3SiCl \xrightarrow[-NH_3,\ -NH_4Cl,\ -NaCl]{NaNH_2,\ C_6H_6,\ \Delta T} (Me_3Si)_2NNa$$

The bulky **bis(trimethylsilyl)amide anion** $(Me_3Si)_2N^-$ is used if the stabilization of low coordination numbers is desired. Thus, in $[(Me_3Si)_2N]_2Co$, the cobalt atom has the unusual coordination number 2; the low melting point of this material points to a molecular lattice. Linear **organopolysilazanes** $(-R_2Si-NR-)_n$ are only obtained with difficulty since the tendency to form cyclosilazanes dominates, six-membered rings being strongly favored.

$$3\ Me_2SiCl_2 + 9\ RNH_2 \xrightarrow{-6\ RNH_3Cl} (Me_2SiNR)_3$$

R = H, alkyl, aryl

a Cyclotrisilazane

Organosilyl derivatives of hydrazine undergo unexpected isomerizations: Whereas in the case of bis(trimethylsilyl)hydrazine the 1,2-isomer is preferred for steric reasons, it is found that a base catalyzed **dyotropic rearrangement** leads to an equilibrium mixture of the 1,1- and 1,2-isomers (West, 1969 R):

$$2\ Me_3SiCl + 3\ N_2H_4 \xrightarrow{-2\ N_2H_5Cl} Me_3SiNH{-}NHSiMe_3$$

$$Me_3SiNH{-}NHSiMe_3 \underset{+H^+}{\overset{-H^+}{\rightleftharpoons}} \left\{ Me_3SiN \cdots H \cdots N^{\ominus} \cdots SiMe_3 \right\} \underset{-H^+}{\overset{+H^+}{\rightleftharpoons}} (Me_3Si)_2N{-}NH_2$$

This **migratory aptitude** is a typical feature of $\mathbf{R_3Si}$ groups (cf. p. 109).

8.1.2 Silicon Organyls of Coordination Numbers 2 and 3 and their Subsequent Products

This topic encompasses the following species:

R_2Si Silylene (Silene)	R_2SiSiR_2 Disilylene (Disilene)	R_3Si^- Silyl anion (silanion)
		$R_3Si\cdot$ Silyl radical
		R_3Si^+ Silicenium ion

Follow-up products with a higher coordination number at silicon, which are closely related to these moieties, are the polysilylenes (= organopolysilanes), the carbosilanes, and the Lewis-base adducts of silicenium ions.

The species R_2E (E = Si, Ge, Sn, Pb) may be regarded as substituted homologues of methylene. They contain the elements of group 14 in the oxidation state E^{II} whose stability increases with increasing atomic number. The organic derivatives generally represent short-lived intermediates which can be investigated in monomeric form under special conditions only (R = alkyl, aryl):

R_2C	R_2Si	R_2Ge	R_2Sn	R_2Pb
Methylene	Silylene	Germylene	Stannylene	Plumbylene

Their identification as intermediates usually involves an analysis of oligomerization- or trapping products or spectroscopic studies in matrix isolation.

Silylenes and Polysilylenes (Organopolysilanes)

Dehalogenation of organochlorosilanes affords **organopolysilanes** which are thought to be formed via silylene intermediates. The degree of polymerization and the ratio linear/cyclic depend on the respective reaction conditions. Examples:

$$Me_2SiCl_2 \xrightarrow{Na/K} (Me_2Si)_n + Me(Me_2Si)_mMe$$

rings: $n = 5, 6, 7$

chains: $m \lesssim 100$

$$3\,Ar_2SiCl_2 \xrightarrow{Li,\ C_{10}H_8} (Ar_2Si)_3$$

$$n\,Me_2SiCl_2 + 2\,Me_3SiCl \xrightarrow{Na/K} Me_3Si(SiMe_2)_nSiMe_3$$

chains: $1 < n < 24$

Larger groups R increase the solubility of the oligomers in organic solvents. From organocyclosilanes **silylenes** can be generated; their identity can be inferred from follow-up reactions (Kumada, 1973):

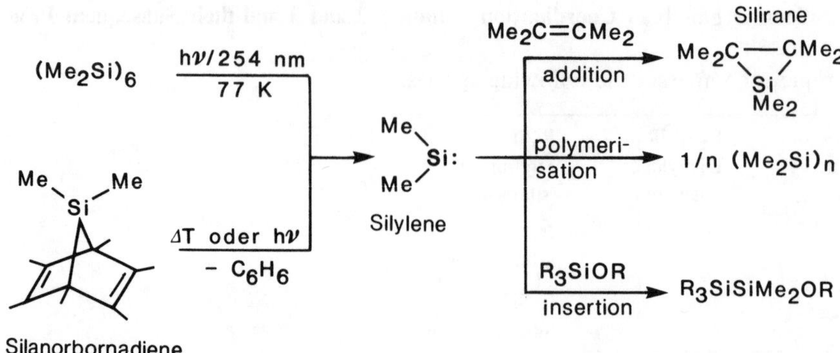

In the context of the reactivity of carbenes and their homologues, their ***molecular and electronic structure*** *as well as the associated* ***spin states*** *are of interest. Experimentally a triplet ground state is found for* Me_2C *(2 unpaired electrons,* $S = 1$*).* Me_2Si, Me_2Ge *and* Me_2Sn *possess a singlet ground state (no unpaired electrons,* $S = 0$*). Whereas the bond angle in* CH_2 *is approximately* 140° *(ESR), in* SiH_2 *it amounts to only* 92° *(IR, Raman; UV). The disparate magnetic and structural features of* CH_2 *and* SiH_2 *may be rationalized by means of differing hybridization ratios. In the* CH_2 *molecule, two* $C(2s, 2p)$ *hybrid orbitals serve for* C—H *bonding, because the angle* HCH *(140°, experiment) lies between* 120° *(expected for* sp^2 *hybrids) and* 180° *(sp hybrid). The remaining two* $C(2s, 2p)$ *hybrid orbitals would then accomodate two non-bonding electrons with parallel spin (*$S = 1$*). In* SiH_2*, on the other hand,* $Si(3p_x)$ *and* $Si(3p_y)$ *orbitals are thought to be used for* Si—H *bonding (90°); the two non-bonding electrons then occupy the* $Si(3s)$ *orbital (*$S = 0$*),* $Si(3p_z)$ *remaining empty. The difference is probably a consequence of the growing energetic separation between ns and np orbitals upon going to heavier elements within a group (limiting case: "inert s electron pair") which renders the participation of s electrons in chemical bonding increasingly unfavorable. In a more refined discussion, the electronic and steric properties of the groups R at* ER_2 *have to be taken into account explicitly.*

A rather special silylene is the compound $(C_5Me_5)_2Si$, **decamethylsilicocene** (Jutzi, 1989):

$$(Me_5C_5)_2SiBr_2 \xrightarrow[-2\ KBr]{K,\ anthracene} (C_5Me_5)_2Si$$

This diamagnetic, thermally stable but highly air-sensitive molecule has the structure of an axially symmetric metallocene.

The cyclic **organopolysilanes** $(R_2Si)_n$ possess singly folded ($n = 4, 5$) or doubly folded ($n = 6$) structures. The **electronic properties** of organopolysilanes are unexpected:

- *In the UV spectra of alkanes, absorption bands appear only for* $\lambda < 160$ nm. *Polysilanes, however, absorb at* $\lambda \leq 350$ nm (*Gilman*, 1964).
- *Contrary to cycloalkanes* $(R_2C)_n$, *cyclosilanes* $(R_2Si)_n$ *may be reduced to radical anions and oxidized to radical cations. The persistent species* $(R_2Si)_n^{\cdot -}$ *and* $(R_2Si)_n^{\cdot +}$ *can be studied by ESR spectroscopy* (*Bock*, 1979 *R*).
- *Upon doping with* AsF_5, *organopolysilanes become semiconducting* (*West*, 1983).

These peculiarities may be crudely explained as follows: Due to similar energy and spatial extension of C(2*s*) and C(2*p*) orbitals, the carbon atoms in alkanes approach ideal sp^3 hybridization and form localized (2*e* 2*c*) bonds along the carbon chain; the σ/σ^* splitting is large. For silicon, on the other hand, s orbital contraction leads to less perfect hybridization, the Si(3*s*) contribution to bonding being reduced. Since Si in the polysilanes is tetravalent but reluctant to fully allocate its 3*s* electrons to bonding, a certain degree of electron deficiency arises which can be alleviated by multicenter bonding along the Si_n backbone. Thus, some one-dimensional metallic character can be assigned to the Si_n chain in polysilanes, which would explain the aforementioned experimental facts (Schoeller, 1987).

Organopolysilanes have recently been the focus of attention for certain technical applications; they have found use as precursors in the manufacture of **ceramic fibres** consisting of **β-silicon carbide**. In a first step, pre-ceramic polycarbosilanes (molecular weight ≈ 8000) are drawn from the melt and spun into fibres. Subsequently, in a two stage thermal process, they are converted into β-SiC, a material with extremely high tensile strength (Yajima, 1975):

$$(Me_2Si)_n \xrightarrow[\text{Argon}]{450^\circ} \quad (Me_2Si)_6 \xrightarrow[\text{Argon}]{400^\circ} \quad \left(\begin{array}{c} H \\ | \\ -Si-CH_2- \\ | \\ Me \end{array} \right)_n \xrightarrow[\text{2. } N_2,\ 1300^\circ,\ -H_2,\ -CH_4]{\text{1. air, } 300^\circ} \beta\text{-SiC}$$

Kumada rearrangement — Carbosilane — "ceramization" — (zincblende structure)

Besides addition, polymerization, and insertion, bonding to a transition metal must be considered as one further carbene-analogous reaction of silylenes. However, whereas the field of transition-metal carbene complexes is highly developed (p. 210) and germylene-, stannylene-, and plumbylene complexes have also been obtained, no silylene-metal complex with the coordination number 3 at Si has yet been identified unequivocally (Tilley, 1990).

A complex bearing two bridging silylene ligands is formed in an unexpected reaction (E. Weiss, 1973):

$$2\ (\eta^5\text{-}C_5H_5)_2TiCl_2 \xrightarrow[\text{DME}]{H_3SiK} 2\ (\eta^5\text{-}C_5H_5)_2Ti(\mu\text{-}SiH_2)_2Ti(\eta^5\text{-}C_5H_5)_2$$

The μ-silylene ligand can be regarded as a doubly metallated silane, a molecule containing Si^{IV}. This may explain why μ-R_2Si complexes, as opposed to complexes containing terminal silylene ligands, are readily accessible.

Molecules with Si$=$E ($p_\pi - p_\pi$) Bonds

Numerous exceptions to the **double bond rule** ("Elements of the third and higher rows avoid forming compounds containing $p_\pi - p_\pi$ multiple bonds") have provoked experiments to synthesize molecules containing $>Si=C<$ and $>Si=Si<$ segments.

Si$=$C ($p_\pi - p_\pi$) bonds were originally proposed to exist in reactive intermediates. Thus, silaethene could be characterized spectroscopically at 10 K in an argon matrix (Guselnikov, 1966):

$$\text{Silacyclobutane } (H_2C-Si(CH_3)_2 / H_2C-CH_2) \xrightarrow[\text{pyrolysis}]{600^\circ} CH_2=CH_2 + \{(CH_3)_2Si=CH_2\} \text{ Silaethene}$$

Pyrolysis of Me_4Si yields a multicomponent mixture of carbosilanes, whereby silaethenes are proposed as reaction intermediates (G. Fritz, 1987 R):

$$(CH_3)_4Si \xrightarrow[700^\circ]{1\text{ min}} CH_4 + \{(CH_3)_2Si=CH_2\} \longrightarrow -Si(CH_3)_2-(CH_2-Si(CH_3)_2)_n-CH_2-$$

The fraction of cyclic carbosilanes amounts to about 15%.

$$\text{cyclo-}(H_2C-SiMe_2-CH_2-SiMe_2) \qquad \text{cyclo-}(H_2C-SiMe_2-CH_2-SiMe_2-H_2C-SiMe_2)$$

Silaethenes also arise from salt eliminations; they immediately dimerize to 1,3-disilacyclobutanes (Wiberg, 1977):

$$Me_2Si(OTs)-C(SiMe_3)_2Li \xrightarrow[-50^\circ]{-LiOTs} \{Me_2Si=C(SiMe_3)_2\} \xrightarrow{\times 2} \text{cyclo-}(Me_2Si-C(SiMe_3)_2-SiMe_2-C(SiMe_3)_2)$$

Ts = Tosylate

A monomeric silaethene, stabilized by metal coordination, has been claimed for the complex $(C_5H_5)(CO)_2HW(\eta^2\text{-}CH_2=SiMe_2)$ (Wrighton, 1983) and characterized by X-ray diffraction in the compound (Me_5C_5) (*i*-Prop$_3$P) HRu(η^2-$CH_2=SiPh_2$) (Tilley, 1988). In a convenient method for the generation of sparingly substituted or unsubstituted silaethenes the retrodiene cleavage of a silabicyclo[2.2.2]octadiene is exploited (Barton, 1972; Maier, 1981):

The first molecule featuring an Si=C ($p_\pi - p_\pi$) bond, which is stable at ambient temperature, was prepared by A. Brook (1981):

Mp. 93°

d(Si=C) = 176 pm
d(Si-C) = 189 pm

$\delta(^{13}C)$ = 214 ppm
^1J(Si=C) 84 Hz

R = Adamantyl

The dimerization or polymerization of this molecule to carbosilanes is prevented by bulky trimethylsilyl- and adamantyl groups.

Apparently, the canonical structure (a) *heavily contributes to the resonance hybrid since, according to* ^{1}H *NMR evidence, rotation around the central* Si–C *bond axis is blocked. A certain contribution of the structures* (b) *and* (c) *is suggested by* ^{13}C *NMR data (resemblance to chemical shifts observed for carbene complexes). The strong deshielding of the carbon nucleus argues against a participation of the silyl ylid canonical form* (d).

Closely related to the silaethene problem is the inclusion of the Si atom into aromatic heterocycles. **Silabenzene** C_5SiH_6 has been generated by means of flash thermolysis and studied in the gas phase (Bock, 1980) and in matrix isolation (G. Maier, 1982):

1. 800°
10^{-4} mbar
$-H_2$
2. matrix isolation Ar, 10 K

Silabenzene

T > 80K → oligomeric products

hν 240 nm ⇅ hν 320 nm

Dewarsilabenzene

In the UV spectrum, silabenzene is identified as an arene which is weakly perturbed by a donor atom.

In analogy to the ready formation of the aromatic anion $C_5H_5^-$ from C_5H_6, deprotonation of **silacyclopentadiene (silole)** C_4SiH_6 to yield its anion would appear to be straightforward. The realization of this idea till now has been frustrated by competing reactions. Whereas heavily substituted siloles are readily accessible and easily characterized,

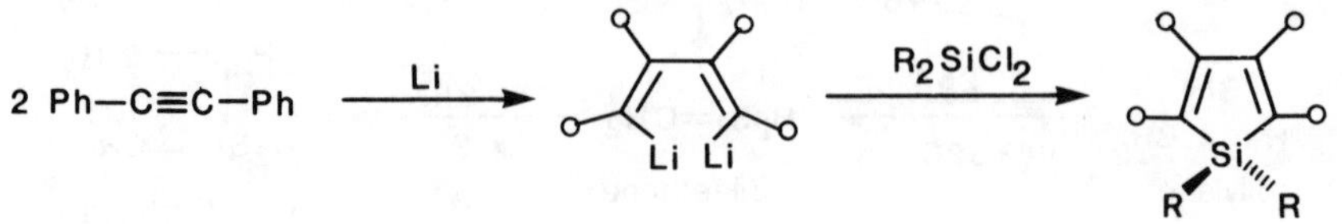

(M. D. Curtis, 1969)

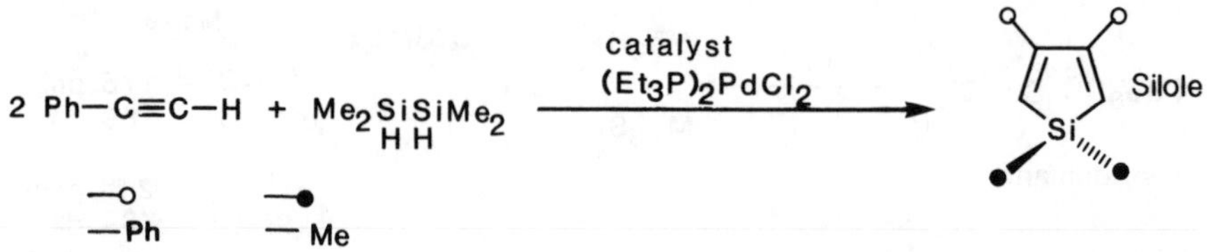

(Kumada, 1975)

the less substituted siloles, which still possess Si—H bonds, tend to form Diels-Alder dimers (T. J. Barton, 1975):

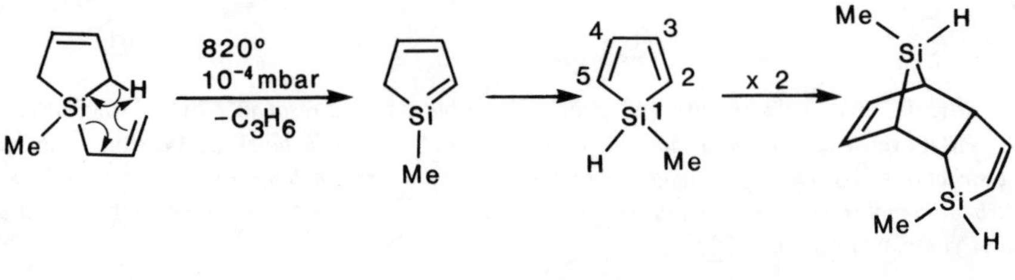

This type of dimerization is suppressed if the positions 3 and 4 bear alkyl substituents (Dubac, 1986).

The preparation of a **silole anion** (silacyclopentadienyl) as yet has only been achieved in perphenylated (Boudjouk, 1984) or in benzanellated form (Gilman, 1958; Kumada, 1983):

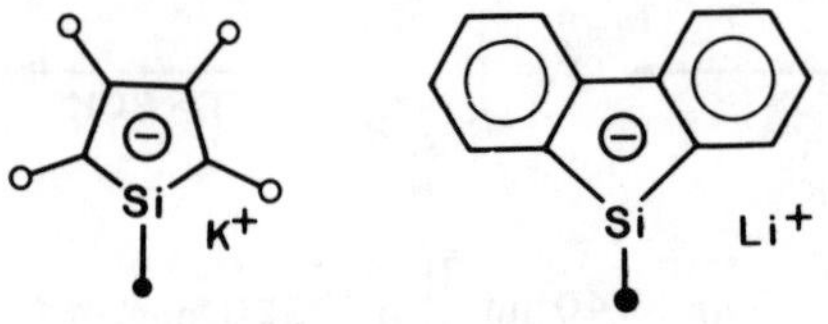

Therefore, at the time of writing, silametallocenes are unknown. The few examples of silole complexes, which have been reported, contain the heterocycle as an η^1- or an η^4-ligand:

$Si=Si\ (p_\pi - p_\pi)$**-bonds** would be expected to be even less stable than $Si=C$ double bonds because, due to the larger distance of the $Si-Si$ σ-bond, $p_\pi - p_\pi$ overlap is unfavorable. A way of "stabilizing" $Si=Si$ and $Si=C$ bonds is their coordination to transition metals:

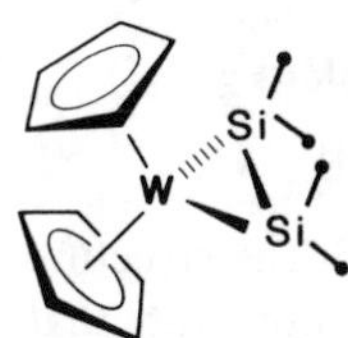

In concordance with the established concepts of bonding in olefin complexes (p. 256), the intra-ligand bond orders in these heteroalkene complexes lie between one and two.

(Berry, 1990) (Tilley, 1988)

The first derivative of **free** disilylene, which is stable at room temperature, was synthesized by West (1981):

$Mes_2Si(SiMe_3)_2$ $\xrightarrow[-\ Me_6Si_2]{h\nu,\ 254\ nm,\ -60°}$ Tetramesityldisilylene 90%

Mes = mesityl

yellow crystals, Mp 178°
d(Si=Si) 215 pm
compare
d(Si−Si) 235 pm

Again it is the steric protection of the vulnerable $Si=Si$ multiple bond by bulky groups which is responsible for the remarkable stability of this molecule. Precursors with sterically less demanding substituents only lead to organopolysilanes.

Disilylenes (and digermylenes) were also prepared by Masamune (1982):

$3\ Ar_2SiCl_2 \xrightarrow{LiC_{10}H_8}$ Hexaarylcyclotrisilane $\xrightarrow{h\nu}$ $Ar_2Si=SiAr_2$ Tetraaryldisilylene

$Ar = 2,6\text{-}Me_2C_6H_3$

The decrease in bond distance in proceeding from $Si-Si$ to $Si=Si$ is comparable to the increment for the pair $C-C$ and $C=C$; it is smaller in relative terms, however. As inferred from the NMR coupling constant $^1J(^{29}Si, ^{13}C)$, the silicon atom in disilylenes is sp^2 hybridized. This judgement is based on the proportionality between the spin-spin coupling constant and the s-character of an intervening bond:

	Me_4Si	$Mes_2Si=SiMes_2$	$(Me_3Si)_2Si=C(OSiMe_3)Ad$
$^1J(^{29}Si, ^{13}C)$ (Hz)	50.2	90.0	84.0
Hybridization at Si	sp^3	sp^2	sp^2

Mes = mesityl Ad = adamantyl

Despite steric shielding, the Si = Si double bond in $Mes_2Si=SiMes_2$ is considerably more reactive than a typical C = C double bond:

$$Mes_2Si{=}SiMes_2 \xrightarrow{H_2O,\ 50°} Mes_2Si(H){-}SiMes_2(OH)$$

$$Mes_2Si{=}SiMes_2 \xrightarrow{R_2C{=}O} \text{cyclo-}[Mes_2Si{-}SiMes_2{-}O{-}CR_2]$$

$$Mes_2Si{=}SiMes_2 \xrightarrow{R{-}C{\equiv}CH} \text{cyclo-}[Mes_2Si{-}SiMes_2{-}CH{=}CR]$$

Silyl anions R_3Si^- are present in solutions of R_3SiLi in ionizing solvents. The structure of silyl anions is pyramidal. Temperature dependent ^{1}H-NMR studies demonstrate that the barrier to inversion is high:

$$Ph_3SiSiPh_3 \xrightarrow{Li,\ THF} Li^+SiPh_3^- \xleftarrow[-LiCl]{Li,\ THF} Ph_3SiCl$$

$$[RR'R''Si:]^{\ominus} \rightleftharpoons [:SiRR'R'']^{\ominus}$$

Barrier to inversion
≥ 100 kJ/mol

Silyl anions R_3Si^- are isoelectronic with phosphanes R_3P and, like the latter, may serve as ligands in transition metal complexes:

$$R_3SiLi + Ni(CO)_4 \longrightarrow Li^+[Ph_3SiNi(CO)_3]^- + CO$$

Mild oxidizing agents convert silyl anions into the respective disilanes. In the case of the anion $(t\text{-}Bu)_3Si^-$, a disilane is thus formed which, compared to the standard value of $d(Si-Si) = 234$ pm, boasts an extraordinarily long Si – Si bond (Wiberg, 1986):

$$2\ (t\text{-}Bu)_3SiK \xrightarrow[heptane]{NO^+BF_4^-} 2\ \{(t\text{-}Bu)_3Si^{\bullet}\} \longrightarrow (t\text{-}Bu)_3Si \underset{pm}{\overset{270}{—}} Si(t\text{-}Bu)_3$$

Incidentally, hexa(*t*-butyl)ethane is unknown.

Silyl radicals $R_3Si\cdot$ are also pyramidal; they are much more reactive than their planar carbon analogues $R_3C\cdot$. Thus, $Ph_3Si-SiPh_3$ shows no inclination to homolytic cleavage:

$$Ph_3Si-SiPh_3 \not\longrightarrow 2\,Ph_3Si\cdot$$

In fact, the bond energy $E(Si-Si)$ is smaller than the value $E(C-C)$ (p. 94) but the repulsion of the substituents, which would promote dissociation, is reduced because of the large Si – Si bond lengths. Furthermore, resonance stabilization of silyl radicals according to

{ phenyl–Si• ⟷ •C6H5=Si ⟷ etc. }

is insignificant due to unfavorable C = Si ($p_\pi - p_\pi$) overlap. Dissociation into radicals has, however, been observed for the sterically congested disilane $Mes_3Si - SiMes_3$ (Neumann, 1984).

Other sources for silyl radicals:

$$Me_6Si_2 + NaOMe \xrightarrow[-Me_3SiOMe]{HMPA} Me_3SiNa$$

$$Me_3SiNa \xrightarrow[-Na^+,\ -C_{10}H_8^{\bullet -}]{+C_{10}H_8 \text{ (naphthalene)}} Me_3Si^{\bullet}$$

$$2\ Me_3SiX \xrightarrow[-2\ NaX]{Na/Hg} (Me_3Si)_2Hg \xrightarrow{h\nu} Me_3Si^{\bullet}$$

Bis(trimethylsilyl)mercury (linear structure, mp: 101 °C), serves admirably in the synthesis of other compounds with **silicon-metal bonds:**

$$Me_3Si{-}Hg{-}SiMe_3 \xrightarrow[-Hg]{2\ Li, THF} 2\ Me_3SiLi$$

$$Me_3Si{-}Hg{-}SiMe_3 \xrightarrow[-Hg]{DME, Mg} (Me_3Si)_2Mg(DME)$$

$$Me_3Si{-}Hg{-}SiMe_3 \xrightarrow[-Hg, -CO]{Fe(CO)_5, h\nu} (Me_3Si)_2Fe(CO)_4$$

Silicenium ions R_3Si^+ as such are still controversial. In cases where their existence has been postulated, tight association with a counter ion could not be excluded, the limiting case being a silanol ester. The latter formulation accounts for the ^{29}Si NMR properties of Me_3SiClO_4 (Olah, 1990).

ClO_4^- NMe_2 $SiPh_2^+$ Fe ClO_4^-

(J. Corey, 1975)

Ph, 174, O—Si, Ph, Ph, O, Cl, O, O

d(Si–C) = 185 pm
∢ C–Si–C = 113.5°
∢ O–Si–C = 105°
(mean values)

(Olah, Bau, 1987 R)

The species R_3Si^+ may in principle be stabilized by means of polarizable (soft) substituents or by the coordination of chelating ligands. In the latter case, because of the increase in coordination number from 3 to 5, the products are referred to as **siliconium ions.**

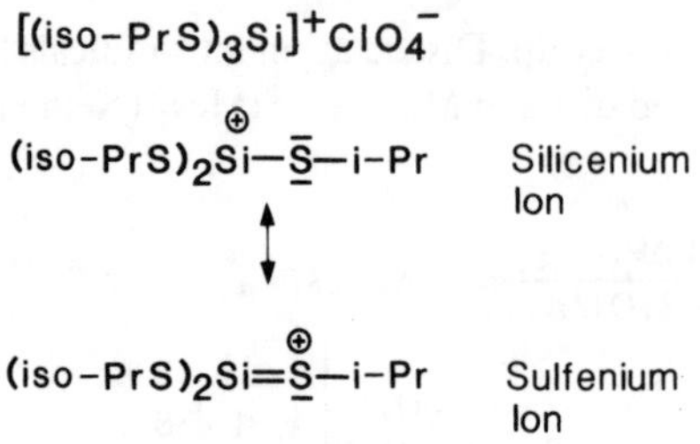

$[(iso\text{-}PrS)_3Si]^+ClO_4^-$

$(iso\text{-}PrS)_2Si{-}\overset{\oplus}{S}{-}i\text{-}Pr$ Silicenium Ion

↕

$(iso\text{-}PrS)_2Si{=}\overset{\oplus}{S}{-}i\text{-}Pr$ Sulfenium Ion

(Lambert, 1983)

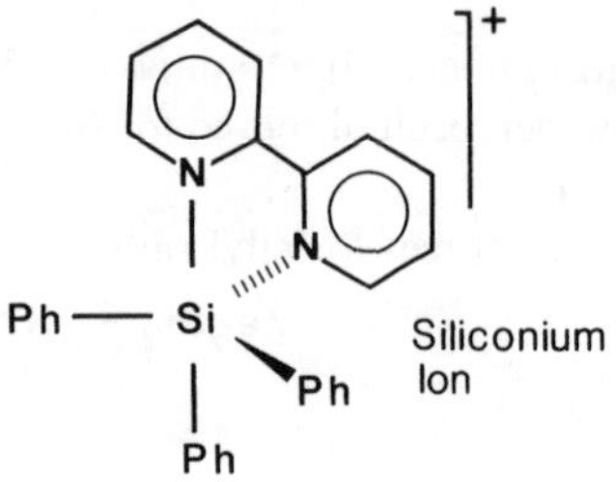

Siliconium Ion

(West, 1963)

As for the low tendency towards formation of R_3Si^+ ions, the strength of the Si–O or Si–halogen bond, respectively, as well as the insignificant resonance stabilization involving canonical structures with Si=C double bonds must be mentioned.
Therefore, **nucleophilic substitution at silicon** does not proceed dissociatively (D, S_N1) via intermediary silicenium ions. Instead, an associative mechanism (A, S_N2) must be assumed (Sommer, 1965 R):

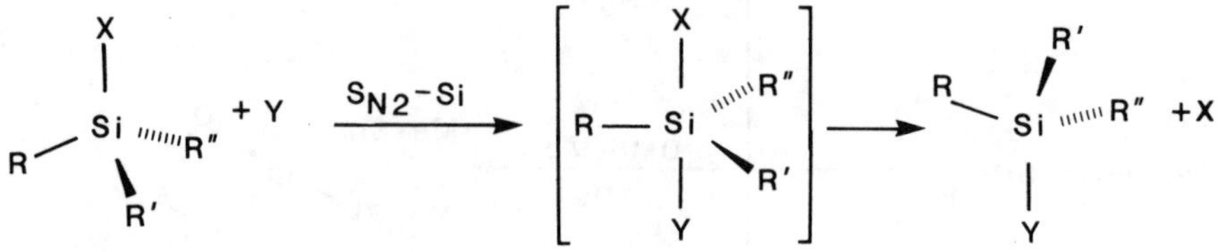

S_N2-Si

(X = *good leaving group, i.e. the conjugate base of a strong acid*)

This is supported by the dependence of the substitution rate on the nature of the attacking nucleophile Y, by the rate decrease in the presence of electron releasing groups R, and by the frequently observed inversion.

An uncritical transfer of diagnostic criteria used in studies of substitution at carbon to those dealing with silicon is inappropriate, because the ready increase of the coordination number at silicon and the rearrangement of the trigonal-bipyramidal intermediates via pseudorotation invalidate arguments, which rest on observations of inversion or retention of configuration.
Thus, swift substitution proceeds even at a bridgehead Si atom, although back-side nucleophilic attack is blocked (Sommer, 1973):

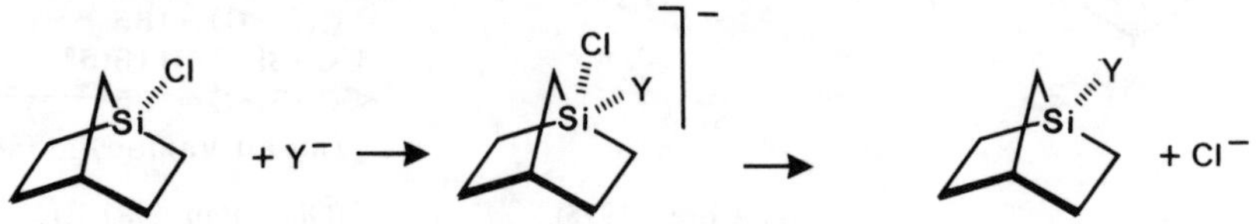

$+ Y^- \rightarrow \;\rightarrow\; + Cl^-$

This front-side attack of the nucleophile is rendered possible by the vacant d orbitals of silicon which stabilize the coordination number 5; it is accompanied by retention of configuration.

Racemization, which is frequently observed during the course of nucleophilic substitution at silicon, may erroneously suggest a dissociative mechanism. Racemization must not, however, be taken as evidence for the intermediary formation of silicenium ions, since a five-coordinate intermediate by means of rearrangements (pseudorotation, PR) can also result in racemization:

8.2 Organogermanium Compounds

Germanium, an element which is settled in the center of the periodic table, possesses typical properties of a semi-metal. For a long time tetraethylgermanium, prepared by Winkler in 1887, remained the only organogermanium compound. Since germanium organyls have as yet found little practical use, the accumulated results are mainly of academic interest. Among the elements Si, Ge, Sn and Pb, Ge most closely matches the electronegativities of H and C. Therefore the polarities of Ge–H and Ge–C bonds and the associated reactivities are strongly influenced by substituents.

8.2.1 Germanium Organyls of Coordination Number 4

PREPARATION

Direct syntheses, as for silicon, are also applicable to germanium. The resulting organogermanium halides may then be converted into other germanium organyls.

$$MeCl + Ge/Cu \longrightarrow Me_nGeCl_{4-n} \quad \boxed{1}$$

$$Me_2GeCl_2 + 2\,RMgX \longrightarrow Me_2GeR_2 + 2\,MgXCl \quad \boxed{4}$$

$$GeCl_4 + 4\,LiR \longrightarrow R_4Ge + 4\,LiCl$$

As compared to organosilicon chemistry, ligand scrambling for organogermanium compounds occurs more readily:

$$GeCl_4 + 3\,Bu_4Ge \xrightarrow[120\,^{\circ}C,\ 5\,h]{AlCl_3} 4\,Bu_3GeCl \quad \boxed{4}$$

This also applies to Ge–C bond cleavage by means of X_2 or HX:

$$Ph_4Ge + Br_2 \longrightarrow Ph_3GeBr + PhBr$$

$$R_4Ge + HCl \xrightarrow{AlCl_3} R_3GeCl + RH$$

$$Me_4Ge \xrightarrow[-SbCl_3]{SbCl_5} Me_3GeCl + Me_2GeCl_2$$

Furthermore, coupling reactions are of synthetic utility:

$$GeCl_4 + 4\,PhBr + 8\,Na \longrightarrow Ph_4Ge + 4\,NaCl + 4\,NaBr$$ 4

$$RGeCl_3 \xrightarrow{Li,\ THF} \text{(RGe)}_6 \quad R = (Me_3Si)_2CH$$

(Ge–Ge 258 pm, 252 pm)

Hexagermaprismane (Sakurai, 1989)

Germacycles are also prepared from Ge^{II} precursors:

$$GeI_2 + \text{butadiene} \longrightarrow \text{1,1-diiodogermacyclopent-3-ene}$$

$$GeI_2 \xrightarrow{2\,RC{\equiv}CR} \text{1,4-digermacyclohexa-2,5-diene } (R_4C_4Ge_2I_4)$$

PROPERTIES AND REACTIONS

Tetraorganogermanes R_4Ge are rather inert chemically; they are cleaved by strong oxidizing agents, often requiring the presence of Friedel-Crafts catalysts. This reaction is employed in the synthesis of ternary and quaternary organogermanes:

$$Ar_4Ge \xrightarrow{Br_2} Ar_3GeBr \xrightarrow{RMgX} Ar_3GeR$$

$$Ar_3GeR \xrightarrow{Br_2} Ar_2GeRBr \xrightarrow{R'MgX} Ar_2GeRR'$$

Note the selective cleavage of the Ge-aryl bond (c.f. p. 97). Like its silicon analog (p. 99) cyclopentadienyl(trimethyl)germane in solution is a **fluxional molecule;** it undergoes **metallotropic 1,2 shifts:**

$$Me_3GeCl + C_5H_5Li \longrightarrow C_5H_5GeMe_3$$

The migration of the Me_3Ge group along the C_5 periphery is faster than that of the Me_3Si group, since the C–Ge bond is weaker than the C–Si bond.

Organopolygermanes result from halogen abstractions using alkali metals or magnesium:

$$2\,Ph_3GeBr \xrightarrow{Li,\ THF} Ph_3Ge-GePh_3$$

$$n\,Me_2GeCl_2 \xrightarrow{Li,\ THF} (Me_2Ge)_n$$

Among the cyclic organogermanes, those with $n = 4$, 5 and 6 are well characterized.

Organohalogermanes R_nGeX_{4-n} like the corresponding silanes tend towards hydrolysis and condensation, albeit with attenuated vigor. The ease of hydrolysis of Ph_3GeX increases in the order $X = F < Cl < Br < I$. **Germanols** are stable in monomeric form if they carry bulky substituents: (*iso*-Pr)$_3$GeOH condenses to the **germoxane** [(*iso*-Pr)$_3$Ge]$_2$O only above 200 °C. The diminished reactivity of R_3GeX, as compared to R_3SiX, reflects the reduced ability of germanium to utilize Ge(4*d*) orbitals in order to stabilize coordination number 5 (cf. p. 115). This lessened inclination toward Ge(4*d*) participation also manifests itself in a comparison of the bond angles EOE in siloxanes and germoxanes:

$Ph_3Si-O-SiPh_3$ (180°) $\qquad$ $Ph_3Ge-O-GePh_3$ (140°)

VB $\qquad$ MO

Organogermanes R_nGeH_{4-n} are conveniently prepared by substitution of X^- by H^-:

$$R_nGeCl_{4-n} \xrightarrow{LiAlH_4} R_nGeH_{4-n}$$

The similar electronegativities (Ge 2.0, H 2.1) result in a low polarity of the Ge−H bond; consequently, their reactivity is low. Thus, organogermanes may even be prepared in aqueous media:

$$MeGeBr_3 \xrightarrow{NaBH_4,\ H_2O} MeGeH_3$$

The reactivity pattern towards lithium organyls argues for hydridic character ($H^{\delta-}$) for Ph_3SiH and for protic character ($H^{\delta+}$) in the case of Ph_3GeH:

$$Ph_3Si-H + RLi \longrightarrow Ph_3Si-R + LiH \quad \text{metathesis}$$
$$Ph_3Ge-H + RLi \longrightarrow Ph_3Ge-Li + RH \quad \text{metallation}$$

The polarity of the Ge−H bond may, however, be reversed by the proper choice of substituents (Umpolung):

$$\overset{\delta+}{Et_3Ge}\leftarrow\overset{\delta-}{H} + {>}C{=}O \longrightarrow H-\overset{|}{\underset{|}{C}}-OGeEt_3 \quad \text{Germylether}$$

$$\overset{\delta-}{Cl_3Ge}\leftarrow\overset{\delta+}{H} + {>}C{=}O \longrightarrow Cl_3Ge-\overset{|}{\underset{|}{C}}-OH \quad \text{Germylcarbinol}$$

The most important reaction of the germanes R_nGeH_{4-n} and X_nGeH_{4-n} is **hydrogermation;** it is catalyzed by transition metal complexes or radical forming reagents and proceeds under milder conditions than hydrosilations:

$$Ph_3GeH + CH_2{=}CHPh \xrightarrow{120\,°C} Ph_3GeCH_2CH_2Ph$$
$$(n\text{-}Bu)_3GeH + HC{\equiv}CR \xrightarrow{H_2PtCl_6} Bu_3GeCH{=}CHR$$

The halogermanes are particularly prone to addition:

$$Cl_3GeH + HC{\equiv}CH \xrightarrow{25\,°C} Cl_3GeCH{=}CH_2 \xrightarrow{Cl_3GeH} Cl_3GeCH_2{-}CH_2GeCl_3$$

Organogermanes are also suited for the preparation of molecules with **germanium-metal bonds:**

$$2\,Et_3GeH + Et_2Hg \longrightarrow Et_3Ge{-}Hg{-}GeEt_3 + 2\,EtH$$

Further modes of formation of Ge–M bonds include

$$Ph_3Ge{-}GePh_3 \xrightarrow{Li,\,THF} 2\,Ph_3GeLi$$

$$Me_3GeBr + NaMn(CO)_5 \longrightarrow Me_3Ge{-}Mn(CO)_5 + NaBr$$

8.2.2 Germanium Organyls of Coordination Numbers 2 and 3 and their Subsequent Products

The chemistry of organogermanium compounds in low coordination states is largely analogous to that of silicon. The increasing preference for the oxidation state E^{II}, exhibited by the heavier elements in group 14, is expressed, i.a., in the formation of **germylenes** through α-elimination of alcohols from alkoxygermanes (Satgé, 1973 R):

$$RGeH_2(OMe) \xrightarrow[-MeOH]{\Delta T} \{RGeH\} \longrightarrow (RGeH)_n$$

Additional sources for germylenes are the following reactions:

$$Et_3Ge{-}GeEt_2Cl \xrightarrow[-Et_3GeCl]{\Delta T} \{Et_2Ge\} \xrightarrow{\text{2,3-dimethylbutadiene}} \text{1,1-diethyl-3,4-dimethylgermacyclopent-3-ene (Ge(Et)}_2\text{)}$$

$$Me_2GeCl_2 \xrightarrow[-\,LiCl]{Li,\,THF} \{Me_2Ge\} \longrightarrow (Me_2Ge)_6 + \underset{\text{polymeric}}{(Me_2Ge)_n}$$

$$Ge(g) + Me_3SiH(g) \xrightarrow[-196°]{\text{cocond.}} \{Me_3SiGeH\}$$

Whereas these simple, highly reactive germylenes can only be identified indirectly from the products of subsequent reactions like polymerization or adduct formation, germylenes with bulky substituents can be isolated in monomeric form at room temperature (Lappert, 1976):

$$[(Me_3Si)_2N]_2Ge + 2\,(Me_3Si)_2CHLi \xrightarrow{Et_2O} \underset{\text{mp: }180\,°C}{[(Me_3Si)_2CH]_2Ge} + 2\,(Me_3Si)_2NLi$$

The simpler germylenes may also be stabilized as ligands in transition-metal complexes – albeit as thermolabile solvates only, displaying coordination number 4 at Ge (Marks,

1971):

$$Na_2[Cr_2(CO)_{10}] + Me_2GeCl_2 \xrightarrow[-78\,°C]{THF} \underset{\underset{THF}{\vdots}}{Me_2GeCr(CO)_5} + NaCl + NaCr(CO)_5Cl$$

The corresponding complexes of stannylenes are more stable (p. 133).
Germylene complexes, which are three-coordinate at Ge, are accessible, if the germylene is sterically demanding (example: $[(Me_3Si)_2CH]_2Ge-Cr(CO)_5$, Lappert, 1977).
In their function as bridging ligands (coordination number 4 at Ge), even the simpler germylenes do not require stabilization by an additional Lewis base (Graham, 1968):

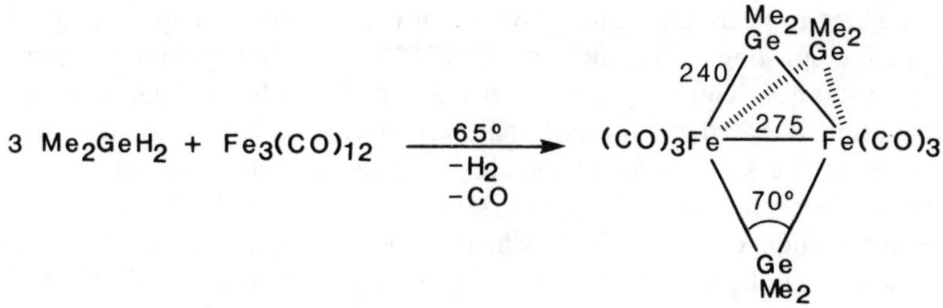

$$3\ Me_2GeH_2 + Fe_3(CO)_{12} \xrightarrow[-H_2,\ -CO]{65°} (CO)_3Fe(\mu\text{-}GeMe_2)_3Fe(CO)_3$$

Fe–Ge 240, Fe–Fe 275, Fe–Ge–Fe 70°

Cyclopentadienyl-substituted germylenes again constitute a special case; they are accessible from Ge^{II} halides (Curtis, 1973):

$$GeBr_2 + 2\,C_5H_5Tl \xrightarrow{THF,\ 20\,°C} (C_5H_5)_2Ge + 2\,TlBr$$

$$GeCl_2 \cdot dioxane + 2\,C_5Me_5Li \xrightarrow{THF} (C_5Me_5)_2Ge + 2\,LiCl + dioxane$$

Monomeric **germanocene** $(C_5H_5)_2Ge$ like stannocene and plumbocene has a tilted sandwich structure; it forms polymers containing bridging C_5H_5 units much more rapidly than the Sn- and Pb analogues. Decamethylgermanocene $(C_5Me_5)_2Ge$ (as well as its Sn congener, p. 136) releases one ligand under the action of HBF_4 to yield the interesting cation $[(C_5Me_5)Ge]^+$ (Jutzi, 1980):

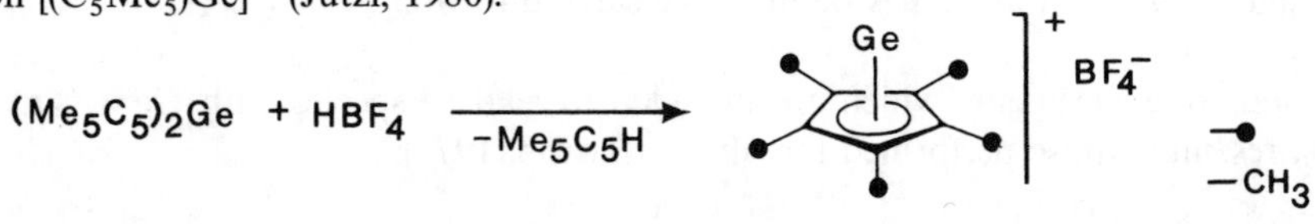

$$(Me_5C_5)_2Ge + HBF_4 \xrightarrow{-Me_5C_5H} [(\eta^5\text{-}C_5Me_5)Ge]^+\ BF_4^-$$

(● = $-CH_3$)

$[(C_5Me_5)Ge]^+$ is isoelectronic and isostructural with $(C_5H_5)In$ (p. 89) and related compounds; together with six π-electrons of the ligand, the central atom $E\,(ns^2)$ attains an octet configuration.
Germyl anions R_3Ge^- originate from organogermyl alkali-metal compounds in highly polar solvents like hexamethylphosphoric trisamide (HMPA) or NH_3 (l); they serve in the formation of germanium-metal bonds:

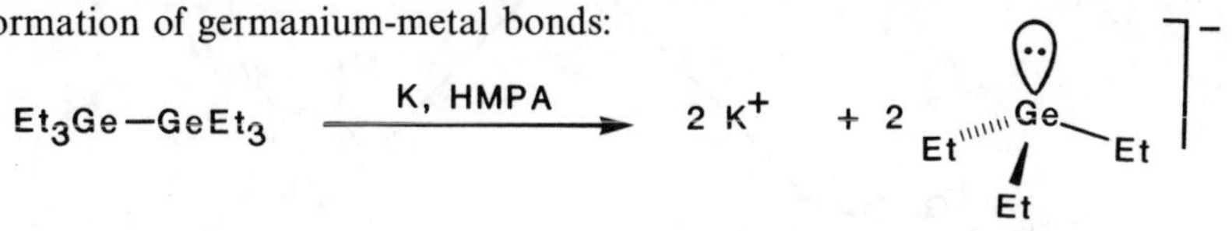

$$Et_3Ge-GeEt_3 \xrightarrow{K,\ HMPA} 2\,K^+ + 2\,[:GeEt_3]^-$$

$$Ph_4Ge \xrightarrow[-40°]{Na,\ NH_3\,(l)} Na^+ + Ph_3Ge^- \xrightarrow{Me_3SnBr} Ph_3Ge-SnMe_3$$

Since, except for sterically crowded species, organodigermanes refrain from homolytic cleavage, **organogermyl radicals** $R_3Ge^{\bullet}$ are generated as follows:

$$t\text{-}Bu_2O_2 \xrightarrow{h\nu} 2\ t\text{-}BuO\cdot \xrightarrow[-t\text{-}BuOH]{R_3GeH,\,-70°} R_3Ge\cdot$$

$$(R_3Ge)_2Hg \xrightarrow[-Hg]{h\nu} R_3Ge\cdot$$

The Ge-centered radicals are identified by means of ESR spectroscopy, taking advantage of the hyperfine splitting by the nucleus ^{73}Ge (7.6%, $I = 9/2$). From the magnitude of the coupling constant $a(^{73}Ge)$, a pyramidal structure of the radicals $R_3Ge\cdot$ can be inferred. This conclusion is based on the fact that isotropic hyperfine coupling constants are proportional to the *s* character of the singly occupied orbital. In this way, the ESR spectrum provides a hint as to the hybridization state of the radical center and therefore to its geometry. Independent evidence which supports nonplanarity of $R_3Ge\cdot$ radicals is the retention of configuration during the homolytic chlorination of chiral germanes (Sakurai, 1971):

$$(Naph)(Ph)(Me)Ge^{*}H + CCl_4 \xrightarrow[80°,\,11h]{t\text{-}Bu_2O_2} (Naph)(Ph)(Me)Ge^{*}Cl + CHCl_3$$

Experiments with chiral silanes point in the same direction.

The kinds of experiments which are intended to lead to species with **Ge=E ($p_\pi - p_\pi$) bonds,** resemble those performed for silicon (Barton, 1973):

$$\text{Et}_2\text{Ge-bicyclo adduct (Cl, Me, 2 CF}_3) \xrightarrow{450°} \{Et_2Ge{=}CH_2\} + \text{C}_6\text{H(Cl)(Me)}_2\text{(CF}_3)_2$$

$\{Et_2Ge{=}CH_2\}$ $\xrightarrow{\times 2}$ 1,1,3,3-tetraethyl-1,3-digermacyclobutane (Et_2Ge, $GeEt_2$)

$\{Et_2Ge{=}CH_2\}$ + 2,3-dimethylbutadiene (Me, Me) → 1,1-diethyl-3,4-dimethyl-1-germacyclohex-3-ene ($GeEt_2$, Me, Me)

The Ge = C double bond is extremely reactive, even if it is part of a 6 π-electron system (Märkl, 1980):

R = t-Bu

t-BuLi, pentane, RT
−t-BuH, −LiBr

Germabenzene

As in the case of silicon, double bonds $\rangle$Ge = Ge$\langle$ may be protected by steric shielding (Masamune, 1984):

$$Ar_2GeCl_2 \xrightarrow[DME]{LiC_{10}H_8} Ar_2Ge(\text{–}GeAr_2\text{–}GeAr_2) \xrightarrow[\text{3-methylpentane}]{h\nu} Ar_2Ge{=}GeAr_2 \quad 90\%$$

221

Ar =

8.3 Organotin Compounds

Due to a multitude of technical as well as synthetic applications, organotin compounds have become the subject of extensive industrial and academic research. A characteristic feature of organotin chemistry, compared to that of the lighter homologues Si and Ge, is the greater structural diversity. This may be traced back to the larger variety of accessible coordination numbers for tin. Thus, two aspects of organotin chemistry, namely the association of organotin halides R_nSnX_{4-n} by means of Sn – X – Sn bridges and the ready formation of stannonium ions $[R_nSn(solv)]^{(4-n)+}$ in donor solvents originate from the ability of tin to increase its coordination number beyond 4. Another distinctive feature is the high reactivity of tin organyls in which C_α is part of an unsaturated system as in Sn-vinyl or Sn-phenyl segments.

Excursion

^{119}Sn-Mössbauer and ^{119}Sn-NMR Spectroscopy

Besides the standard methods of instrumental analysis and X-ray diffraction, these more specialized techniques are also well-suited to solve structural problems in organotin chemistry. Whereas ^{119}Sn-NMR is preferentially applied to species in solution, ^{119}Sn-Mössbauer spectroscopy, operating on solid samples, stands the test in cases where crystalline material for the application of diffraction methods is not available.

The Mössbauer effect rests on the recoil-free emission and resonance absorption of γ-rays. A Mössbauer spectrum furnishes the parameters **isomer shift *IS*** and **quadrupole splitting *QS*.** In their ground- and excited nuclear spin states, atomic nuclei possess different radii and thus differing electrostatic interaction with the surrounding *s* electrons. Therefore, the energy difference between the ground- and the excited state is influenced by the *s* electron density at the nucleus. A comparison of nuclear spin excitation energies for nuclides like ^{119}Sn, which are subjected to different chemical environments, thus provides insight into the respective *s* electron densities. These excitation energies are usually listed as relative values with SnO_2 serving as a standard. They are then named isomer shifts *IS* and for experimental reasons, reported in mms^{-1} whereby a positive *IS* value implies an increased *s* electron density.

If the nuclear spin *I* of the respective nuclide in the ground and/or in the excited state exceeds 1/2, deviations of the electronic environment of the nucleus from cubic, octahedral or tetrahedral symmetry result in a splitting of the Mössbauer signal. This is the quadrupole splitting *QS* which supplies additional structural information.

In the area of organometallics, nuclei suitable for Mössbauer spectroscopy besides ^{119}Sn include ^{57}Fe, ^{99}Ru, ^{121}Sb, ^{125}Te, ^{129}I, ^{193}Ir and ^{197}Au.

The **isomer shifts** *IS* in **^{119}Sn-Mössbauer spectra** cover a range of ± 5 mms^{-1} [typical values: Sn^{IV} -0.5 to 1.8, Sn^{II} 2.5 to 4.3 mm s^{-1}]. The following examples demonstrate how the isomer shift responds to changes in the environment of Sn:

	IS/mms^{-1}	Variation of
$[n\text{-BuSnF}_5]^{2-}$	0.27	ligands
$[n\text{-BuSnCl}_5]^{2-}$	1.03	
$[n\text{-BuSnBr}_5]^{2-}$	1.38	
Ph_4Sn^{IV}	1.15	oxidation state
$n\text{-}Bu_4Sn^{IV}$	1.35	
Cp_2Sn^{II}	3.74	
cis-$Me_2Sn(ox)_2$	0.88	geometry
trans-$Me_2Sn(acac)_2$	1.18	

The **quadrupole splitting *QS*** reflects the stereochemistry at the Sn atom. Therefore, *QS* values may serve to distinguish between different isomeric forms, Scheme I (Bancroft, 1974).

Since these environments also occur in associated organotin compounds like $(R_3SnX)_n$ and $(R_2SnX_2)_n$ the structure of these materials in the solid state is conveniently studied by means of ^{119}Sn-Mössbauer spectroscopy (cf. p.125, 133).

Scheme I

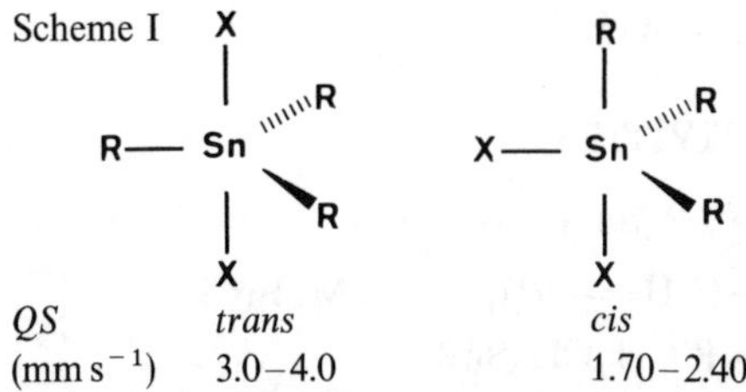

QS $(mm\,s^{-1})$	*trans* 3.0–4.0	*cis* 1.70–2.40	*trans* 3.8–4.2	*cis* 1.7–2.1

In the **^{119}Sn-NMR spectra** of organotin^{IV} compounds, relative to the standard $(CH_3)_4Sn$, chemical shifts δ between $+150$ ppm (low field or high frequency, resp.) and -400 ppm (high field or low frequency, resp.) are observed. In Sn^{II} compounds, the tin nucleus is shielded much more extensively. The following compilation is intended to illustrate a few trends:

	$\delta(^{119}Sn)$/ppm		$\delta(^{119}Sn)$/ppm
Me_3SnCH_2Cl	+4	$(\eta^1\text{-}C_5H_5)_4Sn^{IV}$	−26
$Me_3SnCHCl_2$	+33	$(\eta^5\text{-}C_5H_5)_2Sn^{II}$	−2200
Me_3SnCCl_3	+85		
		Me_3SnH	−104
Me_4Sn	0	Me_2SnH_2	−224
Ph_2SnMe_2	−60	$MeSnH_3$	−346
Ph_4Sn	−137		
$(H_2C{=}CH)_4Sn$	−157	$Me_3SnCl\cdot py$	+25
$Me_2Sn(acac)_2$	−366	$[Me_3Sn(bipy)]BPh_4$	−18

An important application of ^{119}Sn-NMR is the study of association processes in solution which are accompanied by an increase of the coordination number (C.N.) at Sn. The transition from C.N.4 to C.N.5 or C.N.6 is attended by a high field shift which can be used as a diagnostic tool. As an example, consider the monomer/polymer equilibrium for trimethyltin formate (Scheme II).

On the other hand, for inert complexes like $Me_2Sn(acac)_2$ (C.N.6) in which tin has a high coordination number in the molecular unit, the chemical shift is independent of concentration. Additionally, the coupling constants $^1J(^{119}Sn, ^{13}C)$ and $^2J(^{119}Sn, ^1H)$ also respond to changes in the coordination number of Sn, albeit in a less sensitive way.

Scheme II

conc. in $CDCl_3$:	$0.05\ mol\,L^{-1}$	$2.5\ mol\,L^{-1}$
$\delta(^{119}Sn)$:	+152 ppm	+2.5 ppm
Structure:	monomeric	polymeric

8.3.1 Tin Organyls of Coordination Numbers 4, 5 and 6

PREPARATION, STRUCTURE AND REACTIVITIES

$$4\,R_3Al + 4\,NaCl + 3\,SnCl_4 \xrightarrow{PE} 3\,R_4Sn + 4\,NaAlCl_4$$ 4

$$4\,CH_2{=}CH{-}MgBr + SnCl_4 \xrightarrow{THF} (CH_2{=}CH)_4Sn + 4\,MgBrCl$$ 4

$$R'CH{=}CH_2 + R_3SnH \longrightarrow R'CH_2CH_2SnR_3$$ 8

Compounds which are sufficiently CH-acidic are metallated by organotin amides:

$$Me_3SnNMe_2 + HC{\equiv}CPh \longrightarrow Me_3SnC{\equiv}CPh + Me_2NH$$ 6

The **binary tin organyls** R_4Sn are inert towards attack by O_2 or H_2O; usually they are also thermally stable. Me_4Sn decomposes above 400 °C; organotin compounds with unsaturated residues are more labile, a fact which recommends them for synthetic applications:

$$(CH_2{=}CH)_4Sn + 4\,LiPh \longrightarrow 4\,CH_2{=}CHLi + Ph_4Sn$$ 3

(Cyclopentadienyl)trisalkyltin $(C_5H_5)SnR_3$, like the Si- and Ge analogues, is fluxional in solution. Such **metallotropic 1,2-shifts** of R_3Sn groups are also observed at the periphery of larger unsaturated rings:

H SnR3 H SnR3

The thiophilic nature of tin expresses itself in the reaction of Ph_4Sn with sulfur whereby puckered six-membered rings $(Ph_2SnS)_3$ are formed:

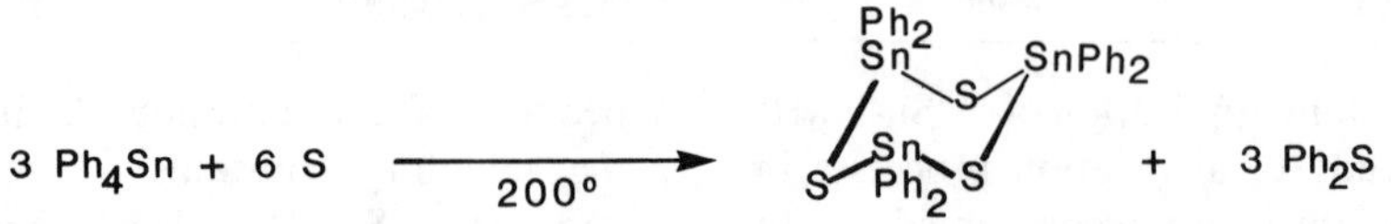

$$3\,Ph_4Sn + 6\,S \xrightarrow[200^\circ]{} (Ph_2SnS)_3 + 3\,Ph_2S$$

Sn−C bonds are also cleaved by dihalogens and by hydrogen halides. In this way, **organotin halides** are obtained:

$$Me_4Sn \xrightarrow[-\,MeBr]{Br_2} Me_3SnBr \xrightarrow[-\,MeBr]{Br_2} Me_2SnBr_2$$

$$Me_4Sn + HX \longrightarrow Me_3SnX + MeH$$

Further methods for the preparation of organotin halides are:

$$Sn/Cu + 2\,MeCl \xrightarrow{200-300\,^\circ C} Me_2SnCl_2 \quad \text{(direct process)}$$ 1

$$SnCl_2 + RCl \xrightarrow{\text{cat. } SbCl_3} RSnCl_3 \quad \text{(oxidative addition)}$$

$$R_4Sn + SnCl_4 \xrightarrow[\text{fast}]{0-20\,^\circ C} R_3SnCl + RSnCl_3 \xrightarrow[\text{slow}]{180\,^\circ C} 2\,R_2SnCl_2$$ 4

$$SnCl_2 + HCl \xrightarrow{Et_2O} HSnCl_3 \cdot Et_2O \xrightarrow{R_2C{=}CHCOOR'} Cl_3SnCR_2CH_2COOR'$$ 8

(hydrostannation)

Organotin halides, -pseudohalides and -carboxylates tend to associate by means of σ-donor bridging:

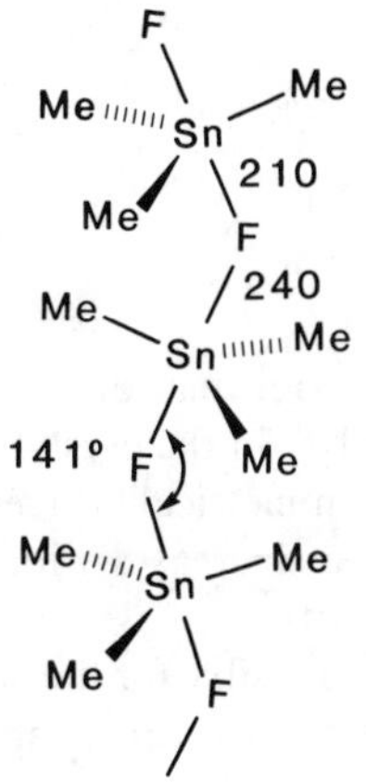

Me_3SnF *(decomp.* 360 °C *before reaching the* mp) *forms chains of trigonal bipyramids, the apical fluoride ions being part of unsymmetrical* Sn – F – Sn *bridges* (*Trotter*, 1964). *Considerably weaker are the bridges* Sn – Cl – Sn *in solid* Me_3SnCl (mp 37 °C, bp 152 °C) *whose structure was studied by Mössbauer spectroscopy.*

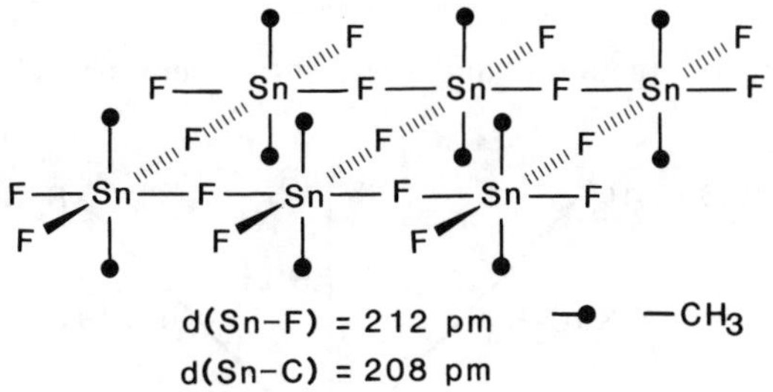

d(Sn–F) = 212 pm — $-CH_3$

d(Sn–C) = 208 pm

Me_2SnF_2 *(decomp.* 400 °C), *like* SnF_4, *crystallizes in a lattice structure whereby two axial terminal methyl groups and four equatorial, bridging fluoride ions generate the coordination number* 6 *at* Sn (Schlemper, 1966).

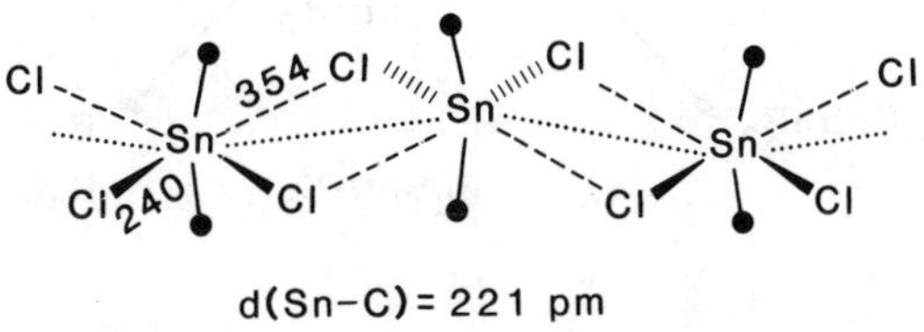

d(Sn–C) = 221 pm

Me_2SnCl_2 (mp 106 °C, bp 190 °C), *however, in the crystal forms zigzag chains of highly distorted edge sharing octahedra with weak chloride bridges* (*Davies*, 1970).

A comparison of the structural data for Me_3GeCN(*s*) and Me_3SnCN (*s*) demonstrates the higher inclination of tin to expand its coordination sphere (Schlemper, 1966):

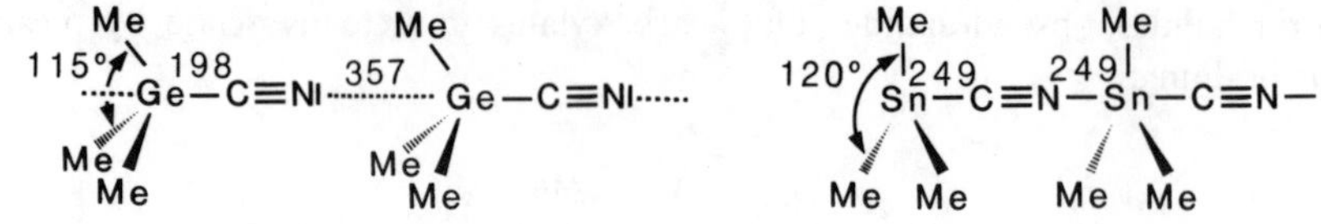

Me_3GeCN *C.N.* (Ge) = 4
compressed tetrahedra

Me_3SnCN *C.N.* (Sn) = 5
trigonal bipyramids

An even greater tendency for high coordination numbers is exhibited by lead, in that Ph_2PbCl_2 – in contrast to Me_2SnCl_2 – in the crystal adopts the coordination number 6 (chains of edge sharing octahedra, symmetrical bridges Pb – Cl – Pb). Thus, the tendency towards a high coordination number increases with the atomic number of the central atom and with ligand electronegativity.
The high tendency to associate is responsible for the fact that chiral trisorganotin halides, as opposed to their C-, Si- and Ge counterparts, rapidly racemize in solution:

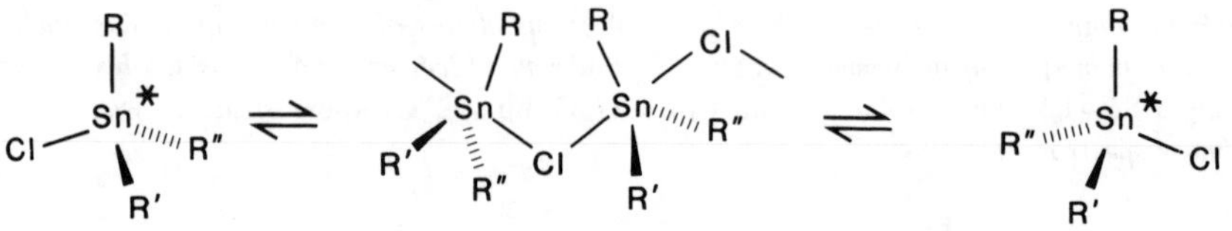

Follow-up reactions of organotin halides lead into several new classes of compounds:

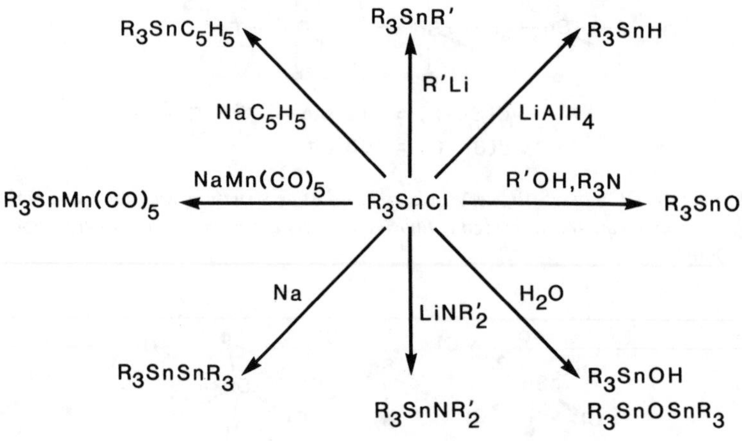

Acid hydrolysis of the $\geq$Sn – Cl bond under conservation of the Sn – C bonds initially yields hydrated **stannonium ions** $R_3Sn(aq)^+$ which can be precipitated with large anions:

$$Me_3SnCl \xrightarrow[NaBPh_4]{H_3O^+_{aq}} [Me_3Sn(OH_2)_2]^+BPh_4^-$$

trigonal bipyramidal cation, H_2O *at axial positions*

In alkaline aqueous media, organotin halides form the corresponding hydroxides; their propensity for condensation increases with decreasing degree of alkylation:

$$R_3SnCl \xrightarrow{OH^-/H_2O} R_3SnOH \xrightarrow[2\times]{Na, C_6H_6} R_3Sn{-}O{-}SnR_3 \xrightarrow{H_2O} 2\ R_3SnOH$$

solution: dimeric
solid state: polymeric

Distannoxane

Stannanol

$$R_2SnCl_2 \xrightarrow{OH^-/H_2O} R_2Sn(OH)_2 \longrightarrow (R_2SnO)_n \begin{cases} \xrightarrow{OH^-} [R_2Sn(OH)_4]^{2-} \\ \xrightarrow{H_2SO_4} R_2SnSO_4 \end{cases}$$

isolable for R = t-Bu

Polystannoxanes

$$\begin{array}{c} R\ \ R \quad R\ \ R \quad R\ \ R \\ -Sn-O-Sn-O-Sn- \\ \uparrow \quad \downarrow \quad \uparrow \quad \downarrow \quad \uparrow \\ -O-Sn-O-Sn-O- \\ R\ \ R \quad R\ \ R \end{array}$$

$$RSnCl_3 \xrightarrow{OH^-/H_2O} \{RSn(OH)_3\} \xrightarrow[-H_2O]{} (RSnOOH)_n$$

Stannonic acid
cross-linked, polymeric

The amphoteric character of the stannoxanes is in marked contrast to the inert nature of siloxanes, which is reflected in the excellent material properties of the latter. The double chain structure of the polystannoxanes follows from ^{119}Sn-Mössbauer spectra.

Organotin alkoxides are prepared from the respective halides and alcohols using an auxiliary base:

$$R_3SnX + R'OH + Et_3N \longrightarrow R_3SnOR' + Et_3NH^+X^-$$

or via transalkoxylation:

$$(R_3Sn)_2O + 2\,R'OH \longrightarrow 2\,R_3SnOR' + H_2O$$

The synthesis of an organotin bis(acetylacetonate) with the coordination number 6 at tin proceeds in a related way:

$$R_2Sn(OMe)_2 + 2\,acacH \longrightarrow \textit{trans}\text{-}R_2Sn(acac)_2 + 2\,MeOH$$

Organotin amides cannot be obtained by an ammonolysis of R_3SnCl; they are prepared via nucleophilic substitution by amide:

$$R_3SnCl + LiNR'_2 \longrightarrow R_3SnNR'_2 + LiCl$$

$$Bu_3SnPh + KNH_2 \xrightarrow{NH_3(\ell)} Bu_3SnNH_2 + PhK$$

Organotin amides have found application in organic synthesis:

$$(Bu_3Sn)_2NEt \xrightarrow{\text{cyclohexanone}} (Bu_3Sn)_2O + \text{cyclohexylidene}{=}N{-}Et$$

$$Bu_3SnNEt_2 \xrightarrow{\text{2-chlorocyclohexanone}} Bu_3SnCl + \text{2-(diethylamino)cyclohexanone}$$

Organotin hydrides, in addition to the halides, are the most versatile reagents. They are easily obtained from the latter:

$$R_nSnX_{4-n} \xrightarrow{LiAlH_4, Et_2O} R_nSnH_{4-n}$$

Organotin oxides or -alkoxides and $NaBH_4$ or sodium amalgam as reducing reagents are also suitable. The stability of the alkylstannanes R_nSnH_{4-n} increases with increasing degree of alkylation: whereas SnH_4 slowly decomposes at room temperature, Me_3SnH is stable indefinitely in the absence of air.
The principal reactions of organotin hydrides are:

hydrostannation $\quad >Sn{-}H + A{=}B \longrightarrow >Sn{-}A{-}B{-}H$

hydrostannolysis $\quad >Sn{-}H + X{-}Y \longrightarrow >Sn{-}Y + HX$

oxidative coupling $\quad >Sn{-}H + H{-}Sn< \xrightarrow{\text{Kat.}} >Sn{-}Sn< + H_2$

The **hydrostannation** reaction (van der Kerk, 1956) serves for the construction of new Sn – C bonds: it may proceed as 1,2- or as 1,4-addition:

$$Ph_3SnH \xrightarrow[80-100^\circ]{H_2C{=}CHPh} Ph_3SnCH_2CH_2Ph$$

$$Ph_3SnH \xrightarrow[80^\circ]{HC{\equiv}CPh} Ph_3SnCH{=}CHPh \text{ (trans)}$$

$$(n\text{-}Bu)_2SnH_2 + HC{\equiv}C{-}CH_2{-}C{\equiv}CH \xrightarrow[80^\circ]{\text{heptane}} \text{1,1-dibutylstannacyclohexa-2,5-diene } (SnBu_2)$$

Under these mild conditions and in the absence of catalysts (Lewis acids or radical forming reagents) polar double bonds like C=O and C=N are not attacked. Therefore, hydrostannation is well-suited for **chemoselective hydrogenations** of activated C=C double bonds (Keinan, 1982):

$$\text{Ph-CH=CH-CHO} + (\text{n-Bu})_3\text{SnH} \xrightarrow[\text{2. } H_2O]{\text{1. cat. } (Ph_3P)_4Pd,\ \text{THF, } 20°} \underset{99\%}{\text{Ph-CH}_2\text{-CH}_2\text{-CHO}}$$

Note the complementary result obtained in the hydroboration of cinnamic aldehyde (p. 61).
In the presence of azoisobutyronitrile (AIBN), 1,4-addition to conjugated dienes occurs:

$$Et_3SnH + CH_2{=}CH{-}CH{=}CH_2 \xrightarrow{AIBN} Et_3SnCH_2{-}CH{=}CH{-}CH_3$$

By appropriate choice of reaction conditions, regioselectivity may be achieved:

$$R_3SnH + CH_2{=}CH{-}C{\equiv}N \begin{cases} \xrightarrow{AIBN} R_3SnCH_2{-}CH_2{-}C{\equiv}N \\ \xrightarrow{\text{uncatalyzed}} CH_3CH(SnR_3){-}C{\equiv}N \end{cases}$$

Compared to hydroboration and hydrosilation, hydrostannations are distinguished by their **mechanistic multiplicity** (Neumann, 1965). The relative weakness and low polarity of the Sn−H bond open up:

- ***radical reaction pathways**, effected by AIBN and/or UV-irradiation and initiated by homolysis of the* Sn−H *bond and*
- ***hydride transfer processes** which proceed in media of high polarity in the presence of Lewis acid catalysts like* $ZnCl_2$ *or* SiO_2 *and which are favored for electrophilic substrates containing polarized multiple bonds.*

Hydrostannolysis as well may follow a homolytic or a heterolytic path. According to the aforementioned low polarity of Sn−H bonds, depending on the nature of the attacking agent organotin hydrides may act as a source of H^-, H^+, or $H\cdot$:

	R_3SnH transfers
$R_3Sn{-}H + MeCOOH \longrightarrow R_3SnOOCMe + H_2$	H^-
$4\,R_3Sn{-}H + Ti(NR'_2)_4 \longrightarrow (R_3Sn)_4Ti + 4\,HNR'_2$	H^+
$2\,R_3Sn{-}H + R'_2Hg \longrightarrow (R_3Sn)_2Hg + 2\,R'H$	$H\cdot$

$$(R_3Sn)_2Hg \xrightarrow[\text{R = Me, } -10\,°C;\ \text{R = Ph, } +100\,°C]{-Hg} R_6Sn_2$$

The homolytic variant of hydrostannolysis constitutes one of the most convenient methods for the conversion R−X → R−H; it proceeds as a chain reaction:

$$R'_3Sn{-}H \xrightarrow{AIBN} R'_3Sn\cdot + H\cdot \quad \text{(Start)}$$

$$\left.\begin{aligned} R'_3Sn\cdot + \mathbf{RX} &\longrightarrow R'_3SnX + R\cdot \\ R\cdot + R'_3SnH &\longrightarrow \mathbf{RH} + R'_3Sn\cdot \end{aligned}\right\} \text{(Propagation)}$$

An example is the selective reduction of geminal dihalides in the presence of other sensitive groups (Kuivila, 1966):

$$\text{Me}_2\text{C}_3(=\text{CH}_2)\text{Br}_2 \xrightarrow[h\nu]{(n\text{-Bu})_3\text{SnH}} \text{Me}_2\text{C}_3(=\text{CH}_2)(\text{Br})\text{H} \xrightarrow[h\nu]{(n\text{-Bu})_3\text{SnH}} \text{Me}_2\text{C}_3(=\text{CH}_2)\text{H}_2$$

Organopolystannanes are of interest in studies concerned with the aptitude of main-group elements to form homonuclear chains. Although the tendency for catenation in group 14 decreases with increasing atomic number, in the case of tin several examples with linear, branched or cyclic structures are known.

In order to secure kinetic stability it is essential, however, to shield the vulnerable Sn – Sn bond [E(Sn – Sn) = 151 kJ/mol, compare: E(Si – Si) = 340 kJ/mol] by means of alkylation or arylation. In the case of the parent stannanes, no unambiguous evidence exists for congeners larger than Sn_2H_6. The preparation of (organo)oligostannanes may be effected by **Wurtz coupling:**

$$\text{Me}_3\text{SnBr} \xrightarrow[\text{NH}_3(\ell)]{\text{Na}} \text{Me}_3\text{Sn}-\text{SnMe}_3 \xrightarrow{\text{Na}} 2\,\text{Na}^+ + 2\,\text{Me}_3\text{Sn}^-$$

$$2\,\text{Me}_3\text{Sn}^- \xrightarrow{\text{SnCl}_4} \text{Sn(SnMe}_3)_4$$

$$\text{Me}_2\text{SnCl}_2 \xrightarrow[\text{NH}_3(\ell)]{\text{Na}} (\text{Me}_2\text{Sn})_6 + \text{ClMe}_2\text{Sn(SnMe}_2)_n\text{SnMe}_2\text{Cl}$$

dodecamethylcyclohexastannane — *chain length varies depending on the reaction conditions, n ≤ 12*

$$\text{Ph}_2\text{SnCl}_2 \xrightarrow[\text{THF}]{\text{NaC}_{10}\text{H}_8} (\text{Ph}_2\text{Sn})_6$$

d(Sn – Sn) = 277 pm (Dräger, 1983);
compare grey α-tin: d(Sn – Sn) = 281 pm

via **thermolysis of organotin hydrides:**

$$\text{Ar}_2\text{SnH}_2 \xrightarrow[-\text{H}_2]{\text{cat. DMF or pyridine}} (\text{Ar}_2\text{Sn})_5 + (\text{Ar}_2\text{Sn})_6$$

$$2\,\text{R}_3\text{Sn}-\text{SnR}_2\text{H} \xrightarrow[\text{cat. amine}]{\Delta T} \text{R}_3\text{Sn(R}_2\text{Sn)}_2\text{SnR}_3$$

or by means of **hydrostannolysis:**

$$\text{RSnH}_3 + 3\,\text{Me}_3\text{SnNEt}_2 \xrightarrow{20\,^\circ\text{C}} \text{R}-\text{Sn(SnMe}_3)_3 + 3\,\text{Et}_2\text{NH}$$

(Organo)oligostannanes are quite stable thermally (example: $Ph_3Sn-SnPh_3$, mp 237°) but considerably more air-sensitive than the peralkylated species R_4Sn. Hexamethyldistannane in air at room temperature slowly forms the stannoxane $(Me_3Sn)_2O$; at its boiling point of 182 °C, Me_6Sn_2 is pyrophoric.
Hexaorganodistannanes have been used to generate transition metal-tin bonds:

$$Me_6Sn_2 \xrightarrow[-2\,Ph_3P]{(Ph_3P)_4Pt} trans\text{-}(Ph_3P)_2Pt^{II}(SnMe_3)_2$$

$$Me_6Sn_2 \xrightarrow{Co_2(CO)_8} 2\,Me_3Sn-Co(CO)_4 \leftarrow Na[Co(CO)_4] + Me_3SnCl$$

These complexes illustrate the iso(valence)electronic nature of R_3Sn^- and R_3P.

8.3.2 Tin Organyls of Coordination Numbers 2 and 3

Organostannyl anions R_3Sn^- are accessible in a variety of ways:

$$\left.\begin{array}{l} R_3Sn-SnR_3 \xrightarrow{Na,\ THF} \\ R_3SnBr \xrightarrow{Na,\ NH_3(\ell)} \\ R_4Sn \xrightarrow{Na,\ NH_3(\ell)} \end{array}\right\} \longrightarrow R_3SnNa \rightleftharpoons R_3Sn^- + Na^+$$

The position of the dissociation equilibrium depends on the nature of R and the solvating power of the medium. In the case of Ph_3SnNa and $NH_3(\ell)$, extensive dissociation is assumed; for Me_3SnLi in $EtNH_2$, on the other hand, electrolytic conductivity is minute. Alkali-metal organostannates(II) R_3SnM complement organodistannanes in the synthesis of transition metal-tin bonds:

$$Ph_3SnLi + Ni(CO)_4 \xrightarrow[-CO]{THF} Li(THF)_n^+[Ni(CO)_3(SnPh_3)]^-$$

Whereas – because of ion solvation – the assignment of coordination number 3 for organostannyl anions R_3Sn^- and -cations R_3Sn^+ (p. 126) is not warranted, it is without question that **organostannyl radicals $R_3Sn\cdot$** do possess this coordination number. The spontaneous formation of $R_3Sn\cdot$ radicals through homolytic cleavage of distannanes only occurs if assisted by bulky substituents (cf. p. 113, 120):

$$Ar_3SnSnAr_3 \xrightarrow{\Delta T} 2\,Ar_3Sn^\bullet \qquad Ar = \text{2,4,6-triethylphenyl}$$

Additional sources for organostannyl radicals are:

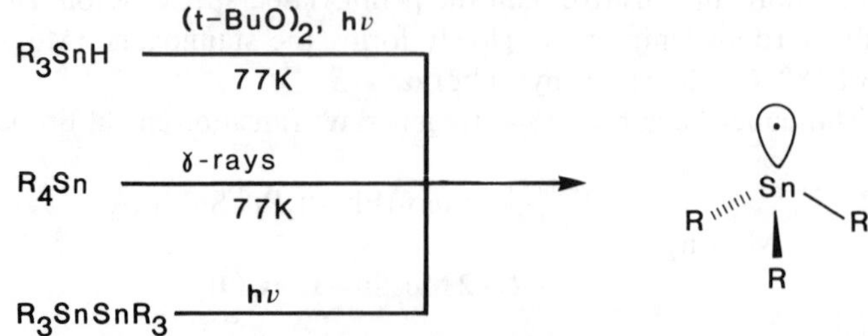

Stannyl radicals $R_3Sn\cdot$, *formally* Sn^{III} *compounds, like* $R_3Ge\cdot$ (*but unlike* $R_3C\cdot$) *are pyramidal* (*Howard*, 1972). *This may be inferred from the large ESR hyperfine coupling constant* $a(^{119}Sn)$ [*example*: $Ph_3Sn\cdot$, $a(^{119}Sn) = 155$ mT] *according to which the singly occupied orbital possesses considerable* Sn(5*s*) *character. In the case of a planar structure and the single occupancy of an* Sn(5*p*) *orbital, a very small coupling constant would be expected which would have to be traced to spin polarization of doubly occupied* Sn(*s*) *orbitals or of* Sn(sp^2) *hybrid orbitals, respectively. A consequence of the pyramidal structure of* $R_3Sn\cdot$ *is the possibility that radical reactions of chiral organostannanes* RR′R″SnH *proceed with retention of configuration. This has been demonstrated in a number of cases.*

The life-time of organotin radicals $R_3Sn\cdot$ strongly depends on the bulkiness of the groups R: the range extends from diffusion-controlled dimerization in the case of $Me_3Sn\cdot$ to a half-life of about one year for $[(Me_3Si)_2CH]_3Sn\cdot$.

The **stannylenes R_2Sn** represent organotin molecules with the coordination number 2. With simple alkyl- or aryl substituents R these stannylenes – like the respective germylenes, silylenes and carbenes – only occur as reactive intermediates which undergo typical follow-up reactions:

R_2SnCl_2 —($NaC_{10}H_8$)→ / $(R_2Sn)_n$ —(hν)→ R_2Sn: (Stannylene)

R_2Sn: —(Me_2SnH_2)→ $HR_2SnSnMe_2H$

R_2Sn: —(R′X)→ $R_2R'SnX$

R_2Sn: —(2,3-dimethylbutadiene)→ 3,4-dimethyl-1,1-R_2-stannacyclopent-3-ene (Sn R_2)

The bulky substituent 2,4,6-*tris*(*iso*-propyl)phenyl stabilizes a distannane which is in thermal equilibrium with the cyclotristannane (Masamune, 1985):

Ar_2SnCl_2 —($NaC_{10}H_8$; 1. −78°, 2. 60°)→ cyclo-$(Ar_2Sn)_3$ $\underset{h\nu,\ -78°}{\overset{\Delta H>0}{\rightleftharpoons}}$ $Ar_2Sn{=}SnAr_2$

Ar = 2,4,6-(iso-$C_3H_7)_3C_6H_2$

The dimeric structure of $Ar_2Sn = SnAr_2$ in solution is proved by the observation of ^{117}Sn satellites ($I = 1/2$) in the ^{119}Sn-NMR spectrum. A stannylene which exists as a monomer in solution at room temperature was devised by introducing the substituent bis(trimethylsilyl)methyl (Lappert, 1976):

$$2\ SnCl_2 + 4\ LiR \xrightarrow[-4\ LiCl]{Et_2O,\,-10°} 2\ :SnR_2 \longrightarrow R_2Sn{=}SnR_2$$

$R = CH(SiMe_3)_2$

solution

solid state

276, 115°

sp2 pz Sn Sn pz sp2

In the same way, the corresponding germylene R_2Ge and plumbylene R_2Pb were obtained.
The bent structure of the stannylene is deduced from the observation of two stretching vibrations, $\nu(Sn-C)_{antisym}$, and $\nu(Sn-C)_{sym.}$, in the IR spectrum. For the red, diamagnetic dimer a "bent double bond" has been proposed in which one doubly occupied sp_xp_y hybrid orbital on each of the stannylene units interacts with the empty p_z orbital of the other stannylene.

Monomeric $[(Me_3Si)_2CH]_2Sn$ is encountered in **transition-metal stannylene complexes**:

$$R_2Sn + Cr(CO)_6 \xrightarrow[-CO]{h\nu} R_2SnCr(CO)_5 \quad (R = CH(SiMe_3)_2)$$

Slightly less plump stannylenes form metal complexes in which the addition of a Lewis base raises the coordination number at tin to 4 (Marks, 1973):

(THF)(t-Bu)$_2$Sn–Fe(CO)$_4$

^{119}Sn-Mössbauer studies point to the oxidation state Sn^{IV} in these base adducts (Zuckerman, 1973).

Stannylene bridges μ-SnR_2 arise from the photolysis of certain transition metal substituted organostannanes (Curtis, 1976):

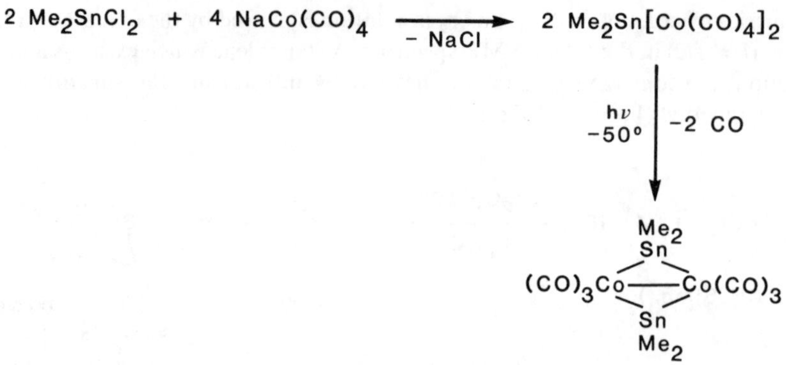

$$2\ Me_2SnCl_2 + 4\ NaCo(CO)_4 \xrightarrow[-\ NaCl]{} 2\ Me_2Sn[Co(CO)_4]_2$$

$$\downarrow h\nu,\ -50°,\ -2\ CO$$

$$(CO)_3Co(\mu\text{-}SnMe_2)_2Co(CO)_3$$

The longest known Sn^{II} organyl is bis(cyclopentadienyl)tin, **stannocene** (E. O. Fischer, 1956):

$$SnCl_2 + 2\,NaC_5H_5 \xrightarrow[-2\,NaCl]{THF} (\eta^5\text{-}C_5H_5)_2Sn$$

mp 105 °C, air- and water-sensitive

The parent stannocene $(C_5H_5)_2Sn$ has a tilted sandwich structure which approaches an axial structure with increasing steric demand from the peripheral substituents:

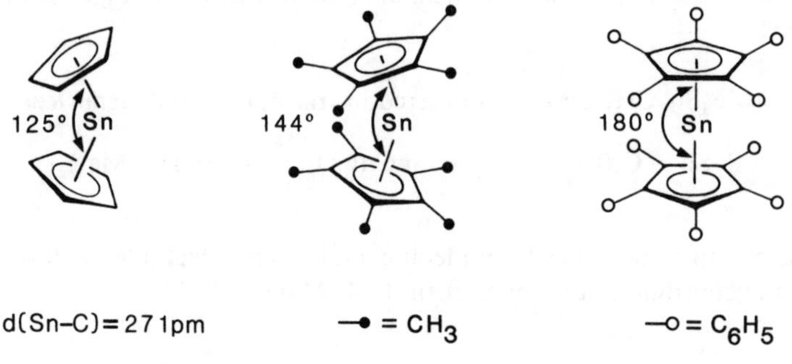

125° Sn — 144° Sn — 180° Sn

d(Sn–C) = 271 pm — ● = CH_3 — ○ = C_6H_5

In a pedestrian discussion of these tilting angles, competition between the attempt to minimize ligand-ligand repulsion and the tendency to place nonbonding metal electrons in hybrid orbitals of high Sn(5*s*) character is assumed. The following limiting cases of hybridization consider only the σ-interaction between atomic orbitals of tin and the a_{1g} molecular orbitals of the ligands C_5H_5 (cf. p. 319):

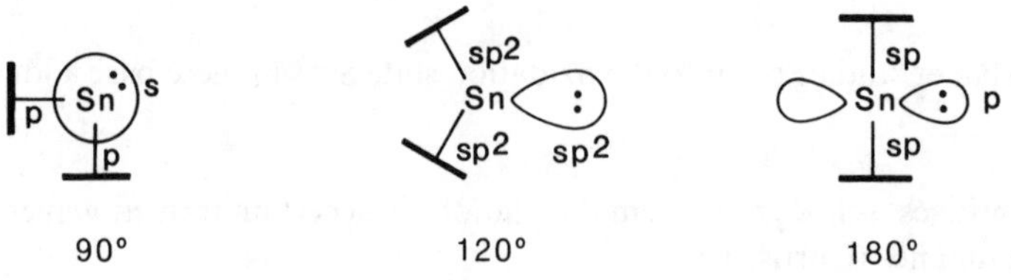

p, Sn, s, p — sp2, Sn, sp2, sp2 — sp, Sn, p, sp

90° — 120° — 180°

In this simple picture, stannocene is best described in terms of Sn(sp^2) hybridization; the sp^2 hybrid orbital not required for metal-ring bonding is occupied by a nonbonding pair of electrons. Additional contributions to metal-ring bonding may stem from π interactions between filled MO's e_1 (C_5H_5) and empty AO's of Sn which possess the appropriate symmetry (cf. p. 90). The high s electron density at the Sn nucleus, which follows from this bonding model, leads to strong magnetic shielding as shown by the appearance of ^{119}Sn-NMR signals at very high field (cf. p. 123). The observation of a coupling $^3J(^{119}Sn, ^1H)$ for $(C_5Me_5)_2Sn$ signifies an appreciable covalent contribution to the metal-ligand bond.

Ligand exchange between stannocene and tin dichloride leads to a half-sandwich complex (Noltes, 1975 R), whose bent structure can be interpreted along the lines given above for stannocene:

$$(\eta^5\text{-}C_5H_5)_2Sn + SnX_2 \longrightarrow 2\ (\eta^5\text{-}C_5H_5)SnX$$

X = Cl, Br

The numerous subsequent reactions of the stannocenes are shaped by the bonding alternatives η^1-C_5H_5 and η^5-C_5H_5, the change in oxidation state $Sn^{II} \rightarrow Sn^{IV}$ and the abstraction of a $C_5H_5^-$ ligand upon attack by a Lewis acid:

$(\eta^5\text{-}C_5H_5)_2Sn^{II}$

MeI, hν → $(\eta^1\text{-}C_5H_5)_2Sn^{IV}(Me)I$

$Fe_2(CO)_9$, −CO → $[(\eta^1\text{-}C_5H_5)_2Sn^{IV}Fe(CO)_4]_2$ (cyclic Sn_2Fe_2 dimer: $(CO)_4Fe$, Sn, Fe$(CO)_4$, Sn)

BF_3, THF → $F_3B\text{—}F\rightarrow Sn^{II}(C_5H_5)_2(\mu\text{-}C_5H_5)Sn^{II}\leftarrow O(THF)$

Contrary to an original assumption, the reaction of stannocene with BF_3 does not lead to the simple adduct $(C_5H_5)_2Sn \cdot BF_3$; instead, one cyclopentadienyl ligand is lost and and the binuclear complex with a bridging C_5H_5 unit is formed (Zuckerman, 1985). In the case of $(\eta^5\text{-}C_5Me_5)_2Sn$, ligand abstraction is effected by tetrafluoroboric acid; the resulting half-sandwich complex remains monomeric (Jutzi, 1979):

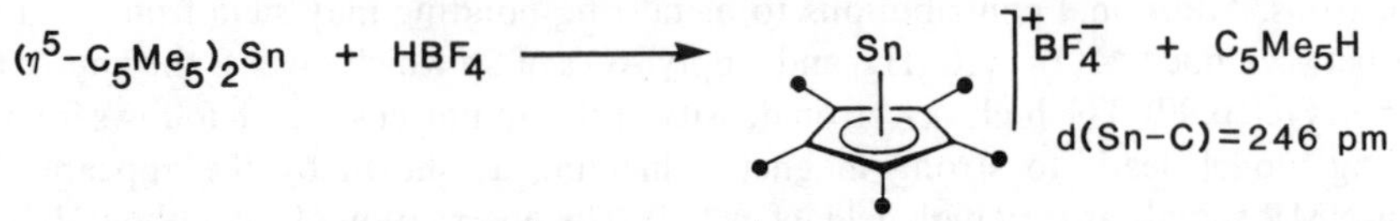

$$(\eta^5\text{-}C_5Me_5)_2Sn + HBF_4 \longrightarrow [(\eta^5\text{-}C_5Me_5)Sn]^+ BF_4^- + C_5Me_5H$$

d(Sn–C) = 246 pm

In the highly symmetric cation $[(\eta^5\text{-}C_5Me_5)Sn]^+$ the six π electrons of the ligand and the two valence electrons of Sn^{II} create an octet configuration at tin.
However, the ion $[(\eta^5\text{-}C_5Me_5)Sn]^+$ may also be regarded as a cluster, and its structure discussed in terms of Wade's rules (p. 68):

Number of vertices	Skeletal bonding electron pairs	Structural type	Geometry
n = 6 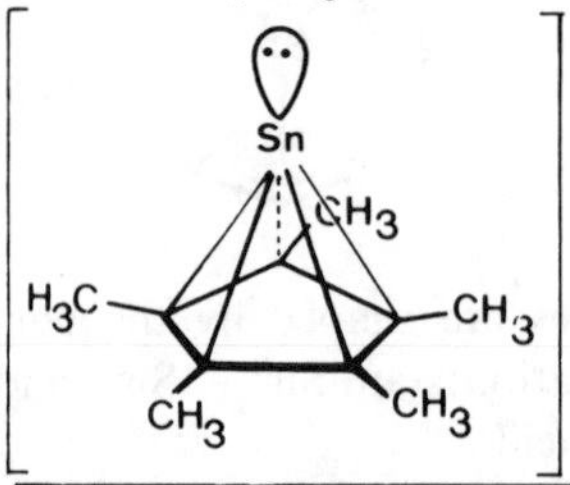	8 (n + 2)	nido	pentagonal bipyramid
	$\frac{1}{2}$ ·Sn(:)$^+$		one vertex unoccupied
	$\frac{15}{2}$ $(:C{-}CH_3)_5$		

As in the case of Ga, In and Tl (C_5H_5M versus $C_6H_6M^+$, p. 90), the question as to the existence of (arene)metal complexes must also be posed for the elements of group 14 ($C_5H_5M^+$ versus $C_6H_6M^{2+}$). In fact, from arenes, $SnCl_2$, and $AlCl_3$ two types of complexes of composition $(ArH)Sn(AlCl_4)_2 \cdot ArH$ and $(ArH)SnCl(AlCl_4)$, respectively, were obtained (Amma, 1979 R). Whereas the former type structurally corresponds to the analogous lead compound (p. 146), the latter type as yet is limited to tin.

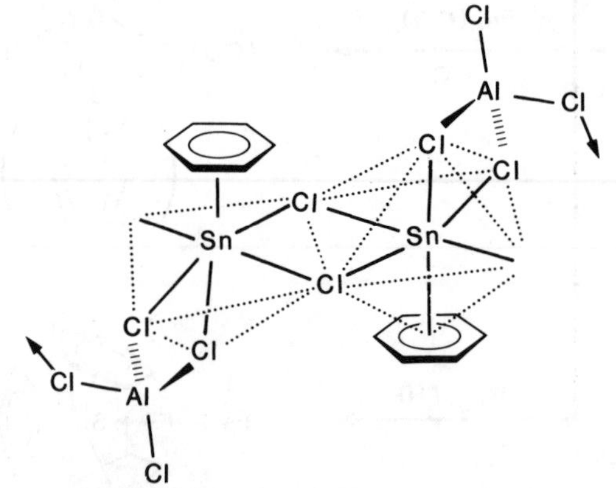

Section of the chain structure of $\mathbf{C_6H_6SnCl(AlCl_4)}$ *in the crystal. The backbone consists of a sequence of four-membered rings* ($Sn_2Cl_2^{2+}$) *and eight-membered rings (formed from one tin atom each of the neighboring* $Sn_2Cl_2^{2+}$ *units and two bridging* $AlCl_4^-$ *ions). The loose coordination of a* C_6H_6 *molecule creates a distorted octahedral environment at tin. Note here again, that the energetically favorable coordination of* Cl^- *ions to* Sn^{II} *is largely preserved.*

Technical Applications of Organotin Compounds

Tin organyls excell in the multitude of areas where they have found practical use. Since 1950 world production figures of organotin compounds have increased fifty-fold; 60% of the total output is used as stabilizers for polyvinylchloride, 30% as biocides and the rest in various special fields.

The toxicity of organotin compounds increases with the degree of alkylation and decreases with increasing chain length of the substituents. Maximal toxicity is therefore reached for materials which can release the ion Me_3Sn^+, compounds of the type $(n\text{-}Bu)_2Sn(SCH_2COO\text{-}iso\text{-}C_8H_{17})_2$ are virtually non-toxic. Plastics stabilizers R_2SnX_2 belong to the latter category. The mode of action of these stabilizers is thought to be the suppression of HCl release during the thermal processing of PVC at 180–200 °C by substituting chlorine atoms at the more reactive regions of the polymer by long chains. After processing, organotin additives also function as UV-stabilizers because of their capacity to terminate radical chain reactions. Another application is the generation of thin SnO_2 coatings on glass, accomplished by the treatment with Me_2SnCl_2 at 400–500 °C.

Compounds of the type R_3SnX serve as biocides in pest control, as fungicides, as antifouling agents in paint protection and, more generally, as disinfectants against gram-positive bacteria [examples: $Ph_3SnOOCH_3$, $(n\text{-}Bu)_3SnOOC(CH_2)_{10}CH_3$, $(n\text{-}Bu_3Sn)_2O$]. The final degradation to non-toxic $SnO_2 \cdot aq$ is an attractive aspect which has helped organotin compounds to find wide acceptance.

8.4 Organolead Compounds

The **technical use** of lead alkyls as gasoline additives, which still consumes about 20% of the annual world production of this metal, is a result of extensive industrial research and has led to the accumulation of a vast amount of factual material. Compared with the lighter homologues C, Si, Ge and Sn, the larger atomic radius of lead favors the realization of the higher coordination numbers 5–8. According to the metallic character of lead, solvated organolead cations are encountered in ionizing media. The tendency towards the formation of homonuclear metal-metal bonds (catenation) in the case of lead has vanished almost completely [$E(\text{Pb}-\text{Pb}) \approx 100$ kJ/mol]. In light of the fact that for purely inorganic compounds Pb^{II} is the favored oxidation state, the readiness of organolead compounds of intermediate oxidation state to disproportionate to Pb^0 and Pb^{IV} is remarkable. In terms of energetics, organolead chemistry is shaped by the **weakness and the polarizability of the Pb–C bond** [$E(\text{Pb}-\text{C}) = 152$ kJ/mol] which paves the way for radical as well as for ionic reactions.

NMR spectroscopy as applied to organolead compounds profits from the existence of the isotope ^{207}Pb (23%, $I = 1/2$). *The chemical shifts in* ^{207}Pb-*NMR spectra cover a range of* 1300 ppm; *the factors which lead to chemical shift differences resemble those for tin organyls. The coupling constants* $^2J(^{207}\text{Pb}, {}^1\text{H})$ *can be retrieved either from* ^{207}Pb-*NMR or from* ^{1}H-*NMR spectra* (^{207}Pb *satellites*). *They lie between* 62 Hz [Me_4Pb] and 155 Hz [$Me_2Pb(acac)_2$], *and provide insight into the* Pb(6*s*) *character in the* $Pb-CH_3$ *bond.*

8.4.1 Pb^{IV} Organyls

PREPARATION IN THE LABORATORY

$$PbCl_2 + 2\,MeMgI \xrightarrow{Et_2O} \{PbMe_2\} + 2\,MgClI$$ 4

$$\{PbMe_2\} \xrightarrow{\times 6} 3\,Me_4Pb + 3\,Pb$$

$$Pb(OAc)_4 + 4\,RMgCl \xrightarrow{THF,\ 5\,°C} R_4Pb + 4\,MgCl(OAc)$$

Complete consumption of the lead introduced into the reaction may be achieved by the addition of methyl iodide:

$$4\,MeLi + 2\,PbCl_2 \xrightarrow{Et_2O} Me_4Pb + 4\,LiCl + Pb$$
$$2\,MeI + Pb \longrightarrow Me_2PbI_2 \quad \text{(oxidative addition)}$$
$$2\,MeLi + Me_2PbI_2 \longrightarrow PbMe_4 + 2\,LiI$$

Mixed Pb^{IV} organyls are obtained as follows:

$$R_3PbCl + LiR' \longrightarrow R_3PbR' + LiCl$$ 4

$$R_3PbH + H_2C{=}CHR' \longrightarrow R_3PbCH_2CH_2R' \quad \text{(hydroplumbation)}$$ 8

Hexaorganodiplumbanes are accessible from alkylations at low temperature:

$$6\,RMgX + 3\,PbCl_2 \xrightarrow[-20\,°C]{Et_2O} R_3Pb{-}PbR_3 + Pb + 3\,MgX_2 + 3\,MgCl_2$$

$$R_3Pb{-}PbR_3 \xrightarrow{\times 2,\ 25\,°C} 3\,R_4Pb + Pb$$

INDUSTRIAL PROCESSES

$$4\,MeCl + 4\,NaPb \xrightarrow[autoclave]{110\,°C} Me_4Pb + 3\,Pb + 4\,NaCl$$ 1

A disadvantage of this **direct synthesis** is the fact that lead is only partially consumed; elemental lead formed has to be recycled. This is circumvented in a process which uses aluminum alkyls in the presence of a catalytic amount of Et_2Cd:

$$6\,Pb(OAc)_2 + 4\,Et_3Al \xrightarrow{HMPA} 3\,Pb + 4\,Al(OAc)_3 + 3\,Et_4Pb$$
$$3\,Pb + 6\,EtI + 3\,Et_2Cd \longrightarrow 3\,Et_4Pb + 3\,CdI_2$$
$$3\,CdI_2 + 2\,Et_3Al \longrightarrow 3\,Et_2Cd + 2\,AlI_3$$

Sum

$$6\,Pb(OAc)_2 + 6\,Et_3Al + 6\,EtI \longrightarrow 6\,Et_4Pb + 4\,Al(OAc)_3 + 2\,AlI_3$$

The role of Et_2Cd is that of an alkyl-transfer agent to Pb; in this function, it is not consumed but regenerated continuously by means of the reaction between the CdI_2 formed and the actual alkylating agent Et_3Al.
The Nalco process (USA) operated electrochemically:

anode (Pb)	$4\,EtMgCl_2^- \xrightarrow[-4\,MgCl_2]{-4e^-} 4\,Et\cdot \xrightarrow{Pb} Et_4Pb$
cathode (reactor made from Mg)	$4\,MgCl^+ \xrightarrow[-2\,MgCl_2]{+4e^-} 2\,Mg \xrightarrow{2\,EtCl} 2\,EtMgCl$
Sum	$2\,EtMgCl + 2\,EtCl + Pb \xrightarrow{THF} Et_4Pb + 2\,MgCl_2$

The commercial importance of these processes is declining; they are of interest, however, as examples of technical ingenuity.

PROPERTIES OF THE ORGANYLS R_4Pb

The simple organoplumbanes Me_4Pb (bp 110 °C/760 mbar) and Et_4Pb (bp 78 °C/10 mbar) are colorless, highly refractive, toxic liquids; Ph_4Pb (mp 223 °C) is a white crystalline solid. Under ambient conditions, these compounds are neither attacked by air or water nor affected by light; at elevated temperatures, they are cleaved into Pb, alkanes, alkenes and H_2. Thermal stability of R_4Pb decreases in the order $R = Ph > Me > Et >$ *iso*-Pr. The thermolysis of Me_4Pb to yield Pb and methyl radicals, and the transport of the lead mirror by means of the radicals present in the gas stream were the subject of a classical experiment (Paneth, 1929).

Tetraethyllead is (was) added to gasoline in a concentration of about 0.1 % *to serve as an antiknock compound. The action of* Et_4Pb *consists in the deactivation of hydroperoxides by* PbO, *as well as in the termination of radical chain reactions of the combustion process by the products of* Et_4Pb *homolysis*:

$$Et_4Pb \longrightarrow Et_3Pb\cdot + Et\cdot \quad (\text{rapid at } T > 250\,°C)$$
$$Et_4Pb + Et\cdot \longrightarrow C_2H_6 + Et_3PbCH_2CH_2\cdot \quad \text{etc.}$$

Another additive, $BrCH_2CH_2Br$, *converts* PbO *into volatile* Pb *compounds which then can exit the combustion chamber. In order to stem further contamination of the environment with toxic lead compounds, the use of methylcyclopentadienyl(tricarbonyl)manganese (p. 335) has been proposed as a substitute. Despite extensive research, the effectivity of tetraethyl lead, in use since* 1922, *has remained unsurpassed. In the long run, the total elimination of lead organyls as fuel additives (banned in the USA since* 1986, *still in use in some European and other countries) appears inevitable.*

The chemistry of the substitution products of binary lead organyls R_nPbX_{4-n} (X = Hal, OH, OR, NR_2, SR, H etc.) differs only moderately from the respective organotin chemistry. Analogous reactions usually proceed under milder conditions for lead as compared to tin.

Organolead halides R_nPbX_{4-n} are accessible through Pb–C cleavage by means of dihalogen or, even better, by hydrogen halides or thionyl chloride:

$$R_4Pb \xrightarrow[-RH]{HX} R_3PbX,\ R_2PbX_2$$

The reaction with HCl is 60 times faster for Ph_4Pb than for Ph_4Sn.
Compounds of the type $RPbX_3$ approach $PbCl_4$ in their instability. They can be prepared via oxidative addition of RI to PbI_2:

$$PbI_2 + MeI \longrightarrow MePbI_3$$ 1

R_3PbX and R_2PbX_2 in solution exist as solvated monomers; in the solid, polymer chains with halide bridges, resembling organotin halides (p. 125 f), are formed. The tendency to associate is more highly developed for the organolead compounds, however. Organolead carboxylates also form polymeric chain structures.

Organolead sulfinates arise through insertion of SO_2 into the Pb–C bond:

$$R_4Pb + SO_2 \longrightarrow R_3Pb{-}R \cdots O{=}S{=}O \longrightarrow R_3Pb{-}O{-}S({=}O){-}R$$

The readiness to increase the coordination number above four is expressed in the existence of numerous complex ions like the following:

$$[R_3PbX_2]^- \qquad [Ph_2Pb(O_2CCH_3)_3]^- \qquad Ph_2PbX_2(TMEDA) \qquad [Ph_3Pb(\mu\text{-}I)_2PbPh_3]^{2-}$$

O⁀O = CH_3COO^- N⁀N = $Me_2N(CH_2)_2NMe_2$ (TMEDA)

Organolead hydroxides $R_nPb(OH)_{4-n}$ are formed in the hydrolysis of the corresponding chlorides:

$$R_3PbCl + 1/2\,Ag_2O_{(aq)} \longrightarrow R_3PbOH + AgCl$$
$$R_2PbCl_2 + 2\,OH^- \longrightarrow R_2Pb(OH)_2 + 2\,Cl^-$$

$R_2Pb(OH)_2$, which prevails in the region $8 < pH < 10$, is **amphoteric.** At $pH > 10$ $[R_2Pb(OH)_3]^-$ is encountered, whereas at $5 < pH < 8$ $[R_2Pb(OH)]_2^{2+}$ and at $pH < 5$ $[R_2Pb(aq)]^{2+}$ are the dominating species. Dications $[R_2Pb(solv)]^{2+}$ are generated from R_2PbCl_2 in other media as well; the group **R_2Pb^{2+}**, like the isoelectronic moieties R_2Te^+ and R_2Hg, has a **linear** structure. The condensation of organolead hydroxides to form

plumboxanes occurs only sluggishly, it requires the continuous removal of water from the equilibrium mixture by distillation or by reaction with sodium metal:

$$2\,R_3PbOH \xrightarrow{Na,\ C_6H_6} (R_3Pb)_2O + NaOH + 1/2\,H_2$$

Conversely, plumboxanes undergo ready solvolytic cleavage; with alcohols, **organolead alkoxides** are obtained which form polymer chains in the solid state:

$$(R_3Pb)_2O \xrightarrow[-H_2O]{R'OH} 1/n\,(R_3PbOR')_n \xleftarrow[-NaCl]{NaOR'} R_3PbCl$$

Polymer chains of $[(CH_3)_3PbOMe]_n$

Organolead alkoxides add to polar C=C double bonds (**oxyplumbation,** Davies, 1967). Methanolysis of this adduct regenerates R_3PbOMe; therefore, in the addition of MeOH to C=C double bonds R_3PbOMe acts as a catalyst:

$$CH_2{=}CH{-}CN \xrightarrow{+Et_3PbOMe} CH_2(OMe){-}CH(PbEt_3){-}CN \xrightarrow[-Et_3PbOMe]{MeOH} CH_2(OMe){-}CH_2{-}CN$$

Lead, like tin, has a high affinity for sulfur; Pb–S bonds are stable in water:

$$2\,R_3PbX + NaHS \xrightarrow{0°,\,H_2O} (R_3Pb)_2S + NaX + HX$$

$$3\,Ph_2Pb(OAc)_2 + 3\,H_2S \longrightarrow (Ph_2PbS)_3 + 6\,HOAc$$

$$R_3PbCl + R'SH + Py \longrightarrow R_3PbSR' + [PyH^+]Cl^-$$

Organolead hydrides R_nPbH_{4-n} clearly demonstrate the gradation of chemical properties displayed by analogous Sn– and Pb compounds. Whereas Me_3SnH is indefinitely stable in an inert environment at room temperature, Me_3PbH decomposes at $-40\,°C$ into Pb, H_2, CH_4 and $Me_3Pb{-}PbMe_3$, decay being accelerated by light:

$$2\,Me_3PbH \xrightarrow{h\nu} 2\,Me_3Pb\cdot + H_2$$
$$2\,Me_3Pb\cdot \longrightarrow Me_3Pb{-}PbMe_3 \quad \text{(i.a.)}$$

Molecules with larger alkyl substituents like Bu_3PbH are somewhat more stable thermally, they all share the high sensitivity to air and light, however. Methods for their preparation include the following reactions:

$$R_3PbBr \xrightarrow{LiAlH_4,\ Et_2O} R_3PbH$$
$$(Bu_3Pb)_2O + Et_2SnH_2 \longrightarrow 2\,Bu_3PbH + (Et_2SnO)_n$$

The treatment of **organoplumbyl anions R_3Pb^-** with proton donors does **not** lead to organolead hydrides:

$$Ph_3PbCl \xrightarrow[-\,NaCl]{Na,\ NH_3(\ell)} NaPbPh_3 \xrightarrow[-\,NaBr]{NH_4Br,\ NH_3(\ell)} NH_4^+[PbPh_3]^-$$

From this and related reactions of the lighter homologues, the following order of decreasing **proton affinity** can be derived:

$$Ph_3Si^- > Ph_3Ge^- > Ph_3Sn^- > Ph_3Pb^- \quad (\text{compare } NH_3 > PH_3 > AsH_3)$$

The high reactivity of the Pb–H bond is apparent from the mild conditions under which the reduction of organo halogen- and carbonyl compounds by means of R_3PbH proceeds:

$$PhH\ (60\%) \xleftarrow[0°C,\ 5d]{PhBr} Bu_3PbH \xrightarrow[20°C,\ \text{instantaneously}]{PhCHO} PhCH_2OH\ (100\%)$$
$$PhCOCH_3 \xleftarrow[0°C,\ 1\,h]{PhCOCH_2Br} Bu_3PbH \xrightarrow[20°C,\ 2\,d]{PhCOCH_3} PhCHOHCH_3$$

For the reactions with organic halides, radical mechanisms are assumed; in the case of the reduction of carbonyl groups, polar mechanisms cannot be excluded. **Hydroplumbation** of alkynes and of conjugated alkenes takes place under mild conditions as well:

$$R_3PbH + HC{\equiv}CR' \xrightarrow{-50°} (R_3Pb)(H)C{=}C(H)(R')$$

$$Bu_3PbH + H_2C{=}CH{-}CN \longrightarrow \underset{Bu_3Pb}{CH_2}{-}CH_2CN + CH_3{-}CH(CN)(PbBu_3)$$

Medium	*Mechanisms*			
Et_2O, 0 °C	radical	92	:	8 %
BuCN, 20 °C	polar	24	:	76 %

As this comparison shows, changing the medium may induce the transition from a radical to a polar reaction mechanism with concomitant change in product composition. The homolytic hydroplumbation does not require the addition of a radical generating reagent, since initiation is effected by the homolysis of the Pb–H bond, which at 0 °C is already appreciable. Bu_3SnH, on the other hand, in the absence of an initiator (5 h, 0 °C) does not cause hydrostannation of $H_2C{=}CHCN$. This once again demonstrates the higher stability of organotin hydrides as compared to their lead counterparts.

8.4.2 Pb^{III}- and Pb^{II} Organyls

Organolead compounds, in which lead is in an oxidation state lower than four, are the organoplumbyl radicals $R_3Pb\cdot$, the organoplumbanes $R(R_2Pb)_nR$ and the plumbylenes R_2Pb. Because of the low Pb–Pb bond energy, compounds with **Pb_n chains** have been well characterized only in a few cases. Besides the aforementioned diplumbanes $R_3Pb{-}PbR_3$, the propane- and neopentane analoga $(c\text{-Hex})_8Pb_3$ (Dräger, 1986) and $(Ph_3Pb)_4Pb$ deserve a mention. The latter is formed in an obscure reaction:

$$Ph_3PbCl \xrightarrow[-\,LiCl]{Li,\,THF} Ph_3PbLi \xrightarrow[0\,°C]{H_2O,\,O_2} (Ph_3Pb)_4Pb$$

red, thermolabile

Hexaaryldiplumbanes, which are easily accessible from $PbCl_2$ and Grignard reagents (van der Kerk, 1970 R), undergo a variety of secondary reactions which lead into many classes of organolead compounds:

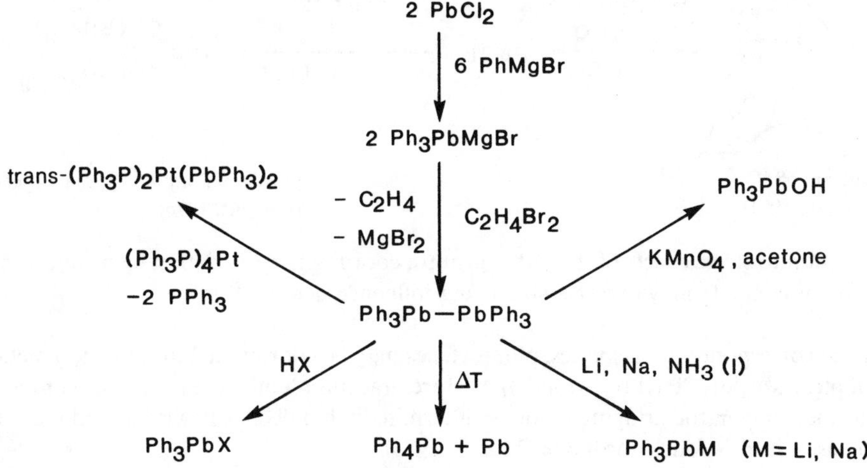

Despite the weakness of the Pb–Pb *bond, evidence for the establishment of a dissociation equilibrium* $R_3Pb{-}PbR_3 \rightleftharpoons 2\,R_3Pb\cdot$ *is absent; hexa(mesityl)diplumbane, dissolved in biphenyl, remains diamagnetic up to the temperature of decomposition* (≈300 °C). *This finding may be traced to small repulsion of the substituents in the dimer which possesses a long* Pb–Pb *bond and to ineffective resonance stabilization of the monomer.*

Organoplumbyl radicals $R_3Pb\cdot$ may, however, be generated in different ways and studied at low temperature by means of ESR spectroscopy:

$$Me_3PbCl \xrightarrow{Na,\,77K} Me_3Pb\cdot \xleftarrow{\gamma,\,77K} Me_4Pb$$

Like the corresponding radicals $Me_3E\cdot$ (E = Si, Ge, Sn), but unlike methyl radicals $R_3C\cdot$, $Me_3Pb\cdot$ has a pyramidal structure (cf. p. 120).

Apart from a few exceptions **Plumbylenes R_2Pb** only occur as reactive intermediates as in the case of (organo)oligoplumbane decomposition:

$$Me_3Pb{-}PbMe_3 \xrightarrow[-Me_4Pb]{RT,\ daylight} \{Me_2Pb\} \xrightarrow{Me_3PbPbMe_3} Me_3Pb{-}PbMe_2{-}PbMe_3 \longrightarrow Pb + 2\,Me_4Pb$$

With particular ligands monomeric diamagnetic plumbylenes may be obtained, however:

$$[2,4,6\text{-}(CF_3)_3C_6H_2]_2Pb \xleftarrow[-2\,LiCl]{(CF_3)_3C_6H_2Li,\ Et_2O} PbCl_2 \xrightarrow[-2\,LiCl]{2\,(Me_3Si)_2CHLi,\ Et_2O} :Pb[CH(SiMe_3)_2]_2$$

Pb 94.5°

yellow, mp 58 °C (Edelmann, 1991)

purple red, mp 44 °C, diamagnetic (Lappert, 1976)

In addition to steric effects of the bulky groups, coordinative saturation by intramolecular Pb–F contacts may excert a stabilizing influence here.

Like the corresponding tin species, plumbylenes may be coordinated to transition metals (example: $[(Me_3Si)_2CH]_2Pb{-}Mo(CO)_5$). More graceful plumbylenes occur as complexing ligands only in the bridging mode or, if terminally bonded, as Lewis base adducts. In both cases, lead is four-coordinate (Marks, 1973):

$$2\,Na_2Fe(CO)_4 + 2\,R_2PbCl_2 \xrightarrow{THF} (CO)_4Fe(\mu\text{-}PbR_2)_2Fe(CO)_4 \underset{}{\overset{B}{\rightleftharpoons}} 2\,(CO)_4Fe{-}PbR_2(\leftarrow B)$$

R = Et, n-Bu, Ph

B = THF, Py etc.

Transition-metal complexes containing bridging plumbylenes were already described by Hein in 1947 (*example*: $[Et_2PbFe(CO)_4]_2$).

Plumbocene, formally also a plumbylene, has been known for a long time (E. O. Fischer, 1956):

$$PbX_2 + 2\,NaC_5H_5 \xrightarrow{\text{DMF, THF}} (\eta^5\text{-}C_5H_5)_2Pb + 2\,NaX$$

(X = Cl, OAc)

yellow, sublimable, mp 138 °C,
air- and water-sensitive,
dipole moment $\mu = 1.2$ *Debye*
($1\,D = 3.33 \times 10^{-30}$ C m)

For isolated plumbocene molecules, a tilted structure similar to that of stannocene follows from dipole-moment measurements in solution as well as from electron-diffraction experiments in the gas phase. In the solid state, on the other hand, a polymeric chain structure is realized (X-ray diffraction, Panattoni, 1966).

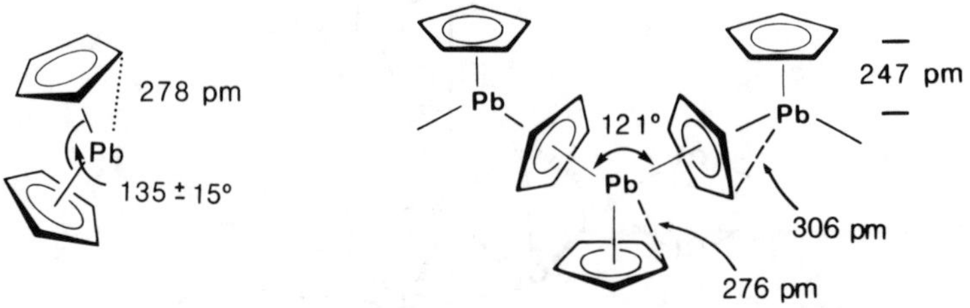

278 pm
Pb
135 ± 15°

Pb
121°
Pb
Pb
247 pm
306 pm
276 pm

Decamethylplumbocene is monomeric even in the crystal (cf. p. 47). Solutions of plumbocene in THF display electrical conductivity, pointing to partial dissociation (Strohmeier, 1962):

$$(C_5H_5)_2Pb \underset{}{\overset{\text{THF}}{\rightleftharpoons}} (C_5H_5)Pb(THF)^+ + (C_5H_5)^-$$

The cation in this equilibrium corresponds to the half-sandwich cations $(C_5Me_5)M^+$ (M = Ge, Sn, p. 119, 136) which, for the lighter homologues of Pb, could be isolated in quantity, albeit with other counterions. In its reactions plumbocene resembles stannocene:

$$(\eta^5\text{-}C_5H_5)_2Pb \xrightarrow{\text{ROH}} Pb(OR)_2 + 2\,C_5H_6$$

$$(\eta^5\text{-}C_5H_5)_2Pb \xrightarrow{\text{HX}} (\eta^5\text{-}C_5H_5)PbX \text{ (bent)}$$

$$(\eta^5\text{-}C_5H_5)_2Pb \xrightarrow{BF_3} (\eta^5\text{-}C_5H_5)_2Pb \cdot BF_3$$

As in the case of $(C_5H_5)_2Pb \cdot BF_3$, the high melting point of $(C_5H_5)PbCl$ (mp 330 °C) suggests a highly associated structure (Puddephatt, 1979):

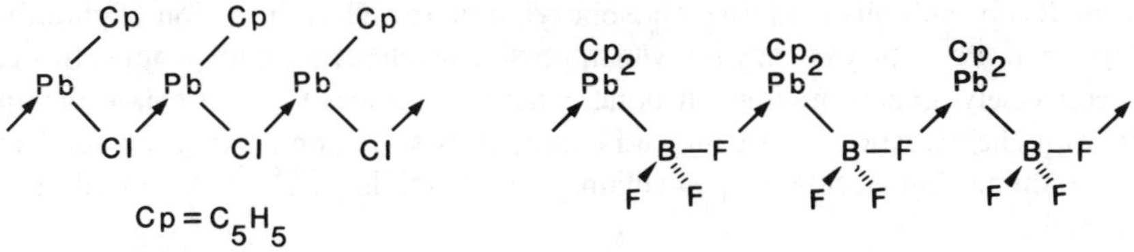

Cp Cp Cp
Pb Pb Pb
Cl Cl Cl

Cp = C_5H_5

Cp_2 Pb B—F F F (×3)

The preference for the lower oxidation state Pb^{II} manifests itself in the disparate behavior of stannocene and plumbocene in the reaction with methyl iodide:

$$CH_3I \xrightarrow{(\eta^5\text{-}C_5H_5)_2Sn} (\eta^1\text{-}C_5H_5)_2Sn(I)CH_3 \quad \text{oxidative addition}$$

$$CH_3I \xrightarrow{(\eta^5\text{-}C_5H_5)_2Pb} (\eta^5\text{-}C_5H_5)PbI + C_5H_5CH_3 \quad \text{substitution}$$

η^6-Coordination of arenes, which is weak for main-group metals (cf. p. 89 f), could also be demonstrated for Pb^{II} (Amma, 1974):

$$PbCl_2 + 2\,AlCl_3 \xrightarrow[\text{2. cryst}]{\text{1. } C_6H_6,\ 80\,°C} [(\eta^6\text{-}C_6H_6)Pb(AlCl_4)_2] \cdot C_6H_6$$

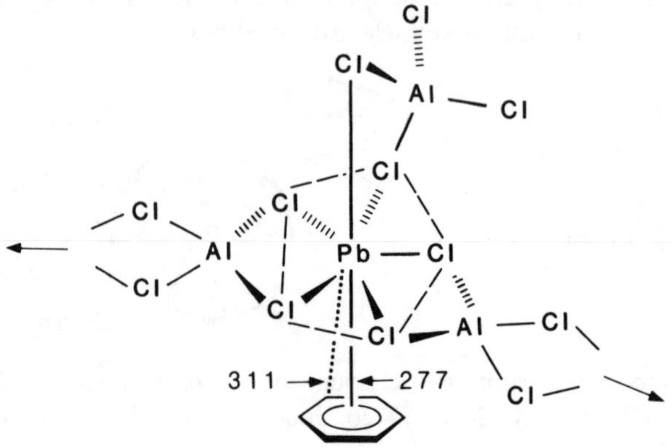

The structure consists of a chain of alternating Pb^{II} *ions and* $AlCl_4^-$ *tetrahedra; the coordination sphere of lead is completed by an additional bidentate* $AlCl_4^-$ *ion and an* η^6-C_6H_6 *ligand. The* Pb–C *distance, which is larger here as compared to plumbocene, signals a rather weak* Pb-*arene interaction. The second* C_6H_6 *molecule of the empirical formula resides outside the coordination sphere of lead.*

Organolead compounds in vivo

In principle, lead organyls share a number of applications with organotin compounds. However, the toxicity of organolead compounds seriously impedes their use as biocides or as stabilizers for plastics. These drawbacks were ignored in the case of gasoline additives and it is only recently that restrictive measures have been adopted.

As in the case of Hg (p. 54f) and Sn (p. 137), maximal toxicity is exhibited by the highly alkylated organolead cations R_2Pb^{2+} and R_3Pb^+. These species are generated via **biomethylation** of inorganic lead compounds and as metabolites of lead organyls in the organism. R_3Pb^+ inhibits oxidative phosphorylation as well as the action of glutathione transferases. R_2Pb^{2+} blocks enzymes which possess neighboring thiol groups. Inorganic Pb^{2+}, conversely, is accumulated in bone, where it replaces Ca^{2+}. **Chelate therapy** is ill-suited for the excretion of organolead compounds since complexing ligands like diethylenetriamine pentaacetate or penicillamine only mobilize Pb^{2+} ions bound in skeletal materials.

9 Organoelement Compounds of the Nitrogen Group (Group 15)

The broadly encompassing term "organoelement chemistry" is justified for group 15 in particular since the organic chemistry of the nonmetal phosphorus, the metalloids arsenic and antimony and the metal bismuth is intimately related. As far as practical and synthetic applications are concerned, the chemistry of As-, Sb- and Bi-organyls is overshadowed by the organometallics of groups 13 and 14. The attention the pharmacological activity of organoarsenic compounds received at the beginning of this century, has dissipated since the discovery of antibiotics. Other conceivable applications are seriously hampered by the high **toxicity of As-, Sb- and Bi organyls.** Addition of organoelement hydrides to unsaturated systems, which as hydroboration, hydroalumination and hydrostannation has proved invaluable in organic synthesis, knows no counterpart in group 15 since the organoelement hydrides of P, As, Sb and Bi are highly labile compounds. The organyls of group 15 are of considerable importance, however, as **σ-donor/π-acceptor ligands in transition-metal complexes.** Another topic of recent interest is the incorporation of P, As, Sb, Bi into **$E{=}E\,(p_\pi-p_\pi)$** and **$E{=}C\,(p_\pi-p_\pi)$ multiple bonds.** Limited practical utility, combined with a large variety of structural and bonding peculiarities have developed the study of group 15 organometallics into what one may call "chemistry for chemists".

The structural multiplicity is a consequence of the fact that in group 15, **two oxidation states E^{III} and E^{V}** both form extensive series of organoelement compounds, whereas in groups 13 and 14 the element organyls of the respective lower oxidation states are species of somewhat exotic nature. Additional features are the variability of the coordination number, the existence of compounds with a mixed ligand sphere (e.g. $R_nE^VX_{5-n,\,n=1-4}$) and the accompanying isomerisms, as well as the formation of oligomers or polymers by means of Lewis acid/base interaction or via condensation reactions. An inconsistent nomenclature (e.g. AsH_3 = arsane, arsine, arsenic hydride) adds to the unwieldiness of the field.

As is the organoelement chemistry of groups 13 and 14, that of group 15 is governed by the decrease in energy of homonuclear (E–E) as well as of heteronuclear bonds (E–C, E–H) in the order E = P > As > Sb > Bi (cf. p. 11). In the same order E–C bond polarity increases.

Structural elucidation of organoantimony compounds has profited from **^{121}Sb-Mössbauer spectroscopy. ^{75}As-NMR** (100%, $I = 3/2$) has as yet only been applied sporadically.

At variance with groups 13 and 14, the organyls of As, Sb, Bi will be discussed jointly; organophosphorus compounds will only be mentioned where interesting parallels to the chemistry of the heavier homologues arise.

9.1 Element^V Organyls (E = As, Sb, Bi)

The following survey lists selected examples of E^V organyls (X = Hal, OH, OR, 1/2 O). Organoelement hydrides are unknown for the oxidation state E^V. For the compounds marked with an asterisk, X-ray structural determinations have been performed.

	As^V	Sb^V	Bi^V
$[R_6E]^-$	$Li^+[As(2,2'\text{-}PhPh)_3]^-$	$Li^+[SbPh_6]^-$	$Li^+[BiPh_6]^-$
R_5E	*Ph_5As Arsorane	*Ph_5Sb Stiborane	Ph_5Bi Bismorane
$[R_4E]^+$	$[Me_4As]^+Br^-$ Arsonium	$[Ph_4Sb]^+[BF_4]^-$ Stibonium	$[Ph_4Bi]^+[BPh_4]^-$ Bismuthonium
R_4EX	Me_4AsOMe	*Me_4SbF, Ph_4SbOH	Ph_4BiONO_2
R_3EX_2	*Me_3AsCl_2 $(Ph_3AsO \cdot H_2O)_2$	*Me_3SbF_2 $Ph_3Sb(OMe)_2$	*Ph_3BiCl_2 Ph_3BiO
R_2EX_3	Me_2AsCl_3 *$Me_2AsO(OH)$ Arsinic acid	*Ph_2SbBr_3 $Ph_2SbO(OH)$ Stibinic acid	
REX_4	$PhAsCl_4$ $PhAsO(OH)_2$ Arsonic acid	$MeSbCl_4$ $PhSbO(OH)_2$ Stibonic acid	

9.1.1 Pentaorganoelement Compounds R_5E

PREPARATION

Due to the strongly oxidizing character of the halides EX_5, the peralkyls ER_5 are not accessible by means of the reaction between EX_5 and RLi or RMgX. Rather, they have to be prepared in a two-step process:

$$Me_3As \xrightarrow{Cl_2} Me_3AsCl_2 \xrightarrow{MeLi,\ Et_2O} Me_5As \quad (Me_5Sb\text{, resp.})$$ 4

$$Ph_3As \xrightarrow{PhI} [Ph_4As]^+I^- \xrightarrow[-LiI]{PhLi} Ph_5As$$

$$Ph_3Sb \xrightarrow{Cl_2} Ph_3SbCl_2 \xrightarrow[-3\,LiCl]{3\,PhLi} Li^+[SbPh_6]^- \quad (\text{"ate complex"})$$

$$Li^+[SbPh_6]^- \xrightarrow[-LiOH,\ -PhH]{H_2O} Ph_5Sb$$

$$Ph_3Bi \xrightarrow[-SO_2]{SO_2Cl_2} Ph_3BiCl_2 \xrightarrow[-MgX_2]{PhMgX} Ph_5Bi$$

These reactions may also be used in the preparation of ternary compounds $R_3R'_2E$. Species of the type $R_4R'E$ are obtained according to:

$$R_5E \xrightarrow[-RX]{X_2} R_4EX \xrightarrow[-MgX_2]{R'MgX} R_4R'E$$

and quaternary compounds $R_3R'R''E$ as follows:

$$R_3E \xrightarrow{R'X} R_3R'EX \xrightarrow[-LiX]{R''Li} R_3R'R''E$$

In additions of alkyl halides R'X to R_3E, the following orders of reactivity are observed:

$$E = As > Sb \gg Bi, \quad R = \text{alkyl} > \text{aryl}$$
$$X = I > Br > Cl, \quad R' = \text{alkyl} > \text{aryl}$$

The addition of aryl halides ArX is promoted by $AlCl_3$.

STRUCTURES AND PROPERTIES

The pentaorganyls are thermally rather stable:

$$Me_5As \xrightarrow{T > 100°} Me_3As + CH_4 + C_2H_4$$

$$Ph_5Sb \xrightarrow{T > 200°} Ph_3Sb + Ph{-}Ph$$

Compared to organoarsanes and -stibanes (R_3As and R_3Sb), the air-sensitivity of organoarsoranes and -stiboranes (R_5As and R_5Sb) is considerably attenuated.
The pentaalkyl derivatives are easily hydrolyzed, however:

$$Me_5As + H_2O \longrightarrow Me_4AsOH + MeH$$

A characteristic feature of the species R_5E is the ready formation of **onium ions**:

$$Ph_5E + BPh_3 \longrightarrow [Ph_4E]^+[BPh_4]^-$$
$$E = As, Sb, Bi$$

and of **ate complexes**:

$$Ph_5E + LiPh \longrightarrow Li^+[EPh_6]^-$$

Of the two complementary geometries of coordination number 5, in the crystal the trigonal bipyramid is adopted by R_5E almost exclusively, an exception being pentaphenylstiborane Ph_5Sb:

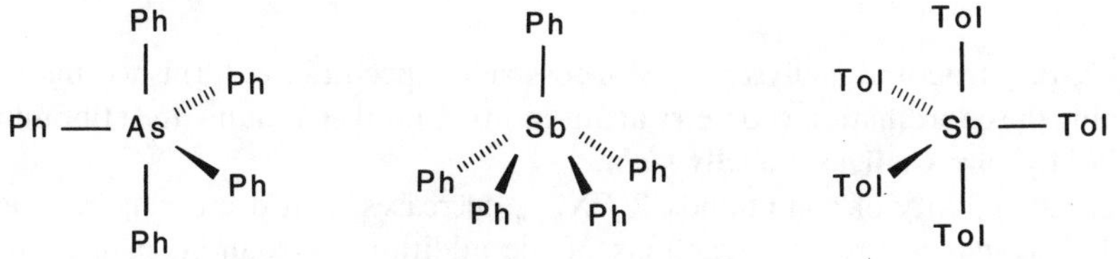

The fact that pentaphenylstiborane, as opposed to trigonal-bipyramidal penta(p-tolyl)stiborane, assumes a square-pyramidal structure demonstrates, however, that the two complementary geometries differ in energy only marginally. The observation that in the ^{1}H-NMR spectrum of $(CH_3)_5Sb$ *even at* $-100\,°C$ *all protons appear equivalent, points in the same direction. Obviously, the exchange between axial and equatorial environments of the trigonal*

bipyramid, which only requires changes in bond angles (**pseudorotation**, *Berry mechanism*), *is a process equipped with a low activation barrier.*
As expected, the onium ions R_4E^+ *have tetrahedral and the ate-complex anions* R_6E^- *octahedral geometry.*

9.1.2 Organoelement Derivatives R_nEX_{5-n}

If arsoranes and stiboranes, in addition to organic residues, contain groups X (X = Hal, OH, OR etc.) which possess lone pairs of electrons, the structural picture becomes more complicated. Methods of preparation for the case X = halogen have already been mentioned. Species R_nEX_{5-n} realize at least four different structural variants which reflect the nature of E, R and X:

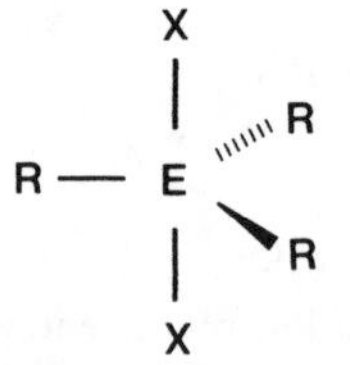

covalent monomeric, *trigonal bipyramids, electronegative groups at axial positions, example*: Me_3AsCl_2

covalent dimeric, *edge sharing octahedra, organic groups at axial position, example*: Ph_2SbCl_3

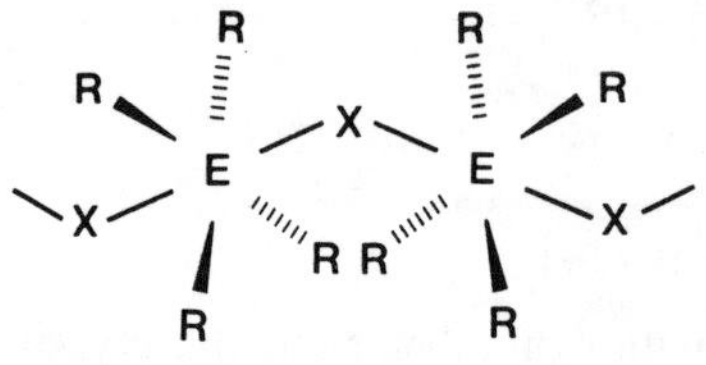

covalent polymeric, *vertex sharing octahedra, halide at bridging position, example*: Me_4SbF

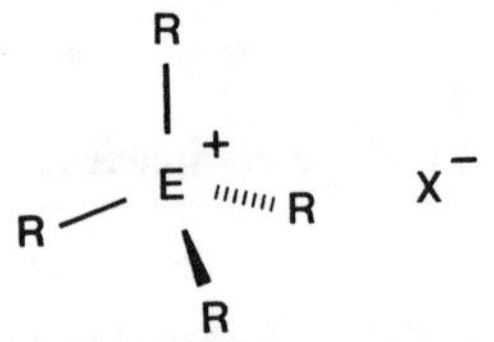

ionic, *tetrahedra, examples*: $[Ph_4As]^+I^-$
$[Me_4Sb]^+[SbCl_6]^-$
$\equiv(Me_2SbCl_3)_2$

Besides X-ray structural analyses, ^{121}Sb-Mössbauer spectra have furnished information concerning the coordination geometry at antimony. Chiral arsonium- and stibonium ions $[RR'R''R'''E]^+$ are **configurationally stable.**
The thermal stability of compounds R_nEX_{5-n} decreases with decreasing *n*. Thermally induced cleavages are reverse reactions of the additions introduced in a preparative context in section 9.1.1:

$$R_3SbX_2 \xrightarrow{\Delta T} R_2SbX + RX$$

$$Ph_2AsCl_3 \xrightarrow[100\,°C]{CO_2} Ph_2AsCl + Cl_2$$

$$Me_2AsCl_3 \xrightarrow[50\,°C]{} MeAsCl_2 + MeCl$$

The more labile organoelement halides may, however, be stabilized as Lewis base adducts (example: $MeSbCl_4 \cdot$ pyridine).
Solvolysis of species R_nEX_{5-n} by water, alcohols, amines etc. leads to several classes of organoelement compounds. Thus, hydrolysis of R_4AsCl yields strongly basic, extensively dissociated **arsonium hydroxides**:

$$R_4AsCl \xrightarrow[-AgCl]{Ag_2O,\ H_2O} [R_4As]^+OH^-$$

The tetraphenylarsonium ion Ph_4As^+ is frequently used to precipitate large anions from aqueous solution. The arsenobetain $Me_3As^{\oplus}CH_2COO^{\ominus}$ has been detected in the abdominal muscle of the rock lobster *Panulirus longipes cygnus George* (Cannon, 1977).
Hydrolysis of the dichloride Ph_3AsCl_2 affords a triphenylarsane oxide hydrate which has an interesting dimeric structure (Ferguson, 1969):

$$2\ Ph_3AsCl_2 \xrightarrow{H_2O} Ph_3As{=}O \cdots (H{-}O{-}H)_2 \cdots O{=}AsPh_3$$

In view of this participation of Ph_3AsO in strong hydrogen bonding, it comes as no surprise that triphenylarsane oxide forms stable adducts with a number of Lewis acids [example: $(Ph_3AsO)_2NiBr_2$].
Hydrogen bonding is also responsible for the dimeric nature of **arsinic acids $R_2AsO(OH)$** which can be prepared by a combination of oxidation and hydrolysis, starting from As^{III}-compounds:

$$PhN_2^+X^- + PhAsCl_2 \xrightarrow[-N_2]{H_2O} Ph_2AsO(OH)$$

$$Me_2AsCl \xrightarrow{H_2O_2} Me_2AsO(OH)$$

$$Me_2As(=O)(O{-}H\cdots O{=})AsMe_2 \text{ (cyclic dimer: } O{-}H\cdots O \text{ bridges)}$$

The dimerization of arsinic acids $R_2AsO(OH)$ *by means of hydrogen bonding resembles that of carboxylic acids.*

Arsinic acids are amphoteric:

$$[Me_2AsO_2]^- \underset{OH^-}{\overset{H_2O}{\rightleftharpoons}} Me_2AsO(OH) \underset{H_2O}{\overset{H_3O^+}{\rightleftharpoons}} [Me_2As(OH_2]^+$$

Stibinic acids $R_2SbO(OH)$, which are polymeric, are accessible analogously to arsinic acids; they are also obtained via hydrolysis of R_2SbCl_3.

Among the organoelement-oxygen compounds of group 15, it is the arsonic- and stibonic acids which have been studied most extensively. This is an outcome of the discovery by Thomas in 1905 that *p*-aminophenylarsonate, *p*-$H_2NC_6H_4AsO(OH)ONa$ (Atoxyl) is effective in the chemotherapy of sleeping sickness by fighting the trypanosomes. Even today, certain substituted arylarsonic acids on a limited scale serve as herbicides, fungicides and bactericides. **Arsonic acids $RAsO(OH)_2$** – like arsinic acids $R_2AsO(OH)$ – are prepared from As^{III}-compounds. The following reaction may formally be described as an oxidative addition of ArX to As^{III} and subsequent hydrolysis:

$$ArN_2^+X^- + As(ONa)_3 \xrightarrow[2.\ H_3O^+_{aq}]{1.\ OH^-_{aq}} ArAsO(OH)_2 \qquad \text{Bart reaction}$$

Arsonic acids are acids of medium strength, they are highly resistant towards oxidation and form polymeric anhydrides $(RAsO_2)_n$. In analytical chemistry, phenylarsonic acid is utilized as a reagent for the **precipitation** of high-valent metal ions (Sn^{IV}, Zr^{IV}, Th^{IV}). **Stibonic acids $RSbO(OH)_2$** which are accessible in a way similar to that for $RAsO(OH)_2$, may be reversibly converted into $RSbCl_4$:

$$PhSbO(OH)_2 + 4\,HCl \rightleftharpoons PhSbCl_4 + 3\,H_2O$$

9.2 Element^{III} Organyls (E = As, Sb, Bi)

Selected examples of E^{III} organyls (X = Hal, OH, OR, 1/2 O, H):

	As^{III}	Sb^{III}	Bi^{III}
R_3E	Ph_3As $(CF_3)_3As$ $(\eta^1\text{-}C_5H_5)_3As$ Arsane (Arsine)	Ph_3Sb $(n\text{-}C_3H_7)_3Sb$ $(CH_2{=}CH)_3Sb$ Stibane (Stibine)	Ph_3Bi Me_3Bi Bismuthane (Bismuthine)
$[R_2E]^-$	$[Me_2As]K$	$[Ph_2Sb]Na$	$[Ph_2Bi]Na$
R_2EX or condensation product	Ph_2AsCl Me_2AsOR $(Ph_2As)_2O$ Me_2AsH	Ph_2SbF $(Me_2Sb)_2O$ Me_2SbH	Me_2BiBr Me_2BiH
REX_2 or condensation product	$PhAsCl_2$ $MeAs(OR)_2$ $(MeAsO)_n$ $MeAsH_2$	$MeSbCl_2$ $MeSb(OR)_2$ $(PhSbO)_n$ $PbSbH_2$	$PhBiCl_2$

For the oxidation state E^{III}, the number of classes of compounds R_nEX_{3-n} is, of course, smaller than for the oxidation state E^V where the variability of n is greater. Furthermore, Lewis acidity is weak for compounds containing E^{III}; therefore, ate-complexes of the type R_4E^- are not encountered.
On the other hand, the species R_3E possess pronounced σ-donor and π-acceptor properties, which has led to their proliferation in coordination chemistry. Another characteristic is the existence of organoelement hydrides R_nEH_{3-n} having no counterpart in E^V chemistry. Selected examples of E^{III} organyls are listed on p. 152.

9.2.1 Trisorganoelement Compounds R_3E

PREPARATION

$$2\,As + 3\,MeBr \xrightarrow[Cu]{\Delta T} Me_2AsBr + MeAsBr_2 \quad \text{(direct synthesis)}$$

$$Me_2AsBr \xrightarrow{PhLi} Me_2AsPh \qquad MeAsBr_2 \xrightarrow{PhLi} MeAsPh_2 \quad \text{ternary organoarsanes} \qquad \boxed{1}$$

$$\underset{E = As,\ Sb,\ Bi}{EX_3} + \underset{(RLi)}{3\,RMgI} \xrightarrow{THF} R_3E + \underset{(LiX)}{3\,MgIX} \qquad \boxed{4}$$

$$AsCl_3 + HC{\equiv}CH \xrightarrow{AlCl_3} \underset{(n = 1,\ 2,\ 3)}{(ClCH{=}CH)_nAsCl_{3-n}} \quad \text{(Insertion)} \qquad \boxed{9}$$

Synthesis of a chelating ligand:

$$Me_2SbCl \xrightarrow[\text{oder } LiAlH_4]{Zn/HCl} Me_2SbH \xrightarrow{Na/THF} Me_2SbNa \xrightarrow{o\text{-}C_6H_4Cl_2} o\text{-}C_6H_4(SbMe_2)_2$$

STRUCTURES AND PROPERTIES

The element trisalkyls R_3E, liquids at room temperature if R is small (Me_3As, bp 50 °C), differ from the pentaalkyls R_5E in their increased air-sensitivity; trisalkylstibanes R_3Sb and -bismuthanes R_3Bi are even pyrophoric. The element aryls Ph_3E, solids at room temperature, are considerably more resistant; their oxidation to Ph_3EO calls for the action of $KMnO_4$ or H_2O_2. The sensitivity towards oxidation increases in the order $R_3As < R_3Sb < R_3Bi$. Water does not attack trisorganoelement compounds R_3E.

Symptoms of poisoning *in rooms which were decorated with wall-paper containing the pigment "Schweinfurter Grün"* $[3\,Cu(AsO_2)_2 \cdot Cu(CH_3COO)_2]$ *were traced back to gaseous organoarsenic compounds generated by the mold Penicillium brevicaule* (Gosio, 1897). *It was not until* 1932 *that Challenger identified trimethylarsane* Me_3As *as the active component of "Gosio gas". His fundamental studies concerning the* ***biomethylation*** *of inorganic ions were subsequently extended to many other elements* (*cf.* Hg *p.* 55, 203; Pb *p.* 146, Se *p.* 171).

Some typical reactions of the element trisalkyls R_3E are presented in the following scheme, using trimethylstibane as an example:

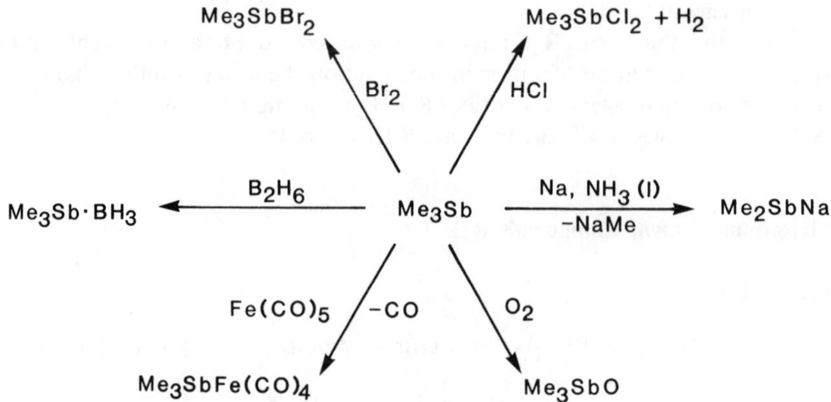

Based on their distinctive σ-donor/π-acceptor character, the species R_3E have found extensive use as ligands for transition metals, complex stability decreasing in the order $R_3P > R_3As > R_3Sb > R_3Bi$.
Trisorganoarsanes, -stibanes, and -bismuthanes R_3E have pyramidal structures; the bond angles C–E–C exceed the angles H–E–H of the respective hydrides EH_3; they decrease according to $R_3As > R_3Sb > R_3Bi$ [example: $(p\text{-}ClC_6H_4)_3E$, angle C–E–C = 102°, 97°, 93° respectively]. Chiral molecules RR′R″E are configurationally stable; for Me(Et)PhAs the inversion barrier amounts to 177 kJ/mol.
As in the case of the species $Me_3M(\eta^1\text{-}C_5H_5)$ (M = Si, Ge, Sn, cf. p. 99), **metallotropic 1, 2 shifts** are also observed for the compounds $Me_2E(\eta^1\text{-}C_5H_5)$ (E = As, Sb). They have the effect that at a characteristic temperature, the ring protons become equivalent on the ^{1}H-NMR time-scale.

$C_5H_5SiMe_3$ (H, $SiMe_3$) + Me_2AsCl ⟶ $C_5H_5AsMe_2$ (H, $AsMe_2$) + Me_3SiCl

$3\ C_5H_5Na + AsCl_3$ $\xrightarrow[0°]{\text{pentane}}$ $As(C_5H_5)_3$ (As, H, H, H) + 3 NaCl

In the triscyclopentadienyl compounds $(\eta^1\text{-}C_5H_5)_3E$ (E = As, Sb, Bi) – *highly sensitive species prone to polymerization – this dynamic process involves all three ligands; for* $(C_5H_5)_3As$ *it is slow at* T < – 30°. *For* $(C_5H_5)_3Sb$ *and* $(C_5H_5)_3Bi$, *the slow exchange region has not yet been reached by lowering the temperature. This concords with the barrier of 1,2-migration decreasing in the order* As > Sb > Bi, *which reflects decreasing* E–C *bond energy.*

9.2.2 Organoelement Derivatives R_nEX_{3-n}

For organoelement^{III} halides a number of special **preparative methods** must be mentioned in addition to the general procedures introduced in section 9.2.1. Partial alkylation requires controlled reaction conditions and sterically demanding groups:

$$SbBr_3 + t\text{-}BuMgCl \xrightarrow{-50\,°C} t\text{-}BuSbBr_2 + MgClBr$$

$$SbCl_3 + 2\,t\text{-}BuMgCl \xrightarrow{0\,°C} (t\text{-}Bu)_2SbCl + 2\,MgCl_2$$

The following transfer of organic residues from silicon compounds to arsenic- or antimony halides is driven by the energy of the silicon-halogen bonds formed:

$$ECl_3 + C_5H_5SiMe_3 \longrightarrow C_5H_5ECl_2 + Me_3SiCl \quad (E = As, Sb)$$

$$2\,PhSi(OEt)_3 + SbF_3 + 10\,HF \xrightarrow{H_2O} Ph_2SbF + 2\,H_2SiF_6 + 6\,EtOH$$

Halomethyl derivatives of arsenic arise from carbene insertion:

$$AsCl_3 + n\,CH_2N_2 \xrightarrow[-N_2]{} (ClCH_2)_nAsCl_{3-n} \quad (n = 1, 2, 3)$$

Compounds of the type $\mathbf{R_2EX}$ are also obtained according to:

$$Me_2PhAs + HI \longrightarrow Me_2AsI + PhH$$

$$R_3E + X_2 \longrightarrow R_3EX_2 \xrightarrow[-RX]{\Delta T} R_2EX$$

Upon the action of BF_3 on $(C_5Me_5)_2AsF$, an **arsenocenium cation** is formed which – like the isoelectronic decamethylgermanocene (p. 119) – has a tilted sandwich structure (Jutzi, 1983):

$$(C_5Me_5)_2AsF + BF_3 \longrightarrow (C_5Me_5)_2As^+BF_4^-$$

$$\sphericalangle\ \text{ring/As/ring} = 143°$$

Compounds $\mathbf{REX_2}$ are rationally prepared via E–C cleavage by hydrogen halide:

$$MePh_2As + 2\,HI \longrightarrow MeAsI_2 + 2\,PhH$$

This selective cleavage of the As–Ph *bond is surprising in view of the relative strengths of* As–C *bonds* [E(As–Ph) = 280 kJ/mol, E(As–Me) = 238 kJ/mol]. *Apparently, it is the group with the superior carbanion stabilization which is preferentially eliminated.*

The reduction of arsonic acids or stibonic acids in the presence of a hydrogen halide also leads to species REX_2:

$$PhEO(OH)_2 \xrightarrow{SO_2,\ HCl} PhECl_2 \quad (E = As, Sb)$$

Like the binary compounds R_3E and EX_3, the organoelement halides R_nEX_{3-n} (X = Cl, Br, I; n = 1,2) are pyramidal; they usually crystallize in molecular lattices. Ph_2SbF is an exception:

In the crystal, $\mathbf{Ph_2SbF}$ *exhibits a chain structure which is caused by strong* Sb—F → Sb *bridges. Four bonding and one lone pair of electrons create a distorted trigonal-bipyramidal environment at antimony* (*Sowerby*, 1979).

(2-Chlorovinyl)dichloroarsane (Lewisite) which is produced from HC≡CH and $AsCl_3$ in the presence of $AlCl_3$, was used as a chemical warfare agent in the First World War. Detoxification is effected by the reaction with 1,2-dithiols:

$$Cl_2As{-}CH{=}CHCl + \underset{\text{SH}\quad\text{SH}\quad\text{OH}}{H_2C{-}CH{-}CH_2} \xrightarrow{-2\,HCl} \text{(}H_2C{-}CH(CH_2OH)\text{ bridged by S, S to As–CH=CHCl)}$$

BAL
British Anti-Lewisite

Other organoarsenic halides, which served similar abhorrent purposes, include Ph_2AsCl (*Clark I*) *and* Ph_2AsCN (*Clark II*).

The **hydrolysis of organoelement^{III} halides** leads to ill-characterized hydroxo compounds, which undergo further reactions:

$$R_2AsX \xrightarrow{H_2O} R_2AsOH + (R_2As)_2O$$

Arsinous acid known in esters R_2AsOR' — Bis(diorganoarsane)oxide (Cacodyl oxide)

analogously: Stibinous acid + Bis(diorganostibane)oxide

$$RAsX_2 \xrightarrow{H_2O} RAs(OH)_2 + 1/n\ (RAsO)_n$$

Arsonous acid known in esters $RAs(OR')_2$ — organoarsane oxide

analogously: Stibonous acid + Organostibane oxide

The polymeric structure of organoarsane oxides $(RAsO)_n$, *which contrasts with the monomeric nature of nitroso compounds RNO, reflects the unfavorable energetics of a* $p_\pi-p_\pi$ *bond in monomeric* R–As=O. *In the case of* $d_\pi-p_\pi$ *interactions, orbital overlap is more effective. The large bond angle* As–O–As (137°) *in bis(diphenylarsane)oxide argues for the participation of canoncial forms with* As=O ($d_\pi-p_\pi$) *bonds*:

$$\left\{ Ph_2As{-}\bar{\underline{O}}{-}AsPh_2 \longleftrightarrow Ph_2As^{\ominus}{=}\bar{O}^{\oplus}{-}AsPh_2 \longleftrightarrow Ph_2As{-}{}^{\oplus}\bar{O}{=}AsPh_2^{\ominus} \right\}$$

Organoelement^{III} hydrides R_nEH_{3-n} are highly reactive, air-sensitive compounds, whose stabilities decrease according to E = As > Sb > Bi, R = alkyl > aryl and $n = 2 > 1$. Due to the low polaritiy of the E–H bonds, they are stable towards hydrolysis, however. A general way of access consists in the substitution of X by H:

$$MeAsCl_2 \xrightarrow{Zn/Cu,\ HCl} MeAsH_2 \quad (bp\ 2\,°C)$$

$$Me_2SbBr \xrightarrow[THF,\ -60\,°C]{LiBH(OMe)_3} Me_2SbH \quad (decomp.\ 30\,°C)$$

$$MeBiCl_2 \xrightarrow[Me_2O,\ -110\,°C]{LiAlH_4} MeBiH_2 \quad (decomp.\ -45\,°C)$$

The reduction of organoelement^V compounds also leads to the respective hydrides:

$$Me_2AsO(OH) \xrightarrow[-ZnCl_2]{Zn,\ HCl} Me_2AsH \quad (bp\ 37\,°C)$$

$$PhAsCl_4 \xrightarrow[-H_2]{LiBH_4,\ Et_2O} PhAsH_2 \quad (bp\ 148\,°C)$$

A few typcial reactions of the compounds R_nEH_{3-n} will be illustrated for the case of $MeAsH_2$:

$$MeAsH_2 \xrightarrow{O_2} (MeAs)_n \xrightarrow{O_2} (MeAsO)_n \xrightarrow{O_2,\ H_2O} MeAsO(OH)_2$$

$$MeAsH_2 \xrightarrow{B_2H_6} MeH_2As \cdot BH_3$$

$$MeAsH_2 \xrightarrow[-H_2]{Na,\ NH_3\ (l)} MeHAsNa$$

Organoelement hydrides of group 15 add to >C=C< multiple bonds; these reactions are of no practical importance, however.

9.3 Chains and Rings containing E–E Single Bonds

A glance at the energies of homonuclear single bonds between elements of group 15 reveals that no great stability accrues to chains and rings of the heavier elements. The trend in bond energies for E_4 molecules is a case in point $E(P-P) = 201$,

$E(As-As) = 146$, $E(Sb-Sb) \approx 120$ kJ/mol. Nevertheless, compounds possessing As_n-rings were already known at the beginning of this century (Bertheim, 1908), and even served a good purpose:

$$HO-C_6H_3(NO_2)-AsO(OH)_2 \xrightarrow[H_2O]{Na_2S_2O_4} \left(HO-C_6H_3(NH_2)-As\right)_n$$

n = 5,6,7 Salvarsan

Salvarsan® is effective against Treponema pallidum, the microorganism which causes syphilis (P. Ehrlich, 1909).

This reaction yields rings of varying size which differ only slightly in energy. Since the discovery of antibiotics, chemotherapeutics based on organoarsenic compounds have, however, lost their importance.
The simplest species containing E–E bonds are the hydrazine analogues R_2E-ER_2. Tetramethyldiarsane, a principal component of Cadet's fuming liquid (p. 1), is prepared more deliberately by means of coupling reactions:

$$Me_2AsH + Me_2AsCl \xrightarrow[-HCl]{} Me_2As-AsMe_2 \quad (bp: 78\,°C)$$

The weakness of the As–As bond accounts for numerous reactions:

$$Me_2As-AsMe_2 \xrightarrow{CF_3C{\equiv}CCF_3} Me_2As(CF_3)C{=}C(CF_3)AsMe_2$$

$$Me_2As-AsMe_2 \xrightarrow{Na,\ NH_3\ (l)} 2\ Me_2AsNa$$

$$Me_2As-AsMe_2 \xrightarrow[-2\,CO]{Mn_2(CO)_{10}} (CO)_4Mn(\mu\text{-}AsMe_2)_2Mn(CO)_4$$

μ-dimethylarsenido bridges

Tetramethyldistibane $Me_2Sb-SbMe_2$, first described by Paneth in 1934, is a pyrophoric compound which decomposes more or less rapidly at room temperature in a closed vessel. Me_4Sb_2 is thermochromic (pale yellow in the solid state at −18 °C, red at +17 °C (mp), pale yellow again in the molten state or in solution). In the crystal, chains with intermolecular Sb–Sb interactions are encountered (Ashe, 1984):

$$\cdots Sb(Me_2) \overset{286}{—} Sb(Me_2) \overset{364}{\cdots} Sb(Me_2) — Sb(Me_2) \cdots \quad (93°)$$

The first well-characterized dibismuthane also is of recent origin (Calderazzo, 1983):

$$2\,Ph_2BiCl \xrightarrow[NH_3(l)]{Na} Ph_4Bi_2$$

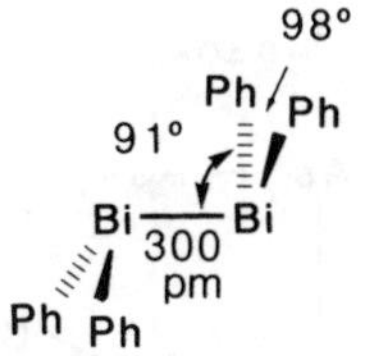

98°
Ph / Ph
91°
Bi — Bi
300 pm
Ph Ph

Tetraphenyldibismuthane Ph_4Bi_2 *is an orange compound which decomposes at* 100 °C. *Judging from the bond angles, the orbitals used by bismuth for bonding to carbon are essentially* 6 p *orbitals; the lone pair at* Bi *therefore almost exclusively possesses* 6 s *character.*

Longer chains between the elements of group 15 result from condensation reactions or from the reduction of compounds which contain RAs^{III}- or RAs^{V} units, the formation of homocyclic rings being strongly favored:

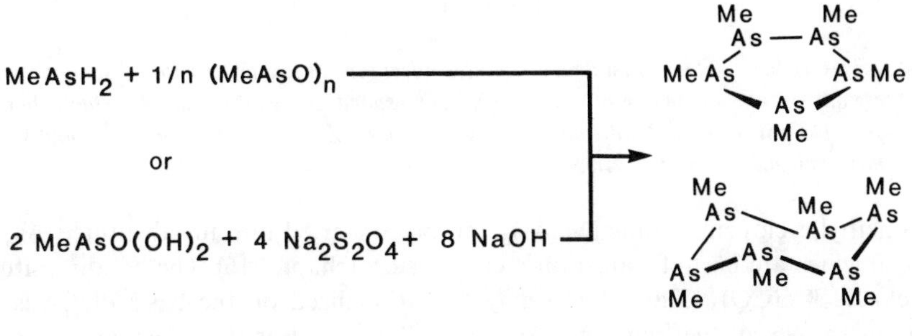

$MeAsH_2 + 1/n\ (MeAsO)_n$

or

$2\ MeAsO(OH)_2 + 4\ Na_2S_2O_4 + 8\ NaOH$ → $(MeAs)_5$ (five-membered ring) / $(MeAs)_6$ (six-membered ring)

Recently, polycyclic organoarsanes have also been prepared. Example (Baudler, 1985):

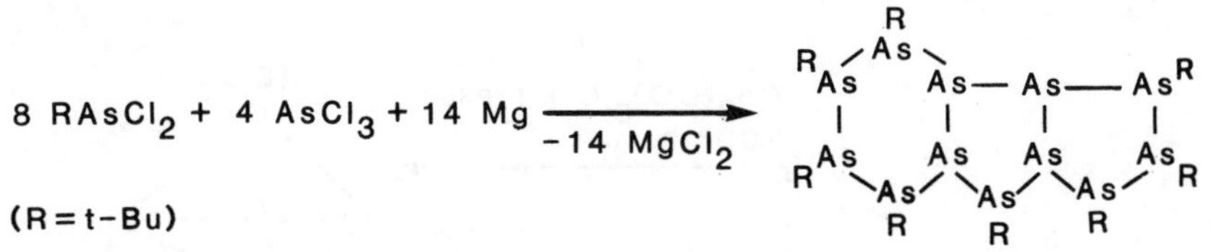

$$8\ RAsCl_2 + 4\ AsCl_3 + 14\ Mg \xrightarrow{-14\ MgCl_2} R_8As_{12}$$

(R = t-Bu)

As for antimony, the weakness of the Sb–Sb bonds in stibacycles $(RSb)_n$, $n = 3-6$, calls for their protection by bulky groups:

$$4\ (t\text{-Bu})_2SbLi \xrightarrow[2.\ CH_3OH]{1.\ I_2} \text{cyclo-}(t\text{-Bu})_4Sb_4$$ (Issleib, 1965)

$$5\ (t\text{-Bu})SbBr_2 \xrightarrow{Mg} \text{cyclo-}(t\text{-Bu})_5Sb_5$$ (Breunig, 1983)

Polyarsinidene $(RAs)_n$ chains of high molecularitiy are obtained from $MeAsH_2$ and $MeAsI_2$ after prolonged reaction times (Rheingold, 1973):

$$n/2\ MeAsI_2 + n/2\ MeAsH_2 \xrightarrow[\text{benzene}]{1-3\ \text{years}} (MeAs)_n + n\,HI$$

or, more quickly, via reduction by R_3Sb:

$$n\ \mathrm{MeAsI_2} + n\ n\text{-}\mathrm{Bu_3Sb} \xrightarrow[25\,^\circ\mathrm{C}]{10\ \mathrm{min}} (\mathrm{MeAs})_n + n\ n\text{-}\mathrm{Bu_3SbI_2}$$

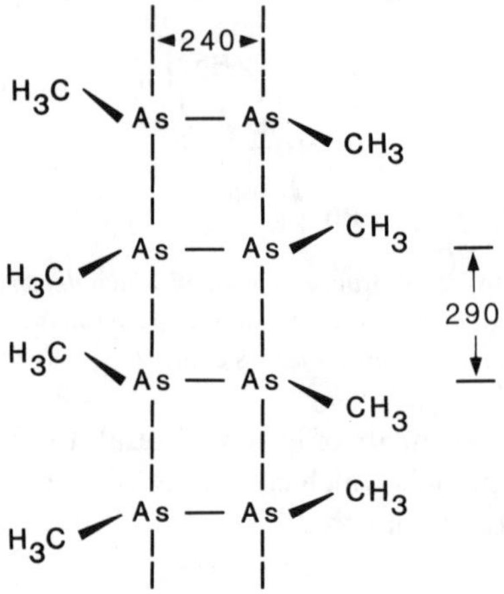

240

H_3C As — As CH_3

H_3C As — As CH_3

290

H_3C As — As CH_3

H_3C As — As CH_3

$(MeAs)_n$ *arises as purple-black crystals and forms a novel type of ladder structure displaying differing* As–As *distances. The distances at the "steps" correspond to regular* As–As *single bonds [cf.* d(As–As) = 243 pm *in* As_4], *the distances along the bars point to a bond order of about* 0.5. *This material has semiconducting properties.*

(Pentamethyl)cyclopentaarsane $(MeAs)_5$ can be converted into unsubstituted **As_n rings** ($n = 3, 5$) which are part of transition-metal clusters (chapter 16). The bonding situation in (cyclo-As_3)$Co(CO)_3$ (Dahl, 1969) may be rationalized on the basis of the isolobal properties (p. 396) of the fragments As⋮ and $(CO)_3$Co⋮. For the triple-decker complex (cyclo-As_5)$[Mo(C_5H_5)]_2$ (Rheingold, 1982), discussions of the bonding condition are more involved:

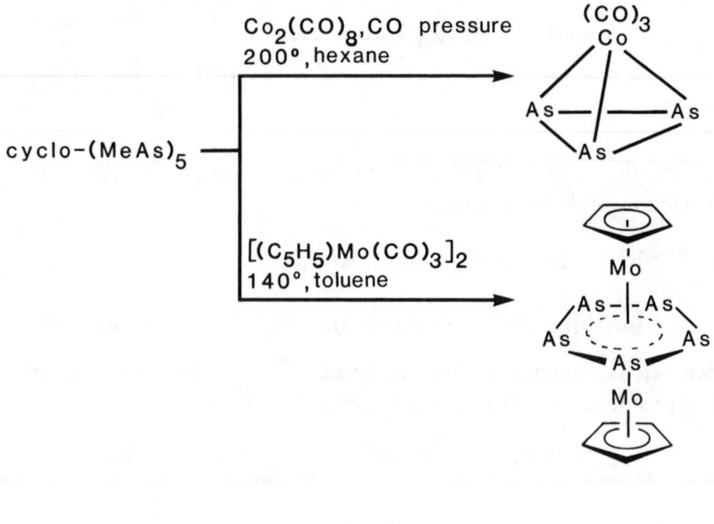

cyclo-$(MeAs)_5$

$Co_2(CO)_8$, CO pressure
200°, hexane
→ $(CO)_3$Co, As, As, As

$[(C_5H_5)Mo(CO)_3]_2$
140°, toluene
→ Mo, As, As, As, As, As, Mo

The bond distance $d(\text{Mo}-\text{Mo}) = 276$ pm argues for a long Mo=Mo double bond which "permeates the As_5 ring". In this context, compare the compounds (cyclo-P_6)[Mo(C_5Me_5)]$_2$ (p. 378) and (cyclo-P_5)Fe(C_5Me_5) (p. 379).

9.4 E=C and E=E Multiple Bonds

Successful strategies for the generation of segments which contain $E=C(p_\pi-p_\pi)$ and $E=E(p_\pi-p_\pi)$ interactions for the heavier elements of group 15 are analogous to those used for group 14: incorporation of the respective multiple bond into conjugated systems, protection of the vulnerable bonding region by bulky groups and coordination to transition metals.

9.4.1 $E=C(p_\pi-p_\pi)$ Bonds

The **heteroarenes C_5H_5E (E = N, P, As, Sb, Bi)** form a complete series of homologous compounds which allows a systematic study of their properties. This field was inaugurated by Märkl (1966) with the synthesis of 2,4,6-triphenylphosphabenzene:

$$2,4,6\text{-}Ph_3C_5H_2O^+\,BF_4^- \xrightarrow[Py]{P(CH_2OH)_3} 2,4,6\text{-}Ph_3C_5H_2P + 3\,CH_2O + H_2O + PyH^+BF_4^-$$

$P(CH_2OH)_3$ acts as a source of PH_3 here. Through variation of the pyrylium salt, a large variety of 2,4,6-derivatives of phosphabenzene has become available.

Substituted derivatives of arsabenzene are also accessible via carbene insertion followed by ring expansion (Märkl, 1972):

$$2,5\text{-}Ph_2C_4H_2AsPh \xrightarrow[-LiPh]{Li,\,DME} 2,5\text{-}Ph_2C_4H_2AsLi \xrightarrow[-LiCl,\ -NaCl,\ -CO_2]{1.\ PhCCl_2COONa\quad 2.\ \Delta T} 2,5\text{-}Ph_2C_4H_2As\text{–}\ddot{C}\text{–}Ph \longrightarrow 2,3,6\text{-}Ph_3C_5H_2As$$

The parent molecules are prepared by means of the following reaction sequence:

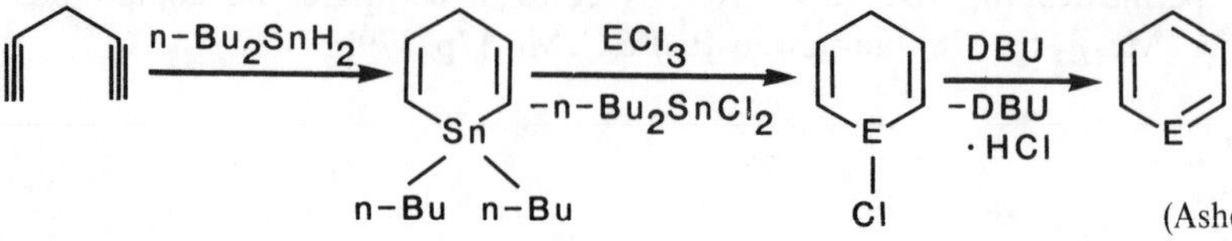

n-Bu$_2$SnH$_2$; ECl$_3$, −n-Bu$_2$SnCl$_2$; DBU, −DBU·HCl

(Ashe, 1971)

E = P, As, Sb, Bi
DBU = Diazabicycloundecene
(a special deprotonating agent)

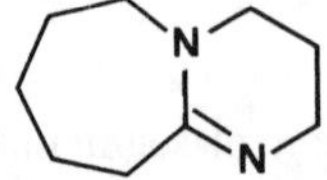

Phosphabenzene (Phosphinine)
Arsabenzene (Arsenine)
Stibabenzene
Bismabenzene

With increasing atomic number of E, the distortion of the six-membered rings increases:

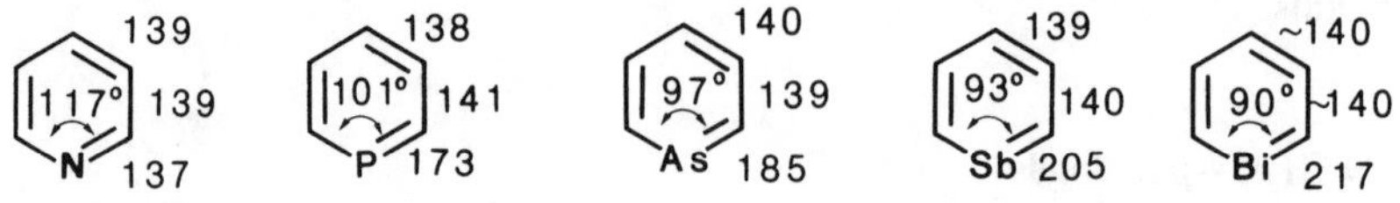

N: 117°, 139, 139, 137; P: 101°, 138, 141, 173; As: 97°, 140, 139, 185; Sb: 93°, 139, 140, 205; Bi: 90°, ~140, ~140, 217

According to spectroscopic data, these heterocycles are aromatic. Chemical evidence concerning their aromaticity is unavailable because the monomeric species C_5H_5E (E = P, As, Sb, Bi) are prone to oxidation and increasingly thermolabile. The phenyl-substituted derivatives are somewhat more stable. Stability decreases in the order P > As > Sb > Bi: arsabenzene is handled comparatively easily, stibabenzene as yet has only been characterized by spectroscopic techniques or as a Diels-Alder adduct. Stibabenzene and bismabenzene at 0 °C engage in a monomer/dimer equilibrium:

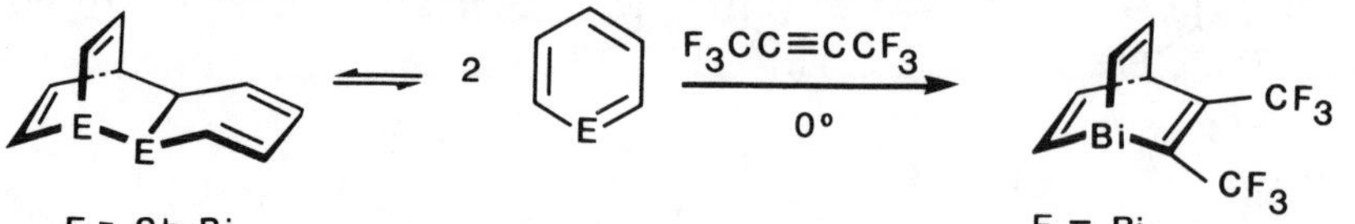

2 ; $F_3CC\equiv CCF_3$, 0° ; CF$_3$, CF$_3$, Bi

E = Sb, Bi   E = Bi

In view of the exceedingly high reactivity of silabenzene (p. 109) the properties of **phosphabenzene** *– e.g. indefinitely stable at room temperature in the absence of air – are surprising. Evidently, $p_\pi - p_\pi$ bonds to phosphorous (>C=P–) are more resistant against addition reactions than $p_\pi - p_\pi$ bonds to silicon (>C=Si<). For the fictitious process*

$$CH_3{-}PH_2 + CH_2{=}SiH_2 \rightleftharpoons CH_2{=}PH + CH_3SiH_3 \quad \text{(Borden, 1987)}$$

quantum-chemical calculations yield the reaction enthalpy $\Delta H = -92$ kJ/mol.

The elements of group 15 have also been incorporated into unsaturated five-membered rings:

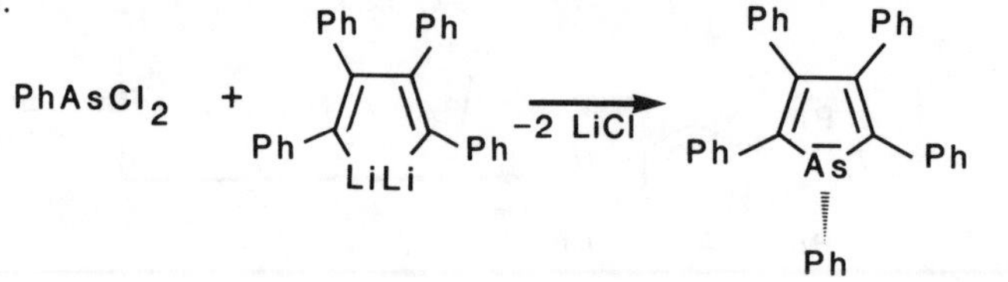

PhAsCl$_2$ + (Ph$_4$C$_4$Li$_2$) —(−2 LiCl)→ Arsole derivative (Ph$_4$C$_4$AsPh)

In phosphole and arsole, P and As retain the pyramidal environment which is familiar from the species R_3E. However, compared to the open-chain, configurationally stable arsanes RR'R''As, the inversion barrier for arsoles is reduced by about 50 kJ/mol:

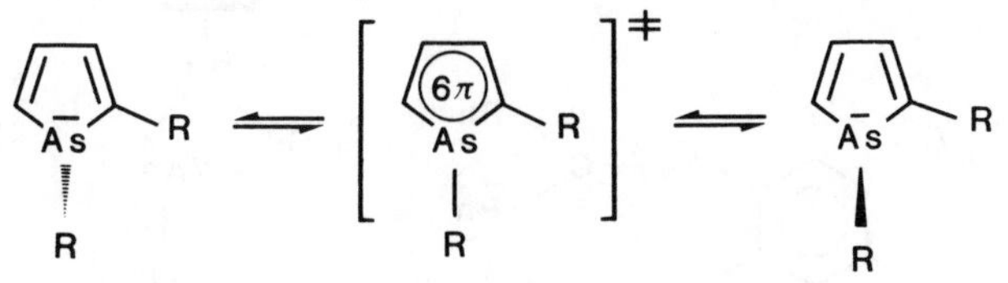

This may be traced to a stabilization of the trigonal planar transition state by means of an As=C ($p_\pi - p_\pi$) interaction. The following survey suggests that with increasing atomic number, the tendency to engage in E=C ($p_\pi - p_\pi$) bonding decreases:

	Pyrrole	Phosphole	Arsole
Configuration at E	planar	pyramidal	pyramidal
Electron pair at E	part of a 6 π-*e*-system	lone pair	lone pair
Character	arene	diolefin	diolefin
Enantiomers isolable?	no	no	yes

The diminishing propensity for a π-contribution to the E–C bond upon going from phosphorus compounds to the arsenic- and antimony analogues is also a feature of the element^V ylids $R_3ECR'_2$.

Arsenic ylids (alkylidene arsoranes) are obtained through deprotonation of arsonium cations:

$$Ph_3AsMe^+Br^- \xrightarrow{NaNH_2,\ THF} Ph_3\overset{\oplus}{As}-\overset{\ominus}{C}H_2$$

$$Me_3AsCH_2SiMe_3^+Cl^- \xrightarrow{BuLi,\ Et_2O} Me_3\overset{\oplus}{As}-\overset{\ominus}{C}HSiMe_3 \xrightarrow{MeOH}$$

$$\left\{ \underset{\textbf{Ylid}}{Me_3\overset{\oplus}{As}-\overset{\ominus}{C}H_2} \longleftrightarrow \underset{\text{Ylene}}{Me_3As=CH_2} \right\}$$

^{13}C-NMR data [high-field shifts, small coupling constant $^1J(^{13}C, ^1H)$ for the CH_2 group] are indicative of sp^3 hybridization at the methylene carbon atom, thereby pointing to dominance of the ylid canonical form. The finding that **methylenearsoranes R_3AsCH_2** are more reactive than methylenephosphoranes is explained by the more pronounced carbanion character of the former as compared to the latter – a consequence of the reluctance of the heavier main-group elements to participate in $p_\pi - p_\pi$ bonding.

Arsaalkenes (alkylidenearsanes) with isolated $As{=}C(p_{\pi}-p_{\pi})$ bonds are stable as monomers only if sterically protected or coordinated to a transition metal:

F. Bickelhaupt (1979) H. Werner (1984)

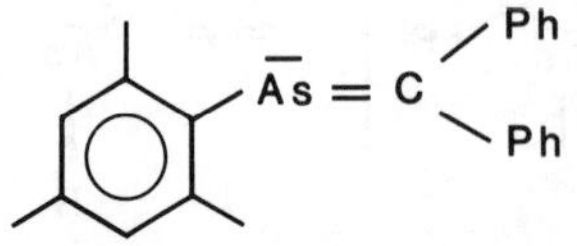

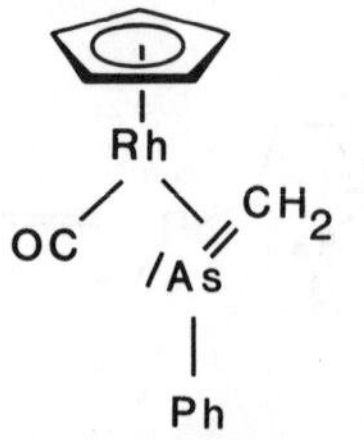

Arsaalkines (alkylidynearsanes) could also be obtained in the presence of bulky substituents or as transition-metal complexes; the first molecule boasting a $-C{\equiv}As$ triple bond was only prepared very recently (Märkl, 1986):

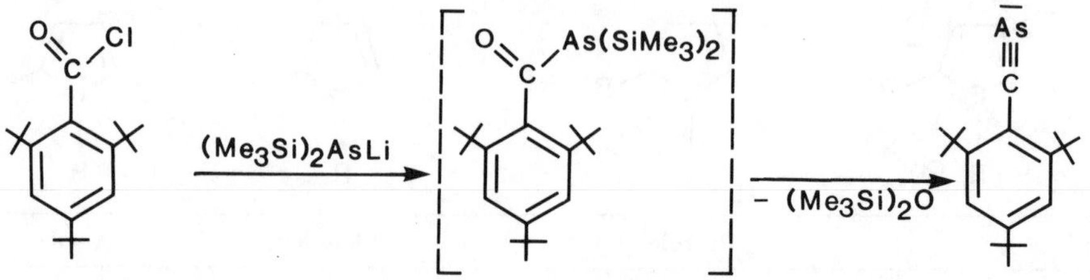

For the sake of completeness, a stable phosphaethyne, which has been known somewhat longer, should be mentioned (Becker, 1981). Interestingly, this phosphaalkyne undergoes thermal cyclooligomerisation to yield a tetraphosphacubane (Regitz, 1989):

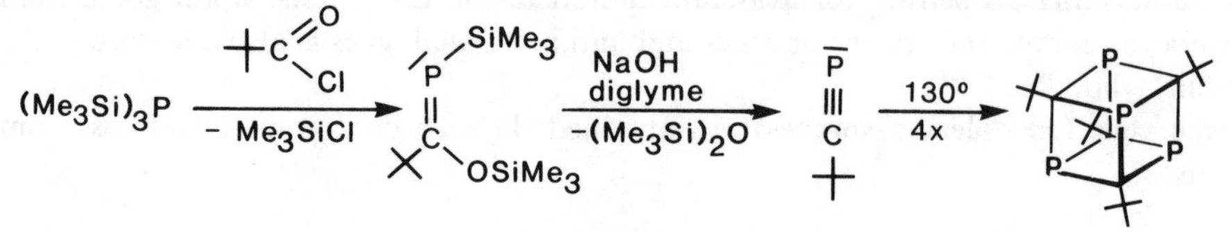

Arsaalkynes can also be generated in the coordination sphere of a transition metal (Seyferth, 1982):

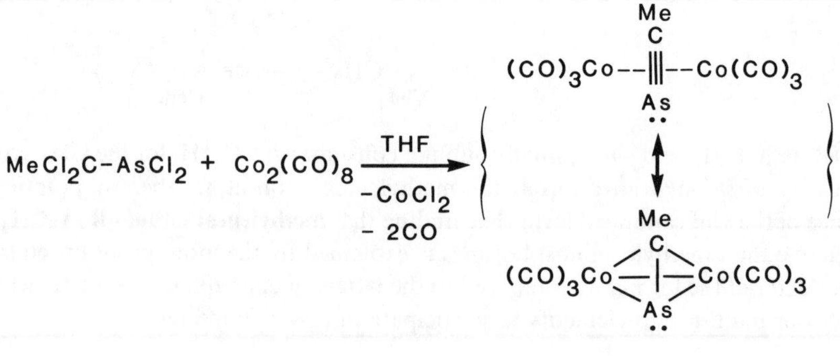

The bonding in the Co_2CAs core resembles the situation in transition-metal carbonyl complexes with bridging alkyne ligands (see chapter 15.2).

9.4.2 E=E ($p_\pi - p_\pi$) Bonds

The realization of molecules containing E=E multiple bonds, which are homologues of the azo compounds RN=NR, also requires steric protection or metal coordination. The element phosphorus led the way:

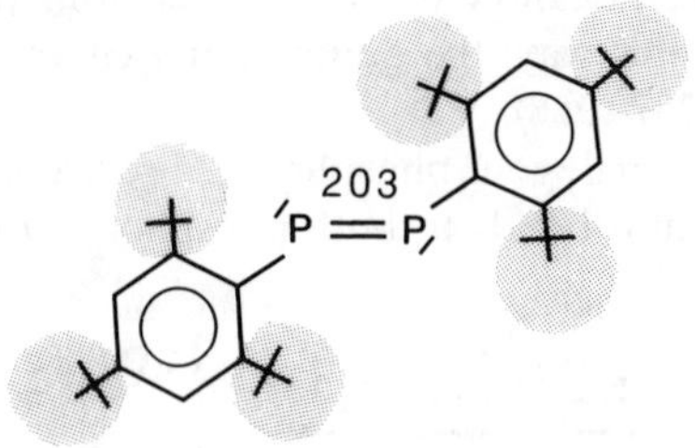

203

Bis(aryl)diphosphene, *Yoshifuji* (1981)
compare P_4: $d(P-P) = 223$ pm

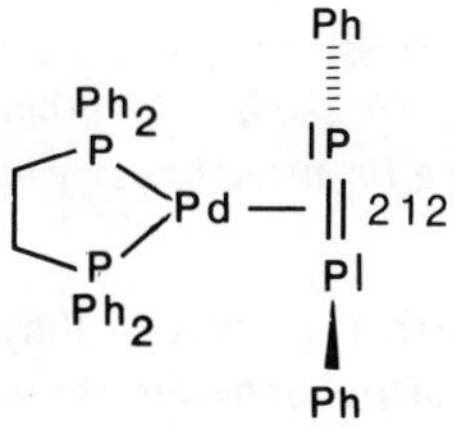

Ph_2 P Pd P Ph_2 — P 212 P, Ph, Ph

t-BuNC, t-BuNC, Ni — N 138 N, Ph, Ph

Diphosphene π-complex (*Chatt*, 1982)

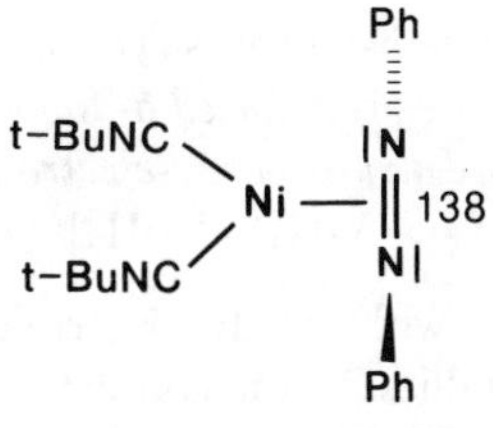

compare: *Diazene π-complex* (*Ibers*, 1972)

Recently, molecules possessing −As=As− and −Sb=Sb− double bonds have been reported (Cowley, 1985):

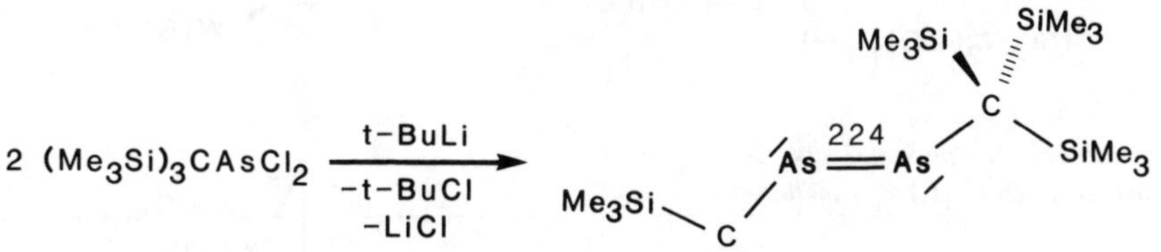

$$2\ (Me_3Si)_3CAsCl_2 \xrightarrow[-t\text{-}BuCl,\ -LiCl]{t\text{-}BuLi} [(Me_3Si)_3C]As{=}As[C(SiMe_3)_3]$$

224

Bis[tris(trimethylsilyl)methyl]diarsene
compare As_4: $d(As-As) = 245$ pm

As of yet, stabilization of $-Sb=Sb-$ double bonds demands both types of protective measure (Cowley, 1985):

$$(Me_3Si)_2CHSbCl_2 \xrightarrow[-NaCl]{Na_2[Fe(CO)_4]} \text{Distibene } \pi \text{ complex: } (Me_3Si)_2CH-\bar{Sb}=\underline{Sb}(Fe(CO)_4)-HC(SiMe_3)_2$$

Distibene π complex

Bonding in the Sb_2Fe segment can be described in analogy to the interaction of olefins with transition metals. In this way, the double bond character of the $Sb=Sb$ bond is partially reduced (see pp. 257–258).
If instead of the stibane, the analogous phosphane is used, a product is obtained in which the $Fe(CO)_4$-fragments form σ-bonds to both phosphorus atoms (Power, 1983):

$$(Me_3Si)_2CHPCl_2 \xrightarrow[-NaCl]{Na_2[Fe(CO)_4]} (CO)_4Fe\leftarrow P((Me_3Si)_2CH)=P(HC(SiMe_3)_2)\rightarrow Fe(CO)_4$$

Diphosphene σ complex

This disparate behavior is typical: lone pairs at phosphorus possess higher p-character and therefore more pronounced σ-donor properties than lone pairs at the heavier atoms of group 15, *which are almost pure s-electrons* [*compare the decreasing Brønsted basicity in the series* $NH_3 \gg PH_3 > AsH_3 > SbH_3$].

Species E_2 with an $E\equiv E$ triple bond contain no further groups – they are not organometallics. In the context of this section it is noteworthy, however, that molecules E_2 (E = As, Sb, Bi) which contrary to N_2 are unstable, can be attached to metal carbonyl fragments (Huttner, 1982):

$$W(CO)_5THF,\ Na_2W_2(CO)_{10} \xrightarrow[E=As,Sb,Bi]{ECl_3} (CO)_5W-E\equiv E\ (\text{with two } W(CO)_5)$$

This remarkable complex may also be regarded as a hetera-[1.1.1]*propellane*

$$(CO)_5W,\ E_2,\ W(CO)_5,\ W(CO)_5$$

10 Organoelement compounds of Selenium and Tellurium (Group 16)

Organometallic chemistry of group 16 is confined to the elements selenium and tellurium, since oxygen and sulfur are typical non-metals, and the strong α-radiation of the nucleus $^{210}_{84}Po$ would rapidly destroy an organic ligand sphere. Since the discovery that organoselenium oxides readily undergo *syn*-eliminations (Reich, Sharpless, 1973), selenium organyls have gained considerable importance in organic synthesis. From the coordination chemists point of view, the stabilization of highly reactive molecules like selenoformaldehyde ($H_2C{=}Se$) or tellurocarbonyl ($C{\equiv}Te$) in transition-metal complexes is of interest (p. 238, 272). Another impetus was provided by the observation that the organic radical cation salt bis(tetramethyl-tetraselenafulvalene)perchlorate at 1.5 K changes from a metallic into a superconducting state (Bechgaard, 1981).
Very recently, thin layers of cadmium telluride CdTe have been produced from mixtures of $(CH_3)_2Cd$ and $(C_2H_5)_2Te$ by means of metal organic chemical vapor deposition (MOCVD). These layers, grown by photoepitaxy, show considerable promise as semiconducting materials in electronic devices (Irvine, 1987 R).

Contrary to sulfur, selenium offers an isotope which is well-suited for NMR spectroscopy: ^{77}Se (7.5%, $I = 1/2$) *features a receptivity which exceeds that of* ^{13}C *by a factor of three and carries no quadrupole moment. Chemical shifts* δ *in* **^{77}Se-*NMR*** *spectra of selenium organyls cover a range of* 2500 ppm, *which is six times the range pertinent to* ^{13}C-*NMR*. *As expected, the values of the coupling constants* $^1J(^{77}Se, ^{13}C)$ *correlate with the s character of the* Se–C *bond* [*Example*: Me**Se**–**C**H_2CH_2Ph 62 Hz, Me**Se**–**C**H=CHPh 115 Hz, Me**Se**–**C**≡CPh 187 Hz].

Organoselenium compounds in their chemical behavior bear strong resemblance to organosulfur compounds. Synthetically important reactions for selenoorganyls usually proceed under milder conditions, a fact which is relevant in natural-product transformations. Compared to sulfur, selenium prefers lower oxidation states. Hence, organoselenium chemistry is dominated by the oxidation states Se^{II} and, to a lesser extent, Se^{IV}. In fact, many important reactions of organoselenium compounds are driven by the elimination of an organoselenium^{II} species from a selenium^{IV} precursor. Selenoles RSeH are more acidic than thiols RSH, yet selenides RSe^- are more nucleophilic than sulfides RS^-. Selenyl halides RSeX are more electrophilic than the corresponding sulfenyl halides RSX. As starting materials for many conversions, organoselenides (organoselenanes) R_2Se and organodiselenides (organodiselenanes) RSeSeR are employed; they are obtained as follows:

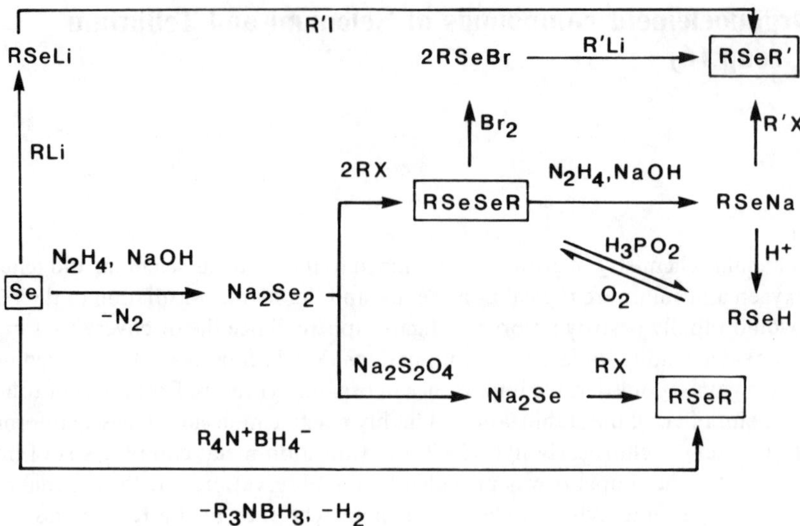

From RSeR and RSeSeR, respectively, a large variety of oxo-, hydroxo- and halo-organoselenium compounds can be prepared. Only a few examples are given here, which illustrate the most popular applications of selenium organyls in organic synthesis:

- β-Ketoselenium oxides via **syn-elimination of RSeOH** form α, β-unsaturated carbonyl compounds (Reich, Sharpless, 1973):

$$CH_3{-}CH_2{-}C(=O){-}R \xrightarrow[\text{2. PhSeBr}, -78^\circ]{\text{1.(iso-Pr)}_2\text{NLi, THF}, -78^\circ} CH_3{-}CH(SePh){-}C(=O){-}R$$

$$CH_3{-}CH(SePh){-}C(=O){-}R \xrightarrow{KIO_4 \text{ or } H_2O_2} CH_2(H){-}CH(Se(=O)Ph){-}C(=O){-}R$$

$$CH_2(H){-}CH(Se(=O)Ph){-}C(=O){-}R \xrightarrow[20^\circ]{-PhSeOH} CH_2{=}CH{-}C(=O){-}R$$

R = H, OR', aryl (alkyl)

This is probably the mildest method for the **preparation of enones** from aldehydes, ketones or esters.

- In a related fashion, the conversion of **epoxides into allyl alcohols** is achieved:

$$RCH_2-\underset{O}{\triangle}-R' \xrightarrow[EtOH]{PhSe^-} RCH_2-CH(SePh)-CHR'(OH) \xrightarrow{H_2O_2} RCH_2-CH(Se(=O)Ph)-CHR'(OH) \xrightarrow[20^\circ]{-PhSeOH} RCH{=}CH-CHR'(OH)$$

This reaction sequence may serve to regenerate a C=C double bond which was protected as an epoxide.

- The proven **oxidation of alkenes** to allyl alcohols by means of **selenium dioxide** is also thought to proceed via organoselenium intermediates (Sharpless, 1973):

Alkene → Allylseleninic acid → {R–CH=CH–CH₂–O–Se–OH} →(H_2O) Allyl alcohol

Alkene Allylseleninic acid Allyl alcohol

Of more academic interest is the question as to the existence of $p_\pi - p_\pi$ bonds between carbon and selenium or tellurium, respectively. Monosubstituted selenocarbonyl compounds oligomerize, whereby the >C=Se double bond is eliminated. Example (Mortillaro, 1965):

$$H_2C = O \xrightarrow[-MgBrOH]{MgBrSeH} \{H_2C = Se\} \longrightarrow (H_2CSe)_3 + (H_2CSe)_n$$

A selenoketone (selenone) with two bulky substituents was shown to be stable as a monomer, however (D. H. R. Barton, 1975):

$$(t\text{-}Bu)_2C = N-N = PPh_3 \xrightarrow[-N_2,\ -PPh_3]{Se,\ 120^\circ,\ Kat.(n\text{-}Bu)_3N} (t\text{-}Bu)_2C = Se$$

blue
$\lambda_{max} = 710$ nm

Selenocarbonyl C≡Se, which is formed from CSe_2 in a high frequency discharge, polymerizes at temperatures as low as − 160 °C; it can only be identified from the products of trapping reactions (Steudel, 1967) or as a ligand in transition-metal complexes (p. 238). In this context, the relative stability of monomeric CSe_2 is noteworthy. The propensity of the group 16 elements to engage in $p_\pi - p_\pi$ bonding to carbon should also be apparent from the extent of conjugation and the derived molecular properties of the heterocyclopentadienes C_4H_4E (E = Se, Te). **Selenophene** (Peel, 1928) and **Tellurophene** (Mack, 1966) are fairly easily accessible and well-characterized compounds:

2 H−C≡C−H + Se —(400°)→ selenophene (143 pm, 137 pm, 185 pm, 87°, Se)

The structural parameters of selenophene stem from the microwave spectrum (Magdesiava, 1969). The Se−C *bond length should be compared with the length of a genuine single bond, d*(Se−C) = 193 pm.

R−C≡C−C≡C−R + Na_2Te —(CH_3OH, 20°)→ 2,5-R₂-tellurophene (R, R, Te)

R = H, CH_2OH, Ph

Cyclohydrotelluration

Judging from chemical evidence (electrophilic substitution dominates over addition) as well as from physical criteria (bond distances, ^{1}H-NMR spectra, diamagnetic susceptibility) aromatic character must be ascribed to selenophene and tellurophene. The following **order of decreasing aromaticity** has been derived (Marino, 1974):

benzene > thiophene > selenophene > tellurophene > furan

The aromatic nature of selenophene C_4H_4Se contrasts with the diolefin character of isoelectronic arsole C_4H_4AsH (p. 163).
Two selenium atoms per five-membered ring are contained in the tetraselenafulvalenes which have already been mentioned with regard to their interesting electrical properties. The corresponding tellurium compounds have also been prepared (Wudl, 1982):

Br, Br —(1. t-BuLi,THF −78°; 2. Te, −15°)→ TeLi, TeLi —($Cl_2C{=}CCl_2$, −80°→25°)→ Te, Te, Te, Te

Tetratellurafulvalene

Recently, tellurium has for the first time been incorporated into a fully unsaturated seven-membered ring (Tsuchiya, 1991):

C≡CH
C≡CH
Na_2Te, $N_2H_4 \cdot aq$
C_6H_6/H_2O
$[(n\text{-}Oct)_3NMe]Cl$
phase transfer catalysis
Te
3-Benzotellurepin
Br_2, hexane
$TeBr_2$
– Te
25°
3d
– $TeBr_2$
ΔT

The thermal lability of this tellurium-containing heteroepin is similar to that of the borepins (p. 64).

Organoselenium Compounds in vivo

For higher organisms, selenium is an essential trace element (Schwarz, 1957); an Se-free diet leads to malfunction of the liver and to hemolytic processes. Since selenium has been detected in the retina, it is likely that this element also plays an important role in vision. The human organism stores 10–15 mg of selenium; a daily uptake exceeding 1 μg Se/g food is toxic but an uptake of less than 0.2 μg Se/g food causes a deficiency syndrome. Selenium plays its biochemical role as a component of glutathione peroxidase, an enzyme which is responsible for the protection of essential SH-groups and for the decomposition of peroxides, thereby acting as an **antioxidant** (Rotruck, 1973). The selenium-containing section of glutathione peroxidase is the amino acid selenocysteine $(HSe)CH_2CH(NH_2)COOH$ (Stadtman, 1984). Other selenium organyls detected in living organisms include selenomethionine $MeS(CH_2)CH(NH_2)COOH$, trimethylselenonium salts $Me_3Se^+X^-$, and dimethylselenide Me_2Se, a component with a garlic-like odor in the breath of organisms which suffer from selenium poisoning. The formation of Me_2Se from inorganic selenium compounds apparently aids in detoxification: Me_2Se is less toxic than the oxoanions SeO_3^{2-} (selenite) and SeO_4^{2-} (selenate) or elemental Se. This contrasts with mercury, where the problems of environmental contamination are aggravated by biomethylation (p. 55).

11 Organometallics of Copper, Silver and Gold (Group 11)

The metals copper, silver and gold clearly belong to the transition elements. The discussion of their chemistry at this particular point follows an established tradition in that it smoothly leads into the organometallic chemistry of the *d*-block elements (Chapters 12–17).

11.1 Copper- and Silver Organyls

The interest which chemists took in organocopper compounds was hitherto mainly confined to their use in organic synthesis. Points of departure were the study of reactions of copper organyls with organic halides (Gilman, 1936) and the observation that Cu^I ions catalyze the 1,4-addition of Grignard reagents to conjugated enones (Kharash, 1941). The intermediacy of organocopper compounds in these additions was demonstrated by House in 1966. Yet, the isolation of pure organocopper species and their characterization by diffraction methods was only achieved during the last decade. Problems arise from the high sensitivity of binary molecules MR (M = Cu, Ag) to H_2O and O_2 and from their lack of solubility in inert solvents. An aggravating factor is the tendency to form non-stoichiometric adducts like $[(CuR)_x \cdot (CuBr)_y]$ or solvates of varying composition, depending on the method of preparation. The organometallic chemistry of **Cu** and **Ag** is exclusively that of the **oxidation state + I**. In this section, only σ-organyls will be treated; π-complexes of the elements Cu, Ag and Au are deferred to sections which review the respective ligands (Chapters 12–17).

PREPARATION

Copper Organyls

$$CuX + LiR \xrightarrow[-LiX]{Et_2O,\ THF} CuR \xrightarrow{LiR} \underset{\text{Organocuprate}}{Li[CuR_2]} \qquad \boxed{4}$$

*The usual designation "organo**cuprate**" for compounds of the composition* $MCuR_2$ *is misleading because – contrary to experimental evidence (p. 175) – it suggests an ionic structure* $M^+[CuR_2]^-$.

$$CuX + Ph_2Zn \longrightarrow CuPh + PhZnX$$

A good source for CuX is the compound $[(n\text{-}Bu_3P)CuI]_4$. Alkynylcopper compounds are prepared by means of metallation:

$$RC{\equiv}CH + [Cu(NH_3)_2]^+ \longrightarrow CuC{\equiv}CR + NH_3 + NH_4^+ \qquad \boxed{6}$$

$$PhC{\equiv}CH + Cu(t\text{-}BuO) \longrightarrow 1/n\,[CuC{\equiv}CPh]_n + t\text{-}BuOH$$

Copper acetylides (copper alkynyls) are inert towards hydrolysis, their association in the solid state is a result of π-complexation of copper to neighboring $-C\equiv CR$ units (p. 208). A particularly stable, sublimable organocopper compound is formed from Cu^I chloride and (trimethyl)methylenephosphorane via transylidation (Schmidbaur, 1973):

$$2\,CuCl + 4Me_3P=CH_2 \longrightarrow [Me_2P^{\oplus}(CH_2-Cu^{\ominus}-CH_2)_2P^{\oplus}Me_2] + 2\,Me_4PCl$$

The inert character of this complex, containing bridging phosphoniodiylid ligands, may be traced back to the absence of β-hydrogen atoms (cf. p. 199).

Silver Organyls

Compared to organocopper compounds, silver organyls are considerably less stable, decomposition setting in at ambient temperature.

$$AgNO_3 + R_4Pb \xrightarrow[-R_3PbNO_3]{} AgR \xrightarrow{20\,°C} Ag + \text{alkanes} + \text{alkenes} \qquad \boxed{4}$$

The corresponding perfluoroalkyl derivates R_FAg are somewhat more stable (cf. p. 204):

$$AgF + CF_2{=}CF-CF_3 \xrightarrow[30\,°C]{MeCN} AgCF(CF_3)_2 \cdot MeCN \qquad \boxed{8}$$

Argentofluorination

Pure phenylsilver was first obtained in 1972; diphenylzinc proved to be a convenient arylating agent here (van der Kerk, 1977):

$$AgNO_3 + ZnPh_2 \xrightarrow{Et_2O,\,0\,°C} AgC_6H_5 + PhZnNO_3$$

In the presence of an excess of LiAr, **organoargentates** are formed:

$$2\,LiAr + AgX \xrightarrow{Et_2O} Li[AgAr_2] + LiX \qquad \boxed{4}$$

Silver alkynyls can be prepared via metallation in the presence of an auxiliary base:

$$RC\equiv CH + AgNO_3 + NH_3 \longrightarrow AgC\equiv CR + NH_4NO_3 \qquad \boxed{6}$$

PROPERTIES OF SELECTED Cu- AND Ag ORGANYLS

Compound	Decomposition/°C	Degree of Association
CuMe	> −15 (explosive)	
AgMe	−50	
CuPh	100	polymer
CuC_6F_5	220	tetramer
AgPh	74	polymer
$CuCH_2SiMe_3$	78 (mp.)	tetramer
$CuC\equiv CR$	200	polymer
$AgC\equiv CR$	100–200	polymer
$Li[CuMe_2]$		dimer

d(Cu--Cu) = 242 pm
∢ C-Cu-C = 163.5°
d(Cu-C) = 202 pm

Structure of **$[CuCH_2SiMe_3]_4$**: Cu^I *adopts its preferred coordination number* 2 *(linear). For the tetramer, this leads to a planar* Cu_4 *ring. The* Cu$^{\diagup C \diagdown}$Cu *bridges may be regarded as* $2e3c$ *bonds which arise from the interaction of* C*(sp*^{3}*)- with* Cu*(sp) hybrid orbitals (Lappert,* 1977*). Compare:* d(Cu–Cu) = 256 pm (*bulk* Cu); r(Cu^+) = 96 pm (*ionic radius*).

$[CuPh]_x$, *which has not yet been obtained as single crystals, is assumed also to be associated by means of* Cu$^{\diagup C \diagdown}$Cu $2e3c$ *bridges.*

Bonding in the Cu$^{\diagup C \diagdown}$Cu bridge probably resembles the situation discussed for Al_2Ph_6 (p. 79); the access to Cu($3d$) electrons may, however, render additional interactions possible:

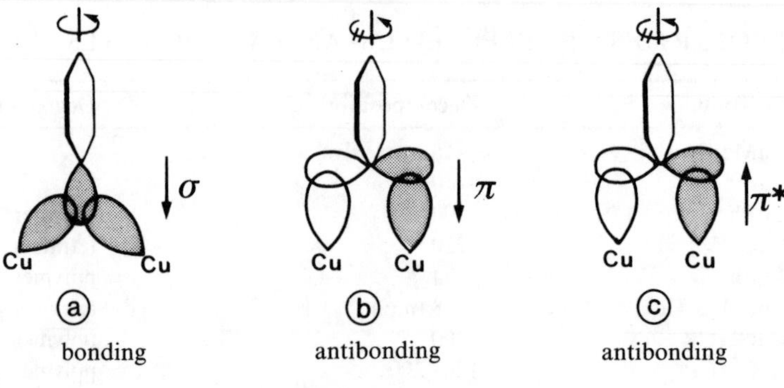

	(a)	(b)	(c)
Cu–Cu interaction:	bonding	antibonding	antibonding
Rotational barrier for the phenyl ring:	no	yes	yes

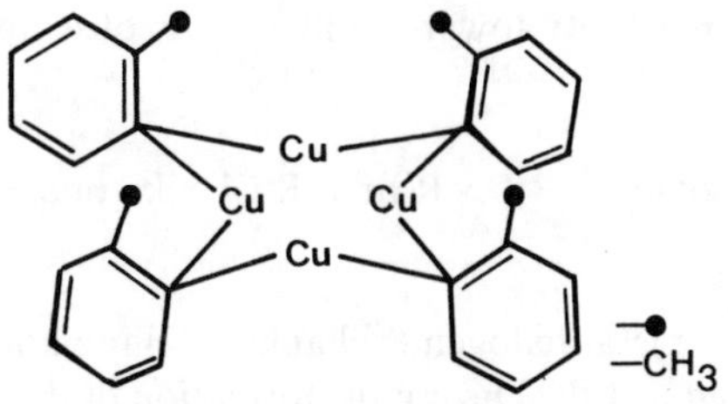

For **[(2-tolyl)Cu]$_4$** ^{1}H-*NMR spectroscopy at low temperature reveals the presence of stereoisomers. Therefore, a barrier to rotation around the* Cu–C *axis must be assumed. This finding argues for contributions of the types* ⓑ *and* ⓒ *to* Cu–C–Cu *bridge bonding* (*Noltes*, 1979).

The dimeric nature of the organocuprates, $[Li(CuMe_2)]_2 \equiv Li_2Cu_2Me_4$, would lead one to assume that their structure, in analogy to Li_4Me_4, is tetrahedral. It is highly likely, however, that a Li_2Cu_2 ring with a trans disposition of the metals is formed instead (R. G. Pearson, 1976):

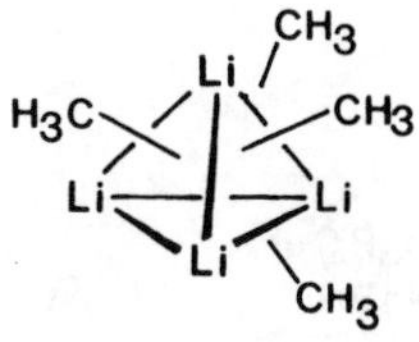

Li_4Me_4:
CH_3 *groups cap*
Li_3 *triangular faces*

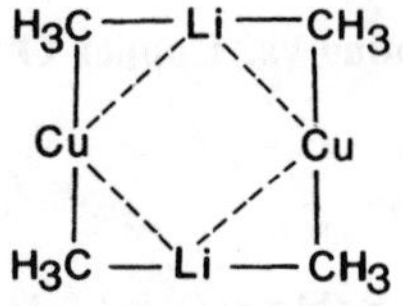

Li_2CuMe_4:
CH_3 *groups bridge*
LiCu *edges*

In solution, ^{7}Li-NMR evidence points to an equilibrium of three species (Lipshutz, 1985):

$$Li_2Cu_2Me_4 \rightleftharpoons LiMe + LiCu_2Me_3$$

APPLICATION OF COPPER ORGANYLS IN ORGANIC SYNTHESIS

① Nucleophilic Substitution (S_N2) at Organic Halides (Cross coupling)

The organocuprate Li[CuR$_2$] (**Gilman reagent**, 1952) is prepared in situ and converted into unsymmetrical coupling products:

$$2\,RLi + CuI \xrightarrow[-LiI]{Et_2O,\ -20\,°C} Li[CuR_2] \xrightarrow[-LiX,\ -CuR]{R'X} R-R'$$

R = Alkyl, Alkenyl, Aryl, Heteroaryl
R′ = Acyl, Alkyl, Alkenyl, Aryl, Heteroaryl

Lithium organocuprates Li[CuR$_2$] are **less nucleophilic** than the parent lithium organyls LiR; therefore, they react **more selectively** with organic substrates. In general, the following order of decreasing reactivity towards $Li[CuR_2]$ is observed:

$$R'C{=O \atop \backslash Cl} > R'C{=O \atop \backslash H} > \text{epoxide} > R'I > R'Br > R'Cl > \text{ketone} > \text{ester} > \text{nitrile} \gg \text{alkene}.$$

Competing reactions like metal/halogen exchange or elimination play a diminished role for $Li[CuR_2]$ as compared to LiR. During the formation of the incipient C–C bond, the configuration of the attacking group R is retained, that of the group R′ in the substrate R′X is inverted. If R′X is an alkenyl halide, its configuration is preserved during the substitution:

Br, Ar + $Li[CuR_2]$ → R, Ar + $Li[RCuBr]$

② **Oxidative Coupling via Copper Organyls**

$BrMg{-}(CH_2)_5{-}MgBr$ → (1. Ph_3PCuI, $-MgBrI$; 2. O_2/THF, $-78°$) cyclopentane 30%

An important application of oxidative coupling is the preparation of **symmetrical bisaryls** and of **conjugated diynes**:

2,2′-dilithiobiphenyl → (2 $CuCl_2$, Et_2O, 20°) tetraphenylene 53% + biphenylene 3% (Wittig, 1967)

In this variant the use of Cu^{II} *obviates the introduction of an additional oxidizing agent.*

The following example portrays the synthesis of [18]annulene, a key compound in discussing the concept of aromaticity (Sondheimer, 1962):

$$3\ HC{\equiv}C{-}(CH_2)_2{-}C{\equiv}CH \xrightarrow[\text{pyridine, }55°]{Cu(OAc)_2}$$

1. t-BuOK
t-BuOH
2. H_2, Pd/C

yield: 1% referred to hexadiyne

③ **Homolytic Thermal Coupling:**

$$2\ Me{-}CH{=}CH{-}CuP(n\text{-}Bu)_3 \xrightarrow{\Delta T} Me{-}CH{=}CH{-}CH{=}CH{-}Me + 2Cu + 2\ P(n\text{-}Bu)_3$$

cis *cis, cis*

In this reaction, the configuration of the alkene is retained (Whitesides, 1966).

④ **Addition to Conjugated Enones (Michael Addition)**

As opposed to organolithium- and Grignard reagents, which preferentially add to the carbonyl function (1, 2-), organocuprates with 1,2-unsaturated carbonyl compounds effect almost exclusively 1,4-addition:

$$\overset{4}{C}H_2{=}\overset{3}{C}H{-}\overset{2}{C}({=}\overset{1}{O}){-}CH_3 + Li[CuR_2] \longrightarrow RCH_2{-}CH{=}C(OLi){-}CH_3 \xrightarrow{H^+} RCH_2{-}CH_2{-}C({=}O){-}CH_3$$

The following example demonstrates the high stereoselectivity which can be attained in additions of organocuprates (House, 1968):

1. $Li[CuMe_2]$
2. H^+

98 : 2

With regard to yield and stereoselectivity, the stoichiometric use of organocuprates is generally superior to the variant which operates with Grignard reagents in the presence of catalytic amounts of $Cu^{I,II}$ salts (p. 172).

⑤ **Addition to Terminal Alkynes (Carbocupration)**

$$\text{RBr} \xrightarrow[\text{2. CuBr}]{\text{1. Mg, Et}_2\text{O}} \text{RCu} \xrightarrow{\text{CH}_3\text{C}\equiv\text{CH}} \text{R(CH}_3\text{)C=CHCu} \xrightarrow{\text{CH}_3\text{I}} \text{R(CH}_3\text{)C=CHCH}_3$$

This synthesis of a terpene is highly stereospecific (99.9% *syn* addition). The actual organocopper compound here probably is the **heterocuprate** MgBr[RCuBr]. If lithium is used instead of magnesium, the **homocuprate** $Li[CuR_2]$ is formed, which at the terminal alkyne effects metallation rather than addition (Normant, 1976 R).

⑥ Cu^I and Ag^I also play an important role in **transition-metal catalyzed valence isomerizations** (p. 415).

11.2 Gold Organyls

The chemistry of gold is defined by the oxidation states Au^I and Au^{III}. Inorganic compounds of Au^I (d^{10}) usually assume the coordination number 2 with a linear geometry and a 14 VE configuration. Less abundant are Au^I complexes, possessing the coordination number 3 (trigonal planar, 16 VE) or 4 (tetrahedral, 18 VE). Au^{III} (d^8) complexes are four-coordinate (square planar, 16 VE), the Au($6p_z$) orbital remaining unoccupied. For purely inorganic complexes of Au^{III}, the coordination numbers 5 and 6 are also encountered.
In organogold chemistry, the cases **Au^I, coordination number 2** (linear) and **Au^{III}, coordination number 4** (square planar) are typical. In this section, only gold σ-organyls will be discussed.

Au^I Organyls

Simple binary compounds AuR are as yet unknown; Au^I organyls are, however, obtained as **adducts** LAuR (L = PR_3, RNC) and as **organoaurates** $M[AuR_2]$, the latter exhibiting low stability (Kochi, 1973):

$$Et_3PAuCl + MeLi \xrightarrow[-LiCl]{} Et_3PAuMe \xrightarrow{MeLi} Li[AuMe_2] \qquad \boxed{4}$$

Organoaurates become significantly more stable if the cation M^+ is converted into a complex ion (Tobias, 1976): [Li(PMDT)] $[AuMe_2]$ only decomposes at 120 °C (PMDT

= pentamethyldiethylenetrisamine). $AuMe_2^-$, like the isoelectronic $M(d^{10})$ species $HgMe_2$, $TlMe_2^+$, and $PbMe_2^{2+}$, has a linear structure.
σ-Cyclopentadienyl Au^I compounds according to their ^{1}H-NMR spectra are *fluxional*:

$$Ph_3PAuCl + C_5H_5Na \xrightarrow[-NaCl]{} Ph_3PAu(\eta^1\text{-}C_5H_5)$$

Au^I σ-organyls are also accessible via carbene insertion into the Au–Cl bond:

$$Et_3PAuCl + CH_2N_2 \longrightarrow Et_3PAuCH_2Cl + N_2$$ 10

Au^I has a high affinity for ylids $R_3P{=}CH_2$ (Schmidbaur, 1975 R):

$$Me_3PAuMe + Me_3P{=}CH_2 \xrightarrow[-\,Me_3P]{} Me_3\overset{\oplus}{P}{-}CH_2{-}\overset{\ominus}{Au}Me$$

$$Me_3PAuCl + 2\,Me_3P = CH_2 \xrightarrow[-Me_3P]{} [(Me_3PCH_2)_2Au]Cl$$

$$\xrightarrow[-2\,Me_4P^+Cl^-]{\times 2,\ \text{trans-ylidation}} [Me_2\overset{\oplus}{P}(CH_2)_2\overset{\ominus}{Au}]_2 \ (\text{Au}\cdots\text{Au } 303\text{ pm}) \xrightarrow{Cl_2} [Me_2P(CH_2)_2AuCl]_2 \ (\text{Au–Au } 260\text{ pm})$$

The oxidation product features an Au–Au bond and therefore must be designated as a Au^{II} compound. Additional evidence, which supports this unusual oxidation state of gold is provided by the ^{197}Au-Mössbauer spectrum and by ESCA data (**e**lectron **s**pectroscopy for **c**hemical **a**nalysis = X-ray photoelectron spectroscopy).

Remarks Concerning the Ylid Ligand:

$$\left\{Me_3P = CH_2 \longleftrightarrow Me_3\overset{\oplus}{P} - \overset{\ominus}{\bar{C}}H_2\right\} \xrightarrow{BH_3} Me_3\overset{\oplus}{P} - CH_2 - \overset{\ominus}{B}H_3$$

$$\xrightarrow{ML_n^+} Me_3\overset{\oplus}{P} - CH_2 - ML_n$$

Ylene Ylid

These reactions are accompanied by a transfer of negative charge from the carbanionic site to acceptors like Lewis acids or coordinatively unsaturated complex fragments ML_n^+.
An additional deprotonation converts the ylid into a **phosphoniodiylid**, which has chelating and bridge forming properties:

$$Me_3P{=}CH_2 \xrightarrow[-\,baseH^+]{base} \left\{ Me_2\overset{\oplus}{P}(\overset{\ominus}{C}H_2)(\overset{\ominus}{C}H_2) \longleftrightarrow Me_2P({=}CH_2)(\overset{\ominus}{C}H_2) \longleftrightarrow Me_2P(\overset{\ominus}{C}H_2)({=}CH_2) \right\}$$

If the ylid itself serves as the base, the proton transfer is termed **transylidation**:

$$(CH_3)_2P(CH_3){=}CH_2 + (CH_3)_2P({=}CH_2)CH_3 \longrightarrow (CH_3)_2\overset{\oplus}{P}(CH_3)CH_3 + (CH_3)_2P({=}CH_2)\overset{\ominus}{C}H_2$$

Coordination modes of phosphoniodiylids:

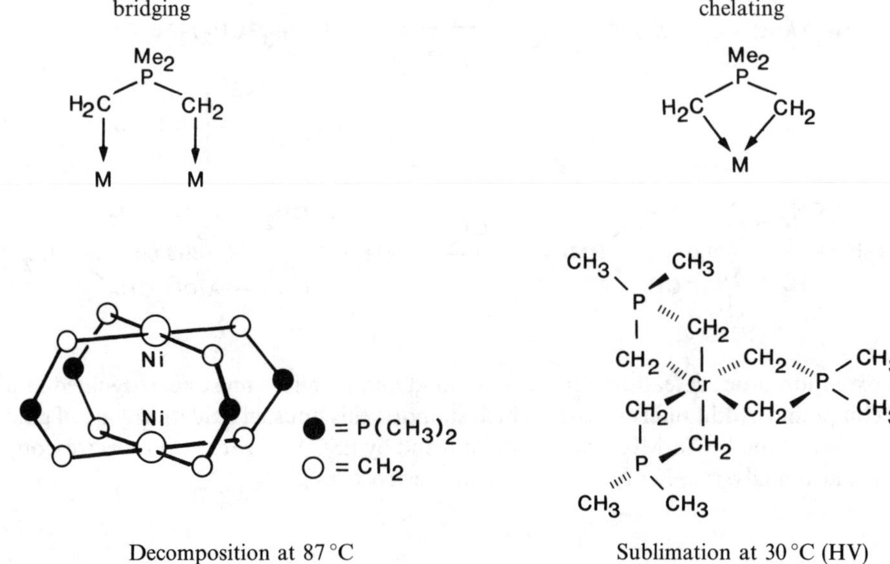

Decomposition at 87 °C

Sublimation at 30 °C (HV)

The bridging variant leads to σ-organometallics, especially of the coinage metals (Cu, Ag, Au), which display surprisingly high stability.

Au^{I}-Carbonyl Complexes

Binary metal carbonyls $Au_m(CO)_n$ are unstable under ambient conditions. The compounds Au(CO) and $Au(CO)_2$, which have been obtained by means of metal-atom ligand-vapor cocondensation (p. 255), decompose above 20 K.
Ternary carbonyl-gold halides are accessible, however:

$$Au_2Cl_6 + 4\,CO \xrightarrow{SOCl_2} 2\,Au(CO)Cl + 2\,COCl_2$$

IR *spectrum*: $\nu_{CO} = 2152\ cm^{-1}$
compare: ν_{CO} (*free*) $= 2143\ cm^{-1}$

The position of the stretching band ν_{CO} suggests that the backbonding contribution Au $\stackrel{\pi}{\rightarrow}$ CO for Au(CO)Cl is insignificant (Calderazzo, 1977). This is what would be expected for an $M(d^{10})$ complex (p. 190). In the first isolable silver carbonyl complex, $Ag(CO)[B(OTeF_5)_4]$, reported by Strauss (1991), the C–O stretching frequency is even higher, $\nu_{CO} = 2204\ cm^{-1}$. Again, CO acts as a Lewis base here only and is exhibiting no π-acidity (see p. 227).

Au^I-Isocyanide and Au^I-Carbene Complexes

Isocyanides RNC as ligands at gold have more to offer than the related ligand CO (cf. p. 240). A considerable number of Au^I-isocyanide complexes are known; these species can be converted into carbene complexes. Neutral $(RNC)Au^I$ species are best obtained via substitution of dimethylsulfide by isocyanide (Bonati, 1973):

$$RNC + (Me_2S)AuCl \longrightarrow (RNC)AuCl + Me_2S$$

Bis(isocyanide)gold[I] cations can be isolated as salts with non-coordinating anions:

$$(RNC)AuCl + RNC + BF_4^- \xrightarrow{\text{acetone}} [RNC{-}Au{-}CNR]BF_4 + Cl^-$$

Conversion into (carbene)gold[I] complexes is effected by the addition of molecules bearing OH- or NH_2 groups (cf. p. 240):

$$(ArNC)AuCl \xrightarrow[65°,\ 2d]{MeOH} (ArNH)(MeO)C{-}AuCl$$

$$[(t\text{-}BuNC)_2Au]PF_6 \xrightarrow[25°,\ 5h]{t\text{-}BuNH_2} [((t\text{-}Bu)NH)_2C{-}Au{-}C(NH(t\text{-}Bu))_2]^+ PF_6^-$$

(McCleverty, 1973)

Au^{III} Organyls

A traditional way of access to organogold[III] compounds is the **auration** reaction (Kharash, 1931):

$$2\ Au_2Cl_6 + 2\ C_6H_6 \xrightarrow[-2H[AuCl_4]]{} [(C_6H_5)ClAu(\mu\text{-}Cl)_2AuCl(C_6H_5)] \qquad \boxed{6}$$

In these binuclear complexes, gold has a square planar environment.

$$Au_2Br_6 + 4\,MeMgBr \longrightarrow Me_2Au(\mu\text{-}Br)_2AuMe_2$$ (Pope, 1907) 4

$$Me_2Au(\mu\text{-}Br)_2AuMe_2 \xrightarrow[H_2O]{+2\,AgNO_3,\ -2\,AgBr} [Me_2Au(OH_2)_2]^+ \xrightarrow{NaN_3} Me_2Au(\mu\text{-}N\text{-}N\equiv N)_2AuMe_2$$

The unit $[Me_2Au(H_2O)_2]^+$ *is surprisingly stable.*

Bis (alkyl)gold^{III} cyanides – as opposed to bis(alkyl)gold^{III} azides – are tetrameric since the ligand $C{\equiv}N^-$ only forms linear bridges. In the ^{1}H-NMR spectrum, four signals are observed instead of the two anticipated signals for the CH_3 protons. Obviously, in addition to form ⓐ (identical orientation of the CN bridges) the disordered form ⓑ is also present:

ⓐ: Me_2Au units linked in a square: Me—Au—C≡N—Au—Me (top), Au—N≡C—Au (left, downward), Au—C≡N—Au (right, downward), Me—Au—N≡C—Au—Me (bottom)

ⓑ: Me—Au—N≡C—Au—Me (top), Au—N≡C—Au (left, downward), Au—C≡N—Au (right, downward), Me—Au—N≡C—Au—Me (bottom)

The inclination of Au^{III} to participate in Au–C–Au ($2e\,3c$) bonding appears to be small; hence, rather than through dimerization to Au_2Me_6 (compare: Al_2Me_6), $AuMe_3$ prefers to achieve coordination number four with the help of additional ligands. The ill-characterized compound trimethylgold, which was obtained in solution by Gilman in 1948, decomposes at T > – 40 °C:

$$Au_2Br_6 + 6\,MeLi \xrightarrow[-65\,°C]{Et_2O} 2\,\text{“}AuMe_3\text{”} + 6\,LiBr$$

The triphenylphosphane adduct Ph_3PAuMe_3, prepared via oxidative addition of MeI to $Li[AuMe_2]$, is stable up to 115 °C (mp, decomp.), however:

$$Li[Au^{I}Me_2] \xrightarrow[Ph_3P]{MeI} Ph_3PAu^{III}Me_3 + LiI \qquad \text{(Kochi, 1973)}$$

$$Ph_3PAu^{III}Me_3 \xrightarrow[-Ph_3P]{MeLi} Li[AuMe_4] \qquad \text{(Tobias, 1975)}$$

The expulsion of the ligand Ph_3P *upon the addition of methyllithium implies that the tetramethylaurate*^{III} *ion is coordinatively saturated with a* 16 VE *configuration. This is a typical feature of* $M(d^8)$ *complexes of the late transition metals (p.* 189*) which we will encounter several times in the following discussion of organotransition-metal chemistry.*

ORGANOMETALLIC COMPOUNDS OF THE TRANSITION ELEMENTS

The full variety of organometallic chemistry presents itself in the realm of transition metals. Contributing factors include:

- *extended possibilities for the formation of metal-carbon bonds. While it is rare for main-group elements to use nd orbitals in addition to ns and np orbitals in chemical bonding, for transition metals, the* ***(n − 1) d, ns and np orbitals*** *must all be regarded as* ***regular valence orbitals.*** *Partial occupation of these orbitals gives transition metals both electron-donor and electron-acceptor properties, which in conjunction with donor/acceptor ligands like* CO, *isonitriles, carbenes, olefins and arenes allows for considerable variation of the metal-ligand bond order* M ⩦ L ***(σ-donor/π-acceptor synergism).*** *This principle is prevalent in large areas of transition-metal organometallic chemistry.*

  Interactions*:

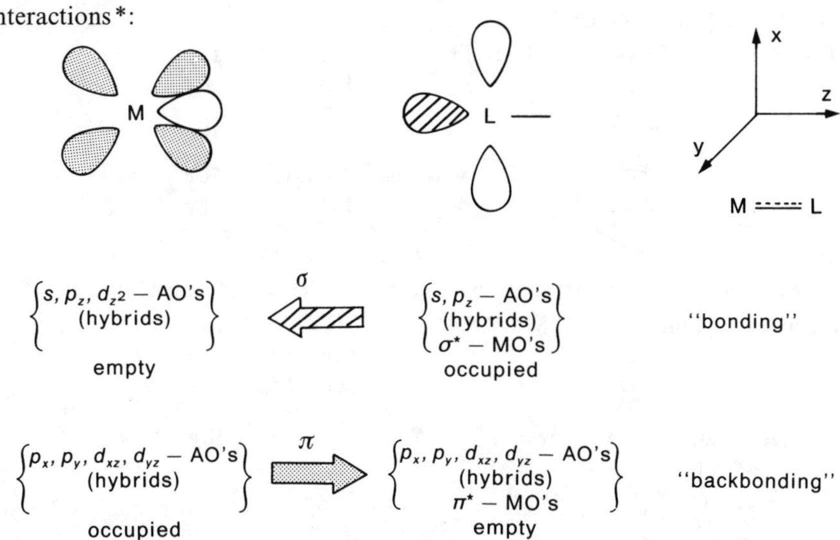

- *various forms of* ***metal-metal multiple bonding*** *with or without additional bridging ligands. This leads to the formation of cluster compounds which are of interest from a structural as well as from a theoretical point of view (see Chapter 16).*

- *the ability to* ***change the coordination number.*** *Together with the lability of metal-carbon σ-bonds this phenomenon offers possibilities for organometallic catalysis (see Chapter 17).*

---

* The shaded areas in this drawing indicate orbital occupation rather than phase.

12 The 18 Valence Electron (18 VE) Rule

The 18 electron rule and the concept of σ-donor/π-acceptor synergism ("bonding and backbonding") represent the most rudimentary basis for discussing structure and bonding in organotransition-metal compounds.

The 18 electron rule (Sidgwick, 1927) is based on the valence bond (VB) formalism of localized metal-ligand bonds; it states that thermodynamically stable transition-metal organometallics are formed when the sum of the metal *d* electrons plus the electrons conventionally regarded as being supplied by the ligands equals 18. In this way, the metal formally attains the electron configuration of the next higher noble gas [18 VE rule, also paraphrased as "inert gas rule", "effective atomic number (EAN) rule"]. When applying the 18 VE rule the following conventions should be considered:

① The intramolecular partitioning of the electrons has to ensure that the total complex charge remains unchanged:

Di(cyclopentadienyl)iron $Fe(C_5H_5)_2$:

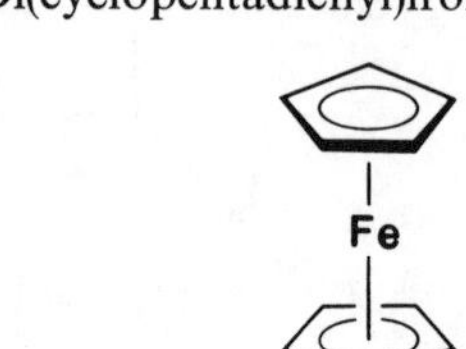

	$2(C_5H_5^-)$	$12e$
	Fe^{2+}	$6e$
		$18e$
or:	$2(C_5H_5\cdot)$	$10e$
	Fe^0	$8e$
		$18e$

② A metal-metal bond contributes *one* electron to the count on each metal:

Decacarbonyldimanganese $Mn_2(CO)_{10}$

$5(CO)$	$10e$
Mn^0	$7e$
$Mn-Mn$	$1e$
	$18e$

③ The electron pair of a bridging carbonyl ligand donates *one electron to each* of the bridged metals:

Noncarbonyldiiron $Fe_2(CO)_9$

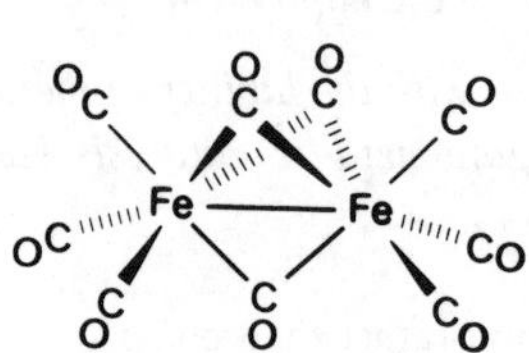

$3(CO)$	$6e$
$3(\mu\text{-}CO)$	$3e$
Fe^0	$8e$
$Fe-Fe$	$1e$
	$18e$

In the following table, ligands commonly encountered in organotransition-metal chemistry are listed together with the respective numbers of electrons relevant to the application of the 18 VE rule:

Neutral	Positive	Negative	Ligand L
1	0	2	alkyl, aryl, hydride, halide (X)
2	–	–	ethylene, monoolefin, CO, phosphane etc.
3	2	4	π-allyl, enyl, cyclopropenyl, NO
4	–	–	diolefin
4	–	6	cyclobutadiene (C_4H_4 or $C_4H_4^{2-}$)
5	–	6	cyclopentadienyl, dienyl
6	–	–	arene, triolefin
7	6	–	tropylium ($C_7H_7^+$)
8	–	10	cyclooctatetraene (C_8H_8 or $C_8H_8^{2-}$)

The 18 VE rule has considerable predictive value in that the composition of many transition-metal complexes may be anticipated from combinations of sets of ligands with transition metals of appropriate *d* electron count. For organometallics of the *f* elements (lanthanides and actinides) this procedure is not applicable (see p. 364).
A perspective view at coordination chemistry (p. 188) suggests a division into three classes:

Class	Number of valence electrons	18 VE rule
I	··· 16 17 18 19 ···	not obeyed
II	··· 16 17 18	not exceeded
III	18	obeyed

How does the nature of the central atom as well as the ligands determine whether a complex belongs to class I, II or III?

GUIDING PRINCIPLES:

– bonding orbitals	should be	occupied
– nonbonding orbitals	may be	occupied
– antibonding orbitals	should not be	occupied

COMMENTS:

Class I: The splitting Δ_0 is relatively small for 3 *d* metals as well as for ligands at the lower end of the spectrochemical series.

- t_{2g} is nonbonding and can be occupied by 0–6 electrons.
- e_g^* is weakly antibonding and can be occupied by 0–4 electrons.

Therefore, 12–22 valence electrons can be accomodated, i.e., the 18 VE rule is not obeyed. Due to their inherently-small splitting Δ_{tetr}, tetrahedral complexes also belong to this class.

Typical transition-metal complexes and their assignment to classes I–III

Class I	n(VE) ≧ 18	
	$n(d)$	n(VE)
TiF_6^{2-}	0	12
VCl_6^{2-}	1	13
$V(C_2O_4)_3^{3-}$	2	14
$Cr(NCS)_6^{3-}$	3	15
$Mn(acac)_3$	4	16
$Fe(C_2O_4)_3^{3-}$	5	17
$Co(NH_3)_6^{3+}$	6	18
$Co(H_2O)_6^{2+}$	7	19
$Ni(en)_3^{2+}$	8	20
$Cu(NH_3)_6^{2+}$	9	21
$Zn(en)_3^{2+}$	10	22

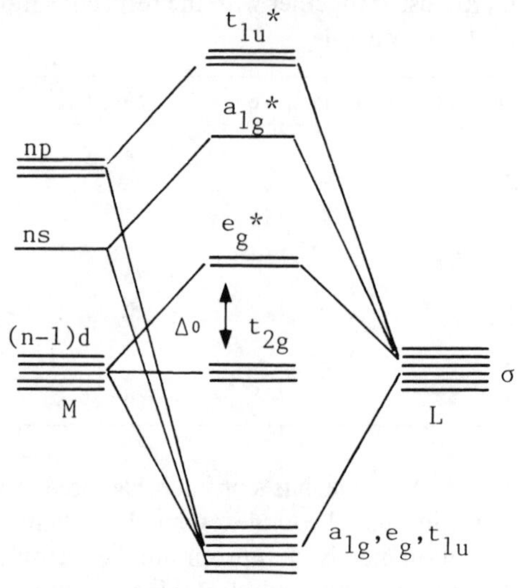

Molecular orbital diagram for an octahedral complex (simplified), σ-bonding only

Class II	n(VE) ≦ 18	
	$n(d)$	n(VE)
ZrF_6^{2-}	0	12
WCl_6	0	12
WCl_6^-	1	13
WCl_6^{2-}	2	14
TcF_6^{2-}	3	15
$OsCl_6^{2-}$	4	16
$W(CN)_8^{3-}$	1	17
$W(CN)_8^{4-}$	2	18
PtF_6	4	16
PtF_6^-	5	17
PtF_6^{2-}	6	18
$PtCl_4^{2-}$	8	16

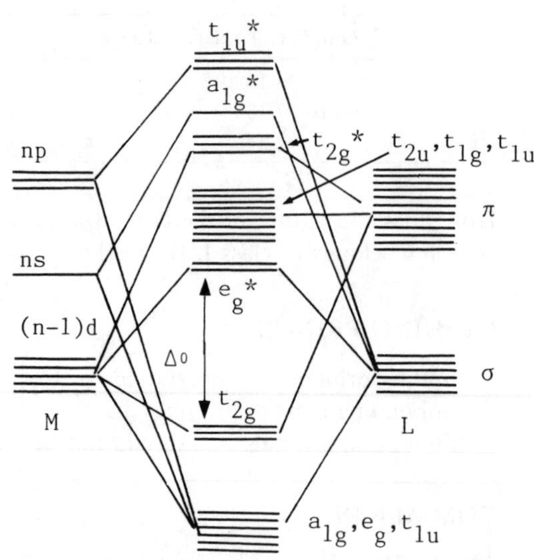

Molecular orbital diagram for an octahedral complex (simplified), σ- and π-bonding

adapted from P. R. Mitchell, R. V. Parish, J. Chem. Ed. 46 (1969) 811.

Class III		n = 18	
	$n(d)$	n(L)	n(VE)
$V(CO)_6^-$	6	12	18
$CpMn(CO)_3$	7	11	18
$Fe(CN)_6^{4-}$	6	12	18
$Fe(PF_3)_5$	8	10	18
$Fe(CO)_4^{2-}$	10	8	18
$CH_3Co(CO)_4$	9	9	18
$Ni(CNR)_4$	10	8	18
$Fe_2(CO)_9$	8	10	18
$[CpCr(CO)_3]_2$	6	12	18

Class II: Δ_0 is larger for $4d$ and $5d$ metals (especially in high oxidation states), and for σ ligands in the intermediate and upper ranges of the spectrochemical series.

- t_{2g} is essentially nonbonding and can be occupied by 0–6 electrons.
- e_g^* is more strongly antibonding and thus no longer available for occupancy.

Consequently, the valence shell contains 18 VE or less. A similar splitting Δ_0 (t_{2g}, e_g^*) is also observed for complexes of the $3d$ metals with ligands possessing extremely high ligand field strength (e.g. CN^-).

Class III: Δ_0 is largest for ligands at the uppermost end of the spectrochemical series (good π-acceptors like CO, PF_3, olefins, arenes).

- t_{2g} becomes bonding due to the interaction with π orbitals of the ligands and should be occupied by 6 electrons.
- e_g^* is strongly antibonding and remains empty.

Therefore the 18 VE rule is obeyed, unless steric reasons prevent the attainment of an 18 VE shell [$V(CO)_6$ (17 VE), WMe_6 (12 VE)].

Organometallic compounds of the transition metals almost exclusively belong to class III.

THE PARTICULAR CASE OF M(d^8) AND M(d^{10}) COMPLEXES

For elements at the end of the transition series ("late transition metals") 16- or 14 VE configurations are favored over 18 VE configurations.

Examples:	M(d^8)	
$[Ni(CN)_4]^{2-}$	16 VE	
$[Rh(CO)_2Cl_2]^-$	16 VE	square planar, C.N. 4
$[AuCl_4]^-$	16 VE	

According to the 18 VE rule, a coordination number (C.N.) of 5 would have been expected. This is found for "earlier transition metals" with a d^8 electron configuration:

$Fe(CO)_5$, $(\eta^4\text{-}C_6H_8)(\eta^6\text{-}C_6H_6)Ru$, $[Mn(CO)_5]^-$.

Examples:	M(d^{10})	
$[Ag(CN)_2]^-$	14 VE	linear, C.N. 2
R_3PAuCl	14 VE	

A qualitative explanation takes into account:

(1) the electroneutrality principle (Pauling, 1948)
(2) the π-acceptor character of the ligands
(3) the change in energetic separation between $(n-1)d$, ns and np orbitals within a transition series

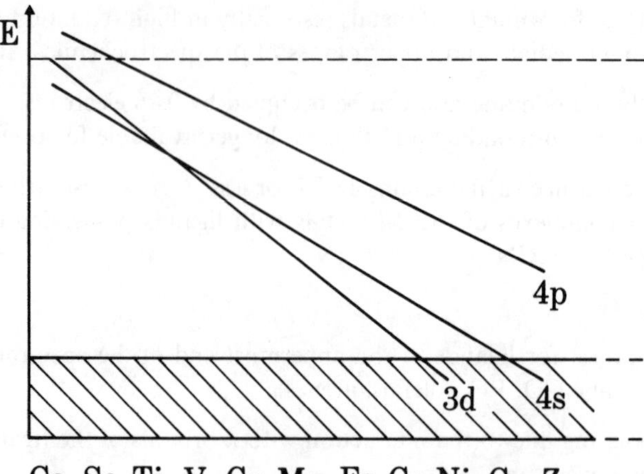

The increase in atomic number is accompanied by a decrease in orbital energies because additional valence electrons only partially screen the higher nuclear charge. For $(n-1)\,d$ *orbitals, the decrease in energy is more pronounced than for ns and np orbitals, as d electrons show less interelectronic repulsion than s and p electrons. Moreover, a positive charge on the metal also serves to increase the separation between s, p and d orbitals. Thus,* Ca^0 *possesses the ground configuration* [Ar] $4s^2$*, while that of the isoelectronic ion* Ti^{2+} *is* [Ar] $3d^2$*. A more refined treatment is offered by F. L. Pilar* (1978)*: "*$4s$ *is* ***always*** *above* $3d$*!".*

ARGUMENTATIVE EXAMPLES BASED ON POINTS (1)–(3) (p. 189)

- $Fe(CO)_5$, $Fe(CNR)_5$
  $Fe^0(d^8)$, 18 VE

The energies of the d, s and p orbitals are still rather close; effective backbonding Fe $\xrightarrow{\pi}$ CO from the relatively large $Fe^0(3d)$ orbitals leads to an equilibration of charge thereby fulfilling the electroneutrality principle.

- $Ni(CO)_4$
  $Ni^0(d^{10})$, 18 VE, C.N. 4

but

$$\underset{Ni^0(d^{10}),\ 18\,VE,\ C.N.\,4}{Ni(PR_3)_4} \overset{25\,^\circ C}{\rightleftharpoons} \underset{Ni^0(d^{10}),\ \mathbf{16\,VE},\ C.N.\,3}{Ni(PR_3)_3 + PR_3}$$

The lower π-acceptor properties of PR_3, compared to CO, and the low energy of the d orbitals in the $Ni^0(d^{10})$ configuration (low π-donor properties) limit the extent of back-bonding, leading to an equilibrium between C.N. 4 und C.N. 3.

- $$\underset{Ni^{II}(d^8),\ 18\,VE,\ C.N.\,5}{[Ni(CN)_5]^{3-}} \rightleftharpoons \underset{Ni^{II}(d^8),\ \mathbf{16\,VE},\ C.N.\,4}{[Ni(CN)_4]^{2-} + CN^-}$$

The equilibrium strongly favors the 16 VE complex, although CN^- is a good π-acceptor and steric crowding is absent. The increase in nuclear charge $Ni^0 \rightarrow Ni^{2+}$ effects a strong

decrease in the energy of $3d$ relative to $4s$ and $4p$. The possibility to attain electroneutrality through $Ni(3d_\pi) \rightarrow CN_\pi$ interaction is limited.
With the ligand Cl^-, which is lower in the spectrochemical series, only the 16-VE complex $[NiCl_4]^{2-}$ is realized.
A further increase in the separation of the $(n-1)$ *d*, *ns* and *np* orbitals is observed when an $M(d^{10})$ configuration *also* has a positive nuclear charge, as in Cu^I, Ag^I, Au^I and Hg^{II}. **14 VE** complexes are now favored with C.N. 2 (linear). Bonding in these complexes is best described in terms of $M(d_{z^2}s)$ rather than $M(sp)$-hybrids (Orgel, 1960 R). The energetically higher $M(p)$ orbitals thus remain uninvolved.

Excursion
Can the VSEPR concept (*v*alence *s*hell *e*lectron-*p*air *r*epulsion, Gillespie-Nyholm model) be applied to transition-metal complexes?

The structures of simple compounds of the main-group elements can often be predicted by counting bonding and non-bonding valence electron pairs. An important role is played by the stereochemically active lone pairs of electrons. The transfer of this principle to transition-metal complexes is generally not possible (see p. 192).

- *The **common occurrence of coordination number 6, independent of the number of valence electrons** (see class* I, *S.* 187) *shows that for transition-metal complexes electrostatic ligand-ligand interaction and the ratio* r_{anion}/r_{cation} *are of structural significance in addition to covalent interactions.*
- ***For 18 VE complexes,** structures are observed which would also have been predicted by VSEPR theory. In the case of* $M(d^4)$, C.N. 7, *the two complementary structures "pentagonal bipyramid" (e.g.* $[Mo(CN)_7]^{5-}$*) and "monocapped trigonal prism" (e.g.* $[(t\text{-}BuNC)_7Mo]^{2+}$*) both occur. For* $M(d^8)$, C.N. 5 *as well, complementary structures are encountered:* $[Ni(CN)_5]^{3-}$ *(square pyramid) and* $Fe(CO)_5$ *(trigonal bipyramid).*
- *For species with **less than 18 valence electrons,** structural predictions by means of VSEPR theory are successful only in the case of* $\mathbf{M(d^{10})}$ *complexes (example:* $Ni(CO)_3$, *trigonal planar).*
- *For species with **less than 18 valence electrons and an incomplete*** $\mathbf{M(d^n)}$ ***valence shell*** $\mathbf{(n < 10)}$, *VSEPR theory is inapplicable. Qualitative predictions can, however, be made as follows:*
  (a) *Start with the corresponding* 18 VE *complex and successively remove the ligands together with their σ-bonding electron pairs.*
  (b) *Leave the structure of the remaining fragment unchanged [example:* $Cr(CO)_6 \rightarrow Cr(CO)_5 \rightarrow Cr(CO)_4 \rightarrow Cr(CO)_3$].

In a phenomenological vein, vacant coordination sites play the same role for transition metals as lone pairs do for main-group elements. This, however, is only true for ligands with modest steric demand.
More subtle explanations are provided by ligand field theory, molecular orbital theory and the angular overlap model.

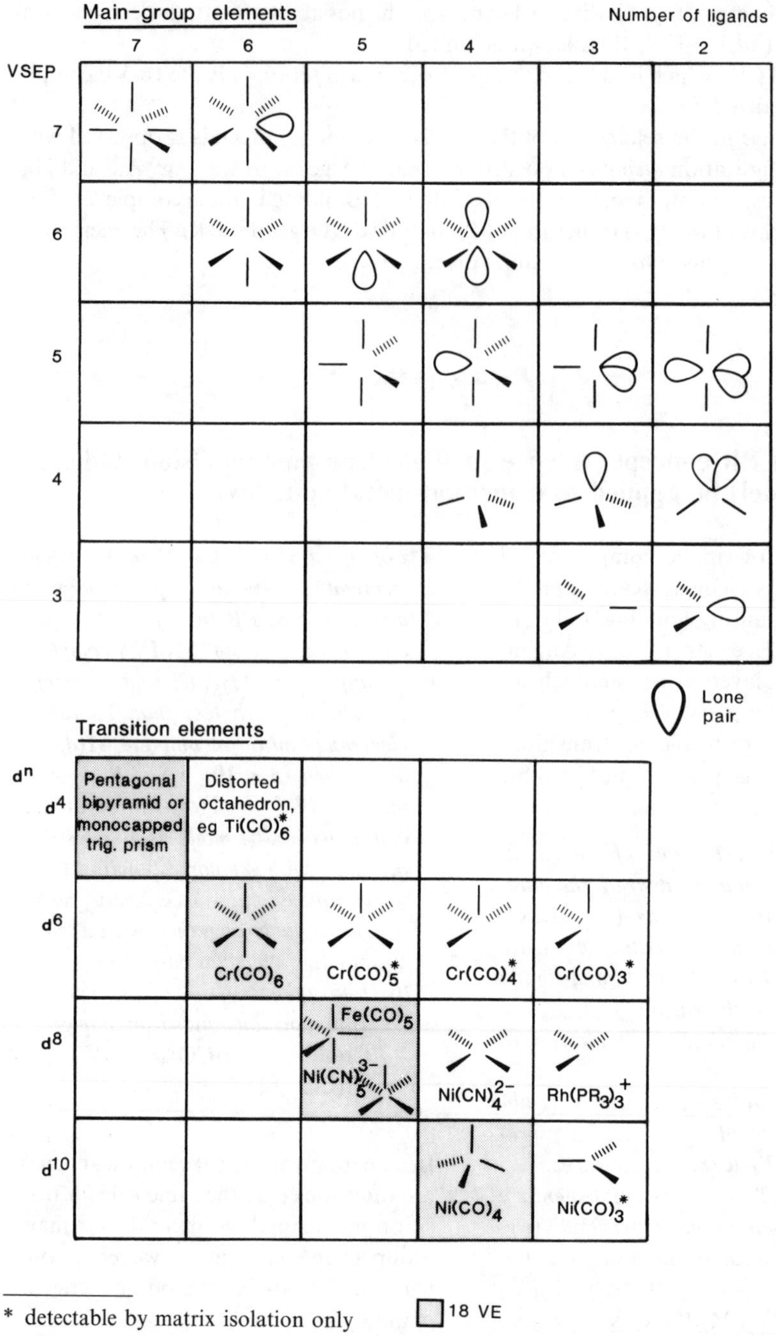

Main-group elements

Number of ligands

7 6 5 4 3 2

VSEP

7

6

5

4

3

Lone pair

Transition elements

d^n

d^4 Pentagonal bipyramid or monocapped trig. prism | Distorted octahedron, eg $Ti(CO)_6^*$

d^6 $Cr(CO)_6$ | $Cr(CO)_5^*$ | $Cr(CO)_4^*$ | $Cr(CO)_3^*$

d^8 $Fe(CO)_5$, $Ni(CN)_5^{3-}$ | $Ni(CN)_4^{2-}$ | $Rh(PR_3)_3^+$

d^{10} $Ni(CO)_4$ | $Ni(CO)_3^*$

adapted from D. M. P. Mingos in COMC **3** (1982) 14

* detectable by matrix isolation only

18 VE

13 σ-Donor Ligands

As mentioned earlier in Chapter 2, it is convenient to classify organometallic compounds of the transition elements according to their respective ligands. The following table lists the more important organic ligands containing at least two carbon atoms, which coordinate to the transition metal through σ bonds:

Carbon Hybridization	Ligand: terminal		Ligand: bridging	
sp^3	M—CR_3	Alkyl	M—M bridged by CR_3	3-Center μ_2-alkyl
			M—M bridged by CR_2	μ_2-Alkylidene
			M_3 capped by CR	μ_3-Alkylidyne
sp^2	M—C_6H_5	Aryl	M—M bridged by aryl	3-Center μ_2-aryl
	M=CR_2	Carbene or alkylidene	M—M bridged by CR	μ_2-Alkylidyne
	M—CR=CR_2	Vinyl		
	M—C(=O)R	Acyl	M—M bridged by C=CR_2	μ_2-Vinylidene
sp	M≡CR	Carbyne or alkylidyne	M—C≡CR→M	$\mu_2(\sigma,\pi)$-Alkynyl
	M—C≡CR	Alkynyl		
	M=C=CR_2	Vinylidene	M—M bridged by C≡CR	3-Center μ_2-alkynyl

Depending on the capability of the ligand to form multiple bonds, the pure σ bond (M – C single bond) can be supplemented by various degrees of π interaction (M ⩦ C multiple bond). The discussions found in this and the following chapters are roughly ordered according to the degree of increasing π-acceptor properties of the ligand, beginning with transition-metal alkyls ("**σ complexes**") and ending with transition-metal complexes of cyclic conjugated π ligands ("**π complexes**").

13.1 Preparation of Transition-Metal Alkyls and -Aryls

Several preparative methods mentioned in Chapter 4 (methods [1] – [12]) for main-group organometallics are applicable to transition metals as well. Additional methods [13] – [15] for transition-metal organometallics use metal complexes in low oxidation states as starting materials (ligand = CO, PR_3, olefins etc.); complexes of this kind are uncommon for main-group elements.

Metal halide + organolithium (Mg, Al) reagent **Metathesis [4]**

$$ZrCl_4 \xrightarrow[Et_2O]{4\ PhCH_2MgCl} Zr(CH_2Ph)_4$$

$$PtCl_2(PR_3)_2 \xrightarrow{Li(CH_2)_4Li} (R_3P)_2Pt(CH_2)_4$$

a metallacycle

Apart from the strongly carbanionic alkyllithium- and Grignard reagents, other main-group organometallics, namely those of Al, Zn, Hg, and Sn, can also be used. Their attenuated alkylating power can be utilized if only partial exchange of halide ligands is desired.

$$TiCl_4 \xrightarrow{Al_2Me_6} MeTiCl_3$$

$$NbCl_5 \xrightarrow{ZnMe_2} Me_2NbCl_3$$

Metal hydride + alkene **Alkene insertion** 8
(Hydrometallation)

trans-$(Et_3P)_2PtClH$ + C_2H_4 ⟶ trans-$(Et_3P)_2PtClC_2H_5$

$CpFe(CO)_2H$ + (butadiene) ⟶ $CpFe(CO)_2{-}CH_2{-}CH{=}CH{-}CH_3$

1,4 addition, cis and trans

The insertion of an olefin into the M–H bond is also the decisive step in certain homogeneous catalytic processes (see Chapter 17).

Metal hydride + carbene **Carbene insertion** 10

$CpMo(CO)_3H$ $\xrightarrow{CH_2N_2}$ $CpMo(CO)_3CH_3$

Carbonylate anion + alkyl halide **Metallate alkylation** 13

$Mn_2(CO)_{10}$ $\xrightarrow{Na/Hg}$ $Na[Mn(CO)_5]$ $\xrightarrow{CH_3I}$ $CH_3Mn(CO)_5$ + NaI

$W(CO)_6$ $\xrightarrow[-3CO]{NaCp}$ $Na[CpW(CO)_3]$ $\xrightarrow{C_2H_5I}$ $CpW(CO)_3C_2H_5$

A common side reaction is the elimination of HX from RX, caused by the nucleophilic carbonylate anions.

Carbonylate anion + acyl halide **Metallate acylation**

$[CpFe(CO)_2]^-$ $\xrightarrow{CH_3COCl}$ $CpFe(CO)_2(COCH_3)$ $\xrightarrow[-CO]{h\nu}$ $CpFe(CO)_2CH_3$

Many metal acyl complexes eliminate one molecule of CO on heating or after photochemical irradiation (see p. 359). This reaction is often reversible.

16 VE metal complex + alkyl halide **Oxidative addition** 14

Many complexes with 16 valence electrons, in which the metal has either a d^{10} or d^8 configuration, react with alkyl halides. This raises the formal oxidation state and the coordination number of the metal by +2. The process is called oxidative addition; it can lead to either cis or trans addition.

16 VE —CH_3I, cis-OA→ 18 VE —$-C_2H_4$→ 16 VE

16 VE —CH_3Cl, trans-OA→ 18 VE —ΔT→ 18 VE

This type of reaction is also observed for 18 VE complexes with a basic metal:

18 VE —CH_3I→ 18 VE → 18 VE

Complexes with high electron density on the central atom and the propensity to increase their coordination number are sometimes referred to as "**metal bases**" (Shriver, 1970 R).

Olefin complex + nucleophile **Addition** 15

—$[OCH_3]^-$→

Upon nucleophilic addition to an η^2-ligand, a π/σ-rearrangement is observed. A further example:

$$\left[(CO)_5Mn \cdots \begin{matrix} CH_2 \\ \| \\ CH_2 \end{matrix}\right]^+ \xrightarrow{NaBH_4} (CO)_5Mn-CH_2-CH_3$$

13.2 Selected Properties of Transition-Metal σ Organometallics

13.2.1 Thermodynamic Stability versus Kinetic Lability

Earlier attempts to prepare binary transition-metal alkyls or -aryls like diethyliron or dimethylnickel showed that such complexes could not be made under normal laboratory conditions, although they possibly existed as solvates at low temperatures.
Known compounds with transitionmetal-carbon σ bonds invariably contained additional ligands, notably η^5-C_5H_5, CO, PR_3 or halides:

$$PtCl_4 + 4\,CH_3MgI \longrightarrow 1/4\,[(CH_3)_3PtI]_4$$

(Pope, Peachey, 1909)

Me_3Pt — I, I — $PtMe_3$, I — $PtMe_3$, Me_3Pt — I

heterocubane type

$$CrCl_3 + 3\,C_6H_5MgBr \xrightarrow{THF} (C_6H_5)_3Cr(THF)_3$$

(Zeiss, 1957)

Further examples: $CpFe(CO)_2CH_3$, $CH_3Mn(CO)_5$ (Cp = η^5-C_5H_5). This led to the assumption that transition-metal TM $\overset{\sigma}{-}$ C bonds generally are weaker than main-group element MGE $\overset{\sigma}{-}$ C bonds, a hypothesis that today is no longer tenable.
A comparison of force constants for the metal-carbon stretching frequency shows that MGE $\overset{\sigma}{-}$ C und TM $\overset{\sigma}{-}$ C bonds can be of comparable strength:

$M(CH_3)_4$	Si	Ge	Sn	Pb	Ti
force constants $k(M-C)/N\,cm^{-1}$	2.93	2.72	**2.25**	1.90	**2.28**

Thermochemical data for transition-metal organometallics point in the same direction and it is currently assumed that the bond energy of TM $\overset{\sigma}{-}$ C bonds lies between 120–350 kJ/mol. It should be noted, however, that the following data are mean values $\bar{E}(M-C)$ rather than individual bond dissociation energies (cf. p. 12). Values found at the lower end of this scale should be considered more representative:

Compound	$\bar{E}(M \overset{\sigma}{-} C)$/kJ/mol	Compound	$\bar{E}(M \overset{\sigma}{-} C)$/kJ/mol
Cp_2TiPh_2	330	WMe_6	160
$Ti(CH_2Ph)_4$	260	$(CO)_5MnMe$	150
$Zr(CH_2Ph)_4$	310	$(CO)_5ReMe$	220
$TaMe_5$	260	$CpPtMe_3$	160

Despite the fundamental importance of $M \overset{\sigma}{-} C$ *bond energies, their quantitative assessment is still a matter of debate. Inconsistencies may arise from the various experimental methods by which these data are obtained (thermochemistry, photochemistry, equilibrium constants, kinetic measurements etc., see Beauchamp, 1990 R). Even if a reliable value for* $\bar{E}(M-C)$ *is available, it remains questionable whether this value can be transferred to other organometallic fragments (Dias, 1987).*

With regard to the TM – C bond, it can nevertheless be stated with confidence that:

- *the* TM – C *bond is weaker than the bond of* TM *to other main-group elements* (F, O, Cl, N)
- *its energy (in contrast to* MGE – C) *increases with increasing atomic number*
- *steric effects must be taken into account* (e.g. $TaMe_5$, WMe_6).

Further, it is now generally recognized that the difficulties in handling binary transition-metal alkyls are not due to particularly low **thermodynamic stability**, but rather to high **kinetic lability** (Wilkinson, 1970). An effective strategy to prepare such compounds must therefore aim at blocking possible decomposition pathways.

One general decomposition mechanism involves **β-elimination**, resulting in a metal hydride and an olefin:

$$L_nM-CH_2(\alpha)-CH(\beta)(R)-H \rightleftharpoons [L_nM\cdots CH_2=CHR\cdots H]^{\ddagger} \rightleftharpoons L_nM(H)(\eta^2\text{-}CH_2=CHR) \rightleftharpoons L_nMH + CHR=CH_2$$

CN = n+1 CN = n+2 decomp.

Experimental proof for this mechanism is provided by the formation of copper deuteride in the following thermolysis:

$$(Bu_3P)CuCH_2CD_2C_2H_5 \longrightarrow (Bu_3P)CuD + CH_2{=}CDC_2H_5$$

β-Eliminations can also be reversible:

$$Cp_2Nb(C_2H_4)C_2H_5 \underset{+C_2H_4}{\overset{-C_2H_4}{\rightleftharpoons}} Cp_2Nb(C_2H_4)H$$

In this context, the facile loss of hydrogen from η^4-cyclopentadiene complexes is readily understood. This reaction proceeds via the intermediate formation of a cyclopentadienyl hydride [*cf.* C–H⇀M (2e 3c) *bonds*, *p.* 267].

The first step can be regarded as an intramolecular oxidative addition:

$$2\ (\eta^4\text{-}C_5H_6)Fe^{0}(CO)_3 \longrightarrow 2\ CpFe^{II}(CO)_2H \longrightarrow [CpFe(CO)_2]_2 + H_2$$

β-Elimination can be suppressed, i.e. TM$\overset{\sigma}{-}$alkyls become **inert**, when

(a) *the formation of an olefin is either sterically or energetically unfavorable*
(b) *the organic ligand is void of hydrogen in the β-position*
(c) *the central atom is coordinatively saturated*

Examples:
(a) **Double bonds to bridgehead carbon atoms are unfavorable (Bredt's rule):**

R =

In binary complexes MR_4, *the norbornyl group stabilizes the rare oxidation states* Cr^{IV}, Mn^{IV}, Fe^{IV} *and* Co^{IV}.

(b) **Absence of β-H:**

C-ligands without β-hydrogens

unidentate

$-CH_3$ $-CH_2C(CH_3)_2-CH_3$ $-CH_2Si(CH_3)_2-CH_3$ $-CH_2-C_6H_5$ $-CH_2C(CH_3)_2-C_6H_5$

bidentate

$(H_3C)_2Si(CH_2)_2$ $(H_3C)_2P(CH_2)_2$ $HC(PR_2CH_2)_2$

The silaneopentyl group ($R = CH_2SiMe_3$) forms rather inert $TM \overset{\sigma}{-} C$ bonds (VOR_3, mp 75 °C, CrR_4, mp 40 °C).
Furthermore, compare the thermal stability of the neopentyl complex $Ti[CH_2C(CH_3)_3]_4$, (mp 90 °C) with the analogous alkyl complexes $Ti(CH_3)_4$ (decomp. −40 °C) and $Ti(C_2H_5)_4$ (decomp. −80 °C). This also holds for the corresponding zirconium complexes; while $Zr(Ph)_4$ cannot be isolated and $Zr(CH_3)_4$ decomposes above −15 °C, tetrabenzylzirconium is stable up to the melting point (132 °C).

(c) **Absence of free coordination sites:**

This aspect is illustrated by the differing properties of $Ti(CH_3)_4$ and $Pb(CH_3)_4$:

$Ti(CH_3)_4$	$Pb(CH_3)_4$
decomp. −40 °C	distills at 110 °C/1 bar (despite a smaller force constant of the $Pb \overset{\sigma}{-} C$ bond, cf. p. 197!)

The decomposition of $Ti(CH_3)_4$ possibly proceeds via a bimolecular mechanism:

$$(CH_3)_3Ti \begin{smallmatrix} CH_3 \\ \\ CH_3 \end{smallmatrix} Ti(CH_3)_3$$

This dimerisation should effect a weakening of the $Ti-C$ $(2e\,2c)$ bond by converting it into a three center bond $Ti \overset{C}{\frown} Ti$ $(2e\,3c)$.

For $Pb(CH_3)_4$, an analogous decomposition pathway would be unfavorable, since the main-group element lead has only high-energy *d* orbitals at its disposal for extending the coordination shell.
Tetramethyltitanium in the form of its bipyridyl adduct (bipy)$Ti(CH_3)_4$ is thermally much more stable (blockage of free coordination sites, suppression of bimolecular decomposition). Bidentate ligands like bis(dimethylphosphino)ethane (dmpe) are particularly suitable. The inert nature of 18 VE $TM \overset{\sigma}{-} C$ alkyls, which contain cyclopentadienyl ligands, is explained in a similar manner. The labile, coordinatively unsaturated species $Ti(CH_3)_4$ contrasts with the relatively inert, sterically shielded molecule $W(CH_3)_6$ (Wilkinson, 1973):

$$WCl_6 \xrightarrow[DME]{CH_3Li} W(CH_3)_6$$ (red crystals, mp 30 °C)
IR: $\nu_{W-C} = 482\ cm^{-1}$
^{1}H-NMR: δ 1.62 ppm, $J(^{183}W, ^1H) = 3$ Hz

Interestingly, the structure of gaseous hexamethyltungsten is trigonal prismatic rather than octahedral (Haaland, 1990).
A further variant of coordinative saturation and thus kinetic stabilization is realized in complexes of the general type CH_3TiX_3 (X = halide, OR, NR_2). Cryoscopic measurements on $CH_3Ti(OR)_3$ show only a small degree of association for $CH_3Ti(OCHMe_2)_3$, while in solution $CH_3Ti(OC_2H_5)_3$ exists almost exclusively as a dimer with alkoxy bridges:

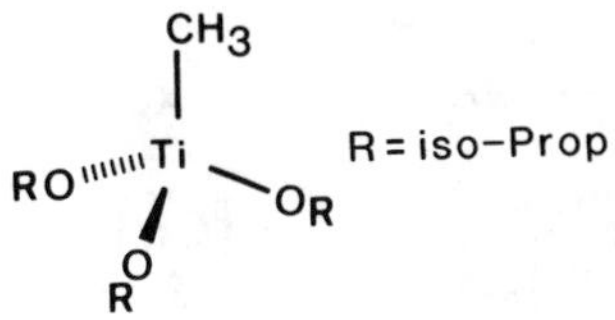

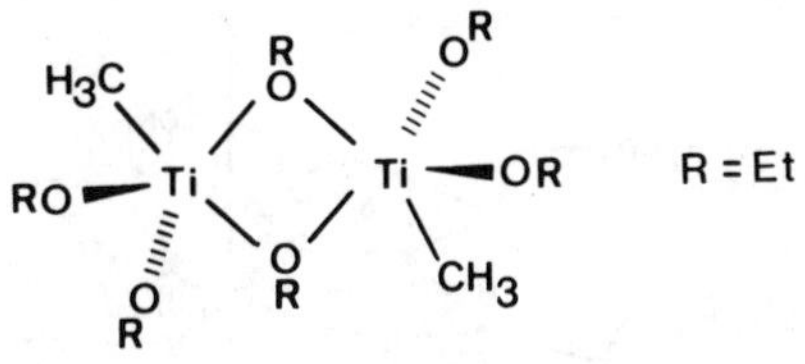

R = iso-Prop

R = Et

In addition, He(II)-photoelectron spectra suggest considerable mixing between empty titanium orbitals and the doubly occupied orbitals of the oxygen atoms. This π-donor effect is also important in other complexes $RTiX_3$, its strength increasing in the order $X = Cl < OR < NR_2$.

If all three criteria (a)–(c) are fulfilled, a compound with remarkable thermal stability results:

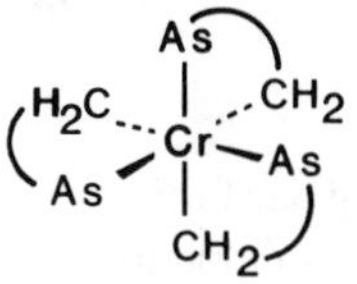

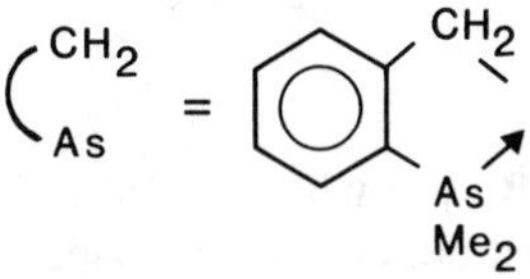

stable up to 350 °C!
(Tzschach, 1970)

$TM \overset{\sigma}{-} C$ bonds can even occur in complexes of the classical Werner type, in so far as these complexes are kinetically inert. This is encountered in the cases of Cr^{III} (d^3), Co^{III} (d^6) and Rh^{III} (d^6).

Examples: $[Rh(NH_3)_5C_2H_5]^{2+}$, $[Cr(H_2O)_5CHCl_2]^{2+}$.

13.2.2 Transition-Metal σ Complexes in vivo

With the possible exception of the iron containing redox enzymes cytochrome P 450 (Ortiz de Montellano, 1982 R), and the nickel containing cofactor F 430, operative in methanogenic bacteria (Jaun, 1991), the **cobalamins** are the only naturally-occurring compounds known to date which at some stage feature TM – C σ bonds. They cobalamines consist of a central cobalt atom, a substituted corrin ring which acts as an equatorial tetradentate ligand, an axial ligand X, and a 5,6-dimethylbenzimidazole base that is linked to the side chain of a pyrrole ring through ribose and phosphate units. The various cobalamins differ in the nature of the ligand X:

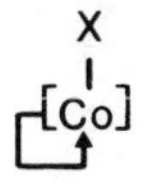

X	CH_3	CN	Ad
Xcobalamins	methylcobalamin	cyanocobalamin (vitamin B_{12})	5′-desoxyadenosylcobalamin (coenzyme B_{12})

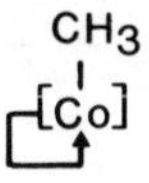

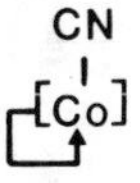

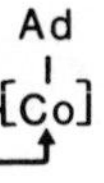

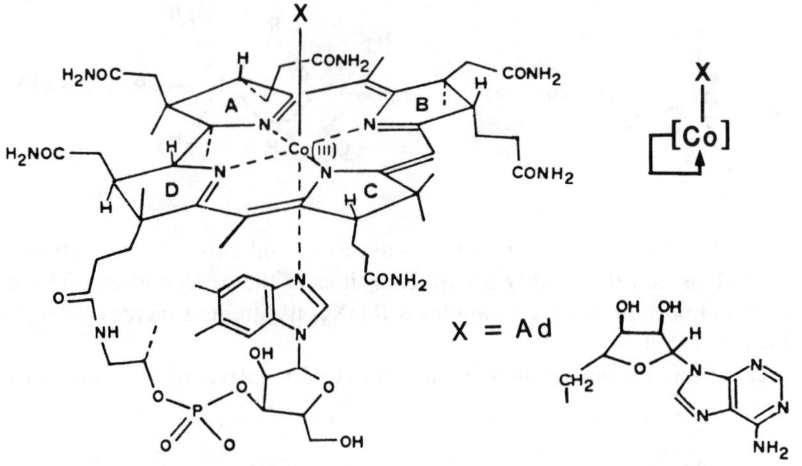

Coenzyme B_{12}, of which the adult human body contains 2–5 mg, is stable in a neutral, aqueous medium and resists oxidation by air. Methylcobalamin (isolable from microorganisms) releases methane and ethane only above 180 °C as result of a homolytic cleavage of the Co–C bond. This homolysis can also be initiated by photochemical irradiation. In the **isomerase reaction**, coenzyme B_{12} catalyses 1,2-shifts of alkyl groups:

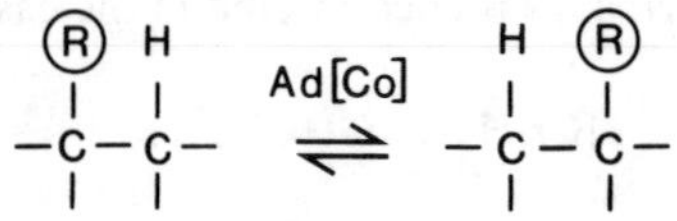

With the intention to clarify the mechanism of this and other reactions, an extensive organometallic chemistry of the cobalamins and their model systems, the cobaloximes, has been developed (Schrauzer, Dolphin, 1976 R).
The chemical versatility of the cobalamins is founded upon the following aspects:

- The largely covalent Co–C bond can be represented by three resonance structures:

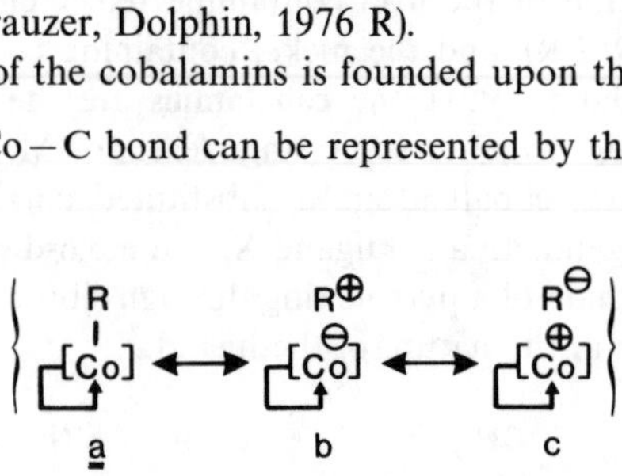

Depending on the reaction conditions and the nature of the attacking reagent, **three possible modes of Co–C bond cleavage** can be envisaged. They are all supported by experimental evidence:

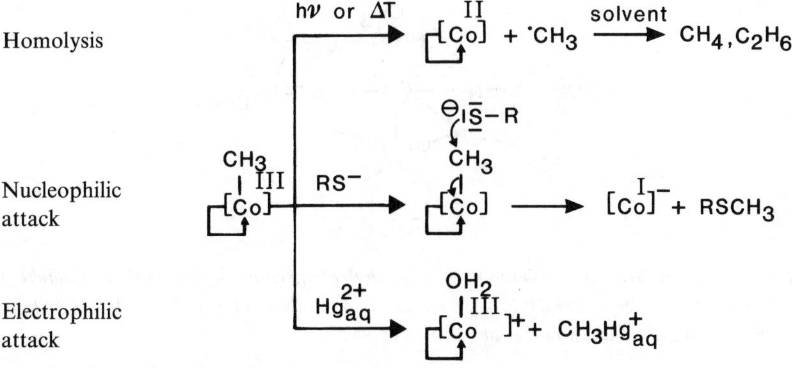

In the latter transalkylation, the methyl group is transferred as a carbanion (cf. p. 54f).

- **Changes of the oxidation state of the cobalt atom** in cobalamins can be induced in vitro by electrochemical means (Savéant, 1982 R) as well as by chemical reducing agents:

$$\underset{\text{purple}}{\underset{B_{12b}}{\overset{OH}{[Co^{III}]}}} \underset{-e^-}{\overset{+e^-}{\rightleftharpoons}} \underset{\text{brown}}{\underset{B_{12r}}{[Co^{II}]}} \underset{-e^-}{\overset{+e^-}{\rightleftharpoons}} \underset{\text{green}}{\underset{B_{12s}}{[Co^{I}]^{\ominus}}} \xrightarrow{RX} \overset{R}{[Co^{III}]} + X^-$$

$$[Co^{I}]^{\ominus} \xrightarrow{H^+} \left\{\overset{H}{[Co^{III}]}\right\} \longrightarrow [Co^{II}] + 1/2\,H_2$$

Vitamin B_{12s} is oxidized by O_2, but metastable under physiological pH conditions. It represents an extraordinarily strong reducing agent. By means of oxidative addition of alkylating agents to B_{12s}, several cobalamin derivatives may be obtained. Hydridocobalamin is unstable in aqueous solution. The reduced form B_{12s} apparently plays an important role in enzyme reactions, catalysed by coenzyme B_{12}. These reactions have not been clarified in detail, but mechanistic patterns of organometallic chemistry may help in proposing reasonable schemes.

Many properties of the natural cobalamins are mimicked by the **cobaloximes**, which are easily prepared from Co^{II} salts, dimethylglyoxime, a base and an alkylating agent (Schrauzer, 1964).

$$Co^{2+}_{aq} + 2\,dmgH_2 \xrightarrow[X^-]{O_2,\ base} XCo(dmgH)_2B + 2\,HB^+$$

$$XCo(dmgH)_2B + CH_3MgX \xrightarrow{THF} CH_3Co(dmgH)_2B + MgX_2$$

($dmgH_2$ = dimethylglyoxime)

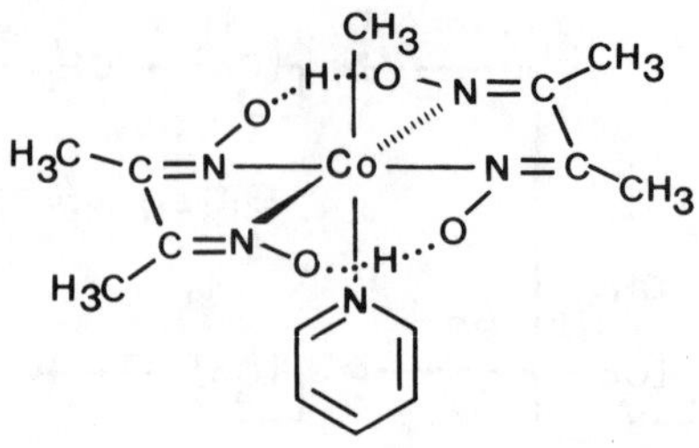

Bis(dimethylglyoximato)(methyl)(pyridine)cobalt, ***methylcobaloxime****. In this synthetic model compound the bond lengths* $d(\text{Co}-\text{N}_{\text{eq}})$ *and* $d(\text{Co}-\text{C})$ *are identical to those of coenzyme* B_{12}*. The ligand field strengths of the original and its model are also very similar.*

13.2.3 Transition-Metal Perfluorocarbon σ Complexes

Transition-metal organometallics R_F-M with perfluorinated ligands, in comparison with the corresponding complexes R_H-M, very often show increased thermal stability which may be traced to a combination of thermodynamic and kinetic factors. Example:

$CF_3Co(CO)_4$	$CH_3Co(CO)_4$
can be distilled without decomposition at 91 °C	decomposition at −30 °C

Due to the high lattice energy of metal fluorides (the likely products of decomposition), R_F-M complexes are thermodynamically less stable than R_H-M species. Therefore, the higher thermal stability of perfluorinated organometallics probably reflects the increased bond dissociation energy $D(M-R_F)$.
Possible reasons for $D(M-R_F) > D(M-R_H)$:

$\overset{\delta-}{R_F} \overset{\sigma}{-} \overset{\delta+}{M}$ *The high partial charge* $M^{\delta+}$ *generates contracted metal orbitals, whereby metal-carbon overlap is increased.*

$Ar_F \overset{\pi}{\leftarrow} M$ *Low-lying π* molecular orbitals in perfluorosubstituted arenes effectively lend themselves to* Ar ← M *backbonding thereby increasing the metal-carbon bond strength.*

The high bond energy of the $M-R_F$ bond may also be responsible for the fact that insertions of CO molecules into M−C bonds are unfavorable:

$$R\text{-}M(CO)_n \xrightarrow{\Delta T} RCO\text{-}M(CO)_{n-1} \quad \textit{readily}$$

$$R_F\text{-}M(CO)_n \xrightarrow{\Delta T} R_FCOM(CO)_{n-1} \quad \textit{rarely and only at elevated temperatures}$$

Suitable methods for the **preparation** of perfluoroalkyl- and perfluoroarylmetal complexes include coupling reactions, oxidative additions and insertions, but, compared to hydrocarbon complexes, certain peculiarities have to be taken into consideration.

Perfluoro**acyl** halides on reaction with carbonylate ions readily form perfluoroacyl-metal complexes, from which the perfluoroalkyl derivatives are easily prepared by decarbonylation:

$$R_FCOCl + Na[Mn(CO)_5] \longrightarrow R_FCOMn(CO)_5 \xrightarrow[-CO]{\Delta T} R_FMn(CO)_5$$

Perfluoro**alkyl** chlorides are unreactive; the corresponding iodides in reactions with metal carbonyl anions display reversed polarity $R_F^{\delta-}-I^{\delta+}$:

$$2\ F_3CI + Na[Mn(CO)_5] \longrightarrow Mn(CO)_5I + NaI + C_2F_6$$

With neutral metal carbonyls, R_FI undergoes oxidative addition, thus behaving like a pseudo-interhalogen compound:

$$R_FI + Fe(CO)_5 \longrightarrow (CO)_4Fe(R_F)(I) + CO$$

Cyclic perfluoroolefins and -arenes are highly susceptible to nucleophilic substitution of F^- by metal carbonyl anions:

$$[CpFe(CO)_2]^- + C_6F_6 \longrightarrow CpFe(CO)_2\overset{\sigma}{-}C_6F_5 + F^-$$

$$[Mn(CO)_5]^- + C_4F_6 \text{ (hexafluorocyclobutene)} \longrightarrow C_4F_5Mn(CO)_5 + F^-$$

In contrast, the less nucleophilic neutral metal carbonyls are subject to oxidative addition in reactions with unsaturated fluorocarbons, the degree of fluorination being maintained:

$$\overset{0}{Fe}(CO)_5 + F_2C{=}CF_2 \longrightarrow (CF_2)_4Fe^{II}(CO)_4 + CO$$

Alternatively, insertion of a perfluoroolefin into a metal-metal bond is observed:

$$Co_2(CO)_8 + F_2C{=}CF_2 \longrightarrow (CO)_4Co{-}CF_2{-}CF_2{-}Co(CO)_4$$

14 σ-Donor/π-Acceptor Ligands

14.1 Transition-Metal Alkenyls and -Aryls

Complexes containing the structural elements

M–C=C	M–C₆H₅	M–C≡C–
-alkenyl (vinyl complexes)	-phenyl (aryl complexes)	-alkynyl (acetylide complexes)

occupy a position between the complexes with pure σ-donor ligands (transition-metal alkyls) and the large class of σ-donor/π-acceptor complexes (L = CO, phosphane etc.). Although alkenyl-, alkynyl- and aryl ligands have empty π^* orbitals which, in principle, are suitable for interaction with occupied d orbitals, structural data suggest that the M–C interaction has little double-bond character. If the M–C distances were shorter than the sum of the single bond covalent radii, then some π-bonding would be indicated. The structural data of the following platinum complexes are a case in point:

	Pt–C	r_{cov} Pt^{II}	r_{cov} C in pm
trans-[PtCl(CH$_2$SiMe$_3$)(PPhMe$_2$)$_2$]	208	131	77 (sp^3)
trans-[PtCl(CH=CH$_2$)(PPhMe$_2$)$_2$]	203	131	67 (sp^2)
trans-[PtCl(C≡CPh)(PPhMe$_2$)$_2$]	198	131	60 (sp)

PPhMe$_2$

Cl — Pt —(203)— CH 135 pm

127° CH$_2$

PPhMe$_2$

In all three examples, the distances Pt–C *are* ***not*** *shorter than the sum of the covalent radii of* Pt^{II} *and* C *(in the respective hybridization state). They therefore give no indication of π-bonding between platinum and carbon. If considerable back bonding* Pt$(d_\pi) \xrightarrow{\pi}$ C(p^*_π) *occurred, one would also expect a lengthening of the* C–C *bond, which is not observed.*

Apart from the usual methods of preparation, which have been shown also to be applicable to transition-metal alkyls (cf. p. 194f), one possible route to **transition-metal σ-vinyl complexes** is the oxidative addition of HX to η^2-alkyne complexes:

$$(R_3P)_2Pt\cdots(RC\equiv CR) \xrightarrow{HCl} (R_3P)_2ClPt-C(R)=C(R)H$$

In this case, the proton enters cis to the metal, which may reflect prior protonation of the metal.

Transition-metal σ-alkenyls can also be obtained by means of nucleophilic addition to cationic alkyne complexes. The *trans* stereochemistry of the product indicates that the nucleophile attacks from the non-coordinated side of the ligand:

$$[CpFe(CO)(PPh_3)(CH_3C\equiv CCH_3)]^+ \xrightarrow{Li_2[Cu(CN)Ph_2]} CpFe(CO)(PPh_3)-C(CH_3)=C(CH_3)Ph$$

Vinyl groups occasionally act as bidentate σ/π-bridging ligands, as the olefinic double bond remains unchanged by σ-coordination. Alternatively, this can be regarded as ferra-allyl coordination to the second iron atom (C. Krüger, 1972):

$$Fe_2(CO)_9 + HBrC=CHBr \longrightarrow (CO)_3Fe(\mu\text{-}Br)(\mu\text{-}CH=CHBr)Fe(CO)_3 \quad (Fe\text{–}Fe\ 252)$$

Compared to the number of metal alkynyls discussed in the next chapter, metal alkenyls are relatively rare.

Transition-metal σ-aryl complexes are not neccessarily more stable than the analogous alkyl complexes. $TiPh_4$ decomposes above 0 °C to give $TiPh_2$ and biphenyl. Exceptional stability is encountered in the square-planar nickel, palladium and platinum complexes $(mesityl)_2M(PR_3)_2$. Contrarily, the corresponding alkyl and phenyl complexes are quite labile.

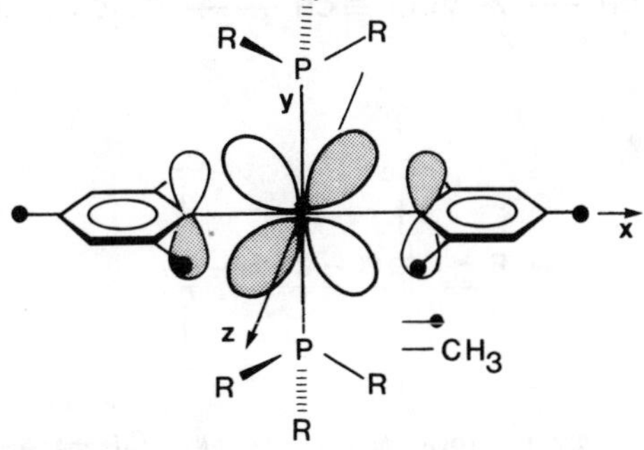

*The **ortho**-substituents prevent rotation of the aryl group, forcing it into a conformation that ensures optimal overlap* $Pt(d_{xy})$-Aryl(π^*). *A further role of the ortho-substituents, in addition to preventing β-elimination, may be to sterically inhibit the attack of ligands along the z-axis.*

14.2 Transition-Metal Alkynyls

Transition-metal alkynyls can be regarded as complexes of the ligand $HC{\equiv}C|^{\ominus}$ which is isoelectronic to CN^-, CO and N_2. In their stoichiometry, color and magnetic properties, metal alkynyls are closely related to the corresponding complex cyanides. The acetylide ligand may also be regarded as a pseudohalide due to its similar behavior in complex formation and precipitation reactions. Therefore, transition-metal alkynyls are often regarded as being part of classical coordination chemistry rather than organometallic chemistry (Nast, 1982 R).

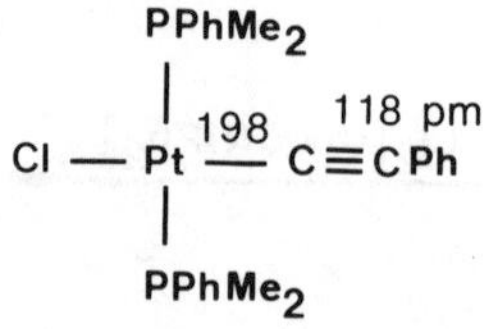

$$Cl-\underset{PPhMe_2}{\overset{PPhMe_2}{Pt}}\overset{198}{-}\overset{118\ pm}{C{\equiv}CPh}$$

As in the corresponding vinyl complex, M—C *as well as* C—C *bond lengths do not indicate substantial π-bonding for the platinum-carbon bond. The π-acceptor character of the alkynyl ligand apparently is of no significance.*

In contrast to CN^- und CO, $RC{\equiv}C|^{\ominus}$ is strongly basic; acetylide complexes are therefore easily hydrolysed. Their preparation requires anhydrous solvents like liquid ammonia:

$$K_3[Cr(CN)_6] + 6\,NaC{\equiv}CH \xrightarrow[-40^{\circ}]{NH_3} K_3[Cr(C{\equiv}CH)_6] + 6\,NaCN$$

Hexa(ethynyl)chromate(III)

Acetylide complexes are generally quite explosive, their thermal stability decreasing in the order $(ArC_2)_nM > (HC_2)_nM > (RC_2)_nM$.
More stable to hydrolysis, but no less explosive than the monomeric alkynyl complexes are the neutral, polymeric metal alkynyls of Cu^I, Ag^I, Au^I:

$$2\,CuI + 2\,KC{\equiv}CH \longrightarrow 2\,CuC{\equiv}CH \xrightarrow{T>45^{\circ}} Cu_2C_2 + HC{\equiv}CH + 2\,KI$$

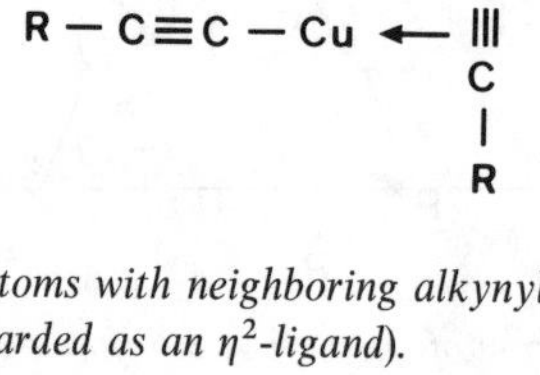

$$R-C{\equiv}C-Cu \leftarrow \|\!|(C{\equiv}C, \text{ with Cu above and R below}); \quad C \uparrow Cu$$

Additional interactions of copper atoms with neighboring alkynyl ligands give each copper atom the coordination number 3 (alkynyl regarded as an η^2-ligand).

$$AuCl_3(aq) \xrightarrow[KBr]{SO_2} [AuBr_2]^- \xrightarrow[NaOAc]{RC\equiv CH} [AuC\equiv CR]_n + H^+$$

COATES, 1962

$$[AuC\equiv CR]_n \xrightarrow{PPh_3} Ph_3P-Au-C\equiv C-R$$

$$[Bu^t-C\equiv C-Au]_4 \xrightarrow{PPh_3} Ph_3P-Au-C\equiv C-R$$

The relative ease of depolymerization of alkynyl-gold complexes by Lewis bases may be associated with the reluctance of gold(I) to exceed coordination number 2. As implied by the weakness of the gold-olefin bond and the non-existence of binary gold carbonyls (Chapter 14.5), the σ-acceptor/π-donor synergism is only weakly established in the "late transition-metal ion" $Au^I(d^{10})$.

Transition-metal σ-alkynyls occur as intermediates in the rearrangement of 1-alkyne to vinylidene ligands (Werner, 1983):

$$(PR_3)_2ClRh(HC\equiv CPh) \rightleftarrows (PR_3)_2ClRh(H)(C\equiv CPh)$$

$$(PR_3)_2ClRh(H)(C\equiv CPh) \xrightarrow{py} (PR_3)_2Cl(py)Rh(H)(C\equiv CPh) \xrightarrow[-C_5H_6,\ -NaCl]{NaCp} (PR_3)_2Rh(py)(C\equiv CPh)$$

$$(PR_3)_2Rh(py)(C\equiv CPh) \xrightarrow[-py,\ -PR_3]{C_5H_6} CpRh(PR_3)(=C=CHPh)$$

The phenylalkyne π complex equilibrates with its σ-alkynyl(hydrido) isomer. Upon addition of pyridine, this equilibrium is shifted, strongly favoring the σ-alkynyl(hydrido) species. The hydrogen migration from metal to carbon is thought to proceed via a two-step intermolecular elimination/addition mechanism (see also vinylidene complexes p. 217).

14.3 Transition-Metal Carbene Complexes

Compounds containing metal-carbon double bonds are generally called metal-carbene complexes. If no heteroatoms are directly bound to the carbene carbon atom, they are referred to as metal-alkylidene complexes. The terms "Fischer carbene complexes" (heteroatom substituted) and "Schrock carbene complexes" (C, H substituted) are also used occasionally. Complexes with metal-carbon triple bonds are designated metal-carbyne or metal-alkylidyne complexes. The first carbene complex was prepared in 1964, the first carbyne complex in 1973. Since then, this class of compounds has developed into an important branch of organometallic chemistry. The role carbene complexes assume as intermediates in olefin metathesis (p. 419) has further contributed to their current relevance. The synthesis of the dichlorocarbene complex (tetraphenylporphyrin)-$Fe(CCl_2) \cdot H_2O$, prepared from CCl_4, is regarded as a model reaction for the degradation of chlorohydrocarbons in the liver as catalyzed by the enzyme system cytochrome P-450 (Mansuy, 1978).

PREPARATION

There are many routes to carbene complexes. The following examples also portray typical reactions of the respective starting materials.

1. Addition of alkyllithium to M–CO (E. O.Fischer, 1964)

$W(CO)_6 \xrightarrow[Et_2O]{LiR} (CO)_5W{=}C(R)O^-Li^+ \xrightarrow{[(CH_3)_3O]BF_4} (CO)_5W{=}C(R)OCH_3$

This reaction is concordant with quantum-chemical calculations (Fenske, 1968), which have shown that the carbon atom in coordinated CO *should bear a larger positive charge than in free* CO. *Nucleophilic attack should therefore be favorable.*

A variant of this procedure is the protonation (alkylation) of neutral acyl complexes (Gladysz, 1983):

$CpRe(NO)(PPh_3)C(O)Me \xrightarrow{CF_3SO_3H} [CpRe(NO)(PPh_3){=}C(Me)OH]^+$

2. Addition of ROH to isocyanide complexes (Chatt, 1969)

$$Cl-Pt(Cl)(PPh_3)-C{\equiv}N-Ph \xrightarrow{EtOH} Cl-Pt(Cl)(PPh_3){=}C(NHPh)(OEt)$$

3. From electron-rich olefins (Lappert, 1977 R)

$$\text{(PhN)}_2C_2H_4C{=}C\text{(NPh)}_2C_2H_4 \xrightarrow[-\,CO]{Fe(CO)_5} (OC)_4Fe{=}C\text{(NPh)}_2C_2H_4$$

4. From carbonyl metallates and geminal dichlorides (Öfele, 1968)

$$Na_2Cr(CO)_5 + Ph_2C_3Cl_2 \xrightarrow[THF,\ -2\ NaCl]{-20^\circ} (CO)_5Cr{=}C_3Ph_2$$

5. Interception of free carbenes (Herrmann, 1975)

$$C_5H_5Mn(CO)_2THF + CH_2N_2 \xrightarrow{-N_2} [(C_5H_5)(CO)_2Mn]_2(\mu\text{-}CH_2) + M-(CH_2{=}CH_2)$$

8% 92%

$$(PPh_3)_3OsCl(NO) \xrightarrow[-PPh_3,\ -N_2]{CH_2N_2} (ON)Cl(PPh_3)_2Os{=}CH_2$$

(Roper, 1983)

6. α-Deprotonation of an M-alkyl group (Schrock, 1975)

intermolecular reaction:

$$[Cp_2Ta(CH_3)_2]^+ \xrightarrow[-\ CH_3OH]{NaOCH_3} Cp_2Ta(=CH_2)CH_3$$

intramolecular reaction:

$$(Me_3CCH_2)_3TaCl_2 \xrightarrow[-LiCl]{LiCH_2CMe_3} (Me_3CCH_2)_2(Cl)Ta(\alpha\text{-CH}\beta\text{-CMe}_3)(H\cdots CH_2CMe_3) \longrightarrow$$

$$(Me_3CCH_2)_2Ta(=CHCMe_3)Cl \xrightarrow[-LiCl]{LiCH_2CMe_3} (Me_3CCH_2)_3Ta{=}C(H)CMe_3$$

$$+ CMe_4$$

7. Hydride abstraction from an M-alkyl group (Gladysz, 1983)

$$CpRe(NO)(L)CH_3 \xrightarrow[CH_2Cl_2]{Ph_3C^+PF_6^-} [CpRe(NO)(L){=}CH_2]^+ + Ph_3CH$$

STRUCTURE AND BONDING

The metal-carbene bond can be described by several resonance forms:

$$\left\{ L_n\overset{\ominus}{M}-\overset{\oplus}{C}(\bar{X})R \longleftrightarrow L_nM{=}C(\bar{X})R \longleftrightarrow L_n\overset{\ominus}{M}-C({=}\overset{\oplus}{X})R \right\}$$

a b c

VB

MO

The relative contributions of the individual resonance forms **a**, **b** and **c** depend on the π-donor properties of the substituents M, X and R at the carbene carbon atom. From

structural studies of various carbene complexes, the following characteristics can be deduced:

- *the coordinated carbene has trigonal planar geometry, carbon hybridization is approximately* sp^2.
- *The bond* M–C*(carbene) is significantly shorter than an* M–C *single bond, but longer than the bond* M–C(CO) *in metal carbonyls. This points to the importance of resonance form* **b** [(M$(d_\pi) \rightarrow$ C(p_π) *contribution*].
- *The bond* C–X (X = *heteroatom) is shorter than a single bond. A contribution of the resonance form* **c** [C$(p_\pi) \leftarrow$ X(p_π) *interaction*] *is thereby implied.*

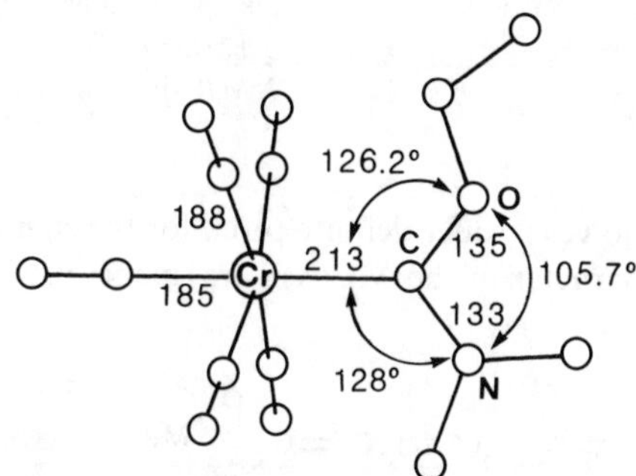

Structure of **$(CO)_5CrC(OEt)NMe_2$**. *The atoms* Cr, C, O, N *are coplanar, the* O–C–N *moiety is staggered with respect to the* $Cr(CO)_5$ *unit (Huttner, 1972).*

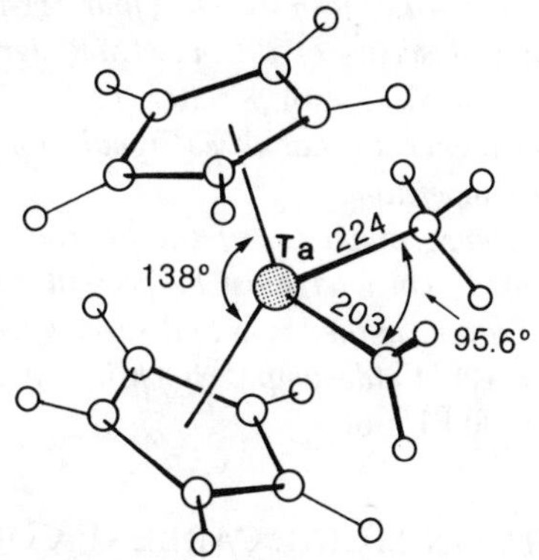

Structure of **$(\eta^5\text{-}C_5H_5)_2Ta(CH_2)CH_3$**. *This complex allows a comparison of bond parameters of the groups* $M{=}CH_2$ *and* $M{-}CH_3$ *in an identical environment (Schrock, 1975).*

Spectroscopic characteristics:

$Cr(CO)_6$: $\nu_{(CO)}$ stretching frequency (Raman): 2108 cm^{-1}
$(CH_3O)CH_3C{=}Cr(CO)_5$: $\nu_{(CO)}$ *trans* to carbene: 1953 cm^{-1}

These **IR data** imply that the carbene carbon atom is a weaker π-acceptor and/or a stronger σ-donor than CO. This is in agreement with the dipole moment

$(CO)_5Cr^{\delta-} \leftarrow C^{\delta+}$ of 4 Debye, whose direction indicates a significant positive charge on the carbene carbon atom (resonance form **a**) and suggests electrophilic reactivity of this site.

^{13}C-NMR	δ (ppm)	reacts as:
$Cp_2(Me)Ta{=}CH_2$	224	Nu
$(t\text{-}BuCH_2)_3Ta{=}C(t\text{-}Bu)H$	250	Nu
$(CO)_5Cr{=}C(N(CH_3)_2)H$	246	El
$(CO)_5Cr{=}C(OCH_3)Ph$	351	El
$(CO)_5Cr{=}C(Ph)Ph$	399	El

The ^{13}***C-NMR*** *signals of coordinated carbene carbon atoms are spread over a large shift range; as with the signals of organic carbenium ions, they appear at low field (compare* Ph_3C^+: δ 212 ppm, Me_3C^+: δ 336 ppm). *The compounds* $(t\text{-}BuCH_2)_3Ta{=}C(t\text{-}Bu)H$ *and* $(CO)_5Cr{=}C(NMe_2)H$ *demonstrate that reliable predictions of nucleophilic or electrophilic behavior cannot be made on the basis of* ^{13}*C-NMR spectra.*

^{1}**H-NMR spectral data,** however, offer definite proof for the significance of M=C as well as C=X double-bond character in the MC(X)R fragment:

$$(CO)_5Cr{=}C(OMe)Me \;(\text{trans}) \rightleftharpoons (CO)_5Cr{=}C(OMe)Me \;(\text{cis}); \quad (CO)_5Cr{=}C(NMe_2)Me$$

In the solid state, the methoxy(methyl)carbene complex is in its ***trans****-conformation. In solution at* −40 °C, *both* ***cis*** *and* ***trans*** *forms occur (four* ^{1}*H-NMR signals). Raising the temperature leads to pairwise coalescence of the* ^{1}*H-NMR signals, caused by the rotation around the* C−O *bond axis* (E_a = 52 kJ/mol, *Kreiter*, 1969). *For the dimethylamino complex, the rotational barrier is higher so that three signals for the non-equivalent methyl groups are observed at room temperature.*

The $M{=}CR_2$ *bond in alkylidene complexes of the Schrock type is particularly rigid; considerable* M=C *double-bond character must be present here* (p. 213). *Thus, the non-equivalence of the methylene protons in the* ^{1}*H-NMR spectrum of* $MeCp(Cp)Ta(CH_2)CH_3$ *remains unchanged up to* $T = 100\,°C$ *(decomp.), which indicates an activation energy for methylene rotation of* $\Delta G^{\ddagger} > 90$ kJ/mol.

REACTIONS OF TRANSITION-METAL CARBENE COMPLEXES

The **heterocarbene complexes**, prepared by methods **1**−**3**, are relatively inert, as the vacant orbital at the carbene carbon atom can be filled by π-interaction with heteroatom lone pairs as well as with occupied metal *d* orbitals. Nevertheless, this **carbene carbon** atom still shows **electrophilic character** (resonance form **a**, p. 212) and reacts with a variety of nucleophiles, e.g. amines and alkyllithium reagents, forming other carbene complexes.

$$(CO)_5Cr{=}C(OCH_3)CH_3 \xrightarrow[-\,CH_3OH]{C_2H_5NH_2} (CO)_5Cr{=}C(NHC_2H_5)CH_3$$

$$(CO)_5W{=}C(Ph)(OCH_3) \xrightarrow[-78^{\circ}]{PhLi} [(CO)_5\overline{W}{-}C(Ph)(Ph)(OCH_3)]^- \xrightarrow[-78^{\circ},\ -CH_3OH]{HCl} (CO)_5W{=}C(Ph)(Ph)$$

(Casey, 1973)

The acidity of a neighboring methyl group is markedly enhanced:

$$(CO)_5Cr{=}C(OCH_3)(CH_3) \xrightarrow[CH_3ONa]{CH_3OD} (CO)_5Cr{=}C(OCH_3)(CD_3)$$

Carbene complexes in general are poor sources of free carbenes. In specific cases they can, however, be used as **carbene transfer agents**. An example is the cationic carbene complex $[CpFe(CH_2)L_2]^+$, which is generated and used *in situ* (Brookhart, 1987 R):

$$CpFe(P\frown P){-}CH_2OCH_3 \xrightarrow[-CH_3OH]{CF_3COOH} [CpFe(P\frown P){=}CH_2]^+ \xrightarrow{CH_2{=}CHOEt} \text{cyclopropyl ethyl ether}$$

cyclopropanation

The following chiral carbene complex reacts with diethyl fumarate under ***chirality transfer*** *to yield an optically active cyclopropane (E. O. Fischer, 1973). This reaction most likely proceeds via initial loss of one CO ligand and coordination of the olefin to the vacant coordination site. Intramolecular cycloaddition then forms a metallacyclobutane intermediate (see Chauvin mechanism, p. 420).*

$$(CO)_4Cr(P^*MePrPh){=}C(OMe)(Ph) + EtO_2C{-}CH{=}CH{-}CO_2Et \xrightarrow{40^{\circ},\ 6h} \text{cyclopropane } (CO_2Et,\ OMe,\ Ph,\ CO_2Et)$$

A further use of carbene complexes in organic synthesis is their reaction with alkynes. This method utilizes the bifunctionality of carbonyl carbene complexes. Pentacarbonyl-[methoxy(phenyl)carbene]chromium reacts under mild conditions with a number of non-heteroatomsubstituted alkynes to form 4-methoxy-1-naphthols, which are π-bonded to $Cr(CO)_3$ fragments. The unsubstituted naphthol ring and the $C(OCH_3)$ unit of the second ring originate from the carbene ligand, while the group C(OH) is derived from a carbonyl ligand (Dötz, 1984 R):

$$(CO)_5Cr{=}C(OCH_3)(C_6H_5) + R_1{-}C{\equiv}C{-}R_2 \xrightarrow[-\ CO]{50^{\circ}} \text{1-OH-2-}R_1\text{-3-}R_2\text{-4-}OCH_3\text{-naphthalene}{-}Cr(CO)_3$$

This cyclization reaction, which can be applied to other carbonyl carbene complexes as well, has been used for the synthesis of certain natural products, e.g. derivatives of Vitamin E and K.
Like organic carbonyl derivatives, carbene complexes can undergo a type of Wittig reaction (Casey, 1973):

$$(CO)_5W{=}C(Ph)_2 + Ph_3P{=}CH_2 \longrightarrow (CO)_5WPPh_3 + H_2C{=}C(Ph)_2$$

compare:

$$O{=}C(Ph)_2 + Ph_3P{=}CH_2 \longrightarrow OPPh_3 + H_2C{=}C(Ph)_2$$

The similarity between these two reactions can be rationalized on the basis of the following isolobal analogy (p. 396):

$$\underset{d^6\text{-}ML_5}{(CO)_5W} \leftarrow\!\circ\!\rightarrow \underset{d^8\text{-}ML_4}{(CO)_4Fe} \leftarrow\!\circ\!\rightarrow CH_2 \leftarrow\!\circ\!\rightarrow O$$

Alkylidene complexes prepared by method **6**, which are void of heteroatom stabilization, are considerably more reactive. They are primarily formed with transition metals in high oxidation states bearing ligands which are weak π-acceptors. In these complexes, the **alkylidene carbon atom** is **nucleophilic** and electrophiles may be added:

$$(Me_3CCH_2)_3Ta{=}C(H)(CMe_3) + O{=}C(R)(H) \longrightarrow \frac{1}{x}[(Me_3CCH_2)_3TaO]_x + (H)(CMe_3)C{=}C(R)(H)$$

Alkylidene complexes therefore react like metal ylides:

$$\left\{Cp_2Ta(CH_3){=}CH_2 \longleftrightarrow Cp_2\overset{\oplus}{Ta}(CH_3)\text{—}\overset{\ominus}{\underline{C}}H_2\right\} \xrightarrow{Al_2Me_6} Cp_2\overset{\oplus}{Ta}(CH_3)\text{—}CH_2\overset{\ominus}{Al}Me_3$$

compare:

$$\left\{Ph_3P{=}CH_2 \longleftrightarrow Ph_3\overset{\oplus}{P}\text{—}\overset{\ominus}{\underline{C}}H_2\right\} \xrightarrow{Al_2Me_6} Ph_3\overset{\oplus}{P}\text{—}CH_2\overset{\ominus}{Al}Me_3$$

A useful preparative application of this analogy is found in **Tebbe's reagent**, an alternative to the classical Wittig reagents (Grubbs, 1980):

$$Cp_2TiCl_2 \xrightarrow[-Me_2AlCl,\ -CH_4]{Al_2Me_6,\ toluene,\ 20°} Cp_2Ti(\mu\text{-}CH_2)(\mu\text{-}Cl)AlMe_2 \xrightarrow[base]{Ph-C(=O)OR} Ph-C(=CH_2)OR$$

The actual ***methylene transfer reagent*** *in this reaction is the alkylidene complex* $Cp_2Ti{=}CH_2$, *formed from Tebbe's reagent by loss of* Me_2AlCl. *For this conversion of an ester into a vinyl ether conventional Wittig reagents are not suitable.*

A variation of this method proceeds via a metallacycle, which is generated from Tebbe's reagent through the reaction with an olefin in the presence of a Lewis base (Grubbs, 1982):

$$Cp_2Ti(\mu\text{-}CH_2)(\mu\text{-}Cl)AlMe_2 \xrightarrow[base,\ -Me_2AlCl]{CH_2{=}C(Me)Pr} Cp_2Ti\text{-metallacyclobutane}(Me)(Pr) \underset{0°}{\overset{-\ CH_2{=}C(Me)Pr}{\rightleftharpoons}} [Cp_2Ti{=}CH_2] \rightarrow \text{subsequent reactions}$$

In contrast to the $\overline{\text{TiCAlCl}}$ reagent, the $\overline{\text{TiCCC}}$ derivative can be handled in air for brief periods. Its use as a methylene transfer reagent does **not** require the presence of base.

Carbene complexes with cumulated double bonds are formed by protonation or by alkylation of σ-alkynyl complexes which are generated *in situ* (Berke, 1980):

$$CpMn(CO)_2(HC{\equiv}CCO_2CH_3) \xrightarrow{3\ RLi} [CpMn(CO)_2{-}C{\equiv}C{-}CR_2{-}O]^{2-}$$

$$R = CH_3:\ \xrightarrow{-(CH_3)_2CO} [CpMn(CO)_2{-}C{\equiv}C]^{2-} \xrightarrow{2\ CH_3OSO_2F} CpMn(CO)_2{=}C{=}C(CH_3)_2$$

Vinylidene complex

$$R = Ph,\ t\text{-}Bu:\ \xrightarrow[-H_2O]{HCl} CpMn(CO)_2{=}C{=}C{=}CR_2$$

Allenylidene complex

The vinylidene ligand is one of the strongest π-acceptors, surpassed in this respect only by the ligands SO_2 and CS.

POST SCRIPTUM

The distribution of the **reactivity patterns "nucleophilic" and "electrophilic" among the two classes of carbene complexes M=C(X)R** may seem paradoxical to the reader: why should Fischer carbene complexes (central atom in low oxidation state, π-donor substituents X) exhibit electrophilic behavior, whereas Schrock complexes (central metal atom in high oxidation state, substituents R void of π-donor character) show nucleophilic behavior? A survey of factors which govern the character of the carbene carbon center shows, however, that a simple answer is not to be expected:

- electron configuration and shielding characteristics of the central metal atom,
- inductive and conjugative effects of the ligands L and of the substituents R and X
- overall charge of the carbene complex.

Furthermore, chemical reactions can proceed either by charge control or by frontier orbital control. In the latter case, a negative partial charge on the carbene carbon atom does not necessarily lead to nucleophilic behavior (Roper, 1986 R).

If, notwithstanding these ambiguities, a qualitative explanation is still desired, one could argue that the **electrophilic nature** of the carbene carbon atom in Fischer complexes is due to the combined effect of the π-acceptor ligands on the metal and the inductive effects of the electronegative heteroatoms on the carbene center. The **nucleophilic nature** of Schrock complexes, however, may be due to the strong M $\xrightarrow{\pi}$ C(carbene) backbonding and the absence of −I substituents on the carbene.

An alternative explanation ascribes electrophilic behavior to those carbene complexes, in which the respective free carbene possesses a singlet ground state and nucleophilic character to those complexes, where the free carbene has a triplet ground state (Hall, 1984). It is best, however, to avoid such strict classifications and discuss the relation between bonding and reactivity in each individual case according to judicious consideration of the various factors involved.

14.4 Transition-Metal Carbyne Complexes

PREPARATION

1. The first examples of this class of compounds emerged serendipitously (E. O. Fischer, 1973):

$(CO)_5M{=}C(OCH_3)R + BX_3 \rightarrow$ ~~$(CO)_5M{=}C(X)R$~~ planned reaction; $\rightarrow$ trans-$X(CO)_4M{\equiv}C{-}R$

M = Cr, Mo, W
X = Cl, Br, I
R = Me, Et, Ph

Substitution always involves the carbonyl group **trans** to the carbene ligand.

2. Another route, applicable only to compounds of the early transition metals in high oxidation states, is **α-deprotonation** of a carbene ligand (Schrock, 1978):

$$CpTa(Cl)_2(=CHR) \xrightarrow[\text{2. } Ph_3P=CH_2,\ -[Ph_3PCH_3]Cl]{\text{1. } PMe_3} CpTa(Cl)(PMe_3)\equiv C-R$$

A related reaction is **α-H elimination**, in which an α-hydrogen atom of a coordinated alkylidene is transferred to the metal atom (Schrock, 1980):

$$Cp^*Ta(Br)_2(=CHCMe_3) \xrightarrow[\text{2. } 2Na/Hg]{\text{1. dmpe}} Cp^*Ta(dmpe)(H)\equiv C-CMe_3$$

---

3. Transition-metal alkylidyne complexes can also be prepared from hexa(*t*-butoxy)-ditungsten and alkynes under mild conditions via a **metathesis reaction** (Schrock, 1982):

$$(t\text{-}BuO)_3W\equiv W(t\text{-}BuO)_3 + RC\equiv CR \longrightarrow 2\ (t\text{-}BuO)_3W\equiv C-R$$

R = alkyl

---

4. A further method is the treatment of a reactive dichlorocarbene complex with an organolithium compound (Roper, 1980):

$$L_3Os(H)Cl(CO) \xrightarrow[-Hg,\ -CHCl_3,\ -L]{Hg(CCl_3)_2} Cl_2(CO)L_2Os=CCl_2 \xrightarrow[-ArCl,\ -2\ LiCl]{2\ ArLi} Cl(CO)L_2Os\equiv C-Ar$$

L = PPh_3
Ar = o-Tolyl

STRUCTURE AND BONDING

$Me-C\equiv Cr(CO)_4(I)$: C–Me 169 pm; Cr≡C 195

∢ Cr – C – Me ~ 180°

The M – C(carbyne) bond is usually shorter than the corresponding distance M – C(carbonyl). The bond axis M ≡ C – R is linear or nearly linear.

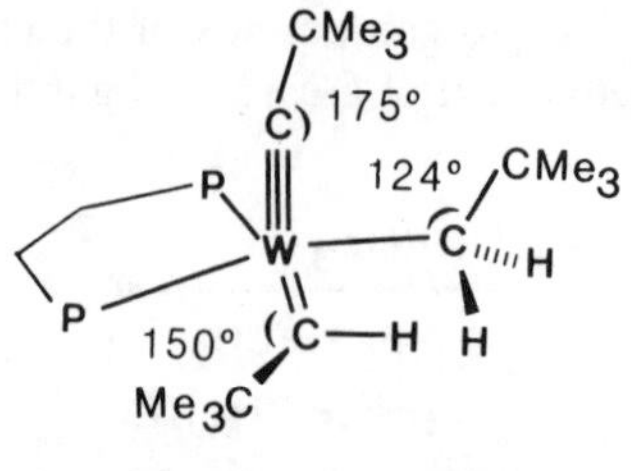

Structural data for the compound **(dmpe)W(CH_2CMe_3) ($CHCMe_3$)$CCMe_3$)** *allow a direct comparison of the lengths of formal* W–C *single,* W=C *double and* W≡C *triple bonds* (*Churchill*, 1979): *d*(W-*alkyl*) = 225 pm; *d*(W-*alkylidene*) = 194 pm; *d*(W-*alkylidyne*) = 178 pm.

The M–C bond in carbyne complexes is best described as a combination of one σ-donor bonding and two π-acceptor backbonding interactions:

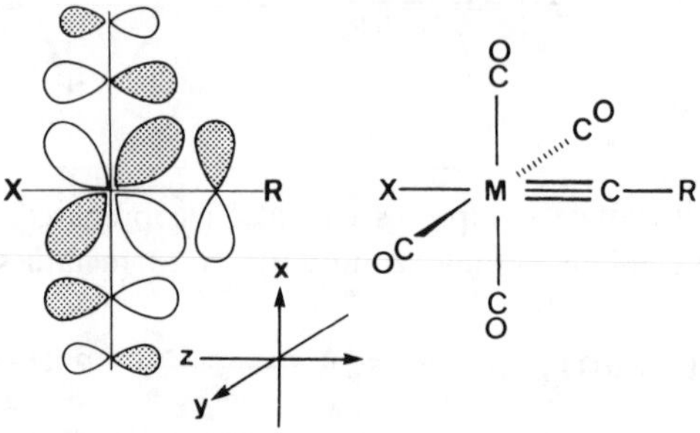

π-interaction (backbonding) of the M(d_{xz}) *orbital with the* p_x *orbital of the carbyne ligand and two* π* *molecular orbitals of two* CO *ligands. The same interaction occurs in the* yz-*plane.*

As in the case of carbene complexes, an electron-rich heteroatom substituent can also contribute to the stabilization of carbyne complexes.

Instead of a triple bond to **one** metal, the ligand RC⋮ can also form three single bonds to **three** metals, thereby becoming the triple-bridging **μ_3-alkylidyne building block** in transition-metal clusters. Details concerning the use of carbene and carbyne complexes in cluster synthesis are deferred to Chapter 16.5.

14.5 Metal Carbonyls

Transition-metal carbonyls are among the longest known classes of organometallic compounds. They are common starting materials for the synthesis of other low-valent metal complexes, especially clusters (see Chapter 16). The carbonyl ligand can not only be substituted for a large number of other ligands (Lewis bases, olefins, arenes), but the

remaining CO groups stabilize the molecule against oxidation or thermal decomposition. Metal carbonyl derivatives play an important role as intermediates in homogeneous catalysis (see Chapter 17).
Carbonyl groups are also useful probes for determining the electronic and molecular structure of organometallic species by spectroscopic methods.

Neutral, binary metal carbonyls:

4	5	6	7	8	9	10	11
Ti	$V(CO)_6$	$Cr(CO)_6$	$Mn_2(CO)_{10}$	$Fe(CO)_5$ $Fe_2(CO)_9$ $Fe_3(CO)_{12}$	$Co_2(CO)_8$ $Co_4(CO)_{12}$ $Co_6(CO)_{16}$	$Ni(CO)_4$	Cu
Zr	Nb	$Mo(CO)_6$	$Tc_2(CO)_{10}$ $Tc_3(CO)_{12}$	$Ru(CO)_5$ $Ru_3(CO)_{12}$ $Ru_6(CO)_{18}$	$Rh_2(CO)_8$ $Rh_4(CO)_{12}$ $Rh_6(CO)_{16}$	Pd	Ag
Hf	Ta	$W(CO)_6$	$Re_2(CO)_{10}$	$Os(CO)_5$ $Os_3(CO)_{12}$	$Ir_4(CO)_{12}$ $Ir_6(CO)_{16}$	Pt	Au

14.5.1 Preparation, Structure and Properties

1. Metal + CO

$$Ni + 4\,CO \xrightarrow[1\text{ bar, }25\,^\circ C]{} Ni(CO)_4$$

$$Fe + 5\,CO \xrightarrow[100\text{ bar, }150\,^\circ C]{} Fe(CO)_5$$

Pure iron (void of surface oxides) reacts with CO at room temperature and a pressure of 1 bar.

2. Metal salt + reducing agent + CO

$$TiCl_4 \cdot DME + 6\,KC_{10}H_8 + 4\,[15]\text{crown-5} + 6\,CO \longrightarrow 2\,[K\{[15]\text{crown-5}\}_2]^+[Ti(CO)_6]^{2-} + 4\,KCl + DME + 6\,C_{10}H_8$$

$$VCl_3 + 3\,Na + 6\,CO \xrightarrow[300\text{ bar}]{\text{diglyme}} [Na(diglyme)_2]^+[V(CO)_6]^- \xrightarrow[-H_2]{H_3PO_4} V(CO)_6$$

$$CrCl_3 + Al + 6\,CO \xrightarrow[140\,^\circ C,\ 300\text{ bar}]{C_6H_6,\ AlCl_3} Cr(CO)_6 + AlCl_3$$

$$WCl_6 + 2\,Et_3Al + 6\,CO \xrightarrow[70\text{ bar}]{C_6H_6,\ 50\,^\circ C} W(CO)_6 + 3\,C_4H_{10}$$

$$2\,Mn(OAc)_2 + 10\,CO \xrightarrow[\text{(iso-Pr)}_2O]{AlEt_3} Mn_2(CO)_{10} + C_4H_{10}$$

$$Re_2O_7 + 17\,CO \longrightarrow Re_2(CO)_{10} + 7\,CO_2$$

$$Ru(acac)_3 \xrightarrow[300\text{ bar, }130\,°C]{CO,\ H_2} Ru_3(CO)_{12}$$

$$2\,CoCO_3 + 2\,H_2 + 8\,CO \xrightarrow[300\text{ bar, }130\,°C]{} Co_2(CO)_8 + 2\,CO_2 + 2\,H_2O$$

3. Miscellaneous methods

$$2\,Fe(CO)_5 \xrightarrow{CH_3COOH,\ h\nu} Fe_2(CO)_9 + CO \qquad \text{(photolysis)}$$

$$Fe(CO)_5 + 2\,OH^- \longrightarrow [HFe(CO)_4]^- + HCO_3^- \qquad \text{(base reaction)}$$

$$3\,[HFe(CO)_4]^- + 3\,MnO_2 \longrightarrow Fe_3(CO)_{12} + 3\,OH^- + 3\,MnO$$

Structures of various binary metal carbonyls are shown on p. 224.

14.5.2 Variants of CO Bridging

The three principal coordination modes of the CO ligand are:

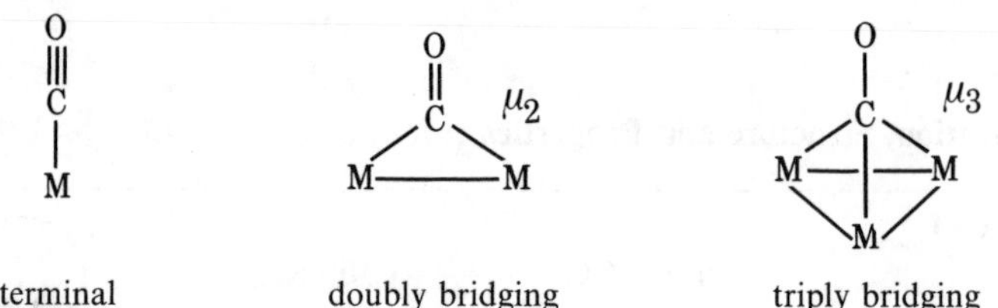

Doubly bridging carbonyl groups are quite common, especially in polynuclear clusters; they appear almost exlusively in conjunction with metal-metal bonds:

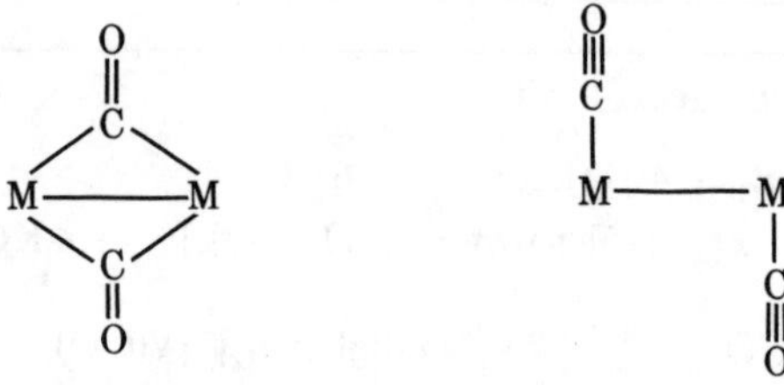

CO bridges often occur in pairs and can be in dynamic equilibrium with the non-bridging mode. For example, octacarbonyldicobalt in solution consists of a mixture of at least two structural isomers (see p. 225 above).

Physical Properties of Selected Metal Carbonyls

Compound	Color	mp. in °C	Symmetry	IR ν_{CO} in cm^{-1}		Miscellaneous
$V(CO)_6$	green-black	70(*d*)	O_h	1976		paramagnetic, $S = 1/2$
$Cr(CO)_6$	white	130(*d*)	O_h	2000		$d(Cr-C) = 192$ pm $\Delta_0 = 32'200\ cm^{-1}$
$Mo(CO)_6$	white	– (subl)	O_h	2004		$d(Mo-C) = 206$ pm $\Delta_0 = 32'150\ cm^{-1}$
$W(CO)_6$	white	– (subl)	O_h	1998		$d(W-C) = 207$ pm $\Delta_0 = 32'200\ cm^{-1}$
$Mn_2(CO)_{10}$	yellow	154	D_{4d}	2044(m) 2013(s) 1983(m)		$d(Mn-Mn) = 293$ pm
$Tc_2(CO)_{10}$	white	177	D_{4d}	2065(m) 2017(s) 1984(m)		
$Re_2(CO)_{10}$	white	177	D_{4d}	2070(m) 2014(s) 1976(m)		
$Fe(CO)_5$	yellow	–20	D_{3h}	2034(s) 2013(vs)		bp 103 °C, highly toxic $d(Fe-C_{ax}) = 181$ pm $d(Fe-C_{eq}) = 183$ pm
$Ru(CO)_5$	colorless	–22	D_{3h}	2035(s) 1999(vs)		unstable; forms $Ru_3(CO)_{12}$
$Os(CO)_5$	colorless	–15	D_{3h}	2034(s) 1991(vs)		very unstable; forms $Os_3(CO)_{12}$
$Fe_2(CO)_9$	gold-yellow	*d*	D_{3h}	2082(m) 2019(2) 1829(s)		$d(Fe-Fe) = 246$ pm
$Co_2(CO)_8$	orange red	51(*d*)	C_{2v} (solid)			$d(Co-Co) = 254$ pm
			D_{3d} (solution)	2112 2071 2059 2044 2031 2001 1886 1857	2107 2069 2042 2031 2023 1991	
$Ni(CO)_4$	colorless	–25	T_d	2057		bp 34 °C, highly toxic $d(Ni-C) = 184$ pm easily decomposes to Ni and 4 CO

Structures of binary metal carbonyls

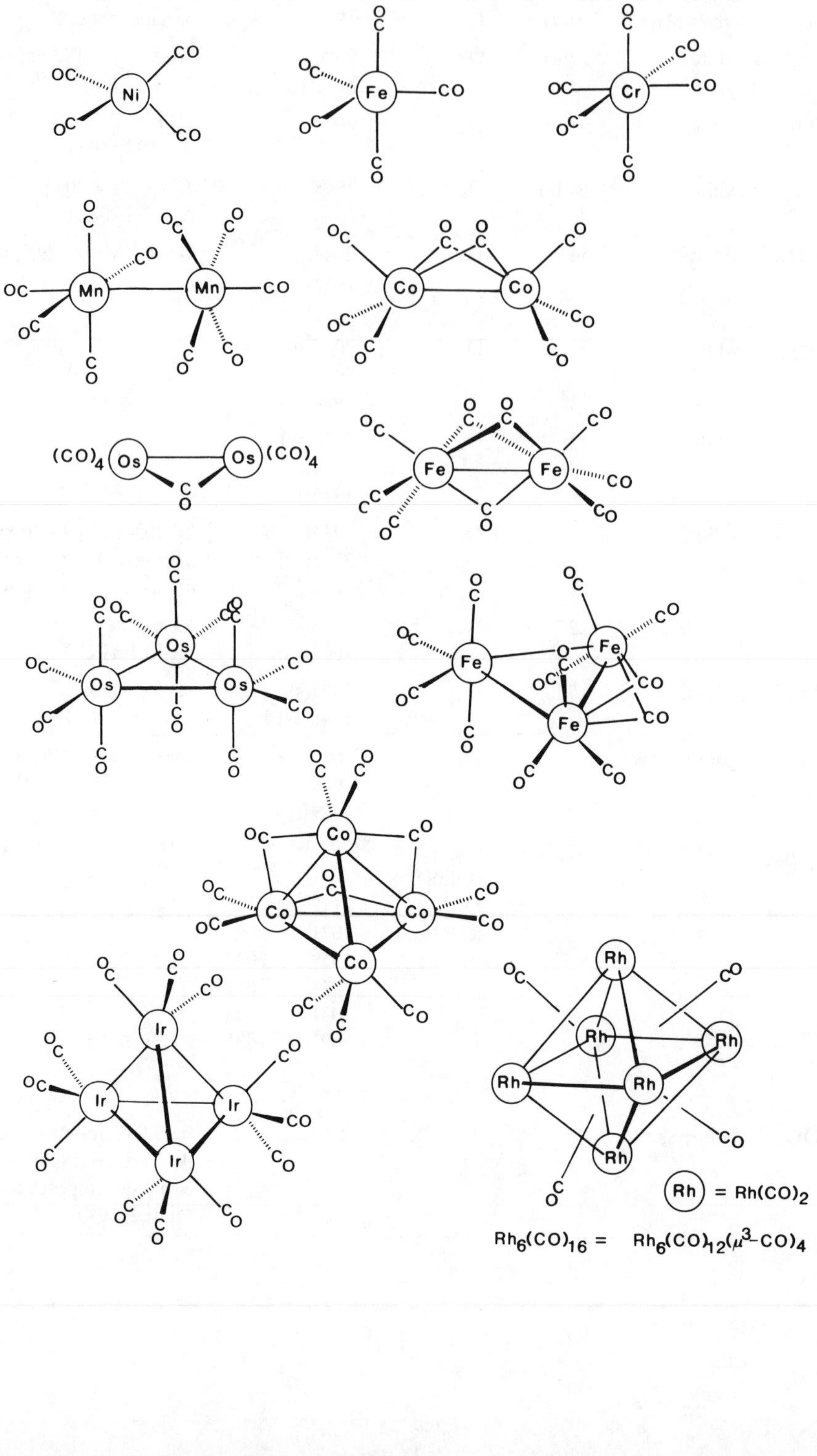

(Rh) = $Rh(CO)_2$

$Rh_6(CO)_{16} = Rh_6(CO)_{12}(\mu^3\text{-}CO)_4$

D_{3d} 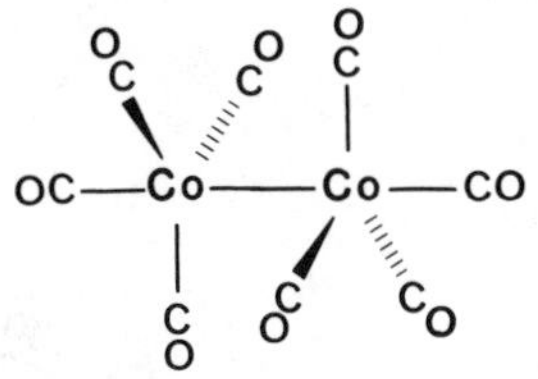 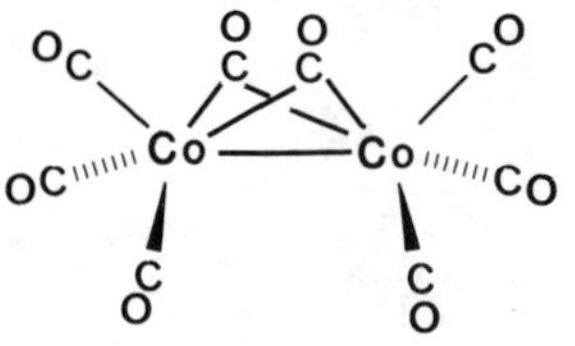 C_{2v}

Larger metals prefer the unbridged form (compare the structures of $Fe_2(CO)_9$ vs $Os_2(CO)_9$ or of $Fe_3(CO)_{12}$ vs $Os_3(CO)_{12}$). Presumably, in the case of larger metal atoms, the bond distance M–M and the angle M–C–M of a CO-bridge are incompatible. Apart from the symmetrical bridge, unsymmetrically **semibridging** μ_2-CO groups are also occasionally encountered. They can be regarded as being intermediate between terminal and bridging carbonyls.

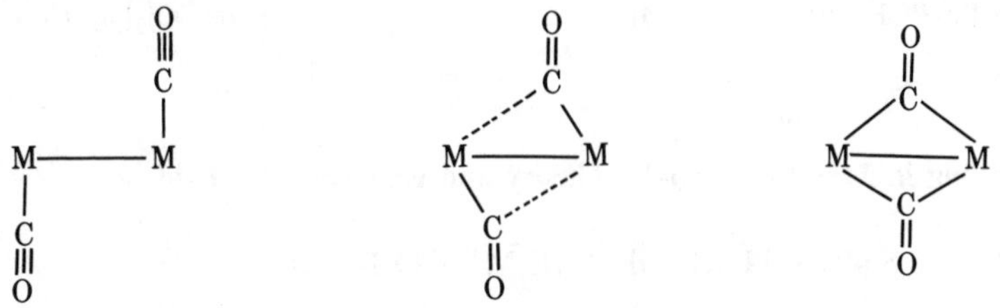

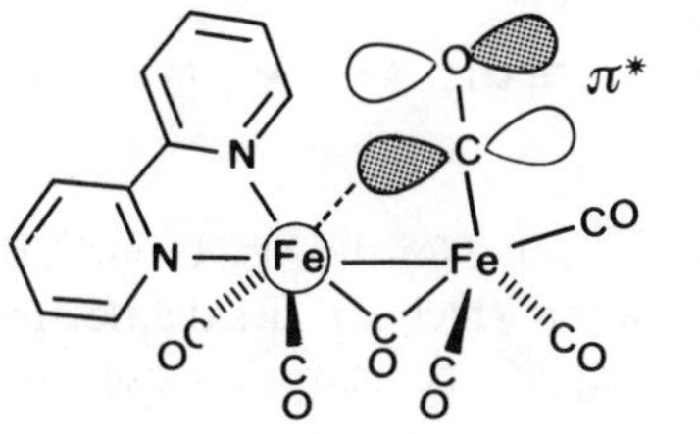

Substitution of two CO *groups in* $Fe_2(CO)_9$ *for 2,2′-bipyridyl (bipy) leads to considerable structural change in the bridging region. The superior σ-donor and inferior π-acceptor character of bipy as compared to* CO *would entail an uneven charge distribution over the* Fe *atoms. The higher negative charge density on* (Fe) *may, however, be reduced by means of a* ***semibridging* CO** *unit which towards* (Fe) *exhibits π-acceptor character exclusively, the σ-donor action being restricted to* Fe.

In a further unsymmetrical form, the ***σ/π*-bridge**, CO acts as a four- or six-electron donor displaying side-on coordination which is uncommon in mononuclear carbonyl complexes (this coordination mode is reminiscent of vinyl complexes, p. 207):

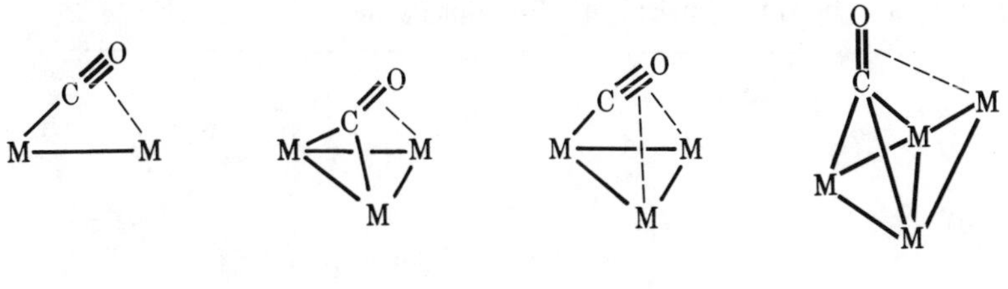

Examples for σ/π-bridging CO:

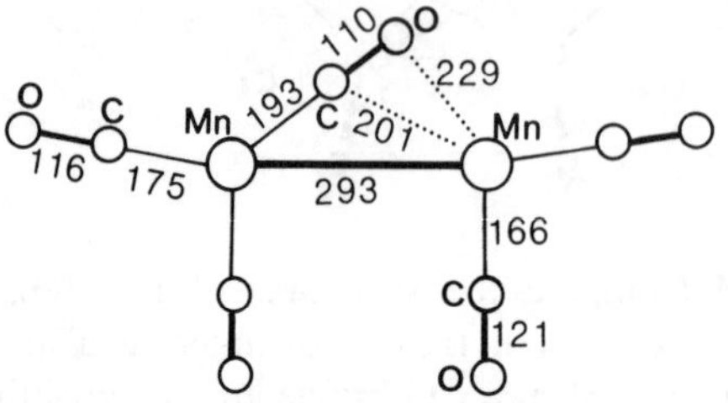

The $Mn_2(CO)_5$ *core in* $(\mu\text{-}Ph_2PCH_2PPh_2)_2Mn_2(CO)_5$

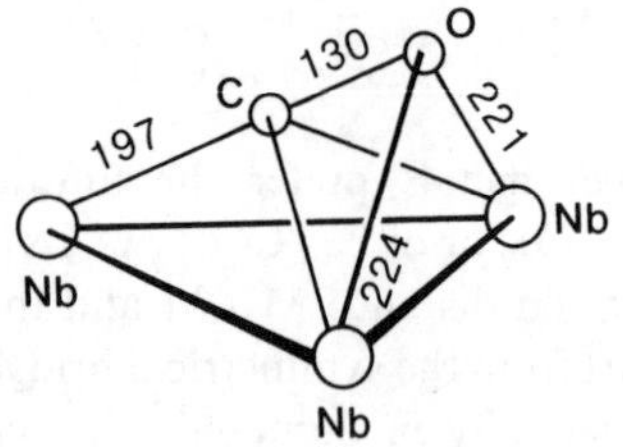

The $Nb_3(\mu\text{-}CO)$ *core in* $[CpNb(CO)_2]_3(\mu\text{-}CO)$

14.5.3 Bonding in Metal Carbonyls, Theory and Experimental Evidence

ELECTRONIC STRUCTURE OF THE M–CO BOND

The description of the transition metal-CO bond as a resonance hybrid

$$\left\{ \overset{\ominus}{\underline{M}} - C \equiv \overset{\oplus}{O}| \longleftrightarrow M = C = \overline{\underline{O}} \right\}$$

(Brockway, Pauling, 1935)

leads to a bond order between 1 and 2 for the M–C bond, between 2 and 3 for the C–O bond and to less charge separation which satisfies the electroneutrality principle. Because of its prevalence in organometallic chemistry, a closer look at the carbonyl ligand is warranted, however.

The character of the frontier orbitals of CO may be derived from the following experimental evidence:

Species	Configuration	d(C–O)/pm	ν_{CO}/cm^{-1}	Conclusion
CO	$(5\sigma)^2$	113	2143	
CO^+	$(5\sigma)^1$	111	2184	5σ is weakly antibonding
CO*	$(5\sigma)^1(2\pi)^1$	S 124	1489	2π is strongly antibonding
		T 121	1715	

(Johnson, Klemperer, 1977) S = singlet state, T = triplet state

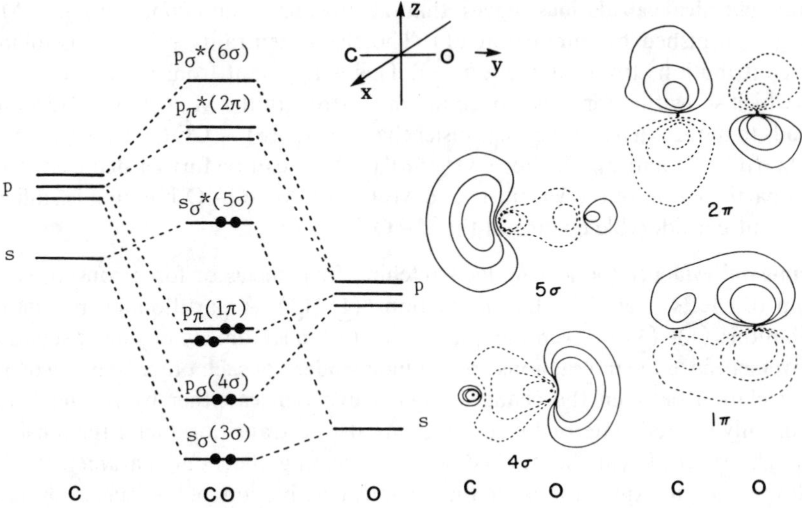

*This rudimentary **energy-level diagram** for* CO *can be refined by inclusion of s, p_y-mixing. Even in its simplest form, the nature of the HOMO (σ^*, antibonding) and the LUMO (π^*, antibonding) is depicted correctly. The symbols in brackets refer to the energetic sequence of the molecular orbitals, which results if all atomic orbitals, including* C(1s) *and* O(1s)*, are considered.*

*Based on quantum-chemical calculations, **contour lines** for the* MO's *of free* CO *may be drawn. They portray the shape of the ligand molecular orbitals which are relevant for* M–C–O *bonding. Solid and broken lines indicate opposite phase. The absolute values* (0.3, 0.2 *and* 0.1) *decrease with increasing radial distance.*

These data for CO in the ground state, the electronically excited state (CO*) and as the cation (CO^+) contribute to an understanding of the changes this molecule undergoes on coordination to a transition metal. Consider a single moiety M–C–O:

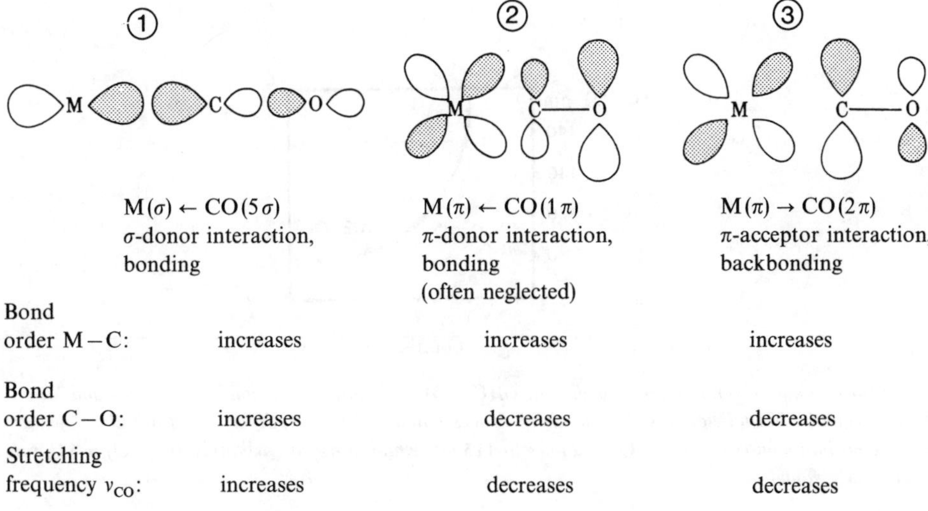

	①	②	③
	M(σ) ← CO(5σ) σ-donor interaction, bonding	M(π) ← CO(1π) π-donor interaction, bonding (often neglected)	M(π) → CO(2π) π-acceptor interaction, backbonding
Bond order M–C:	increases	increases	increases
Bond order C–O:	increases	decreases	decreases
Stretching frequency ν_{CO}:	increases	decreases	decreases

Quantum-chemical calculations suggest that the most important contribution to M–CO bonding is furnished by interaction ①. The electroneutrality principle is fulfilled by minor contributions from interaction ③. This is apparent from the contour-line diagrams which show that for bonds of comparable strength the M $d(\pi) \rightarrow$ CO(2π) interaction leads to more extensive charge transfer than the M$(d\sigma) \leftarrow$ CO(5σ) interaction. From the data d(C–O) and ν_{CO} for free CO* $(5\sigma)^1(2\pi)^1$, it can be further deduced that even a small participation of the vacant ligand MO 2π in M–C–O bonding is sufficient to bring about considerable change in the C–O bond order.

Experimental evidence (bond lengths, stretching frequencies or force constants) reflects the sum of effects caused by the interactions ①–③. A partitioning into individual contributions ①–③ or even the question as to the legitimacy of such a separation is controversial since the interactions are not independent of each other (synergism, p. 185). The correlation between the change a ligand experiences either by raising it into an electronically excited state or by coordinating it to a transition metal remains a useful concept, however. It can be applied when discussing the bond parameters of olefin complexes and in explanations of the chemical activation of molecules by complex formation.

The transition from a fragment M–C–O to a real molecule, e.g. M(CO)$_6$, does not raise any new aspects concerning the nature of the M–CO bond, but is an application of the MO approach to larger molecules. The separation of the interactions into σ (octahedron: a_{1g}, e_g, t_{1u}) and π (octahedron: t_{2g}) is facilitated by symmetry considerations.

EXPERIMENTAL EVIDENCE FOR MULTIPLE BONDING M⋯C⋯O

The most conclusive indications of the bond order in metal carbonyls are provided by **crystal structure determinations and by vibrational spectra.** As backbonding from the metal to CO increases, the M–C bond should become stronger (shorter) and the C–O bond correspondingly weaker (longer).

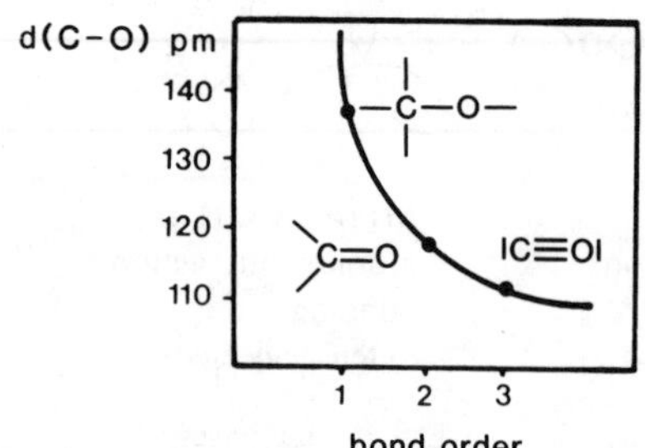

For the discussion of bond orders, the ***distance d*(C–O)** *is ill-suited since bonds of orders* 2 *and* 3 *vary only marginally in their lengths. For bonds of orders* 1 *and* 2, *d*(C–O) *differs substantially, however. In free carbon monoxide, d*(C–O) *amounts to* 113 pm *while in metal carbonyls d*(C–O) = 115 pm *is generally found.*

More revealing is an inspection of the ***distance d*(M−C);** here the progression from M−C to M=C is accompanied by a shortening of 30–40 pm. However, a quantitative assessment is again difficult, as reference values d(M−C, single bond) often are unavailable for zerovalent metals.

Estimation of d(M−C, *single bond*):
The covalent radius for Mo^0 *can be inferred by subtracting the covalent radius of nitrogen* (sp^3, 70 pm) *from the* Mo−N *distance* (231 pm). *This difference* (161 pm) *added to the covalent radius of carbon* (*sp*, 72 pm) *leads to the distance d*(Mo−C, *single*) = 233 pm. *The smaller value d*(Mo−C) = 193 pm *found in* $(R_3N)_3Mo(CO)_3$ *therefore indicates significant double-bond character for the molybdenum-carbon bond.*

The discussion of chemical bonding in metal carbonyls profits from the consideration of **infrared spectra:** To a first approximation, C−O stretching vibrational frequencies, in contrast to M−C stretching frequencies, can be regarded as being independent from other vibrations in the molecule. Therefore a relation between CO stretching vibrational frequency and CO bond order can be established.

(a) Bonding modes of the carbonyl group

The change in vibrational stretching frequencies ν_{CO} is quite characteristic:

	Free	Terminal	μ_2-CO	μ_3-CO
	$\|C\equiv O\|$	M−C−O	M₂C=O	M₃C−O
ν_{CO} (cm^{-1})	2143	1850–2120	1750–1850	1620–1730

Example: $Cr(CO)_6$ ν_{CO} = 2000 cm^{-1}, k = 17.8 N cm^{-1}
Bond order C−O ≈ 2.7.

(b) Charge on the complex

		ν_{CO}/cm^{-1}
d^{10}	$Ni(CO)_4$	2060
	$Co(CO)_4^-$	1890
	$Fe(CO)_4^{2-}$	1790
d^6	$Mn(CO)_6^+$	2090
	$Cr(CO)_6$	2000
	$V(CO)_6^-$	1860
	CO_{free}	2143

Increasing negative charge leads to expansion, increasing positive charge to contraction of the metal d orbitals with attendant increase or decrease in the overlap M(*d*, π)−CO(π*), *respectively. The charge of the complex therefore influences the extent of backbonding, which manifests itself in the value of* ν_{CO}.

In the adduct $H_3B^{\ominus}-{}^{\oplus}CO$, boron can be regarded as a central atom void of π-donor character. Backbonding effects are therefore absent and the CO stretching frequency is shifted to a **higher** wave number ($\nu_{CO} = 2164\ cm^{-1}$), in accordance with the nature of the donor orbital CO(5σ), which is slightly antibonding with respect to the intra-ligand bond (p. 226).

(c) Symmetry of the molecule

The number and intensities of carbonyl stretching bands in the vibrational spectra largely depend on the local symmetry around the central atom (p. 231). Quite often, the symmetry of a metal carbonyl complex is determined simply by counting the number of infrared bands. The expected number of IR-active bands can be derived by means of group theory.

(d) π-Acceptor and σ-donor properties of other ligands

Ligands in mutual *trans* position compete for the electrons of a particular metal *d* orbital. Two CO groups therefore weaken each other's bond to the same central atom. By replacing a CO group with a ligand which is a weaker π-acceptor, the M−CO bond in the *trans* position is strengthened and the C−O bond weakened.

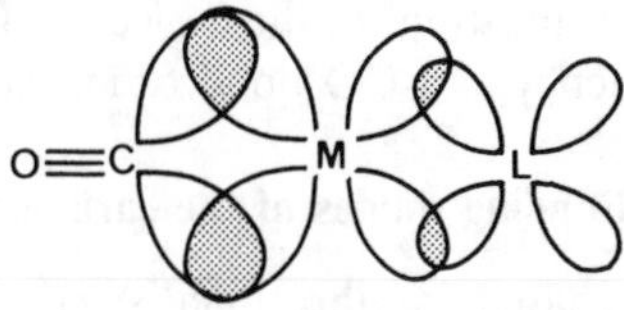

This notion is exemplified by a comparison of CO stretching frequencies in a series of complexes of the general composition $L_3Mo(CO)_3$ (Cotton, 1964):

Complex	ν_{CO} cm^{-1}
$(PF_3)_3Mo(CO)_3$	2055, 2090
$(PCl_3)_3Mo(CO)_3$	1991, 2040
$[P(OMe)_3]_3Mo(CO)_3$	1888, 1977
$(PPh_3)_3Mo(CO)_3$	1835, 1934
$(CH_3CN)_3Mo(CO)_3$	1783, 1915
$(dien)Mo(CO)_3$	1758, 1898
$(Py)_3Mo(CO)_3$	1746, 1888

In this series, PF_3 is the better π-acceptor (backbonding into *d* orbitals of phosphorus, electron-withdrawing substituents); pyridine and the tridentate ligand diethylenetriamine (dien) have poor or no π-acceptor character and behave as σ-donors exclusively.
Based on the ν_{CO} stretching frequencies in complexes with a mixed ligand sphere, the following **order of π-acceptor strength** can be proposed:

$NO > CO > RNC > PF_3 > PCl_3 > PCl_2R > PClR_2 > P(OR)_3 > PR_3 > RCN > NH_3$.

Number and Modes of IR-active bands (ν_{CO}) *in carbonyl complexes, depending on the local symmetry of* $M(CO)_n$

Complex	Number and Modes of IR-active Bands ν_{CO}	Point Group
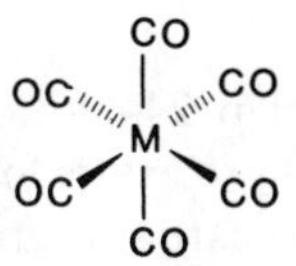 $M(CO)_6$	1 T_{1u}	O_h
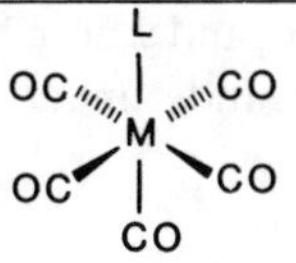 $LM(CO)_5$	3 $2A_1 + E$	C_{4v}
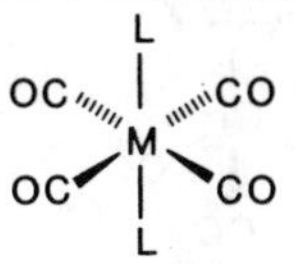 trans-$L_2M(CO)_4$	1 E_u	D_{4h}
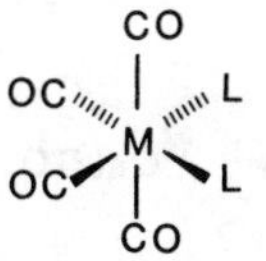 cis-$L_2M(CO)_4$	4 $2A_1 + B_1 + B_2$	C_{2v}
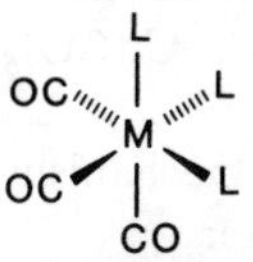 fac-$L_3M(CO)_3$	2 $A_1 + E$	C_{3v}
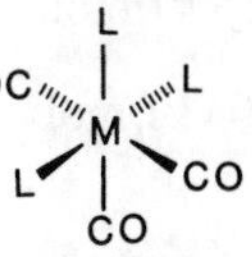 mer-$L_3M(CO)_3$	3 $2A_1 + B_2$	C_{2v}
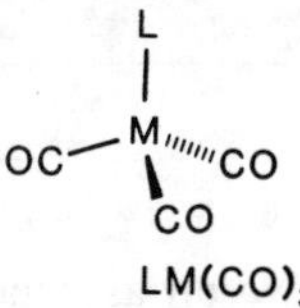 $LM(CO)_3$	2 $A_1 + E$	C_{3v}
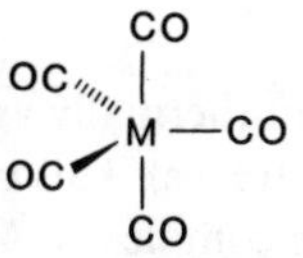 $M(CO)_5$	2 $A''_2 + E'$	D_{3h}
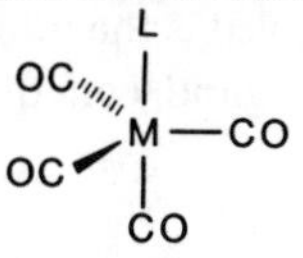 $LM(CO)_4$	3 $2A_1 + E$	C_{3v}
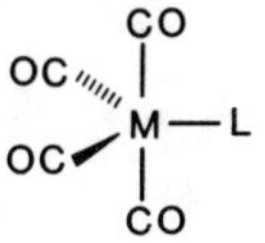 $LM(CO)_4$	4 $2A_1 + B_1 + B_2$	C_{2v}
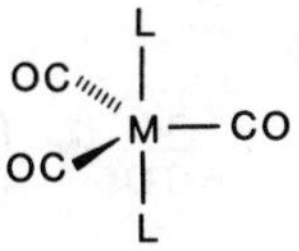 trans-$L_2M(CO)_3$	1 E'	D_{3h}
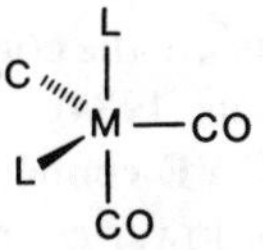 cis-$L_2M(CO)_3$	3 $2A' + A''$	C_s
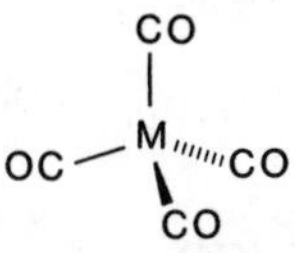 $M(CO)_4$	1 T_2	T_d
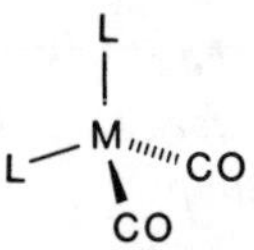 $L_2M(CO)_2$	2 $A_1 + B_1$	C_{2v}

14.5.4 Principal Reaction Types of Metal Carbonyls

CO is such a common ligand that a chapter covering reactions of metal carbonyls would necessarily cover a substantial part of organometallic chemistry. Here, only a small survey will be given, with references to other relevant sections of the text.

① Substitution

CO groups can be substituted thermally as well as photochemically (cf. p. 244) for other ligands (Lewis bases, olefins, arenes). This general method is a standard procedure for the synthesis of low-valent metal complexes. A complete substitution of all carbonyl groups is rarely accomplished.
One variant frequently employed is the intermediate introduction of a ligand whose bond to the metal is labile. This ligand can then be replaced by other ligands under mild conditions:

$$Cr(CO)_6 \xrightarrow[\text{reflux}]{CH_3CN} Cr(CO)_3(CH_3CN)_3 \xrightarrow{C_7H_8} (C_7H_8)Cr(CO)_3$$

$$Cr(CO)_6 + \text{cyclooctene} \xrightarrow[-CO]{h\nu} (\text{cyclooctene})Cr(CO)_5 \xrightarrow{+L} LCr(CO)_5$$

Other **M(CO)$_x$ transfer agents** are the complexes (THF)Mo(CO)$_5$, (CH$_2$Cl$_2$)Cr(CO)$_5$ and (cyclooctene)$_2$Fe(CO)$_3$ (Grevels, 1984).
Carbonyl substitutions at **18 VE complexes** proceed by a dissociative mechanism **(D)**, that is *via* an intermediate of lower coordination number (or its solvate):

$$Cr(CO)_6 \xrightarrow[-CO]{\text{slow}} \{Cr(CO)_5\} \xrightarrow[+L]{\text{fast}} LCr(CO)_5 \qquad D(Cr-CO) = 155\ \text{kJ/mol}$$

$$\underset{18\ \text{VE}}{Ni(CO)_4} \xrightarrow[-CO]{} \underset{16\ \text{VE}}{\{Ni(CO)_3\}} \xrightarrow[+L]{} \underset{18\ \text{VE}}{LNi(CO)_3} \qquad D(Ni-CO) = 105\ \text{kJ/mol}$$

Ni(CO)$_4$ reacts faster than Cr(CO)$_6$, the activation enthalpies ΔH^{+} of these reactions being virtually identical to the bond energies D(M−CO).

*For reasons yet unexplained, the rate of substitution is highest for the **central** element in each group:*

Cr < **Mo** > W
Co < **Rh** > Ir
Ni < **Pd** > Pt

Carbonyl metal complexes with a **17 VE shell** undergo substitution *via* an associative mechanism **(A)**; it can be shown that the formation of a 19 VE intermediate is accompanied by a gain in bond energy (Poë, 1975):

$$\underset{17\ \text{VE}}{V(CO)_6} \xrightarrow[+L]{\text{rate-determining}} \underset{19\ \text{VE}}{LV(CO)_6} \xrightarrow[-CO]{\text{fast}} \underset{17\ \text{VE}}{LV(CO)_5}$$

The 17 VE complex $V(CO)_6$ reacts 10^{10} times faster than the 18 VE complex $Cr(CO)_6$! The accelerated substitution of a 17 VE as compared to an 18 VE complex is utilized in **electron transfer (ET) catalysis.** This principle will be explained for the case of ET catalysis initiated by oxidation (Kochi, 1983):

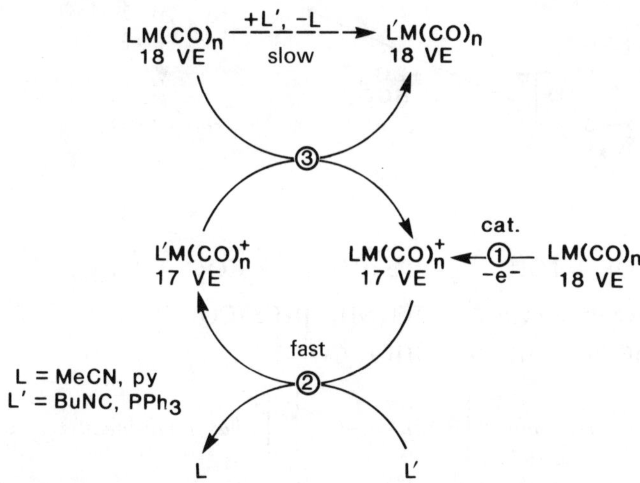

The labile species $LM(CO)_n^+$ *must be generated chemically or electrochemically in catalytic amounts only, step* ①. *A necessary requirement is that the redox potentials allow step* ③ *to proceed spontaneously. ET catalyses can also be initiated by reduction* (18 VE → 19 VE).

② Addition of nucleophiles to η-CO

- Formation of carbene complexes (p. 210)
- Formation of carbonylmetallates (p. 234)
- Formation of formyl complexes:

$$Fe(CO)_5 + Na^+[(MeO)_3BH]^- \xrightarrow[-(MeO)_3B]{} Na^+\left[(CO)_4Fe-C{\overset{\displaystyle =O}{\underset{\displaystyle \diagdown H}{}}}\right]^-$$

(Casey, 1976)

③ Disproportionation

$$3\,Mn_2(CO)_{10} + 12\,py \xrightarrow[-10\,CO]{120\,^\circ C} 2\,[Mn(py)_6]^{2+} + 4\,[Mn(CO)_5]^-$$

(Hieber, 1960)

This type of disproportionation is also observed for $Co_2(CO)_8$; it can be effected photochemically (Tyler, 1985 R).

④ Oxidative decarbonylation (p. 237, 336)

14.5.5 Carbonylmetallates and Carbonyl Metal Hydrides

The classical route to metal carbonyl anions (**carbonylmetallates**) is the reaction of metal carbonyls with strong bases (base reaction, W. Hieber, 1895–1976, the pioneer of metal carbonyl chemistry). In this reaction, the base OH^- attacks the carbonyl carbon atom (cf. the analogous reaction with alkyllithium reagents, p. 210) and the initial addition product decomposes, presumably by β-elimination, to give the carbonylmetallates:

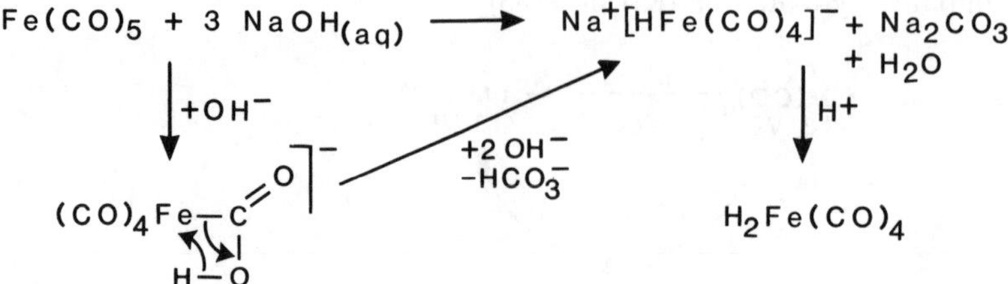

$$Fe_2(CO)_9 + 4\,OH^- \longrightarrow [Fe_2(CO)_8]^{2-} + CO_3^{2-} + 2\,H_2O$$

$$3\,Fe(CO)_5 + Et_3N \xrightarrow{H_2O} [Et_3NH]^+\,[HFe_3(CO)_{11}]^-$$

$$Cr(CO)_6 + BH_4^- \longrightarrow [HCr_2(CO)_{10}]^-$$

The initially formed formyl complex $\left[(CO)_5Cr{-}C{\lesssim}^{O}_{H}\right]^-$ *has a half-life of* $\tau_{1/2} \approx 40$ min *at room temperature* (*Casey*, 1976).

Carbonylmetallates are formed by most metal carbonyls. Their protonation generally (but not always) leads to metal carbonyl hydrides (Examples: $Mn(CO)_5^-$, $Fe(CO)_4^{2-}$, $Co(CO)_4^-$).

Other methods to generate **transition-metal carbonyl hydrides:**

$$Co_2(CO)_8 + 2\,Na \longrightarrow 2\,Na^+[Co(CO)_4]^- \xrightarrow{H^+} 2\,HCo(CO)_4$$

$$Fe(CO)_5 + I_2 \xrightarrow[-CO]{} Fe(CO)_4I_2 \xrightarrow[THF]{NaBH_4} H_2Fe(CO)_4$$

$$Mn_2(CO)_{10} + H_2 \xrightarrow[200\text{ bar, }150\,^\circ C]{} 2\,HMn(CO)_5$$

Properties of some metal carbonyl hydrides:

Complex	mp °C	Decomp. °C	IR ν_{M-H}	^{1}H-NMR δ ppm	pK_a	Acidity comparable to
$HCo(CO)_4$	−26	−26	1934	−10	1	H_2SO_4
$H_2Fe(CO)_4$	−70	−10		−11.1	4.7	CH_3COOH
$[HFe(CO)_4]^-$					14	H_2O
$HMn(CO)_5$	−25	stab. RT	1783	− 7.5	7	H_2S

The term "hydride" for these complexes is based on the assignment of formal oxidation states (hydrogen has a higher electronegativity than most transition metals, cf. p. 8) and should not be equated with chemical reactivity. The properties of transition-metal hydrides vary considerably; they range from hydridic through inert to acidic in character.

M – H functioning as a hydride donor:

$$C_5H_5Fe(CO)_2H + HCl \longrightarrow C_5H_5Fe(CO)_2Cl + H_2$$

M – H functioning as a proton donor:

$$HCo(CO)_4 + H_2O \longrightarrow H_3O^+ + [Co(CO)_4]^-$$

Structural elucidation of metal hydrides by means of X-ray diffraction is difficult (the atomic scattering factors are proportional to the square of the atomic number; light atoms have high amplitudes of thermal vibration). Neutron diffraction is preferable. Typical values for the bond parameters in transition-metal hydrides:

$$d(M-H) = 150-170 \text{ pm} \qquad E(M-H) \approx 250 \text{ kJ/mol}$$

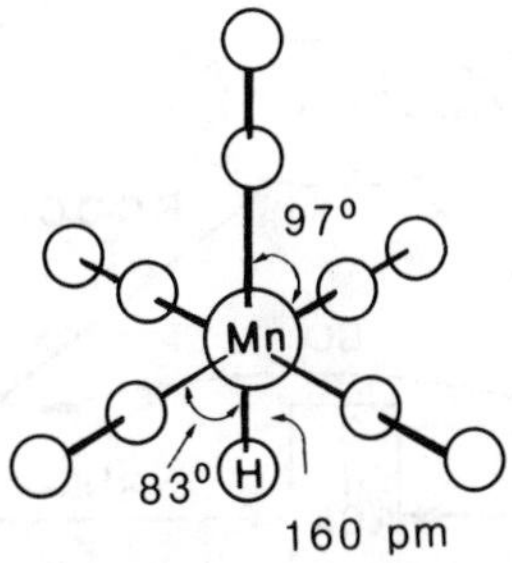

97°
Mn
83° H
160 pm

The hydride ligand occupies a definite position in the coordination polyhedron. **$Mn(CO)_5H$** *has the structure of a slightly distorted octahedron. The* Mn – H *distance approximates the sum of the covalent radii. With increasing bulk the ancillary ligands (e.g.* PPh_3*) may dominate the overall geometry of the complex, however.*

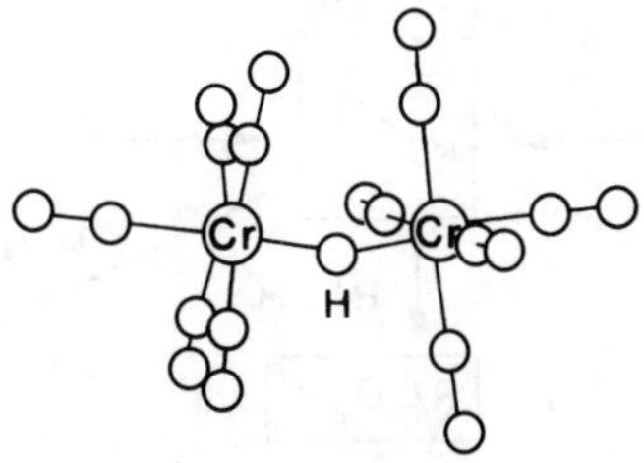

Cr
Cr
H

The localization of a bridging hydride ligand (M–H–M) is especially difficult because of the proximity of two strongly scattering metal atoms. The position of the hydride ligand in **$[Et_4N]^+ [HCr_2(CO)_{10}]^-$** *was determined by neutron diffraction.*

The bonding situation in the M–H–M bridge is similar to that in the moiety B–H–B ("open" $2e\,3c$ bond) encountered in boranes.

M – H units are easily detected by **large high-field shifts of their ^{1}H-NMR signals** (see also p. 306). This shielding cannot be directly related to structure and reactivity of these

complexes, though. The absence of a high-field NMR signal ($0 > \delta > -50$ ppm) does not necessarily imply that an M–H moiety is absent. More reliable indications of the presence of direct M–H bonds are found in the **coupling constants,** e.g. $^1J(^{103}Rh, ^1H) = 15-30$ Hz, $^1J(^{183}W, ^1H) = 28-80$ Hz, $^1J(^{195}Pt, ^1H) = 700-1300$ Hz.
In addition to **protonation**, other important reactions of carbonylmetallates include **alkylation** and **silylation:**

$$[Mn(CO)_5]^- + CH_3I \longrightarrow CH_3Mn(CO)_5 + I^-$$

$$[C_5H_5W(CO)_3]^- + (CH_3)_3SiCl \longrightarrow C_5H_5W(CO)_3Si(CH_3)_3 + Cl^-$$

Carbonylmetallates have also found application in organic synthesis ($Na_2Fe(CO)_4$, "**Collman's reagent**"). Organic halogen compounds can be functionalized in many ways via intermediary reaction with disodium tetracarbonylferrate:

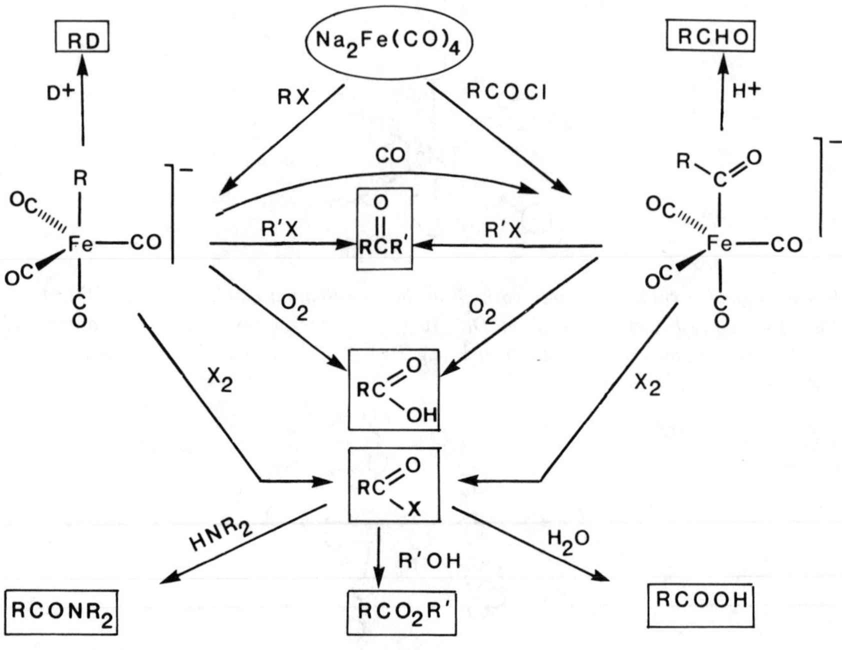

Advantages: *high yields* (70–90 %), *other functional groups need not be protected since the organoiron anions do not add to carbonyl or nitrile groups.*

Limitations: $Fe(CO)_4^{2-}$ *is a strong base and, with tertiary halogen compounds, effects* HX-*elimination.*

Examples:

$$CH_3(CH_2)_5COCl \xrightarrow[2)\ HOAc]{1)\ Na_2Fe(CO)_4,\ THF,\ CO} CH_3(CH_2)_5CHO \quad 98\%$$

$$NC(CH_2)_{8}Br \xrightarrow[2)\ n\text{-}C_7F_{15}COCl]{1)\ Na_2Fe(CO)_4,\ THF,\ CO} NC(CH_2)_{8}COC_7F_{15} \quad 72\%$$

14.5.6 Carbonyl Metal Halides

Carbonyl metal halides are known for most transition metals. They are formed by reaction of metal carbonyls with halogens or by reaction of metal salts with CO:

$$Fe(CO)_5 + I_2 \longrightarrow Fe(CO)_4I_2 + CO$$

$$Mn_2(CO)_{10} + I_2 \longrightarrow 2Mn(CO)_5I \xrightarrow[-2CO]{120^\circ C} [(OC)_4Mn(\mu\text{-}I)_2Mn(CO)_4]$$

$$[(OC)_4Mn(\mu\text{-}I)_2Mn(CO)_4] \xrightarrow{C_5H_5N\ (py)} Mn(CO)_4I(py) \xrightarrow{py} fac\text{-}Mn(CO)_3I(py)_2$$

*The formation of the **fac**-substitution product results from the **trans** effect of the* CO *ligands.*

The late transition metals copper, palladium, platinum, and gold, while not forming stable binary carbonyls, do give carbonyl metal halides:

$$2\,PtCl_2 + 2\,CO \longrightarrow [Pt(CO)Cl(\mu\text{-}Cl)]_2$$

The reversible fixation of CO to Cu(I) serves to remove carbon monoxide from gases:

$$2\,[CuCl_2]^- + 2\,CO \rightleftharpoons [Cu(CO)Cl]_2 + 2\,Cl^-$$

$$Cu^+/NH_4OH + CO \longrightarrow [Cu(NH_3)_2CO]^+ \xrightarrow[-NH_4^+]{H^+} Cu^+ + CO$$

The corresponding complex [Cu(en)CO]Cl of the chelating ligand ethylenediamine (en) can be isolated, although it is rather unstable.

Carbonyl metal halides are suitable precursors for the formation of **metal-metal bonds**:

$$Mn(CO)_5Br + [Re(CO)_5]^- \longrightarrow (CO)_5Mn-Re(CO)_5 + Br^-$$

14.6 Thio-, Seleno-, and Tellurocarbonyl Metal Complexes

Although the molecules CS, CSe, and CTe as monomers are unstable at room temperature, they can be generated in the coordination sphere of a transition metal. The most common source for CS is CS_2, which itself is first coordinated as an η^2-ligand to the metal. The η^2-CS_2 unit is then attacked by triphenylphosphane to yield the η^1-thiocarbonyl complex and triphenylphosphanesulfide (Wilkinson, 1967):

$$RhCl(PPh_3)_3 \xrightarrow[\text{MeOH},\ -PPh_3]{CS_2} [(Ph_3P)_2Rh(\eta^2\text{-}CS_2)]^+Cl^- \xrightarrow[-Ph_3PS]{+PPh_3} trans\text{-}RhCl(CS)(PPh_3)_2$$

$$CpMn(CO)_2(THF) \xrightarrow{CSe_2} CpMn(CO)_2(CSe_2) \xrightarrow[-Ph_3PSe]{+PPh_3} CpMn(CO)_2(CSe)$$

Other sources for CS are ethylchlorothioformate or thiophosgene (Angelici, 1968):

$$[CpFe(CO)_2]^- \xrightarrow{EtOC(S)Cl} CpFe(CO)_2C(S)OEt \xrightarrow[-EtOH]{HCl} [CpFe(CO)_2(CS)]^+$$

$$2\,Cr(CO)_6 \xrightarrow[\text{Na/Hg, THF},\ -2CO]{+2e^-} [Cr_2(CO)_{10}]^{2-} \underset{-2e^-}{\overset{+2e^-}{\rightleftharpoons}} 2\,[Cr(CO)_5]^{2-} \xrightarrow[-2\,Cl]{Cl_2CS\ \text{(Thiophosgene)}} (CO)_5CrCS$$

(Angelici, 1973)

An additional approach is the substitution of two chlorine atoms in a dichlorocarbene complex by E^{2-} (Roper, 1980):

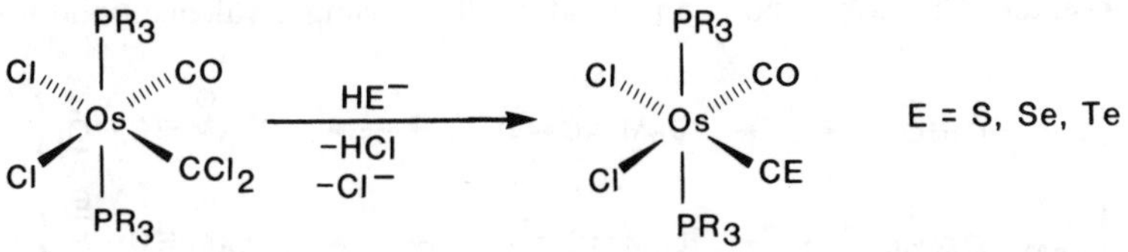

E = S, Se, Te

The M−C−S segment is usually linear, the M−CS bond being shorter and therefore stronger than an M−CO bond in the same coordination sphere:

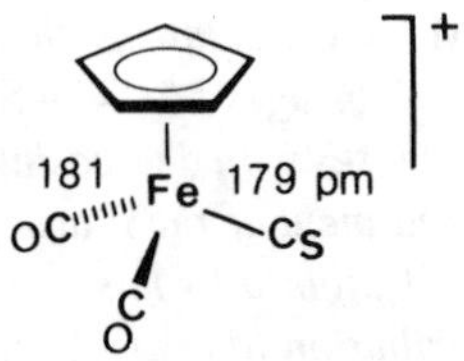

(Angelici, 1987)
$\nu_{CS} = 1348\ cm^{-1}$

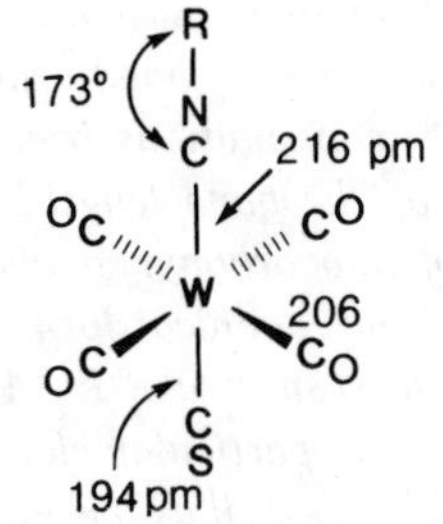

(Angelici, 1976)
$\nu_{CS} = 1240\ cm^{-1}$
(R = cyclohexyl)

Bridging CS ligands are also known; apparently CS is even more inclined to serve as a bridging ligand and in this function more versatile than CO:

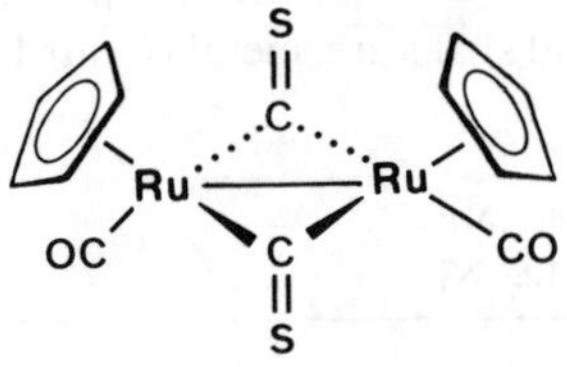

(Angelici, 1977)
$\nu_{CS} = 1031\ cm^{-1}$

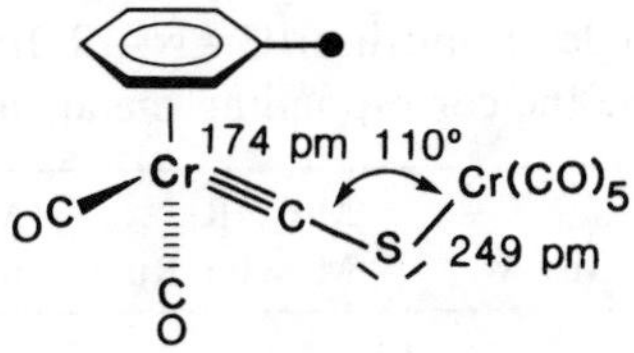

(Lotz, 1986)
$\nu_{CS} = 1156\ cm^{-1}$
[$\nu_{CS} = 1220\ cm^{-1}$
in $(PhMe)Cr(CO)_2CS$]

Note the preference of μ-CS bridging over μ-CO bridging in $[(C_5H_5)Ru(CO)CS]_2$ and the bridging mode M−C−S−M in $(PhMe)Cr(CO)_2CSCr(CO)_5$. For the ligand CO, bridging of the type M−C−O−M has only been observed in a few exceptional cases (Trogler, 1985).
The valence electron formula used above is in agreement with the bond angle CSCr and with the very short Cr≡C bond, which is reminiscent of the bond length in carbyne complexes (p. 219).

Transition-metal thiocarbonyl complexes exhibit ν_{CS} frequencies in the range of 1160 to 1410 cm^{-1} (free CS in matrix isolation: $\nu_{CS} = 1274\ cm^{-1}$). Obviously, metal coordination can either weaken or strenghten the CS bond. Compared to CO, **CS** is thought to be a **better π-acceptor**. This conclusion can be rationalized using a valence bond argument:

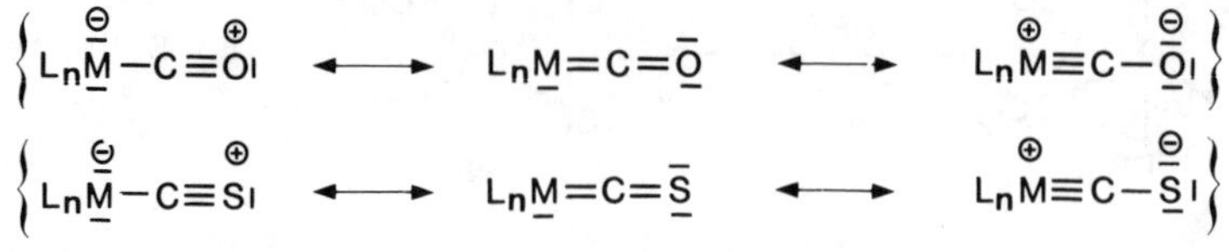

$$\left\{ L_nM^{\ominus}-C\equiv O^{\oplus} \longleftrightarrow L_nM=C=O \longleftrightarrow L_nM^{\oplus}\equiv C-O^{\ominus} \right\}$$

$$\left\{ L_nM^{\ominus}-C\equiv S^{\oplus} \longleftrightarrow L_nM=C=S \longleftrightarrow L_nM^{\oplus}\equiv C-S^{\ominus} \right\}$$

a b c

M $\overset{\pi}{\rightarrow}$ C backbonding leads to an increase of the metal-ligand $d_\pi - p_\pi$ and a decrease of the intra-ligand $p_\pi - p_\pi$ bond order (canonical forms **b, c**). *According to the "double-bond rule" (p. 108) contributions b and c should be more significant for* CS *than for* CO. *A comparison of the bond lengths d*(M–CS) *versus d*(M–CO) *confirms this notion. The propensity of thiocarbonyl complexes to form "end to end" bridges* M–C–S–M *reflects the weight of the canonical form* **c**(S); *this propensity increases with decreasing frequencies* ν_{CS} *in the infrared spectra of* L_nMCS. *In a more refined treatment of the π-acceptor/σ-donor ratio for* CS, *the particular electron density at the metal atom also has to be considered (Andrews, 1977) as well as the fact that the π-donor contribution* M(π) ← 1(π) (②, *p. 227) is more important for* CS *complexes than for* CO *complexes.*

14.7 Isocyanide Complexes (Metal Isonitriles)

Substitution of the oxygen atom in carbon monoxide |C≡O| for the group NR leads to isocyanides (isonitriles) |C≡N–R. In composition and structure, metal isonitriles closely resemble the corresponding metal carbonyls.

$M(CNR)_6$	$M(CNR)_5$	$M_2(CNR)_8$	$M(CNR)_4$
M = Cr, Mo, W	M = Fe, Ru	M = Co	M = Ni

However, a more detailed analysis (major contributions by L. Malatesta) reveals **important differences between CNR and CO:**

- *In contrast to* CO, *isonitriles possess a considerable dipole moment*: $\mu_{CNPh} = 3.44$ Debye (negative end on carbon), $\mu_{CO} = 0.1$ Debye.
- CNR *can displace* CO *as a ligand*:

$$Ni(CO)_4 + 4\,CNPh \longrightarrow Ni(CNPh)_4 + 4\,CO$$

- *Some metal isonitriles as yet have no metal carbonyl counterparts* *Examples*: $[Pt(CNR)_4]^{2+}$, $[M(CNR)_4]^{+}$ (M = Cu, Ag, Au).
- *Although documented, the ability of* CNR *to acts as bridging ligand is not as pronounced as for* CO [*example*: $(RNC)_3Co(\mu\text{-}(CNR)_2)Co(CNR)_3$].

- *Compared to metal carbonyls, metal isonitriles have a greater tendency to occur in higher oxidation states* (M^I, M^{II}) *and a reduced ability to stabilize low oxidation states* (M^0, M^{-I}).
- *Metal coordination of* CO *without exception results in a lowering of the stretching frequency* ν_{CO}. *For isonitriles, shifts of* ν_{CN} *to higher as well as to lower wave numbers are observed upon coordination*:

L = p-$CH_3C_6H_4NC$	L	L_4Ni	L_4Ag^+
ν_{CN}/cm^{-1}	2136	2065	2186 (sh)
			2177
		2033	2136 (w)

Compared with CO, isonitriles **CNR** display **stronger σ-donor** and **weaker π-acceptor** character, the latter being influenced by the nature of the group R.

Contrary to what might be anticipated from the contribution of canonical form b, backbonding does not lead to a bent structure for the isonitrile ligand.

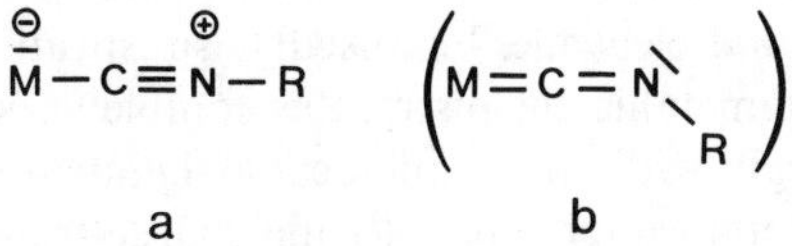

$$\overset{\ominus}{M}-C\equiv\overset{\oplus}{N}-R \qquad \left(M=C=N\diagdown_{R}\right)$$

a b

Whereas strict adherence to the octet rule (a VB *concept) for b would necessitate placing a lone pair of electrons on the nitrogen atom with concomitant ligand bending, the use of vacant antibonding orbitals* π* *of linear* CNR *for the backbonding interaction circumvents a "violation" of the octet rule. Note that the concept of antibonding molecular orbitals and their engagement in chemical bonding is alien to the valence bond formalism.*

From the large number of reactions of metal isonitriles, only the addition of compounds with active hydrogen (alcohols, phenols, amines, hydrazines) are mentioned here:

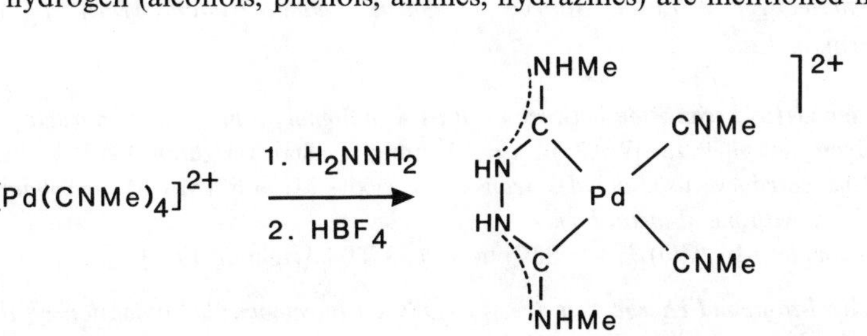

$$[Pd(CNMe)_4]^{2+} \xrightarrow[\text{2. } HBF_4]{\text{1. } H_2NNH_2} \left[\text{(cyclic bis(carbene) Pd complex: HN–NH bridging two C(NHMe) carbene ligands, plus two CNMe)}\right]^{2+}$$

This type of reaction constitutes a general route for preparing aminocarbene complexes (cf. Chapter 14.3).

The fact that the chemistry of metal isonitriles has long been overshadowed by that of metal carbonyls certainly is not due to a lack of reactivity, but rather to the repulsive smell of the free ligands. Interestingly, the radioactive isonitrile complex $[^{99m}Tc(CN-CH_2CMe_2OMe)_6]^+$ has recently been approved for cardiac imaging (Cardiolite™, Du Pont Merck Pharmaceutical Co., 1991).

Excursion
Photochemical Reactivity of Organotransition Metal Complexes

Absorption of a photon converts a molecule into an excited electronic state, which constitutes a new chemical species with altered molecular and electronic structure and therefore different chemical behavior. Compared to the respective ground state, molecules in an electronically excited state may display different bond angles (formation of strained systems, p. 415), as well as drastic changes of the acidity constants and in the redox potentials. Most importantly, specific bond weakening in an excited state may cause greatly enhanced chemical reactivity. Thus, thermal substitution of CO in (η^5-C_5H_5)Mn(CO)$_3$ requires high temperatures, whereas photochemically it proceeds at room temperature.

A thorough discussion of a photochemical reaction ideally includes a description of **photophysical aspects** (nature and life-time of the electronically excited state), **photochemical primary processes** (bond cleavage, bimolecular substitution, redox process) and of **synthetic utility**. In organometallic chemistry, this completeness is hardly ever achieved. Already, the problems begin with an unequivocal assignment of the excitation since the differentiation between metal-centered ($d-d$) and charge-transfer (CT) transitions, dear to the practitioner of ligand field spectroscopy, finds only limited applicability to organotransition-metal complexes, where covalent bond character dominates. Here **both** orbitals involved in an electronic transition usually contain metal- **and** ligand contributions, rendering criteria of assignment like solvatochromism or substitution effects inappropriate.

In the simplified MO *scheme of an octahedral complex, the various types of electronic excitation are defined. Correlation lines lead to the basis orbitals which dominate in the respective molecular orbitals. Source*: *N. Sutin, Inorg. React. Meth.* ***15*** (1986) 260, *J. J. Zuckerman, Ed.*

LF ***Ligand-field transition*** *between orbitals with dominant metal (nd) character.*
Example: (*piperidine*)W(CO)$_5$, $\lambda = 403$ nm, $\varepsilon = 3860$ (*Wrighton*, 1974).
This category also includes transitions σ(M–M) $\rightarrow$ σ^*(M–M) *in oligonuclear complexes with metal-metal bonds.*
Example: $Mn_2(CO)_{10}$, $\lambda = 336$ nm, $\varepsilon = 33\,700$ (*Wrighton*, 1975).

MLCT ***Metal-to-ligand charge transfer****, formally an intramolecular oxidation of the metal and reduction of the ligand(s).*
Example: (*4-formyl-pyridine*)W(CO)$_5$, $\lambda = 470$ nm, $\varepsilon = 6470$ (*Wrighton*, 1973).

LMCT ***Ligand-to-metal charge transfer.***
Example: $[Cp_2Fe]^+$, $\lambda = 617$ nm, $\varepsilon = 450$ (*H. B. Gray*, 1973 *R*).

IL ***Intraligand transition****, electronic excitation* $n \rightarrow \pi^*$ *or* $\pi \rightarrow \pi^*$, *respectively, between orbitals which are predominantly centered on a ligand, counterpart to* LF*-transition.*
Example: (*trans-4-styrylpyridine*)W(CO)$_5$, $\lambda = 316$ nm, $\varepsilon = 16300$ (*Wrighton*, 1973).

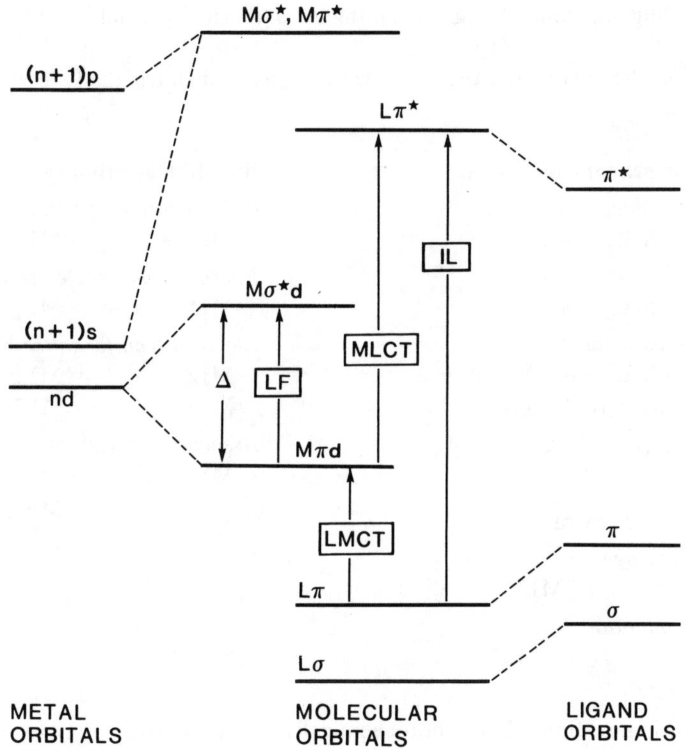

Additional excitations are:

IT — ***Intervalence transition,*** M → M′ *charge transfer in mixed valence oligonuclear complexes. Example*: *biferrocenyl*$^+$, $[(C_5H_5)Fe^{II}(C_5H_4-C_5H_4)\text{-}Fe^{III}(C_5H_5)]^+$, $\lambda = 1900$ nm, $\varepsilon = 550$ (*Cowan*, 1973 *R*).

MSCT — ***Metal-to-solvent charge transfer.*** *Example*: Cp_2Fe *in* CCl_4, $\lambda = 320$ nm. *Irradiation at the frequency of the* MSCT *band causes photo-oxidation of the substrate according to*

$$Cp_2Fe + CCl_4 \rightarrow \{Cp_2Fe^{\dot{+}}CCl_4^{\dot{-}}\}^* \rightarrow Cp_2Fe^{\dot{+}} + Cl^- + \dot{C}Cl_3$$

For the photochemical primary processes which follow the initial absorption of light, the life-times of the electronically excited states are of critical importance. The photochemically active state of a molecule is not necessarily generated directly by the absorption of a photon:

The intensive red color of $[Fe(bipy)_3]^{2+}$ *is due to an* MLCT *transition. However, the* MLCT *state is extremely short-lived* ($\tau \leq 10^{-11}$ s) *and rapid relaxation to a longer lived chemically active* LF *state occurs* ($\tau = 10^{-9}$ s, *Sutin*, 1980). *Note that direct generation of an* LF *state* (*d* – *d transition*) *would be Laporte-forbidden.*

Electronic excitations are usually accompanied by vibrational excitations (vertical transition, Franck-Condon principle). Excessive vibrational energy is rapidly dissipated into

the surrounding medium, though, resulting in a thermally equilibrated excited state ("thexi-state").
Depending on the energy and the life-time of the excited states, the following processes may succeed:

Unimolecular primary processes:

Radiationless deactivation

$$ML_6^* \longrightarrow ML_6 + \text{heat}$$

Luminescence

$$ML_6^* \longrightarrow ML_6 + h\nu$$

Dissociation, association

$$ML_6^* \xrightarrow[-L]{} ML_5 \xrightarrow{A} ML_5A$$

Dissociation, oxidative addition

$$ML_6^* \xrightarrow[-L]{} ML_5 \xrightarrow{A-B} L_5M{<}^{A}_{B}$$

Isomerization

cis/trans or *d/l* rearrangement

Homolytic cleavage

$$L_5M-ML_5^* \longrightarrow 2\,ML_5^{\bullet}$$

Reductive elimination

$$L_4M(H)_2^* \longrightarrow L_4M + H_2$$

Bimolecular primary processes:

Collisional deactivation (thermal)

$$ML_6^* + A \longrightarrow ML_6 + A + \text{heat}$$

Energy transfer (electronic)

$$ML_6^* + A \longrightarrow ML_6 + A^*$$

Electron transfer (after MLCT)

$$ML_6^* + A \longrightarrow ML_6^+ + A^-$$

$$ML_6^* + B \longrightarrow ML_6^- + B^+$$

Association (after MLCT)

$$ML_6^* + A \longrightarrow ML_6A$$

$$(\longrightarrow ML_5A + L)$$

In the following sections, four photochemically initiated reaction types, which have gained preparative importance, will be discussed.

1. PHOTOCHEMICAL SUBSTITUTION AT METAL CARBONYLS

This is the best known and the most frequently executed photoreaction in organometallic chemistry. Examples:

$$W(CO)_6 + PPh_3 \xrightarrow{h\nu} W(CO)_5(PPh_3) + CO$$

$$Fe(CO)_5 + \text{(butadiene)} \xrightarrow{h\nu} (\text{butadiene})\cdots Fe(CO)_3 + 2\,CO$$

$$CpV(CO)_4 + PhC{\equiv}CPh \xrightarrow{h\nu} CpV(CO)_{3-n}(PhC{\equiv}CPh)_n + n\,CO \qquad n = 1, 2$$

$$CpMn(CO)_3 \xrightarrow[THF,\,-CO]{h\nu} CpMn(CO)_2THF \xrightarrow[20\,°C,\,-THF]{L} CpMn(CO)_2L$$

The rate constant for the dissociation $Cr(CO)_6 \rightarrow Cr(CO)_5 + CO$, which preceeds the entrance of a new ligand L, is increased upon photochemical excitation $Cr(CO)_6 \xrightarrow{h\nu} Cr(CO)_6^*$ by a factor of 10^{16}! This effect may be traced to the ligand-field transition $t_{2g}(\pi) \rightarrow e_g(\sigma^*)$ which leads to the depopulation of an M–CO bonding- and the population of an M–CO antibonding MO (see p. 188). In metal carbonyl complexes $M(CO)_mL_n$ with a mixed coordination sphere, photochemical excitation causes dissociation of that ligand which is most weakly bonded in the ground state as well. This will be the ligand at the lowest position respectively in the spectrochemical series:

$$M(CO)_5THF \xrightarrow{h\nu} M(CO)_5 + THF$$

For this reason, the weakly bonded ligand THF can only be introduced once. Among ligands which form bonds of comparable strength, competitive reactions are observed:

$$CO + M(CO)_4L \xleftarrow{h\nu} M(CO)_5L \xrightarrow{h\nu} M(CO)_5 + L$$

Thus, in the presence of an excess of trimethylphosphite, quantitative photochemical conversion of $Mo(CO)_6$ into $Mo[P(OMe)_3]_6$ can be achieved (Poilblanc, 1972).
In (η^6-arene)$Cr(CO)_3$, the arene as well as the carbonyl groups may be replaced:

$$(\eta^6\text{-arene})Cr(CO)_3 \xrightarrow{h\nu} \{(\eta^6\text{-arene})Cr(CO)_2\} \begin{cases} \xrightarrow[+CO]{\text{arene}'} (\eta^6\text{-arene}')Cr(CO)_3 \\ \xrightarrow{+L} (\eta^6\text{-arene})Cr(CO)_2L \end{cases}$$

Mo- and W-complexes undergo CO substitution less readily.
If suitable free ligands are absent, the gap in the coordination sphere, generated through photochemical dissociation of CO, may be closed by dimerization:

$$2\ Re(CO)_5Br \xrightarrow[CCl_4,\ -2\ CO]{h\nu} (CO)_4Re(\mu\text{-}Br)_2Re(CO)_4$$

(Wrighton, 1976)

$$2\ C_5H_5Co(CO)_2 \xrightarrow[-2\ CO]{h\nu} (C_5H_5)Co(\mu\text{-}CO)_2Co(C_5H_5)$$

(Brintzinger, 1977)

The gap may also be filled internally via $\sigma(\eta^1) \rightarrow \pi(\eta^3)$ rearrangement:

$$(\eta^1\text{-}C_3H_5)Mn(CO)_5 \xrightarrow{h\nu} (\eta^3\text{-}C_3H_5)Mn(CO)_4 + CO$$

(M. L. H. Green, 1964 R)

Another possibility of coordinative saturation is oxidative addition, as exemplified by hydrosilation (Graham, 1971):

$$Fe^0(CO)_5 \xrightarrow[-CO]{h\nu} \{Fe(CO)_4\} \xrightarrow{HSiCl_3} (CO)_4Fe^{II}(H)(SiCl_3)$$

Whereas the previous examples of carbonyl substitution featured **dissociative activation,** the electronic peculiarities of the ligand **NO** permit an **associative mechanism** which is initiated by a preceding photochemical conversion of NO^+ (linear coordination) into NO^- (bent coordination) (Zink, 1981):

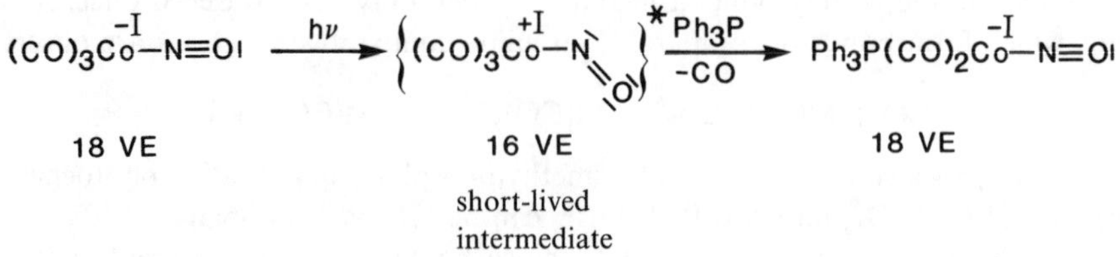

$$\underset{18\ VE}{(CO)_3\overset{-I}{Co}{-}N{\equiv}O|} \xrightarrow{h\nu} \underset{\substack{16\ VE \\ \text{short-lived} \\ \text{intermediate}}}{\left\{(CO)_3\overset{+I}{Co}{-}N{=}O\right\}^*} \xrightarrow[-CO]{Ph_3P} \underset{18\ VE}{Ph_3P(CO)_2\overset{-I}{Co}{-}N{\equiv}O|}$$

2. PHOTOCHEMICAL REACTIONS WITH CLEAVAGE OF METAL-METAL BONDS

Oligonuclear organotransition metal compounds may hold metal-metal bonds of the orders 1–4 (chapter 16). In the simplest case, two 17 VE fragments are linked by means of an M–M single bond. Examples: $[(C_5H_5)Mo(CO)_3]_2$, $Mn_2(CO)_{10}$, $Co_2(CO)_8$.

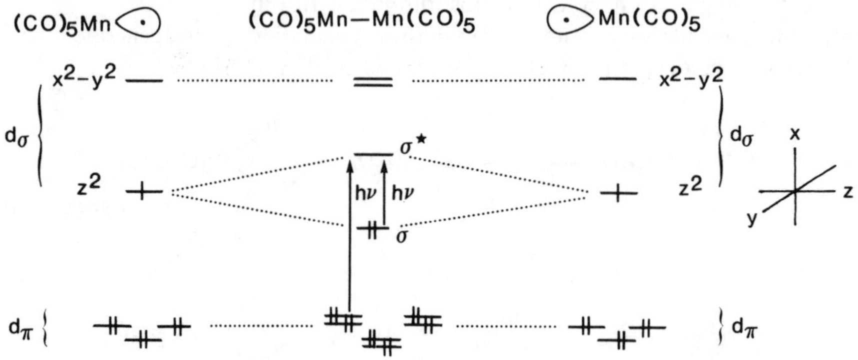

$(CO)_5Mn\cdot$ $(CO)_5Mn{-}Mn(CO)_5$ $\cdot Mn(CO)_5$

x^2-y^2 $d\sigma$ σ^* z^2 $h\nu$ $h\nu$ σ $d\pi$ x y z

In these dinuclear species, the M–M single bond is the weakest link. Photochemical excitation $\sigma \rightarrow \sigma^*$ reduces the bond order to zero and **homolytic cleavage** yields two organometallic radicals which undergo subsequent reactions:

$$Mn_2(CO)_{10} \xrightarrow{h\nu} 2\,Mn(CO)_5^{\cdot} \xrightarrow{CCl_4} 2\,Mn(CO)_5Cl$$

From a study of competitive reactions for the halogen abstractions, the following reactivity sequence was derived (Wrighton, 1977):

$$Re(CO)_5^{\cdot} > Mn(CO)_5^{\cdot} > CpW(CO)_3^{\cdot} > CpMo(CO)_3^{\cdot} > CpFe(CO)_2^{\cdot} > Co(CO)_4^{\cdot}$$

This series parallels the trend in the energetic splitting $E(\sigma, \sigma^*)$ which manifests itself in the optical spectra of the respective dimers.

Other photolytic and follow-up reactions which are synthetically useful include:

$$[CpNi(CO)]_2 \xrightarrow{h\nu} 2\,CpNi(CO)^{\cdot} \xrightarrow{R_2S_2} 2\,CpNi(CO)SR$$

$$[CpMo(CO)_3]_2 \xrightarrow{h\nu} 2\,CpMo(CO)_3^{\cdot} \xrightarrow[-2\,CO]{2\,NO} 2\,CpMo(CO)_2NO$$

Through co-photolysis heterometallic dinuclear complexes may be prepared:

$$Re_2(CO)_{10} + [CpFe(CO)_2]_2 \xrightarrow{h\nu} 2\,(CO)_5ReFe(CO)_2Cp$$

A few seemingly simple photochemically induced substitions have been shown to proceed via an initial cleavage of the metal-metal bond (Wrighton, 1975):

$$Mn_2CO_{10} + PPh_3 \longrightarrow Mn_2(CO)_9(PPh_3)$$

$$\downarrow h\nu \qquad\qquad \uparrow$$

$$2\,Mn(CO)_5^{\cdot} \xrightarrow[-CO]{\Delta,\ PPh_3} Mn(CO)_5^{\cdot} + Mn(CO)_4PPh_3^{\cdot}$$

The 17 VE radical $Mn(CO)_5^{\cdot}$ is substitution-labile; replacement of CO by PPh_3 takes an **associative** path, that is without preceding expulsion of CO (T. L. Brown, 1985).
In $[CpFe(CO)_2]_2$, however, the dimeric structure is maintained during the substitution process (T. J. Meyer, 1980):

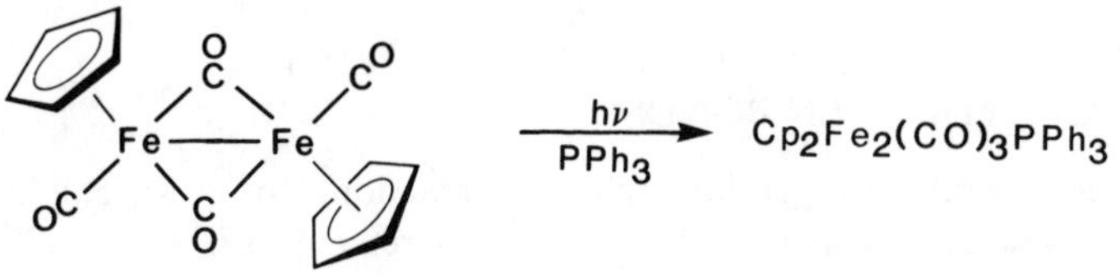

$$[CpFe(CO)_2]_2 \xrightarrow[PPh_3]{h\nu} Cp_2Fe_2(CO)_3PPh_3$$

This also applies to $\mu(\eta^2:\eta^2\text{-}Et_2C_2)\,[CpMo(CO)_2]_2$ (Muetterties, 1980):

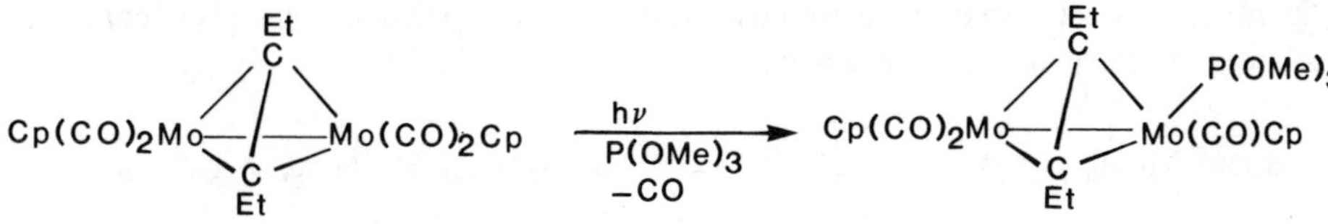

$$Cp(CO)_2Mo(\mu\text{-}EtC{\equiv}CEt)Mo(CO)_2Cp \xrightarrow[P(OMe)_3,\ -CO]{h\nu} Cp(CO)_2Mo(\mu\text{-}EtC{\equiv}CEt)Mo(CO)Cp\,P(OMe)_3$$

Obviously, bridging ligands in addition to metal-metal bonds obviate homolytic cleavage. If the cluster $Ir_4(CO)_{12}$ is photolyzed in the presence of an alkyne, the aggregation Ir_4 is preserved, the metal frame changing from tetrahedral to planar geometry, however. The entrance of an electron-rich alkyne abates the electron deficiency of the Ir_4-unit with concomitant **reduction of metal-metal connectivity** (B. F. G. Johnson, 1978):

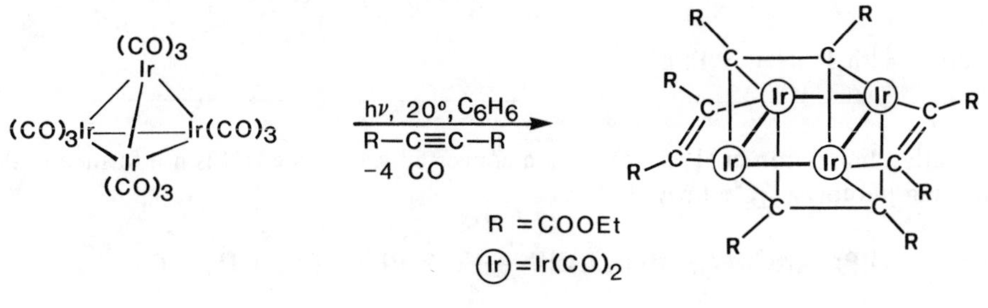

$$Ir_4(CO)_{12} \xrightarrow[R-C{\equiv}C-R,\ -4\,CO]{h\nu,\ 20°,\ C_6H_6} Ir_4(CO)_8(RC_2R)_4$$

R = COOEt
(Ir) = $Ir(CO)_2$

On the other hand, the thermodynamic stability of certain organometallic clusters may also lead to the **generation of new metal-metal bonds** under photochemical conditions:

$$2\,Co_2(CO)_8 \xrightarrow{h\nu} Co_4(CO)_{12} + 4\,CO$$

Occasionally, photoreactions of dinuclear metal carbonyls are accompanied by disproportionation:

$$[CpMo^{I}(CO)_3]_2 + Cl^- \xrightarrow[CH_3CN]{h\nu} CpMo^{II}(CO)_3Cl + [CpMo^{0}(CO)_3]^-$$
(T. J. Meyer, 1974)

$$Mn_2(CO)_{10} + 3\,NH_3 \xrightarrow[\text{pentane}\;-2\,CO]{h\nu} [fac\text{-}Mn^{I}(CO)_3(NH_3)_3]^+\,[Mn^{-I}(CO)_5^4]^-$$
(Herberhold, 1978)

Mechanistically, these reactions may follow the sequence homolysis → substitution → electron transfer.

3. PHOTOCHEMICAL REACTIONS WITH CLEAVAGE OF METAL-HYDROGEN BONDS

While transition-metal hydrogen bond cleavage reactions do not strictly belong to the realm of organometallic chemistry, they are closely related to it.

- In the case of **mononuclear di- and oligohydrides,** the primary reaction of the photochemically excited molecule usually is **reductive elimination of H_2** (Geoffroy, 1980 R). The electron-rich and sparsely coordinated species which are formed in this way are highly reactive. They combine even with relatively inert partners, which is of considerable preparative significance (see p. 440):

$$(dppe)_2MoH_4 \xrightarrow[-2\,H_2]{h\nu} \{(dppe)_2Mo\} \xrightarrow{2\,N_2} trans\text{-}[(dppe)_2Mo(N_2)_2]$$
(Geoffroy, 1980)

$$Cp_2WH_2 \xrightarrow[-H_2]{h\nu} \{Cp_2W\} \xrightarrow{C_6H_6} Cp_2W\genfrac{}{}{0pt}{}{H}{Ph}$$
(M. L. H. Green, 1972)

$$(dppe)_2ReH_3 \xrightarrow[-H_2]{h\nu} \{(dppe)_2ReH\} \xrightarrow{CO_2} (dppe)_2Re\genfrac{}{}{0pt}{}{O}{O}CH$$
(Geoffroy, 1980)

dppe = $Ph_2PCH_2CH_2PPh_2$

Evidently, elimination of H_2 proceeds in a concerted way, since HD is not formed in the following photolysis (Geoffroy, 1976):

$$(PPh_3)_3IrClH_2 + (PPh_3)IrClD_2 \xrightarrow{h\nu} 2\,(PPh_3)_3IrCl + H_2 + D_2$$

- **Oligohydridometal clusters** like $Re_3H_3(CO)_{12}$ resist photochemical H_2 elimination. Here, the hydride ligands are in metal bridging positions, which impedes the concerted elimination of an H_2 molecule. Instead, the photochemistry of these hydrido clusters is governed by metal-metal bond cleavage and by carbonyl substitution.
- In **monohydrido complexes** as well, photoinduced ligand substitution is the usual reaction, homolytic cleavage of transition-metal hydrogen bonds being the exception. This is plausible because the concerted loss of an H_2 molecule from a dihydride is a low energy reaction path, whereas the cleavage of a single metal-hydrogen bond in a monohydride is a high energy one. Photoinduced dissociation of a metal carbonyl hydride can be studied by means of IR spectroscopy in matrix isolation (Orchin, 1978):

$$HCo(CO)_4 \xrightarrow[\text{Ar, 14 K}]{h\nu\ (\lambda = 310\ \text{nm})} HCo(CO)_3 + CO$$

In the presence of potential ligands, photosubstitution is effected:

$$HRe(CO)_5 \xrightarrow[-CO]{h\nu} HRe(CO)_4 \xrightarrow{PBu_3} HRe(CO)_4PBu_3$$

This reaction is much slower than photosubstitution of $Re_2(CO)_{10}$, which was shown to proceed via $Re(CO)_5^{\cdot}$ radicals. Thus for $HRe(CO)_5$, Re–H homolysis as a primary process may confidently be excluded (T. L. Brown, 1977).

4. PHOTOCHEMICAL REACTIONS WITH CLEAVAGE OF TRANSITION METAL-CARBON BONDS

The study of photochemically initiated reactions of organometallics containing transition metal–carbon σ bonds is complicated by the fact that these molecules often have access to thermal pathways of decomposition like β-hydride elimination (p. 198). If one turns to thermally more stable complexes with a mixed ligand sphere (example: $CpMo(CO)_3CH_3$), the photochemical reactivity of the metal-alkyl bond is complemented by the possibility of reactions in other segments of the molecule. This diversity clouds the mechanistic picture. Even more than in the preceding sections preparative utility will be stressed here.

- An early application of photochemical excitation in organometallic chemistry was the synthesis of π-allyl complexes via **$\sigma \rightarrow \pi$ rearrangement** (M. L. H. Green, 1963):

Mo(CO)3 (η¹-allyl) —hν, -CO→ [Mo(CO)2 (η¹-allyl)] → Mo(CO)2 (η³-allyl)

18 VE 16 VE 18 VE

The primary photoreaction here is loss of a CO ligand, followed by the change in the bonding mode of the allyl group from η^1 (σ-complex) to η^3 (π-complex). According to this pattern, (η^5-pentadienyl)metal moieties can also be generated (p. 290).

- **Photodealkylation,** a method for the formation of reactive organometallic intermediates, has found extensive preparative application. Example (Alt, 1984 R):

$$Cp_2Ti(CH_3)_2 \xrightarrow[-2\,CH_4]{h\nu} \{Cp_2Ti\}$$

16 VE → 14 VE

$\{Cp_2Ti\} \xrightarrow{CO} Cp_2Ti(CO)_2$

$\{Cp_2Ti\} \xrightarrow{S_8} Cp_2TiS_5$

$\{Cp_2Ti\} \xrightarrow{PhC\equiv CPh}$ Cp_2Ti(tetraphenyltitanacyclopentadiene)

Cp = C_5H_5 —o = —Ph

A remarkable aspect is the loss of CH_4: labeling studies indicate that the hydrogen required is furnished by the C_5H_5 ligand. Only traces of C_2H_6 are observed; therefore, the intermediacy of $CH_3^{\bullet}$ radicals is unlikely. Apparently, the photolysis of molecules with TM$\overset{\sigma}{-}$C bonds differs fundamentally from that of organometallic dihydrides, where reductive elimination of H_2 dominates. In the absence of potential ligands, photodealkylations may be succeeded by dimerization of the organometal fragments (Wrighton, 1982):

$$CpRu(CO)_2C_2H_5 \xrightarrow[-CO]{h\nu} \{CpRu(CO)(H)(C_2H_4)\} \xrightarrow{CO} [CpRu(CO)_2]_2 + H_2 + C_2H_4$$

Here as well, the initial step is photochemical loss of CO. An olefin(hydrido)complex intermediate is implied by the reaction products H_2 and C_2H_4. A prerequisite for the β-elimination mechanism to occur is the initial generation of a vacant coordination site. If in addition to σ-bonded alkyl groups, the central metal atom carries only multidentate ligands, photochemical excitation may lead to bond weakening for one of these ligands, reducing its hapticity (e.g. $\eta^5 \rightarrow \eta^3$). The photochemical formation of $(\eta^5\text{-}C_5H_5)_3Th$ from the thermally stable compound $(\eta^5\text{-}C_5H_5)_3Th(i\text{-}Pr)$ is thought to follow this pattern (Marks, 1977):

$$(\eta^5\text{-}C_5H_5)_3Th(i\text{-}Pr) \xrightarrow[5°]{h\nu} \{(\eta^5\text{-}C_5H_5)_2(\eta^3\text{-}C_5H_5)Th(H)(C_3H_6)\}$$

$$\downarrow \Delta$$

$$(\eta^5\text{-}C_5H_5)_3Th + C_3H_6 + C_3H_8$$

• The **Fischer-Müller reaction,** a versatile method for the synthesis of arene- and olefin complexes is also based on the photochemical cleavage of $M \overset{\sigma}{-} C$ bonds (M = V, Cr, Fe, Ru, Os, Pt; R = *i*-Pr). Example (E. O. Fischer, 1962):

$$FeCl_3 \xrightarrow[Et_2O,\ -70°]{(i\text{-}Pr)MgBr} (i\text{-}Pr)_3Fe(solv) \xrightarrow[\text{1,3-cyclohexadiene}]{h\nu} (C_6H_6)Fe(C_6H_8) + C_3H_6 + C_3H_8 + H_2$$

(i-Pr)₃Fe(solv): not isolated; via $(i\text{-}Pr)_2Fe(H)(C_3H_6)$

Once again, the formation of C_3H_6 and C_3H_8 (1 : 1) suggests a *β*-hydride elimination mechanism with an olefin(hydrido)complex as an intermediate.

Despite numerous practical applications, a mechanistic understanding of the photochemistry of organometallics is still in its infancy. Significant advances are expected from the application of ultra-fast methods like laser flash photolysis, which explores the picosecond range (Peters, 1986 R). A particularly attractive aspect of photochemistry is the close symbiosis between spectroscopy and chemistry. After all, electron spectroscopic characterization of the educts in their ground states and photochemical generation of the new reactive species often are identical processes.

15 σ, π-Donor/π-Acceptor Ligands

A common feature of the extensive class of **π-complexes** is the fact that the L → M donor- as well as the L ← M acceptor interaction utilizes ligand orbitals which – with regard to the intra-ligand bond – have π-symmetry. The ligand-metal bond in π-complexes always contains an L ← M π-acceptor component; the L → M donor contribution, as will be discussed for the various ligands, can have σ-symmetry (monoolefins) or σ- and π-symmetry (oligoolefins, enyl ligands, arenes and heteroarenes).

15.1 Olefin Complexes

15.1.1 Homoalkene Complexes

Olefin complexes are widespread among the transition metals. These complexes play an important role in reactions that are catalyzed by organotransition-metal compounds, such as hydrogenation, oligomerization, polymerization, cyclization, hydroformylation, isomerization, and oxidation (see Chapter 17).
Conjugated oligoolefins form particularly stable complexes, as do nonconjugated diolefins with a sterically favorable arrangement of the double bonds (chelate effect).

PREPARATION

1. Substitution Reactions

$$K_2[PtCl_4] + C_2H_4 \xrightarrow[\text{60 bar}]{\text{dilute HCl}} K[C_2H_4PtCl_3]\cdot H_2O + KCl$$

This complex was prepared for the first time by Zeise in 1827 *by boiling* $PtCl_4$ *in ethanol. The first preparation starting from ethylene was described by Birnbaum* (1868). *With* $SnCl_2$ *as a catalyst, this reaction proceeds in a few hours, given a* C_2H_4 *pressure of* 1 *bar.*

$AlCl_3$ facilitates the substitution of Cl^- by incorporating it into the $[AlCl_4]^-$ counterion:

$$Re(CO)_5Cl + C_2H_4 \xrightarrow{AlCl_3} [Re(CO)_5C_2H_4]AlCl_4$$

The use of $AgBF_4$ leads to the precipitation of silver halide and the introduction of the noncoordinating anion BF_4^- (Caution: $AgBF_4$ is also a good oxidizing agent!):

$$CpFe(CO)_2I + C_2H_4 + AgBF_4 \longrightarrow [CpFe(CO)_2(C_2H_4)]BF_4 + AgI$$

Thermal ligand substitution (Reihlen, 1930):

$$Fe(CO)_5 + \text{butadiene} \xrightarrow[20\ bar]{135°} (\eta^4\text{-butadiene})Fe(CO)_3 + 2\ CO$$

Photochemical ligand substitution proceeds at low temperatures (E. O. Fischer, 1960):

$$CpMn(CO)_3 + CH_2{=}CH{-}CHO \xrightarrow{h\nu} CpMn(CO)_2(\eta^2\text{-}CH_2{=}CH{-}CHO)$$

Metal-induced ligand isomerization (Birch, 1968):

$$Fe(CO)_5 + \text{1-methoxycyclohexa-1,4-diene} \xrightarrow{140°} (\eta^4\text{-1-methoxycyclohexa-1,3-diene})Fe(CO)_3 + 2\ CO$$

A simple method to separate isomeric olefins by recrystallization of their silver nitrate adducts is based on the following equilibrium:

$$AgNO_3 + \text{olefin} \underset{}{\overset{EtOH}{\rightleftharpoons}} [Ag(\text{olefin})_2]NO_3$$

Coordinatively unsaturated complexes can add olefinic ligands without replacement of another group (**addition**):

$$IrCl(CO)(PPh_3)_2 + R_2C{=}CR_2 \rightleftharpoons (Ph_3P)_2Ir(CO)(Cl)(\eta^2\text{-}R_2C{=}CR_2)$$

16 VE 18 VE

2. Metal Salt + Olefin + Reducing Agent

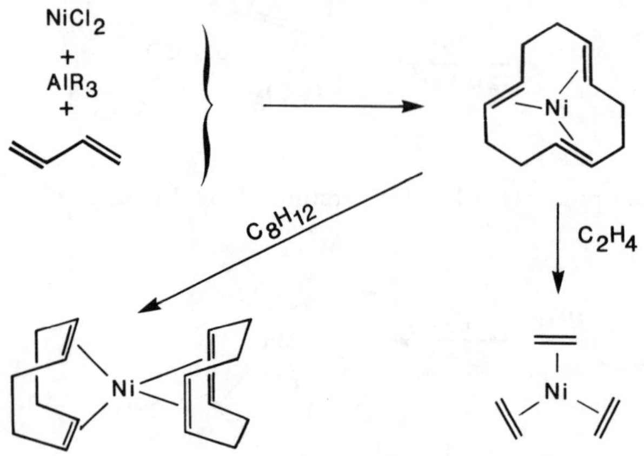

(*Trans-trans-trans-cyclododecatriene*)Ni⁰ *The high reactivity of this complex is expressed in the term "naked nickel".* (*Wilke*, 1963 *R*)

$Ni^0(COD)_2$, *a good source of* Ni^0 *for further reactions.*

The first binary metal-ethylene complex, colorless, stable up to 0 °C (*Wilke*, 1973).

$$PtCl_2(1,5\text{-COD}) + C_8H_8Li_2 + 1,5\text{-COD} \xrightarrow{Et_2O} Pt(1,5\text{-COD})_2 \xrightarrow{C_2H_4} Pt(C_2H_4)_3$$

Trisethyleneplatinum is only stable under a C_2H_4 atmosphere (Stone, 1977).
A special variation of this reaction type is the reductive cleavage of metallocenes (Jonas, 1980):

$$Cp_2Co \xrightarrow[-20°C]{K/C_2H_4} CpCo(C_2H_4)_2 + KC_5H_5$$

$CpCo(C_2H_4)_2$ *is a useful reagent for the transfer of the half-sandwich unit* CpCo.

$$RhCl_3 + C_2H_4 \xrightarrow[C_2H_5OH/H_2O]{} [(C_2H_4)_2Rh(\mu\text{-}Cl)]_2$$

In this reaction, Rh(III) is reduced by an excess of ethylene.

3. Butadiene Transfer by Means of Magnesium Butadiene (cf. p. 41)

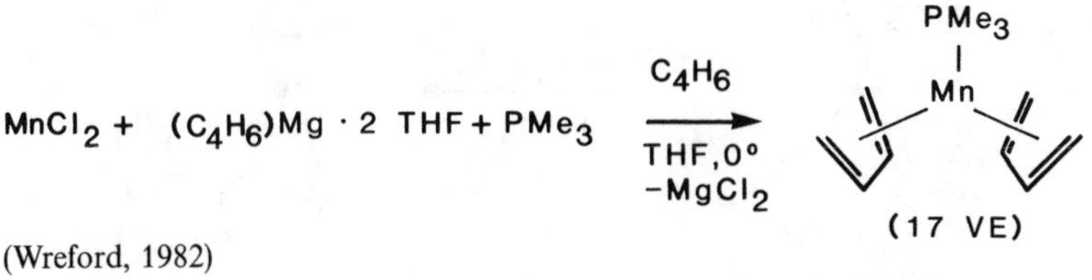
MnCl2 + (C4H6)Mg · 2 THF + PMe3
C4H6
THF, 0°
-MgCl2
PMe3
Mn
(17 VE)

(Wreford, 1982)

This procedure is also applicable to other transition-metal halides.

4. Metal-Atom Ligand-Vapor Cocondensation (CC)

In the cocondensation method (sometimes referred to as metal-vapor synthesis) metal vapors are condensed with the gaseous ligand on a cooled surface or into ligand solutions of low vapor pressure. On warming to room temperature, metal complexes are formed in competition with metal aggregation (Skell, Timms, Klabunde, M. L. H. Green). For a number of fundamental organometallics, cocondensation techniques currently represent the only way of access. Drawbacks to this method include the frequently low yields and the large amount of cooling agent required.

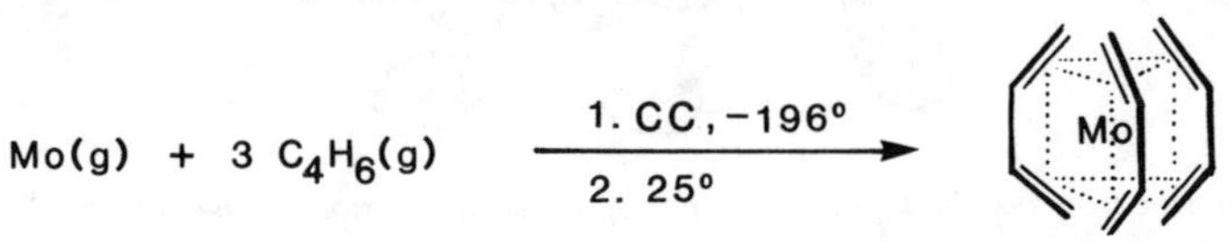
Mo(g) + 3 C4H6(g)
1. CC, -196°
2. 25°
Mo

(Skell, 1974)

trigonal-prismatic coordination

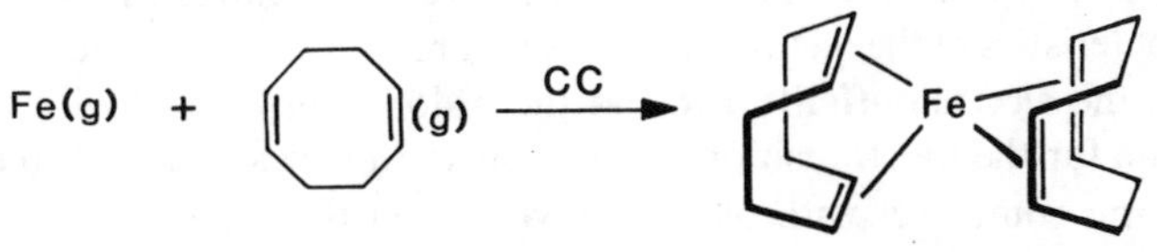
Fe(g) +
(g)
CC
Fe

(Timms, 1974)

decomposes above $-20\,^{\circ}C$,
starting material for other Fe^0*-complexes.*

5. Transformation of Enyl Complexes

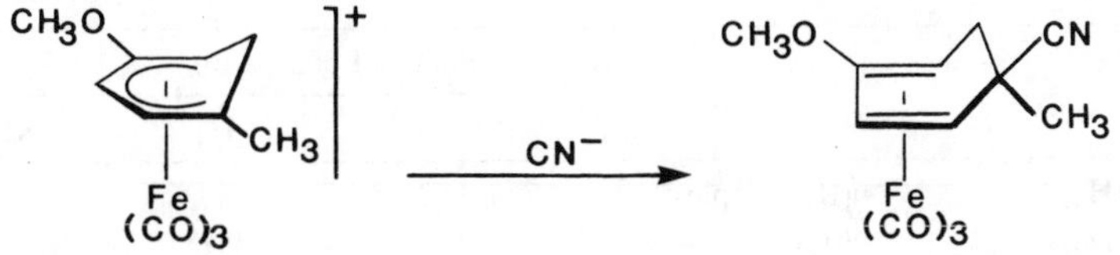
CH3O
+
CH3
Fe
(CO)3
CN⁻
CH3O
CN
CH3
Fe
(CO)3

Such nucleophilic additions are generally regio- as well as stereoselective; the nucleophile attacks trans to the metal.

6. Hydride Abstraction from Alkyl Complexes

$[CPh_3]BF_4$

$-HCPh_3$

STRUCTURE AND BONDING OF MONOOLEFIN COMPLEXES

The coordination of a monoolefin to a transition metal provides the simplest example of a metal π-complex. The qualitative bonding description (Dewar, 1951; Chatt, Duncanson, 1953) is similar to that for the M–CO moiety as far as the donor-acceptor synergism is concerned.

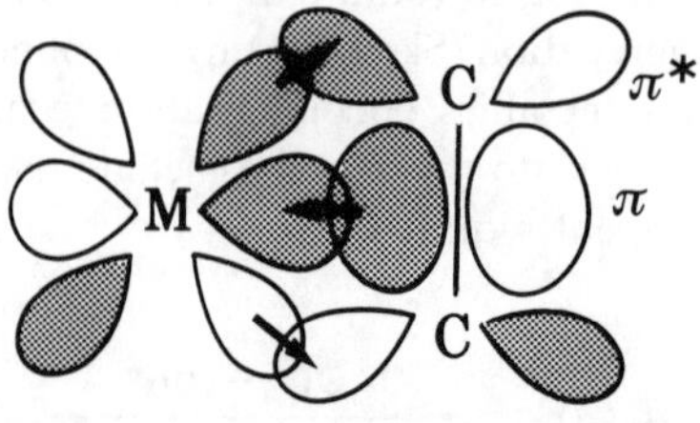

The donor component (from the ligand's viewpoint) is the interaction of the filled, π-bonding orbital of ethylene with vacant metal orbitals, the acceptor component that of filled metal orbitals with the vacant π^*-antibonding orbital of ethylene (shading indicates orbital phases.)

The tendency to form olefin complexes is therefore also controlled by the σ-acceptor/π-donor characteristics of the metal. Under the assumption that the σ-acceptor properties correlate with the **electron affinity** ***EA*** and the π-donor properties with the **promotion energy** ***PE,*** data for the free transition-metal atoms or ions can be used to evaluate their tendency to form complexes with olefins or with other donor/acceptor ligands (p. 257).

The ability of an olefin to function as a Lewis base as well as a Lewis acid in its bonding to a transition metal helps to fulfill the electroneutrality principle. In this context, consider the results of quantum-chemical calculations on some model compounds (Roos, 1977):

		Calculated charge on		
		Ni	C_2H_4	NH_2 or NH_3
$(C_2H_4)Ni(NH_2)_2$	"Ni(II) complex"	+0.83	+0.02	−0.43
$(C_2H_4)Ni(NH_3)_2$	"Ni(0) complex"	+0.58	−0.78	+0.11

Despite the different formal oxidation states, this quantitative treatment of the bonding leads to very similar charges on the central atom!

High electron affinity (EA) of the metal favors the M $\overset{\sigma}{\leftarrow}$ olefin contribution, low promotion energy (PE) the M $\overset{\pi}{\rightarrow}$ olefin contribution to chemical bonding: **Ni^0** *is a good π-donor,* Hg^{2+} *a good σ-acceptor;* Pd^{2+} *is a good π-donor* ***and*** *a good σ-acceptor. The donor/acceptor behavior of the metal towards the olefin is also influenced by ancillary ligands.*
Source: R. S. Nyholm, Proc. Chem. Soc. (1961) 273

Atom or Ion	Electr. Config.	*PE* in eV	*EA* in eV
Ni(0)*	d^{10}	1.72	1.2
Pd(0)*	d^{10}	4.23	1.3
Pt(0)*	d^{10}	3.28	2.4
Rh(I)	d^8	1.6	7.31
Ir(I)	d^8	2.4	7.95
Pd(II)	d^8	3.05	18.56
Pt(II)	d^8	3.39	19.42
Cu(I)	d^{10}	8.25	7.72
Ag(I)	d^{10}	9.94	7.59
Au(I)	d^{10}	7.83	9.22
Zn(II)	d^{10}	17.1	17.96
Cd(II)	d^{10}	16.6	16.90
Hg(II)	d^{10}	12.8	16.90

$$PE\begin{cases} nd^{10} \rightarrow nd^9\ (n+1)\,p^1 \\ nd^8 \rightarrow nd^7\ (n+1)\,p^1 \end{cases}$$

$$EA\begin{cases} nd^{10} \rightarrow nd^{10}\ (n+1)\,s^1 \\ nd^8 \rightarrow nd^8\ (n+1)\,s^1 \end{cases}$$

* Ground-state configurations:
Ni^0 $d^8 s^2$,
Pd^0 d^{10},
Pt^0 $d^9 s^1$

The MO *treatment of the* MC_2 *fragment consists of symmetry-matched combinations of metal and ligand orbitals:*

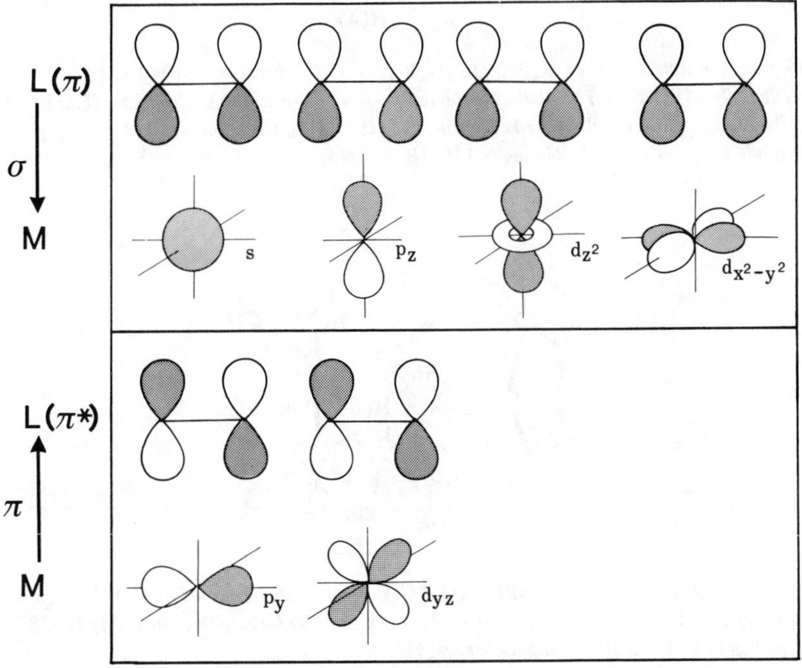

Both the M $\overset{\sigma}{\leftarrow}$ olefin donor and the M $\overset{\pi}{\rightarrow}$ olefin acceptor part of the interaction serve to weaken the intra-ligand C–C bond. In principle, this is evident from the phase relations in the L (π) und L (π^*) molecular orbitals, verification being provided by a comparison of the $\nu_{C=C}$ vibrational frequencies of free and coordinated ethylene:

Complex	$\nu_{C=C}/cm^{-1}$	Complex	$\nu_{C=C}/cm^{-1}$
$[(C_2H_4)_2Ag]BF_4$	1584	$(C_2H_4)Fe(CO)_4$	1551
$[(C_2H_4)_2Re(CO)_4]PF_6$	1539	$[CpFe(CO)_2C_2H_4]PF_6$	1527
$[C_2H_4PdCl_2]_2$	1525	$K[PtCl_3(C_2H_4)]\cdot H_2O$	1516
$CpMn(CO)_2(C_2H_4)$	1508	$[C_2H_4PtCl_2]_2$	1506
$CpRh(C_2H_4)_2$	1493	**C_2H_4, free**	**1623**

An important structural aspect is the **loss of planarity of the olefin upon coordination to a transition metal.** In substituted ethylenes C_2X_4, this deformation increases with increasing electronegativity of X.

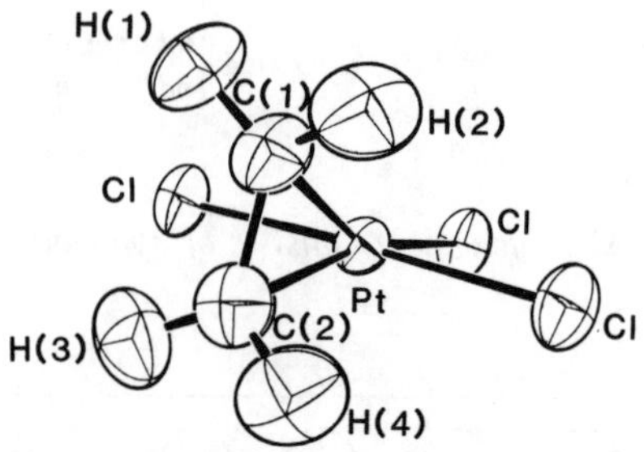

Structure (neutron diffraction) of **$K[PtCl_3(C_2H_4)]$**. *The* C–C *bond length is* 137 pm, *similar to the uncomplexed olefin* (135 pm). *The olefin is oriented perpendicular to the* $PtCl_3$*-plane. The* Pt–Cl *bond trans to the olefin is slightly elongated. The atoms* H(1)–H(4), C(1) *and* C(2) *are not coplanar. The dihedral angle between the* CH_2 *planes is* 146° (*Bau*, 1975).

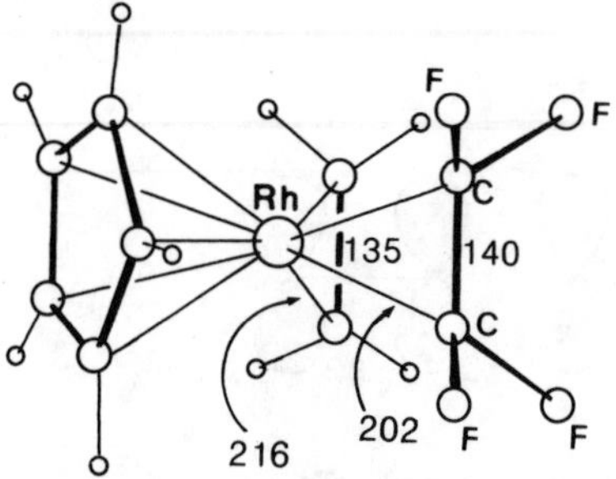

Struture (X-ray diffraction) of **$(C_5H_5)Rh(C_2H_4)(C_2F_4)$**. *The carbon atoms of* C_2F_4 *are closer to the metal than those of* C_2H_4. *The dihedral angle between the two* CH_2*-planes in* C_2H_4 *is* 138°, *that of the* CF_2*-planes in* C_2F_4 *is* 106° (*Guggenberger*, 1972).

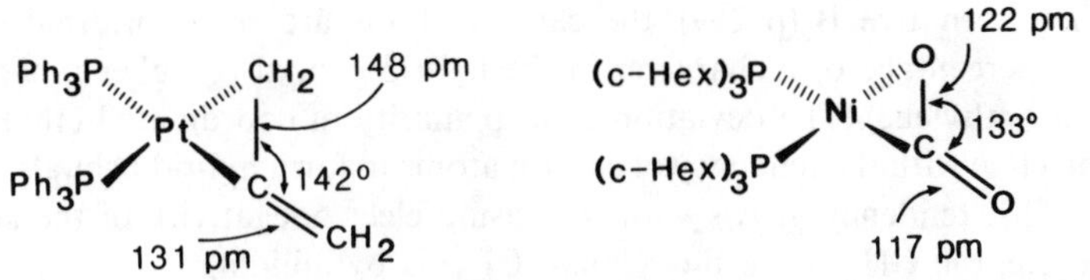

The coordination of allenes leads to considerable distortion of the C_3*-unit, which is linear in the free state. A similar distortion is experienced by* CO_2*, coordinated side-on as a "diheteroallene" ligand in complexes of the type* $(R_3P)_2Ni(\eta^2\text{-}CO_2)$ *(Darensbourg, 1983 R).*

There is a certain structural similarity between the fragment (η^2-alkene)M and an epoxide, warranting a metallacyclopropane description of the former:

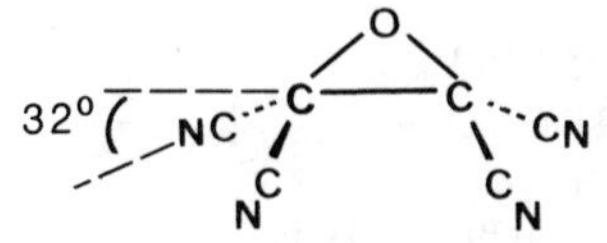

Epoxide

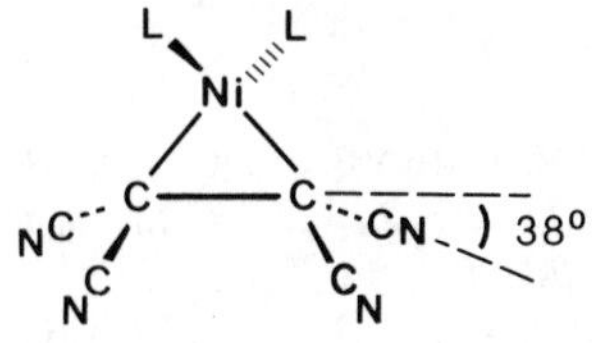

(η^2-Tetracyanoethylene)Ni complex (nickelacyclopropane)

At first sight, such a formulation seems to bear little resemblance to the Dewar-Chatt-Duncanson model. It can be shown, however, that the actual bonding situation is well in accordance with a description somewhere between these two borderline cases (R. Hoffmann, 1979):

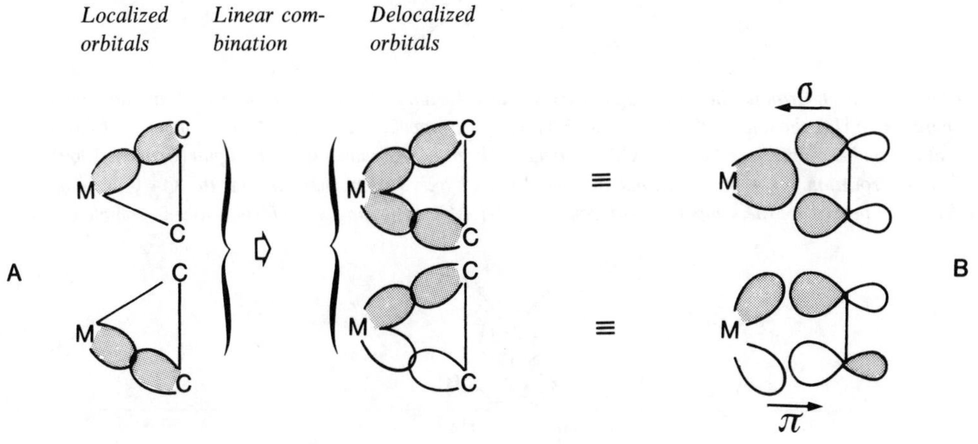

Metallacyclopropane description:
two localized 2e3c MC σ-bonds, alkene acting as a bidentate ligand (non-planar)

Dewar-Chatt-Duncanson model:
One delocalised 2e3c MC_2 *σ-bond, one delocalised 2e3c* MC_2 *π-bond, alkene acting as a "monodentate" ligand (planar).*

While in the alternative **B** (p. 259), the carbon atoms are sp^2-hybridized (planar η^2-ethylene), the percentage of p character in the hybrid orbital is higher in alternative **A** (non-planar η^2-ethylene). The deviation from planarity of coordinated ethylene should therefore correlate with the tendency of carbon atoms to form hybrid orbitals with higher p character. This tendency grows with increasing electronegativity of the substituents (**Bent's Rule**, e.g. the $CH_3\cdot$ radical is planar, $CF_3\cdot$ is pyramidal).
The conformation relative to the M-olefin axis depends on the coordination number of the metal as well as on the number of valence electrons:

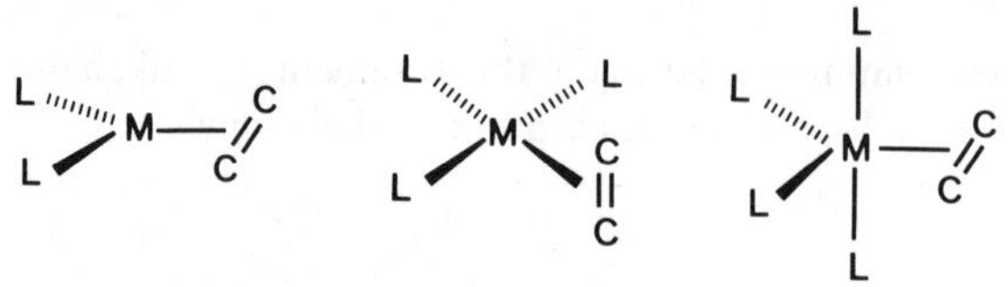

	C.N. 3, 16 VE	C.N. 4, 16 VE	C.N. 5, 18 VE
	L_2M (*alkene*)	L_3M (*alkene*)	L_4M (*alkene*)
	L_2M (*alkyne*)		
Examples:	$(PPh_3)_2NiC_2H_4$	$K[PtCl_3(C_2H_4)]$	$(PPh_3)_2IrBr(CO)TCNE$
	$Pt(C_2H_4)_3$		TCNE = *tetracyanoethylene*

Hindered ligand rotation of coordinated alkenes (and alkynes) often occurs in a temperature range which is suitable for routine ^{1}H- and ^{13}C-NMR studies. This is demonstrated by the temperature-dependent ^{1}H-NMR spectrum of the fluxional molecule $C_5H_5Rh(C_2H_4)_2$ (p. 261).

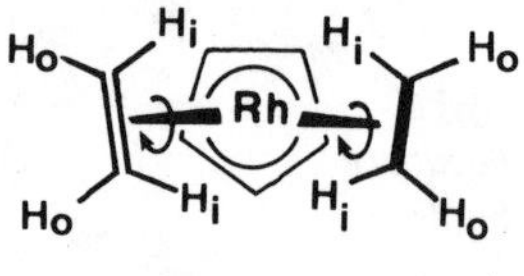

On fast rotation around the metal-ligand axis (the rotational frequency is higher than the signal separation in Hz), *the inner* (H_i) *and the outer* (H_0) *protons appear equivalent. At low temperatures, this rotation can be "frozen", and an* AA'XX' *spectrum is observed instead of a single signal* (*Cramer*, 1969). *That this rotation takes place around the metal-ligand axis rather than around the* C–C *axis of ethylene is proved by the temperature-dependent* ^{1}H-*NMR spectrum of a chiral ethylene complex*:

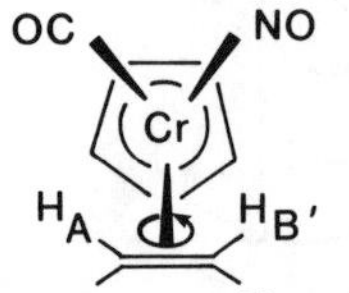

Upon raising the temperature, the spectrum of $C_5H_5CrCO(NO)C_2H_4$ *converts from an* ABCD- *to an* AA'BB' *type; two pairs of diastereotopic protons are therefore retained. An additional rotation around the ethylene* (C–C) *axis would, however, result in an* A_4-*type spectrum* (*Kreiter*, 1974).

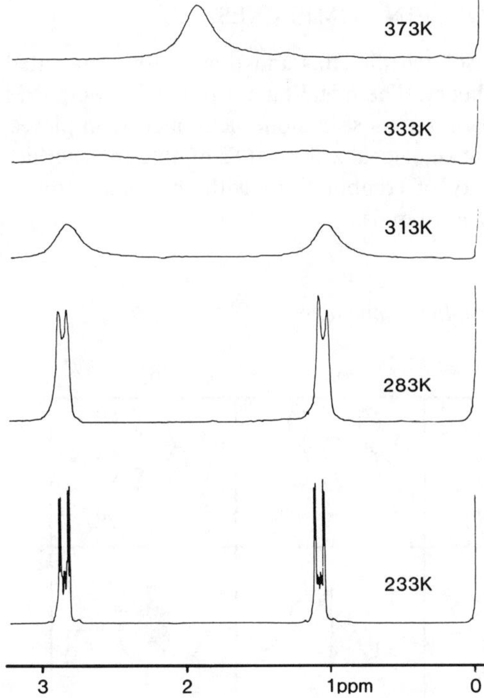

^{1}H-*NMR spectrum of* $C_5H_5Rh(C_2H_4)_2$ (200 MHz).

If the ligands C_2H_4 and C_2F_4 are present in the same molecule (see structure p. 258), the rotation around the $M-C_2F_4$ axis ceases at a higher temperature than that around the $M-C_2H_4$ axis, a consequence of the different rotational barriers for the two ligands. Apparently the backbonding contribution $M \xrightarrow{\pi}$ olefin, mainly responsible for the rotational barrier, is larger for C_2F_4 than for C_2H_4.
Inspired by the large variety of bridging modes observed for the ligands CO and $RC{\equiv}CR$, one may ask whether **alkenes** as well may assume a **bridging role**. The first example of ethylene bridging two metal-metal bonded centers was presented by Norton (1982):

$$Na_2[Os_2(CO)_8] \xrightarrow[\text{THF, 0°}]{ICH_2CH_2I} [Os_2(CO)_8(\mu_2\text{-}CH_2CH_2)]$$

153 pm
H₂C—CH₂
OC—Os—Os—CO (each Os bearing further CO ligands)

$^1J(^{13}C, ^{13}C) = 34.0$ Hz
$^1J(^{13}C, ^1H) = 125.3$ Hz

The C−C bond length and the scalar couplings (p. 300) imply sp^3 hybridization at the carbon atoms and therefore the designation diosmacyclobutene is appropriate. The complex $[Os_2(CO)_8([\mu_2 - 1{:}2\eta\text{-}C_2H_4)]$ may serve as a model for chemisorbed ethylene (Norton, 1989).

STRUCTURE AND BONDING OF DIOLEFIN COMPLEXES

Distinguishing between compounds with isolated diolefins and those with conjugated diolefins is justified in terms of bonding theory. The metal-ligand bonds in complexes with nonconjugated diolefins closely correspond to those in monoolefin-metal complexes. Conjugated di- and oligoolefins, however, have delocalized π-MO's of differing energies and symmetries. They offer a larger variety of combinations with the metal atomic orbitals, resulting in more stable metal-carbon bonds.

Metal-ligand interactions in butadiene complexes:

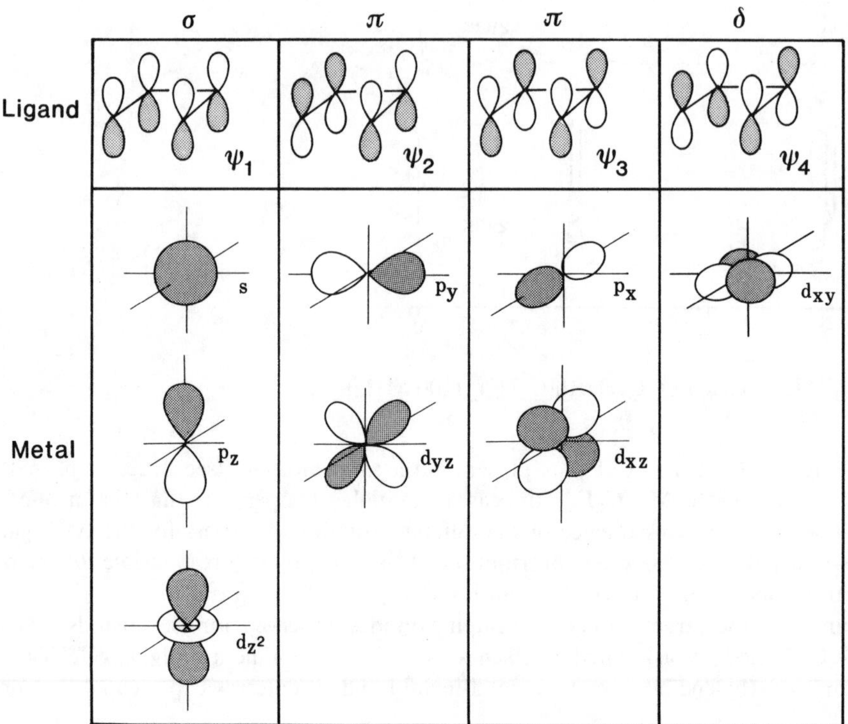

In an extension of the Dewar-Chatt-Duncanson model, the bonding situation in $(\eta^4\text{-}C_4H_6)M$ can be described by the components $\psi_1 \xrightarrow{\sigma} M$, $\psi_2 \xrightarrow{\pi} M$, $\psi_3 \xleftarrow{\pi} M$ and $\psi_4 \xleftarrow{\delta} M$. Both interactions $M \overset{\pi}{-} L$ should result in a lengthening of the terminal C–C bonds and a shortening of the internal C–C bond. This is a consequence of the depopulation of ψ_2, the population of ψ_3 and the nodal characteristics of these ligand orbitals. In this context, it should be mentioned that electronic excitation of free butadiene induces similar changes in bond lengths:

The following examples illustrate the influence of transition-metal coordination on the ligand parameters:

Apart from bond lengths, coupling constants $^1J(^{13}C, ^{13}C)$ in ^{13}C-NMR spectra also argue for a change of bond orders within the η^4-1,3-diene ligand. For the terminal carbon atoms, a hybridization state intermediate between sp^2 and sp^3 can be assumed on the basis of the observed coupling constants $^1J(^{13}C, ^1H)$. This may be described by the two canonical forms:

The relative weight of these two resonance forms can be discussed on the basis of the M – C distances. While complexes of the "late" transition metals merit a description as **π-complexes**, corresponding compounds of the "early" transition metals are better designated as **metallacyclopentenes**. In such generalizations, however, the nature of the substituents on the ligands should not be disregarded (cf. η^2-C_2X_4, p. 259).

"***Metallacyclopentene***"
(η^4-*dimethylbutadiene*)$ZrCp_2$
$M-C_1$ (*terminal*) = *short*
$M-C_2$ (*internal*) = *long*

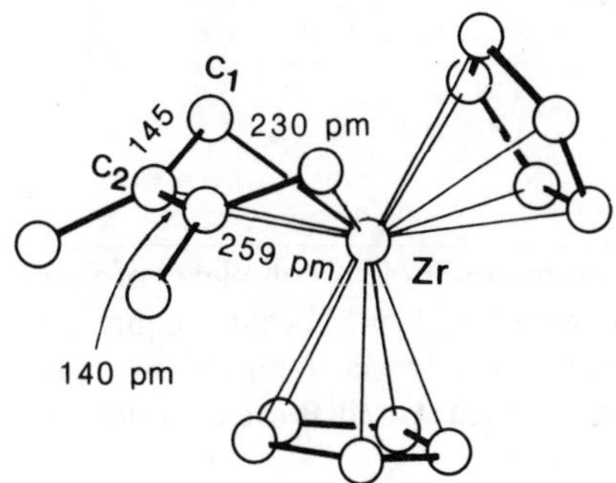

"***π-Complex***"
(η^4-*cyclohexadiene*)$_2$Fe(CO)
$M-C_1$ (*terminal*) = *long*
$M-C_2$ (*internal*) = *short*

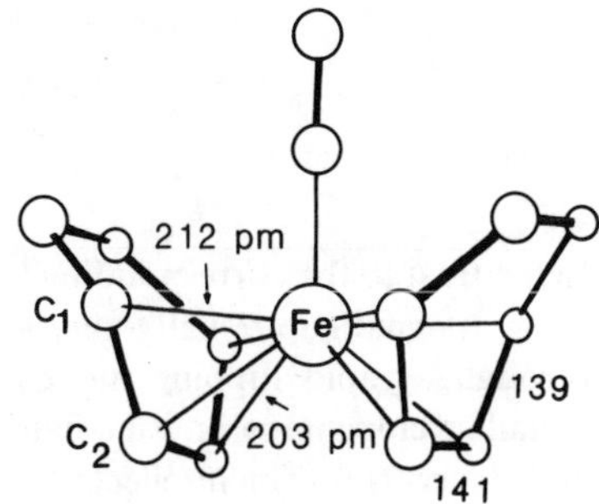

Further structural characteristics of (1,3-diene)Fe(CO)$_3$ complexes can be deduced from extensive X-ray diffraction studies and from spectroscopic data. This is discussed for the following representative example, the structure of tricarbonyl[3-6η-(6-methylhepta-3,5-dien-2-one)]iron (Prewo, 1988):
(Guide to three-dimensional viewing: p. 71)

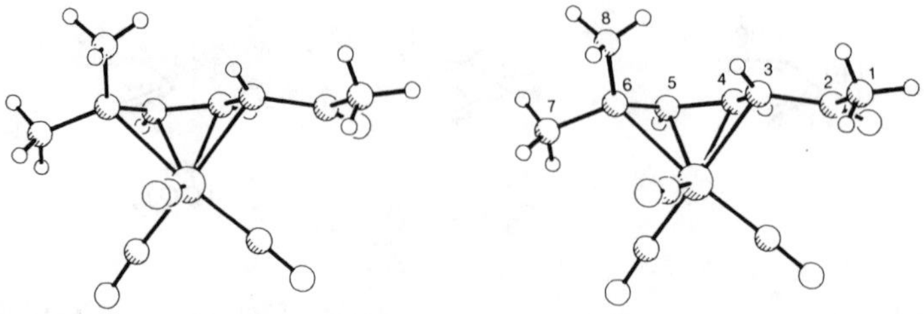

- *The terminal substituents* [*e.g.* C(7) and C(8)] *are not coplanar with the coordinated carbon atoms* C(3)–C(6). *The deviation from planarity is* 16.6° *for the bond* C(6)–C(7) *and* 52.4° *for the bond* C(6)–C(8). *This is due to a strong* ***distortion of* C(6)–C(8)** *relative to* C(3)–C(6), *one result being that the overlap of the* C(6) p_π*-orbital with* Fe(3*d*)*-orbitals is altered.*
- (1,3-*diene*)Fe(CO)$_3$ *complexes are* ***conformationally labile****. Crystal structures disclose a nearly square-planar geometry, with the apical and two basal positions occupied by carbonyl groups. For a species of low symmetry like the iron carbonyl complex shown above, three signals for the three nonequivalent carbonyl groups would be expected in the* ^{13}*C-NMR spectrum. At room temperature, however, only one signal is usually observed, indicating a rapid exchange process within the* Fe(CO)$_3$ *moiety. Upon lowering the temperature, the slow exchange region can be reached* (*Takats*, 1976). *The experimental data are most reasonably explained by an exchange process involving rotation of the* Fe(CO)$_3$ *moiety about its* ***pseudo-*C$_3$** *axis, whereby the olefin retains its planarity.*
- (1,3-*diene*)Fe(CO)$_3$ *complexes are* ***configurationally stable****. Prochiral diolefins form chiral complexes which can be resolved into enantiomers.*

In contrast to the corresponding (diene)Fe(CO)$_3$ compounds, diolefin complexes of zirconium are **configurationally nonrigid.** The two cyclopentadienyl rings become equivalent through a rapid flipping movement of the diolefin (possibly via a purely σ-bonded metallacyclopentene intermediate). If R$_1$ and R$_2$ differ, then **A** and **B** are enantiomeric and the rearrangement effects racemization (Erker, 1982):

A **B**

Among the organozirconium complexes, one also finds the only monomeric complexes in which butadiene adopts the ***s-trans* configuration** (Erker, 1982):

hν

Δ

In (η^5-C_5H_5)Nb(η^4-C_4H_6)$_2$, according to its 500 MHz ^{1}H-NMR spectrum, the *cis* and *trans* forms of butadiene even coexist in the same molecule (Yasuda, 1988):

CpNbCl$_4$ + (Mg—CH$_2$CH=CHCH$_2$)$_n$ ⟶ Nb (cis, trans) + bis(cis) isomer

Trimethylenemethane, a constitutional isomer of butadiene, is only stable if coordinated to a transition metal (Emerson, 1966):

$$H_2C{=}C(CH_2Cl)_2 \xrightarrow[-\ FeCl_2]{Fe_2(CO)_9}$$

C$_2$ C$_1$ 212 pm 193 pm Fe

The different Fe–C *distances reflect the reluctance of trimethylenemethane to sacrifice its planarity and thereby its* ***intra-****ligand π-conjugation.*

REACTIONS OF NONCONJUGATED OLEFIN COMPLEXES

The **thermodynamic stability** of these complexes is strongly influenced by the nature of the olefin:

- *Electron-withdrawing* ***substituents*** *increase the stability whereas electron-donating substituents decrease it.*
- *In cases where* ***cis-trans isomerism*** *is possible, the more stable complex is invariably formed by the* ***cis****-olefin.*
- *Complexes of* ***ring-strained*** *cycloalkenes like cyclopropene,* ***trans****-cyclooctene and norbornene display surprising stability.*
- *As a consequence of the* ***chelate effect****, particularly high stability is shown by complexes in which isolated olefins are part of a carbocycle of favorable geometry. Examples:*

The **redox behavior** of olefin complexes is not easily generalized:

$$(C_5H_5)M(CO)_2(\text{Olefin}) \xrightarrow[]{H_2,\,25\,°C}\!\!\!\!\!/\;\; \text{no reaction}$$
$$M = Mn, Re$$

but

$$[C_2H_4PtCl_2]_2 \xrightarrow{H_2,\,25\,°C} 2\,Pt + 4\,HCl + 2\,C_2H_6$$

The most characteristic reaction of monoolefin complexes is **ligand substitution**, occuring in many cases under mild conditions, and providing an important method for the synthesis of metal complexes of low thermal stability:

$$[(C_6H_{11})_3P]_2Ni(C_2H_4) + O_2 \longrightarrow [(C_6H_{11})_3P]_2Ni(O_2) + C_2H_4$$

(Wilke, 1967) (first O_2-nickel complex; stable up to $-5\,°C$)

$$(Ph_3P)_2Pt(C_2H_4) + C_{60} \xrightarrow{\text{toluene}} (Ph_3P)_2Pt(\eta^2\text{-}C_{60}) + C_2H_4$$

C_{60} *designates buckminsterfullerene, the soccerball-shaped new allotrope of carbon. The carbon-carbon double bonds of* C_{60} *react like those of very electron-deficient alkenes* (*Fagan*, 1991).

Also of synthetic interest is the **behavior** of some metal-olefin complexes **towards nucleophilic reagents.** This reaction has been extensively studied for dimeric (olefin)palladium chlorides:

$$[C_2H_4PdCl_2]_2 \xrightarrow{NaOAc} CH_2{=}CHOAc$$

(Synthesis of vinyl acetate)

$$[C_2H_4PdCl_2]_2 \xrightarrow{C_2H_5OH} CH_2{=}CHOC_2H_5 \longrightarrow CH_3CH(OC_2H_5)_2$$

(Synthesis of vinyl ethers and acetals)

Complexes of the general type $[CpFe(CO)_2(olefin)]^+$ react with a variety of nucleophiles via a $\pi - \sigma$ rearrangement of the metal-ligand bond to give stable, neutral metal alkyls (Rosenblum 1974 R):

REACTIONS OF DI- AND OLIGOOLEFIN COMPLEXES

Metal-coordinated di- and oligoolefins are quite unreactive compared to the respective free olefin; they can neither be catalytically hydrogenated nor do they undergo Diels-Alder reactions. However, provided they are stable to oxidation, they can react with **electrophiles** such as strong acids. If the proton prefers addition to a coordinated carbon atom rather than to the metal, the valence shell has to be filled by an additional two-electron ligand like carbon monoxide:

An alternative way to compensate this electron deficit at the metal is the formation of a C−H⇀M bridge. Protonated diolefin complexes (form **A**) exist in a dynamic equilibrium with forms **B** and **C**. At low temperatures, for M = Co, Rh form **B** and for M = Ir form **C** dominates (Salzer, 1987):

18 VE
M = Co, Rh, Ir

A
16 VE

B
M = Co, Rh
C−H⇀M bridge

C
18 VE
M = Ir
metal hydride

The C_5Me_5 congener of the rhodium complex is a fluxional molecule, whose dynamic ^{13}C-NMR spectrum is discussed on p. 304).

Such $2e\,3c$ bonds between the C−H entity of a ligand and the central metal atom, for which M. L. H. Green has proposed the term **"agostic"**, strongly resemble the familiar B⌒H⌒B bridges in borane chemistry. Agostic interactions model the attack of an unsaturated fragment L_nM at a C–H bond; they contribute to our understanding of C–H activation at transition-metal centers (p. 439). C–H⇀M bridges have mainly been detected in complexes in which C–H bonds are positioned β to coordinated allyl and diolefin units (Brookhart, 1988 R).

The **existence of C–H⇀M bridges** can be deduced from

- *structural data (especially those obtained from neutron diffraction)*
- *chemical shifts to high-field in the* ^{1}*H-NMR spectrum* ($\delta = -5$ to -15 ppm)
- *reduced coupling constants* [^1J(C, H) = 75–100 Hz]
- *vibrational frequencies at low wave numbers* (ν_{CH} = 2700–2300 cm^{-1}).

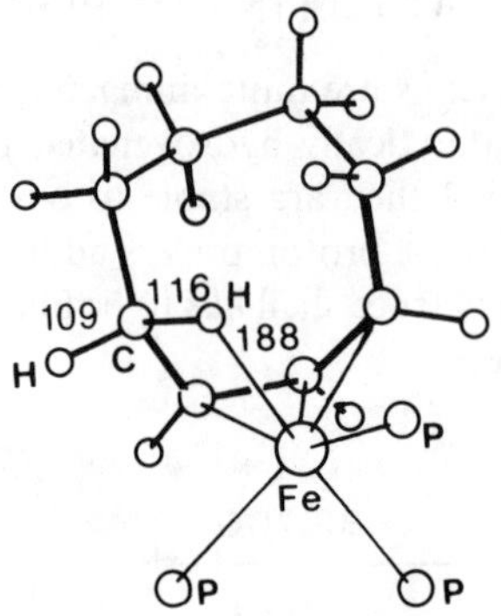

A typical example of a C–H⇀M *bridge is seen in the structure of* **{(η^3-C_8H_{13})Fe[P(OMe)$_3$]$_3$}BF_4** *(Stucky, Ittel, 1980), obtained by means of neutron diffraction at* 30 K. *Note the strongly differing* C–H *bond lengths at the methylene group engaged in an agostic interaction.*

Friedel-Crafts acylations of (diene)Fe(CO)$_3$ complexes are also initiated by an electrophilic attack at the terminal carbon atom of the ligand (Pauson, 1974):

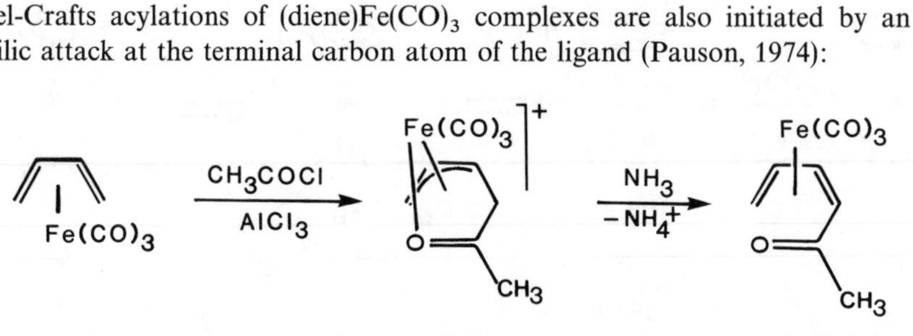

Besides electrophilic additions, **nucleophilic additions** to coordinated olefins are of growing synthetic interest. At -78 °C, alkyllithium reagents attack (isoprene)Fe(CO)$_3$ exclusively at an internal unsubstituted carbon atom. Since this addition is kinetically controlled and reversible, warming to room temperature favors the formation of the thermodynamically more stable allyl complex. The anionic complexes are only stable in solution, so that their structures have to be deduced from product analysis after aqueous quenching (Semmelhack, 1984):

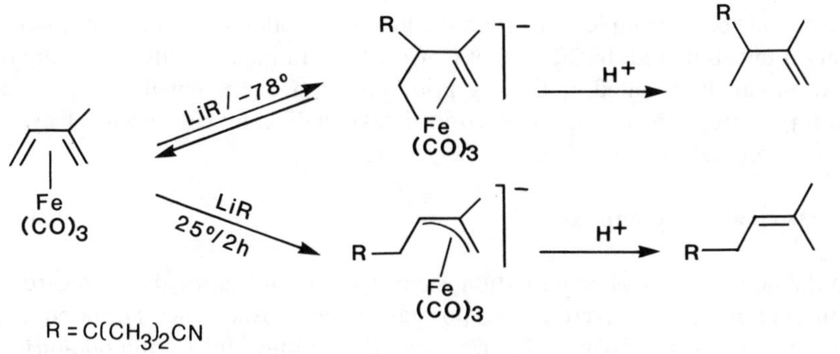

R = C(CH$_3$)$_2$CN

Metal-coordinated oligoolefins can undergo **valence isomerization** in a manner similar to the free olefins; the 'Woodward-Hoffmann rules' of organic chemistry are not applicable to these organometallics, however. Instead, the course of the reaction is mainly influenced by the respective organometallic fragment and its coordinative preference. Bicyclo[6.1.0]nonatriene and its metal complexes may serve as an illustrative example:

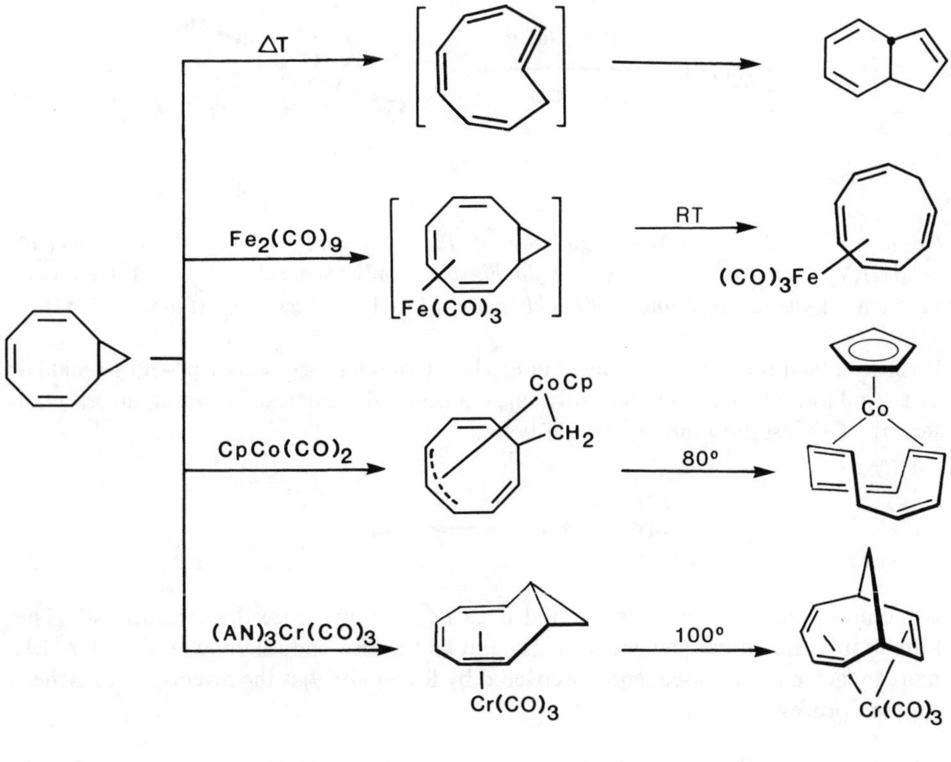

The iron and cobalt complexes contain the ligand cyclononatetraene, a molecule of low stability if uncomplexed. In the case of $CpCo(CO)_2$, an intermediate displaying σ, π-coordination can be isolated, indicating prior insertion of the cobalt complex into the cyclopropane ring. The valence isomerization takes a different course when bicyclononatriene is coordinated to a $Cr(CO)_3$ moiety.

15.1.2 Heteroalkene Complexes

From the point of view of coordination chemistry it seems reasonable to regard organic carbonyl compounds as heteroalkenes (formaldehyde = oxaethene). The bonding modes η^1-vinyl and η^2-alkene of olefins then correspond to terminal (η^1-O) and side-on (η^2-C, O) coordination of carbonyl groups, respectively:

$(C_5H_5)_2V$ —[$H\text{-}(OCH_2)_n\text{-}OH$ (paraformaldehyde)]→ $Cp_2V^{II}(\eta^2\text{-}CH_2O)$ vanada-oxirane

The IR vibrational frequency for this η^2 (C, O) formaldehyde complex ($\nu_{CO} = 1160\ cm^{-1}$) *is strongly shifted to lower wave numbers* (*for free* H_2CO: $\nu_{CO} = 1746\ cm^{-1}$) (Floriani, 1982).

$(C_5H_5)_2VCl$ —[1. $H_2O, NaBPh_4$; 2. Me_2CO]→ $[Cp_2V^{III}\text{-}O{=}CMe_2]^+\ BPh_4^-$

On the other hand, η^1(O) *coordination of the keto group to the electron-deficient* $[(C_5H_5)_2V]^+$ *ion weakens the* C=O *double bond only moderately* ($\nu_{CO} = 1660$ *versus* $1715\ cm^{-1}$), *the reason being that backbonding* M $\xrightarrow{\pi}$ L *is absent here* (*Floriani*, 1981).

Because of their possible involvement in Fischer-Tropsch processes (see p. 430), formaldehyde- and formyl complexes currently enjoy considerable interest. However, an essential step, the CO-insertion into an M–H bond,

$$L_nM(H)\text{-}CO + L' \longrightarrow L_nM(L')\text{-}C(H){=}O$$

is thermodynamically unfavorable and does not usually proceed spontaneously (The formation of free formaldehyde from H_2 and CO is also endothermic, $\Delta H^0 = +26$ kJ/mol). In fact, a reaction sequence discovered by Roper (1979) is the reversal of a Fischer-Tropsch process:

$$(Ph_3P)_3Os(CO)_2 + CH_2O \xrightarrow{-PPh_3} (CO)_2(PPh_3)_2Os(\eta^2\text{-}CH_2O) \xrightarrow[\text{solid}]{75°} (CO)_2(PPh_3)_2Os(H)(CHO) \xrightarrow[\text{solution}]{40°} (CO)_2(PPh_3)_2Os\text{—}CO + H_2$$

The formation of trimeric (η^2-formaldehyde)zirconocene from zirconocene dihydride and carbon monoxide does proceed in the proper Fischer-Tropsch fashion (Erker, Krüger, 1983):

$$\frac{3}{x}(Cp_2ZrH_2)_x \xrightarrow[\text{toluene, }25°]{CO,\,150\text{ bar}} [Cp_2Zr(\eta^2\text{-}CH_2O)]_3 \longrightarrow 3\,Cp_2ZrCl_2 + 3\,CH_3OH \xrightarrow{-6\,Cl^-,\ +6\,H^-} \frac{3}{x}(Cp_2ZrH_2)_x$$

zirconaoxirane

The work to formally split H_2 into H^+ and H^- – required for the formation of Cp_2ZrH_2 from Cp_2ZrCl_2 – has to be invested prior to the CO insertion described above, however. A typical example of the concept "stabilization of reactive molecules by complex formation" is supplied by the isolation of compounds containing **chalcogeno formaldehydes** $H_2C{=}E$ (E = S, Se, Te) as ligands. These species are generated in the coordination sphere of the metal from the respective precursor ligands (Roper, 1977, see p. 272 above).

*The ambidentate nature of chalcogeno formaldehydes becomes apparent in their function as **bridging ligands** in that the π-electron pair of the* C=E *double bond as well as the lone pair on* E *engage in metal-ligand bonding* (E = S, Se, Te) (*Herrmann, Herberhold*, 1983).

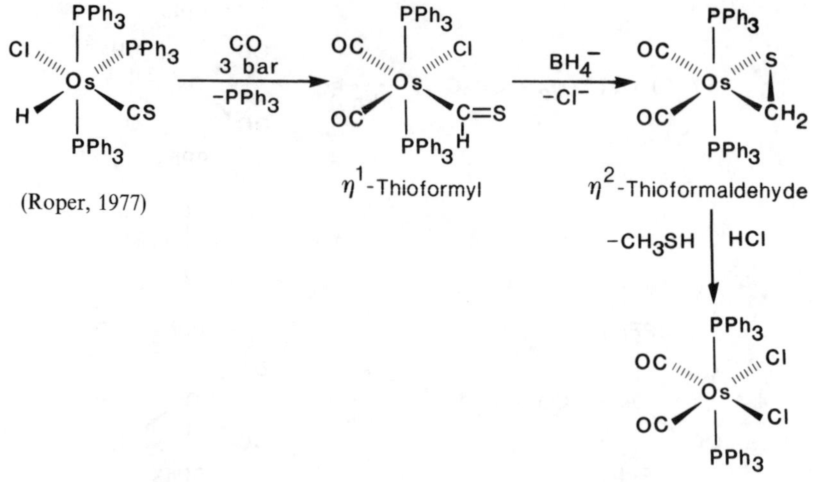

CO
3 bar
$-PPh_3$

BH_4^-
$-Cl^-$

η^1-Thioformyl

η^2-Thioformaldehyde

$-CH_3SH$ HCl

(Roper, 1977)

Another class of highly reactive small molecules is that of the phosphaalkenes (cf. p. 161f). An achievement of recent vintage is the preparation of the first **η^4-1-phosphabutadiene** complex of a transition metal. The reactivity of this ligand is apparent from the fact that for its stabilization, the coordination of the π-electron system as well as of the phosphorus lone pair to carbonyl tungsten fragments is essential (Mathey, 1987).

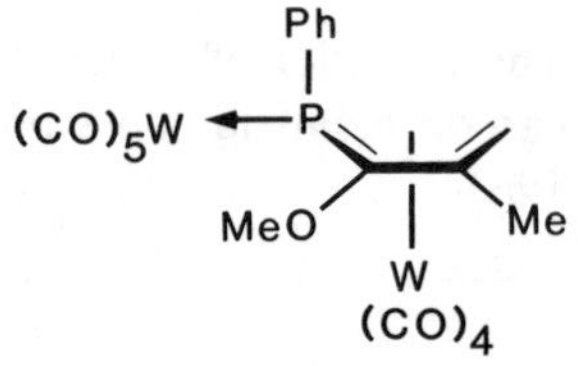

η^2-CO_2 metal complexes which may formally be regarded as coordination compounds of the ligand dioxaallene, have been alluded to on p. 259. They may play a role in transition-metal catalyzed carbon-carbon coupling between CO_2 and organic substrates (**"CO_2 activation"**, Braunstein 1988 R).

15.2 Alkyne Complexes

The most important aspects pertaining to the organometallic chemistry of alkynes concern the **cyclooligomerization of acetylenes** and the amazing **product variety** found in the reactions between acetylenes and metal carbonyls. This is largely due to the following facts:

- *The alkyne ligand can coordinate like two orthogonal alkene units.*
- *The alkyne ligand can formally occupy one or two (resp. three or four) coordination sites. The assignment of its denticity is closely associated with descriptions of the bond type:*

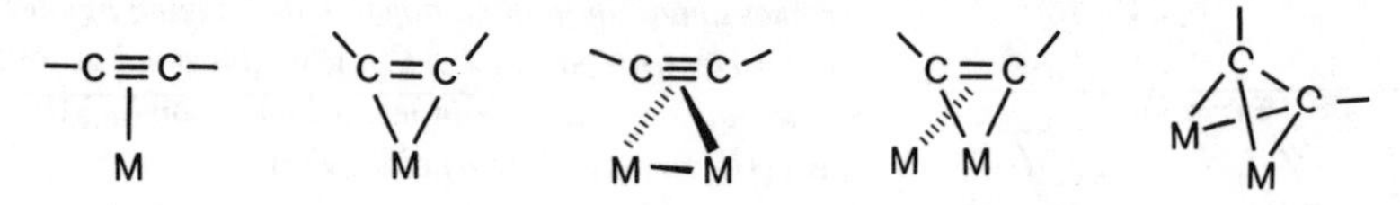

monodentate bidentate bidentate tridentate quadridentate

- *Alkyne ligands can be complexed in dimerized or trimerized form.*
- *Alkyne oligomerization can also involve the incorporation of carbonyl groups and/or the metal atom.*

15.2.1 Homoalkyne Complexes

ALKYNES AS FORMAL MONO- OR BIDENTATE LIGANDS:

$Na_2[PtCl_4]$ —(1. t-Bu_2C_2, EtOH; 2. RNH_2; R=4-MeC_6H_4)→ RH_2N–Pt(Cl)(Cl)–(t-BuC≡C-t-Bu): 165°, 124 pm

(Chatt, 1961) $\nu_{CC} = 2028\ cm^{-1}$

$Cp_2Ti(CO)_2$ + PhC≡CPh —(heptane, vacuum, 25°C, 3h, -CO)→ $Cp_2Ti(CO)(PhC_2Ph)$: 128 pm, 146°

(Floriani, 1978) $\nu_{CC} = 1780\ cm^{-1}$

$(Ph_3P)_2Pt$–(R_2C=CR_2) + C_2Ph_2 ⟶ $(Ph_3P)_2Pt(PhC_2Ph)$: 140°, 132 pm

(Grim, 1967) $\nu_{CC} = 1750\ cm^{-1}$

The C–C bond lengths in transition-metal coordinated alkynes cover almost the whole range between $d(C{\equiv}C)_{free} = 120$ pm and $d(C{=}C)_{free} = 134$ pm. Notable here is the correlation between $d(CC)_{coord.}$, the CCR bond angle and the stretching frequency ν_{CC} ($\nu_{C\equiv C,\ free} = 2190$–$2260\ cm^{-1}$, depending on the substituents). In fact, it seems appropriate to describe $(PPh_3)_2Pt(Ph_2C_2)$ as a **metallacyclopropene** since the alkyne formally occupies two coordination sites.

As with alkene complexes, the stability of alkyne complexes increases with increasing electron-withdrawing character of X in XC≡CX. Thus, halogen-substituted acetylenes, which are highly explosive compounds in the uncomplexed state, may be stabilized by complex formation (Dehnicke, 1986):

$$WCl_6 + C_2Cl_2 \cdot OEt_2 + C_2Cl_4 \xrightarrow[2.\ -Et_2O]{1.\ -C_2Cl_6} [WCl_4(C_2Cl_2)]_2 \xrightarrow{PPh_4Cl} [WCl_5(C_2Cl_2)]^- \; PPh_4^+$$

144° 128 pm

This example shows that alkynes can also be coordinated to transition metals in higher oxidation states. Such complexes might also contribute to the understanding of WCl_6-catalyzed alkyne polymerization (Masuda, 1984 R).

Even cyclohexyne, unstable as a free molecule, is obtained in metal-coordinated form. This stabilization is due to the distortion from linearity which alkynes experience on coordination, and which in this case relieves ring strain (Whimp, 1971):

$$C_6H_8Br_2 + Pt(PPh_3)_3 \xrightarrow[-PPh_3]{Na/Hg} (C_6H_8)Pt(PPh_3)_2$$

127° 129 pm

No less remarkable is the coordination of benzyne C_6H_4, also not stable as a free molecule (Schrock, 1979):

$$Cp^*M(CH_3)_3(C_6H_5) \xrightarrow[-CH_4]{120^0}$$

M = Nb,Ta

Ta 141 137 136 140 141 136 pm

The aromatic ring of the η^2-benzyne unit exhibits C–C bond-length alternation (symmetry D_{3h}); the bond length between the two coordinated carbon atoms is not significantly different from those of the other two short C–C bonds. The formulation of this complex as an aryne π-complex or as a metallabenzocyclopropene remains largely a matter of semantics.

Very recently, even benzdiyne C_6H_2 (tetradehydrobenzene) could be stabilized by means of coordination to two Ni(0) centers. The structural parameters of this unusual complex indicate extensive localization of the π electrons in the region of the coordinated carbon atoms (Bennett, 1988):

(c-Hex)₂P / Ni — [arene] — Ni / P(c-Hex)₂ ; 131 pm ; 141 pm

ALKYNES AS BRIDGING LIGANDS

$Co_2(CO)_8$ + PhC≡CPh ⟶ $(CO)_3Co$—$Co(CO)_3$ (PhC≡CPh bridging; C–C 146 pm; Co–Co 247)

$[CpNi(CO)]_2$ + RC≡CR ⟶ (RC≡CR bridging Ni–Ni, each Ni bearing Cp)

The classification of these and similar oligonuclear compounds as acetylene complexes is somewhat arbitrary; they could just as well be regarded as metal-carbon clusters (cf. Chapter 16).

The coordination of an alkyne to a $Co_2(CO)_6$ moiety considerably reduces the reactivity of the C≡C triple bond, allowing selective reactions at the functional groups of the coordinated alkyne to be carried out. For example, propargyl alcohols are converted by strong acids into **metal-stabilized propargyl cations.** Spectroscopic data speak for delocalization of the positive charge over the whole (alkyne)$Co_2(CO)_6$ unit. Such complexes can be used as selective electrophilic alkylating agents in reactions with ketones, enol acetates and arenes (Nicholas, 1987 R).

HC≡C—CH_2OH ($Co_2(CO)_6$) —(HBF_4, Ac_2O)⟶ $[HC≡C{=}CH_2\ (Co_2(CO)_6)]^+$ BF_4^- —(OAc enol acetate)⟶ ketone–CH_2CH_2–C≡CH ($Co_2(CO)_6$)

The free alkynes can be generated by oxidative decomplexation.

A further interesting reaction of the cobalt cluster (alkyne)$Co_2(CO)_6$ is the conversion μ_2-alkyne $\rightarrow$ μ_3-alkylidyne:

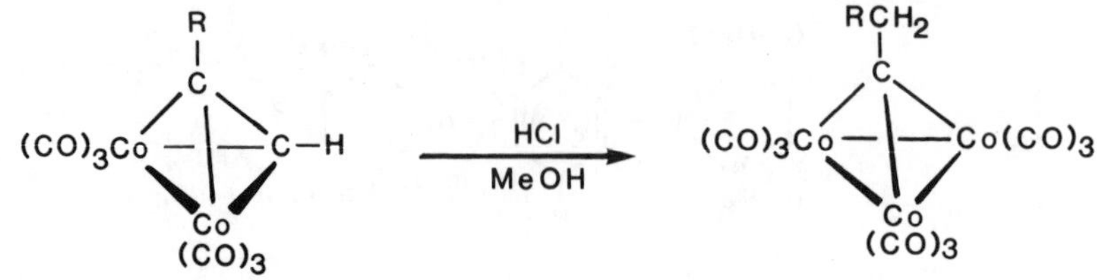

Two edge-sharing metallatetrahedrane clusters build the framework of the binary iron alkyne complex $(Me_3SiC_2SiMe_3)_4Fe_2$ which forms from the precursor (η^6-toluene)ironbis(ethylene), the latter being accessible through metal-atom ligand-vapor cocondensation (Zenneck, 1988):

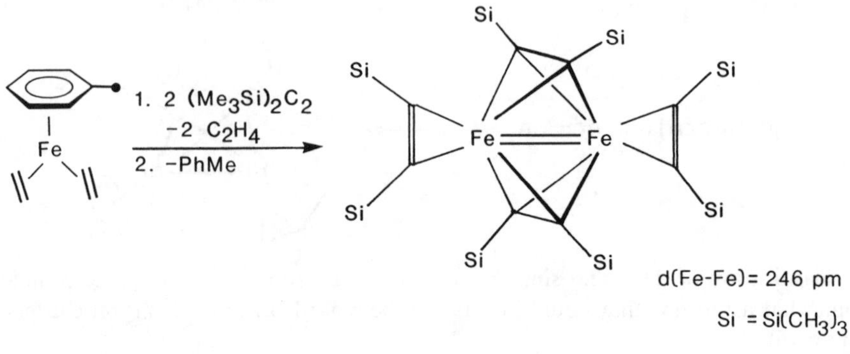

d(Fe-Fe) = 246 pm

Si = $Si(CH_3)_3$

OLIGOMERIZATION OF ALKYNES

Reactions of alkynes with metal compounds can lead to di- tri- or tetramerization of the organic ligand. Not in all cases is an isolable metal complex formed, though (see Chapter 17). The reaction with palladium salts yields cyclobutadiene complexes in addition to alkyne complexes and benzene derivatives (Maitlis, 1976 R):

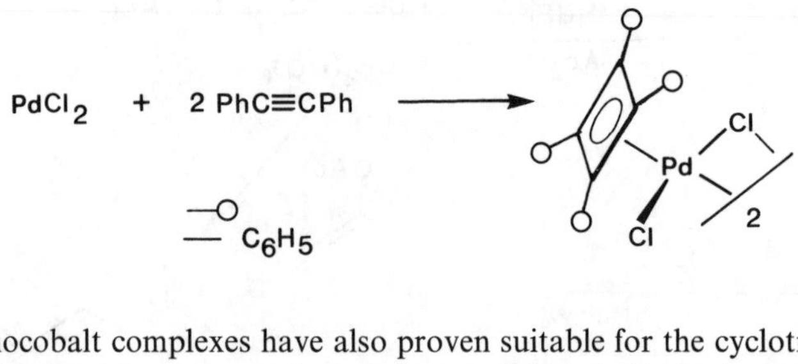

$PdCl_2$ + 2 PhC≡CPh →

—O = C_6H_5

Certain organocobalt complexes have also proven suitable for the cyclotrimerization of alkynes, the isolation of arene complexes being possible in individual cases (Jonas, 1983):

Co + 3 $CH_3C{\equiv}CCH_3$ → Co 20 VE

$=^{\bullet}$ CH_3

$Co_2(CO)_8$ + $2\,{}^tBuC{\equiv}CH$ + $HC{\equiv}CH$ →

$\downarrow Br_2$

This binuclear complex with a ***"fly-over" ligand****, formed by alkyne trimerization, contains* Co(d^7) *centers which are* η^1*-bonded to terminal carbons and* η^3*-bonded to internal allyl units. Carbonyl ligands and a* Co–Co *bond complete the* 18 VE *shell. The sterically encumbered hexaisopropylbenzene was also prepared for the first time by* $Co_2(CO)_8$*-catalyzed alkyne trimerization (Arnett,* 1964).

Of greater importance is the cobalt-catalyzed cotrimerization of alkynes and nitriles discovered by Yamazaki (1973) (see p. 375). Whereas this reaction has been extended by Bönnemann (1985 R) into a versatile pyridine synthesis, Vollhardt (1984 R) has placed the emphasis on synthesizing homocyclic systems:

2 $HC{\equiv}CH$ + $R{-}C{\equiv}N$ $\xrightarrow{CpCo(COD)}$ 2-R-pyridine

COD = 1,5-Cyclooctadiene

(Bönnemann, 1974)

$(CH_2)_n(C{\equiv}C{-}H)_2$ + $R^1{-}C{\equiv}C{-}R^2$ $\xrightarrow{CpCo(CO)_2}$ $(CH_2)_n$-annulated benzene (R^1, R')

n = 2–5

(Vollhardt, 1974)

The best catalysts for the cyclotetramerization of acetylenes to give cyclooctatetraene derivatives are labile Ni(II) complexes like $Ni(acac)_2$ or $Ni(CN)_2$ (Reppe, 1940f). This reaction forms the basis of an industrial process to produce cyclooctatetraene (see p. 421). Cyclooligomerizations can also be accompanied by incorporation of one or more molecules of CO:

Fe(CO)5 + RC≡CR
R = H
C=O
Fe (CO)3
R = Ph
R = Me
R = H
O= =O
Fe (CO)3
Fe (CO)3
O
Fe (CO)3

Among the possible products are dimeric complexes in which one $Fe(CO)_3$ moiety is π-bonded, the other becoming part of a metallacyclopentadiene ("ferrole"):

HC≡CH + Fe(CO)5
(CO)3 Fe
(CO)3 Fe
(CO)3 Fe
(CO)3 Fe

The two $Fe(CO)_3$ units are presumably linked by a metal-metal bond (249 pm), and, as labelling experiments have shown, are subject to intramolecular exchange. Chiral derivatives of these ferracyclopentadienes undergo rapid thermal racemization.

A related cyclization is the **Pauson-Khand reaction**, in which substituted cyclopentenones are formed in a single step from an alkyne, an alkene and CO, with $Co_2(CO)_8$ acting as a catalyst (Pauson, 1985 R).

+ H–C≡C–H + CO
10% Co2(CO)8
60 - 70°
O

In most cases, however, this reaction is carried out by first isolating the (alkyne)$Co_2(CO)_6$ complex, followed by its stoichiometric addition to the alkene. The Pauson-Khand reaction excells in high regio- and stereospecificity. Thus, in the above example the exo product is formed exclusively. The CO group usually ends up adjacent to the bulkier substituent of the alkyne:

+ CH3–C≡CH
Co2(CO)6
CH3
O

15.2.2 Heteroalkyne Complexes

The heteroalkynes RC≡E (E = P, As, Sb see p. 164) are related to the heteroalkenes $R_2C{=}E$ (E = S, Se, Te) in that they can also function as π ligands (Nixon, 1981):

$(Ph_3P)_2Pt(C_2H_4)$ → (t-BuC≡P; 25° C_6H_6; $-C_2H_4$) → $(Ph_3P)_2Pt(\eta^2\text{-}P{\equiv}C\text{-}t\text{-}Bu)$ [132°]

This type of coordination results in a considerable lengthening of the C–P bond from 154 pm (free) to 167 pm (η^2-bonded), a consequence of Pt $\overset{\pi}{\rightarrow}$ ligand backbonding. The well-known alkyne bridging mode is also displayed by phosphaalkynes:

$Co_2(CO)_8$ → (t-BuC≡P; THF, T < 25°; − CO) → $(CO)_3Co(\mu\text{-}t\text{-}BuC{\equiv}P)Co(CO)_3$

This side-on coordination of the phosphaalkyne is in contrast to the usual terminal coordination of nitriles (azaalkynes), for which side-on coordination is exceptional (Wilkinson, 1986). Recently, however, an example has been reported in which a phosphaalkyne coordinates via the terminal phosphorus atom (Nixon, 1987):

$Mo(N_2)_2(P\frown P)_2$ → (2 AdC≡P; − 2 N_2) → $Mo(P{\equiv}CAd)_2(P\frown P)_2$

P⌒P = 1,2-Diethylphosphinoethane

Ad = Adamantyl

As expected, the P≡C triple bond is affected only marginally by this bonding mode.

15.3 Allyl and Enyl Complexes

Unsaturated hydrocarbons C_nH_{n+2} with odd numbers of carbon atoms can be regarded as neutral ligands with an odd number of π electrons or as anionic or cationic ligands possessing an even number of valence electrons:

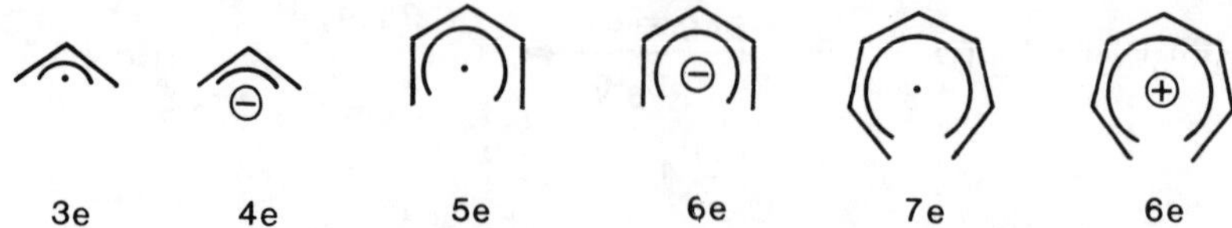

The neutral radicals are highly reactive, whereas the anions in the form of Li-, K-, or Mg salts are stable and characterizable, although very air- and moisture-sensitive (see p. 23). In their transition-metal complexes, allyl- and enyl units are remarkably inert, as evident from thermal robustness and stability towards hydrolysis.

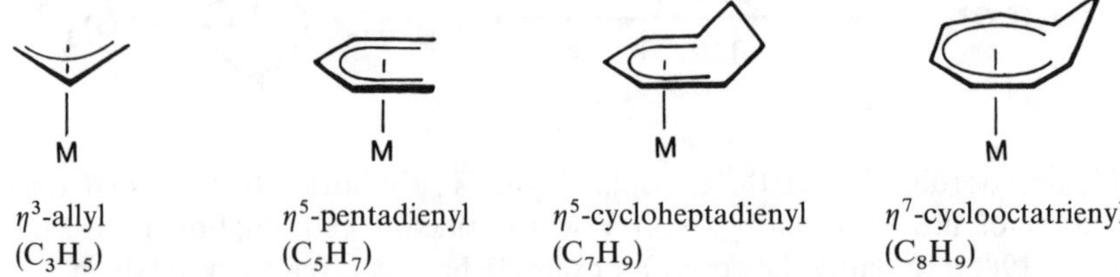

η^3-allyl (C_3H_5) η^5-pentadienyl (C_5H_7) η^5-cycloheptadienyl (C_7H_9) η^7-cyclooctatrienyl (C_8H_9)

Only the binary allyl complexes are rather labile; they tend to release the metal via dimerization of the organic ligand and thus become important intermediates for homogeneously catalyzed cycles (Chapter 17). As far as bonding and reactivity are concerned, no basic differences exist between enyl- and oligoolefin complexes. As a matter of fact, their mutual interconversion is one of the most important features of their synthesis and reactivity.

15.3.1 Allyl Complexes

The first synthesis and accurate description of an η^3-allyl metal complex was reported by Smidt and Hafner (1959), who prepared $[(C_3H_5)PdCl]_2$ from the reaction of palladium chloride with allyl alcohol. Since then, the structural element (η^3-allyl)M has become widespread among transition-metal complexes, sometimes appearing as part of larger ligands. The routes to η^3-allyl metal complexes therefore are manifold.

PREPARATION

The following synopsis of preparative methods can be divided into the reaction types:

- *Replacement of* X^- *by allyl*$^-$ *(metathesis reaction)* **1**
- *Rearrangement* σ*-allyl* (η^1) $\rightarrow$ π*-allyl* (η^3) **2, 3**
- *Conversion* π*-olefin* (η^2 *or* η^4) $\rightarrow$ π*-allyl* (η^3) **4–7**

1. Metal Salt + Main-group Organometallic (Metathesis)

$$NiBr_2 + 2\,C_3H_5MgBr \xrightarrow[-10\,°C]{Et_2O} Ni(C_3H_5)_2$$

$$Co(acac)_3 + 3\,C_3H_5MgBr \longrightarrow Co(C_3H_5)_3 \quad \text{decomp.} > -55\,°C$$

$$ZrCl_4 + 4\,C_3H_5MgCl \xrightarrow[-78\,°C]{Et_2O} Zr(C_3H_5)_4$$

This is a general route to the binary allyl complexes. Synthesis and isolation must be performed at low temperatures, as complexes of this type are very thermolabile (Wilke, 1966 R).

$$4\,PdCl_2 + (C_3H_5)_4Sn \xrightarrow[-SnCl_4]{4\,PPh_3} 4\,(C_3H_5)PdCl(PPh_3)$$

2. Carbonyl Metallate + Allyl Halide

$$Na[Mn^{-I}(CO)_5] + \text{CH}_2{=}\text{CHCH}_2\text{Cl} \longrightarrow \eta^1\text{-CH}_2{=}\text{CHCH}_2\text{Mn}^{I}(CO)_5 \xrightarrow{-CO} \eta^3\text{-C}_3\text{H}_5\text{Mn}^{I}(CO)_4$$

The σ/π rearrangement with attendant loss of CO *can be initiated thermally* (80 °C) *or photochemically.*

$$[CpMo^{0}(CO)_3]^- \xrightarrow[-Cl^-]{PhCH_2Cl} CpMo^{II}(CO)_3CH_2Ph \xrightarrow[-CO]{h\nu} CpMo^{II}(CO)_2(\eta^3\text{-CH}_2C_6H_5)$$

(allyl π-complex of the benzyl anion)

Metal Carbonyl + Allyl Halide

$$CpCo^{I}(CO)_2 + C_3H_5I \longrightarrow [CpCo^{III}(\eta^3\text{-}C_3H_5)(CO)]^+I^- + CpCo^{III}(\eta^3\text{-}C_3H_5)I$$

3. Metal Hydride + Diolefin

$HCo(CO)_4$ + butadiene $\xrightarrow{-CO}$ anti-η^3-$CH_3C_3H_4Co(CO)_3$ + syn-η^3-$CH_3C_3H_4Co(CO)_3$

via $[(CO)_4Co\text{–}CH_2CH{=}CHCH_3]$ $\xrightarrow{-CO}$

This reaction is initiated by a 1,4-**hydrocobaltation** of butadiene. The syn-complex is the more stable isomer.

4. Metal Salt + Olefin + Base

$Na_2[PdCl_4]$ + 1-methylcyclopentene $\xrightarrow{Na_2CO_3}$ $[(\eta^3\text{-allyl})Pd(\mu\text{-}Cl)]_2$

A mechanism for this reaction was proposed by Trost (1978):

propene + $PdCl_4^{2-}$ $\xrightarrow{-Cl^-}$ η^2-complex $[PdCl_3(C_3H_6)]^-$ $\xrightarrow{-Cl^-}$ η^3-allyl $PdCl_2H$ $\xrightarrow{-HCl}$ $[(\eta^3\text{-}C_3H_5)Pd(\mu\text{-}Cl)]_2$

5. Metal Salt + Allyl Halide

$Na_2[PdCl_4]$ + allyl chloride $\xrightarrow{CO(cat.)/H_2O/MeOH}$ $1/2\,[C_3H_5PdCl]_2$

The formation of an η^3-allyl complex from allyl chloride formally is a disproportionation which yields organic keto products in addition to the coordinated allyl anion (Jira, 1971). The reaction can be accelerated by introducing other reducing agents (e.g. CO, $SnCl_2$). This measure becomes unnecessary, however, when one starts from Pd^0 (Klabunde, 1977):

$$2\ C_6H_5CH_2Cl\ (g) + 2\ Pd^0(g) \xrightarrow[OA]{CK} [(\eta^3\text{-benzyl})Pd(\mu\text{-}Cl)]_2$$

6. Conversion η^2-Allyl Alcohol → η^3-Allyl:

CH_2OH HBF$_4$ Mn OC CO

7. Electrophilic or Nucleophilic Addition to Olefin Complexes

H^+/CO Ir CO

The coordinatively unsaturated 16 electron complex generated by protonation accepts a Lewis base.

R^- Mo OC CO η^4 η^3 R

8. Hydrogenation of a Complex with Excess Electrons

20 VE Ni Na/Hg EtOH Ni 18 VE H_2 H_2

9. Dimerization of an Allene

$$Fe_2(CO)_9 + 2\ CH_2{=}C{=}CH_2 \xrightarrow[-\ 3\ CO]{} (CO)_3Fe{-}Fe(CO)_3$$

STRUCTURE AND BONDING

Metal-ligand interactions in allyl complexes:

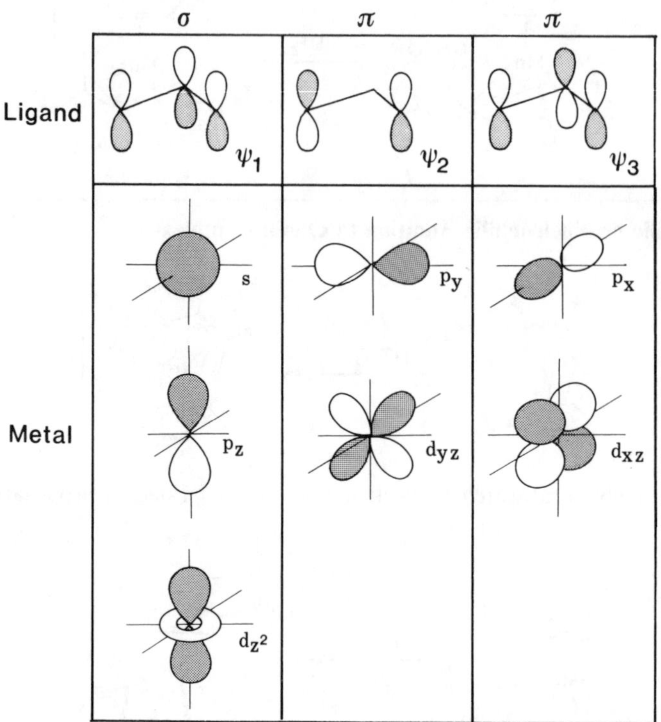

In the ligand $C_3H_5^-$ ψ_1 and ψ_2 are doubly occupied. The metal-ligand bond can be described by the components $\psi_1 \xrightarrow{\sigma} M$, $\psi_2 \xrightarrow{\pi} M$ and $\psi_3 \xleftarrow{\pi} M$. These overlap characteristics induce an electronic rotational barrier within the (η^3-allyl)M unit (see p. 287).

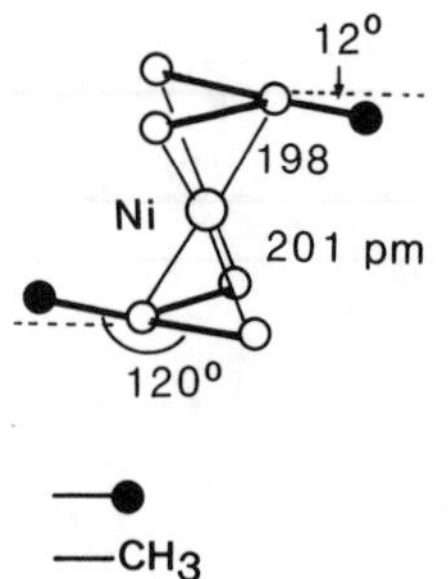

The structure of the prototypical ***bis(2-methylallyl)nickel*** *in the solid state is characterized by a parallel disposition of the two allyl ligands and slight bending of the* C$-$CH$_3$ *bond towards the metal (Dietrich,* 1963*).*

Bis(η^3-allyl)nickel (Wilke, 1961*) was the first binary allyl metal complex to be isolated.*

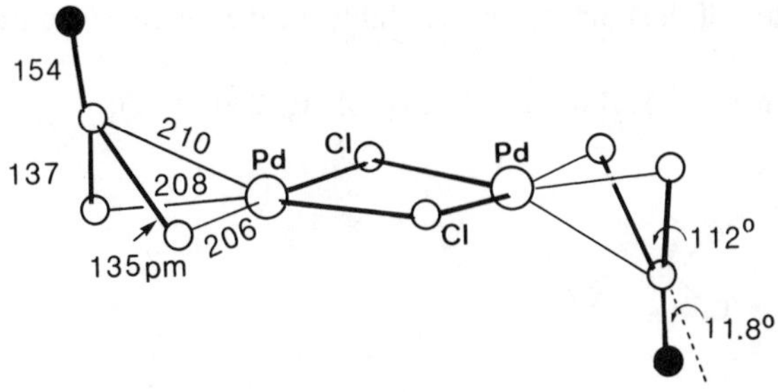

Structure of $[C_4H_7PdCl]_2$. *The* Pd- *and* Cl *atoms lie in one plane which intersects the plane of the three allyl carbon atoms at an angle of* 111.5°.

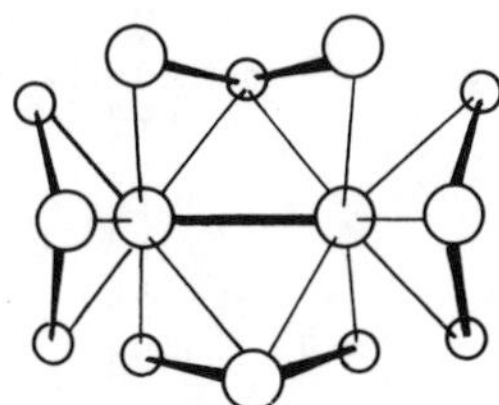

Structure of the binuclear complex $(C_3H_5)_4Mo_2$. *Two allyl groups function as bridging ligands. This complex is fluxional in solution, whereby the bridging and the terminal allyl groups experience ligand scrambling* (*Cotton*, 1971).
$d(Mo-Mo) = 218$ pm

In the **^{1}H-NMR spectra** of allyl complexes, *syn*- and *anti*-protons of the terminal CH_2 groups are generally nonequivalent. Frequently, however, dynamic behavior is observed in solution which renders the terminal protons equivalent on the NMR time scale. The most likely mechanism to explain this phenomenon is a rapid $\pi-\sigma-\pi$ conversion, which at the σ-stage allows free rotation around the C–C- and the M–C bonds. In this manner, *syn*- and *anti*-protons can exchange their positions:

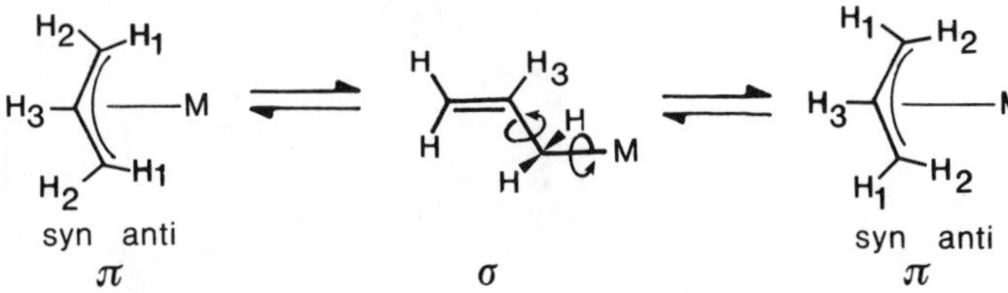

Often, the presence of a Lewis base (e.g. a nucleophilic solvent) is sufficient to trigger this dynamic behavior. This will be illustrated by the ^{1}H-NMR spectra of $[C_3H_5PdCl]_2$. In $CDCl_3$ (upper trace), the spectrum of the π-allyl ligand is that of an A_2M_2X system, while in the solvent d^6-DMSO at 140 °C (lower trace) an A_4X pattern with seemingly equivalent terminal hydrogen atoms is observed. Presumably, upon attack of solvent molecules, the dimeric structure of the complex is reversibly cleaved with accompanying $\pi - \sigma - \pi$ conversions (Chien, 1961).

^{1}H-NMR spectrum of $[C_3H_5PdCl]_2$ (**$CDCl_3$**, 25 °C, 200 MHz):

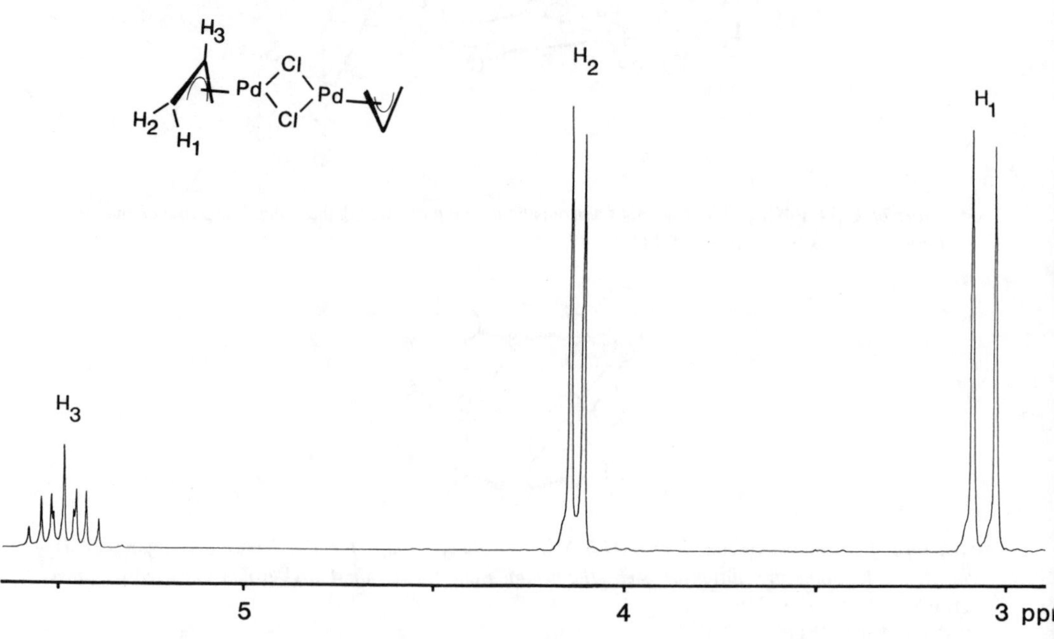

^{1}H-NMR spectrum of $[C_3H_5PdCl]_2$ (**d^6-DMSO**, 140 °C, 200 MHz):

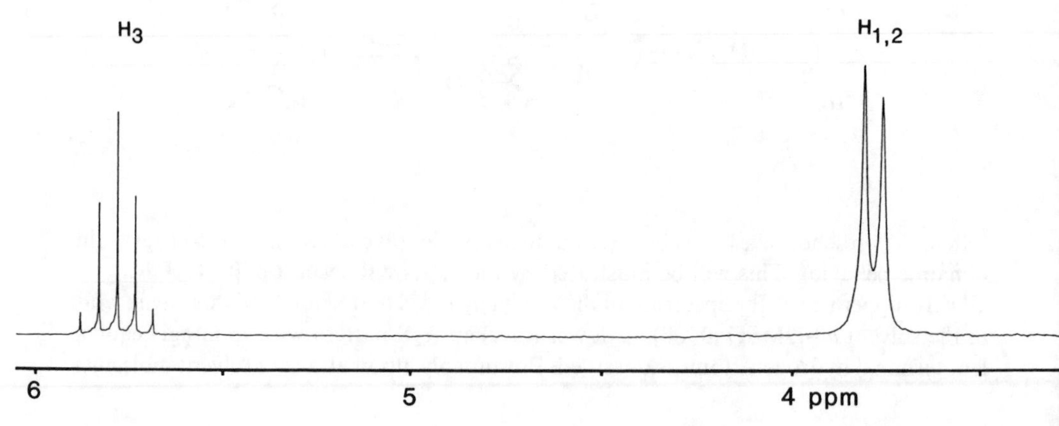

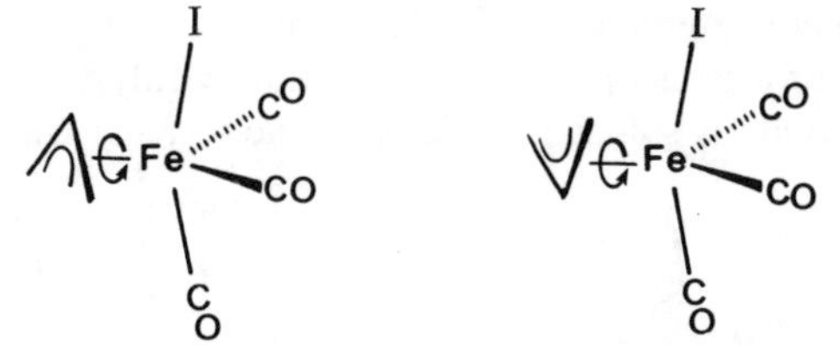

During the rotation of the allyl ligand around the (η^3-C_3H_5)$-$M *axis* ***without*** $\pi-\sigma-\pi$ *conversion the terminal syn- and anti-protons necessarily remain nonequivalent; they experience changes in the shielding, however. As this rotation at room temperature is slow on the NMR time scale, both isomers are observable* (*Nesmeyanow*, 1968).

Structural fluxionality of a different type is found in binuclear complexes in which two organometallic moieties are synfacially bonded to the bridging ligand cycloheptatriene. For the compound (μ-C_7H_8)[C_5H_5Rh]$_2$, low-temperature NMR results suggest the simultaneous presence of the bonding modes η^3-allyl and η^1-alkyl + η^2-alkene. On raising the temperature, the two enantiomers interconvert, coalescence of the respective NMR signals eventually being reached (Lewis, 1974).

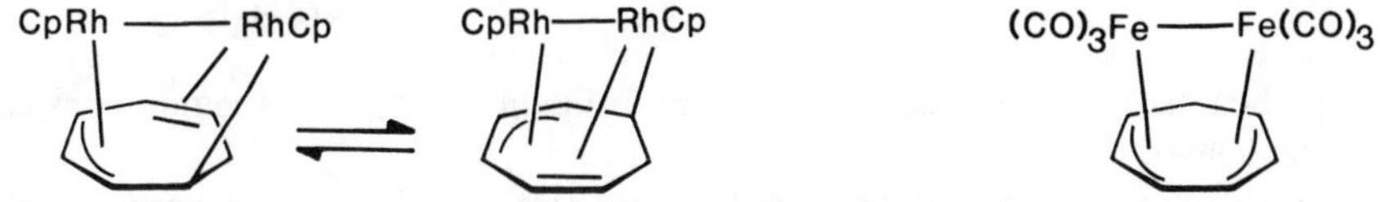

The isoelectronic complex (μ-C_7H_8)[$Fe(CO)_3$]$_2$ even at low temperature exhibits the bis(η^3-allyl) structure (Cotton, 1971).

REACTIONS OF ALLYL COMPLEXES

Besides the significance of allyl complexes as intermediates in homogeneous catalysis (e.g. oligomerization reactions, p. 422), they have become increasingly important as **selective electrophilic substrates** in organic synthesis.

Thus, the cations [(allyl)$Fe(CO)_4$]$^+$, readily obtained by protonation of (η^4-diene)$Fe(CO)_3$ complexes in the presence of CO, are highly electrophilic. They are attacked at the terminal carbon atom by a variety of nucleophiles to generate unstable (η^2-olefin)$Fe(CO)_4$ species which easily decompose releasing the respective olefin (Whitesides, 1973):

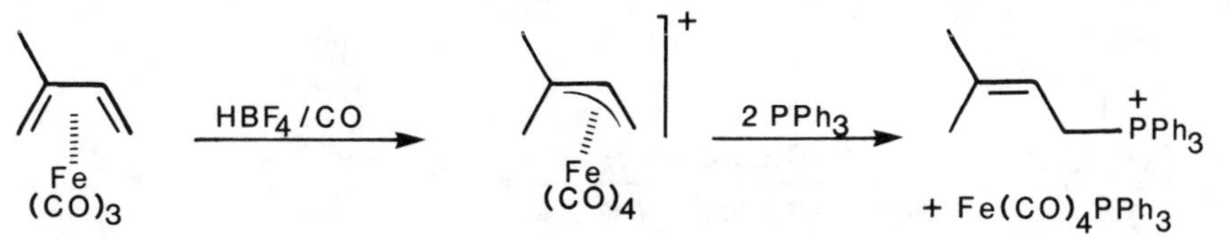

In the case of the dimeric complex $[(allyl)PdCl]_2$, the reaction usually requires the presence of a phosphane or a nucleophilic solvent, because only the cationic allyl complex formed as an intermediate is sufficiently electrophilic to react with other nucleophiles:

OCH3 PdCl4 2− Base OCH3 Pd Cl 2 PPh3 THF/20° OCH3 2 Pd PPh3 PPh3 + HN(CH3)2 −H+ OCH3 + Pd0 + 2 PPh3 N(CH3)2

Nucleophilic attack always occurs from an **antifacial** direction, i.e. at the non-complexed side of the ligand.

For economical reasons, palladium complexes are rarely used in stochiometric amounts. Instead, Pd^0-catalysed reactions are generally preferred, η^3-allyl complexes playing an essential role as intermediates (Tsuji, 1986 R). Nucleophilic substitution at allyl acetates may serve as an example:

X OA ① L Pd X L X = OAc L = PPh3 ② σ → π PdL4 −2L PdL2 + X− Pd L L Nu +X− ③ Nu−

Pd^0-catalyzed allylic substitutions proceed stereospecifically with retention of configuration at the α-allyl carbon atom (Trost, 1977 R):

Me H OAc NaCH(CO2Me)2 Pd(PPh3)4, PPh3, THF Me H CH(CO2Me)2

This is due to the fact that the acetate group leaves *trans* to palladium and the new nucleophile again enters from the *trans*-direction.
In **chiral allyl complexes**, even the terminal carbon atoms of symmetrical allyl ligands are nonequivalent. Selective attack by a nucleophile at one of these centers and subsequent oxidative decomplexation can generate chiral olefins in high enantiomeric purity (Faller, 1983):

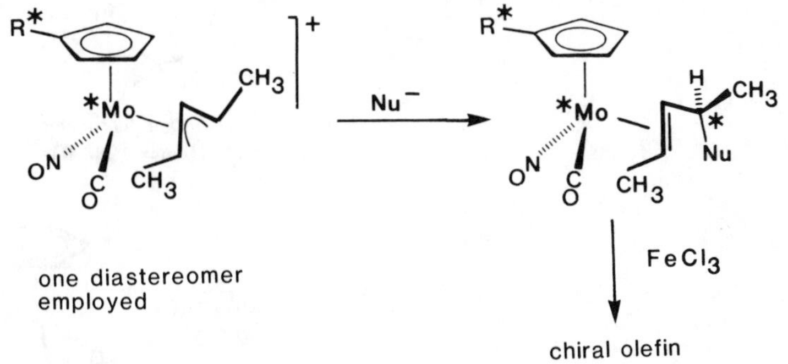

Olefin isomerization, catalyzed by transition metals, presumably also proceeds *via* allyl complex intermediates. By metal insertion into a CH_2 group adjacent to a coordinated olefin, an allyl hydride complex is formed, which can then rearrange to give the isomeric olefin complex:

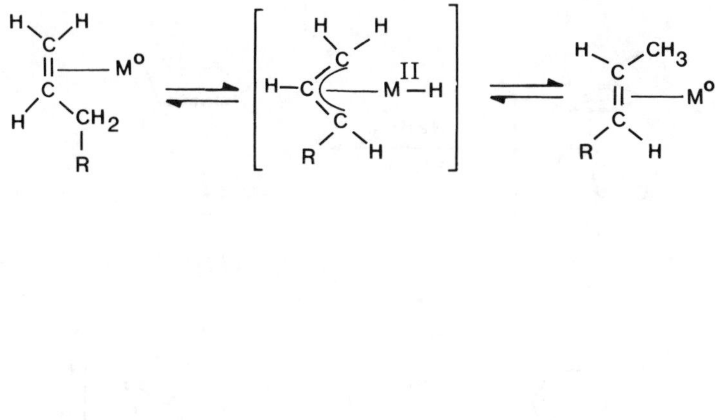

15.3.2 Dienyl and Trienyl Complexes

η^5-Dienyl and η^7-trienyl complexes can be regarded as **vinylogues of allyl complexes**. The stabilization of enyl ligands by complex formation facilitates their use as carbocation or carbanion analogues in subsequent reactions. The preparation of dienyl and trienyl complexes in many ways parallels the synthesis of allyl complexes.

PREPARATION

1. Metal Halide + Main-Group Organometallic (Metathesis)

2 K^+ + $FeCl_2$ →

"open ferrocene"
(Ernst, 1988 R)

$(CH_3)_3Sn$ + $Mn(CO)_5Br$ —($-Me_3SnBr$, $-2CO$)→

"open cymantrene"

2. Transfer of H^+ or H^-

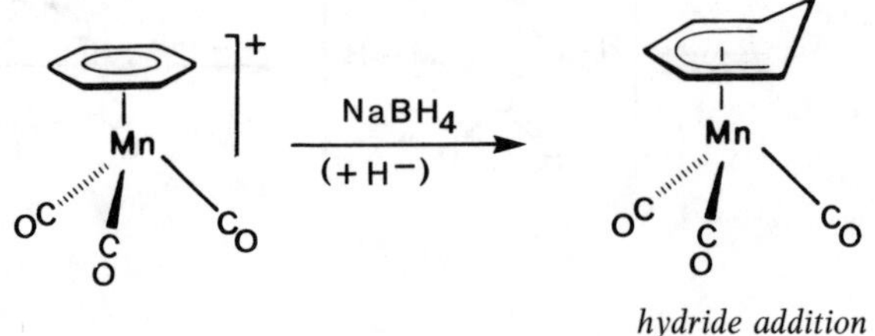

hydride addition

Fe $(CO)_3$ —$Ph_3C^+BF_4^-$ (− H^-)→ Fe $(CO)_3$ +

hydride abstraction

$Ph_3C^+BF_4^-$ *is the classical reagent for hydride abstraction. It is usually applied to cyclic systems.*

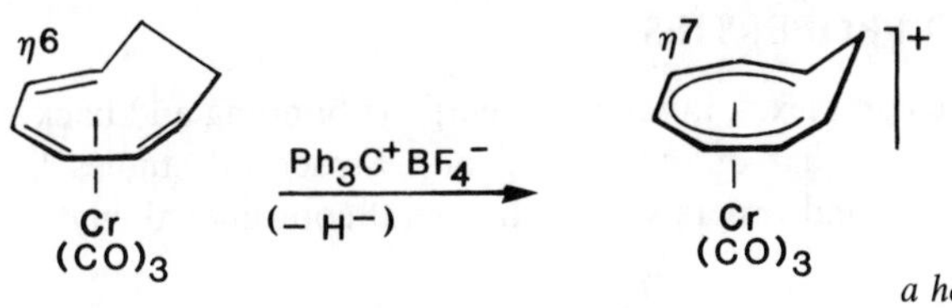

a homotropylium complex

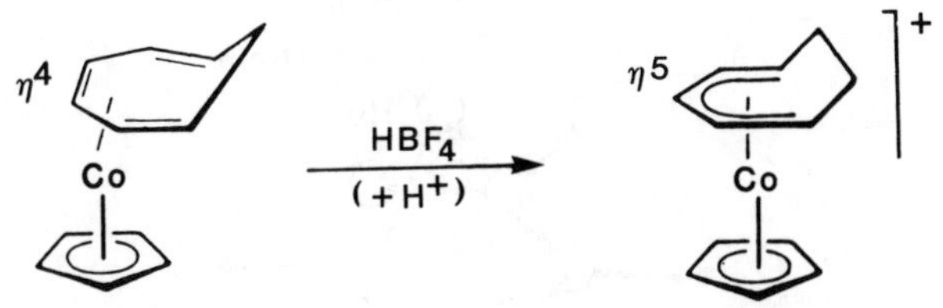

protonation of an uncoordinated double bond

Alternatively, protonation can occur at an α-OH or α-OCH_3 group of an (η^4-1,3-diene)M unit, water or methanol then being released to yield the dienyl complex:

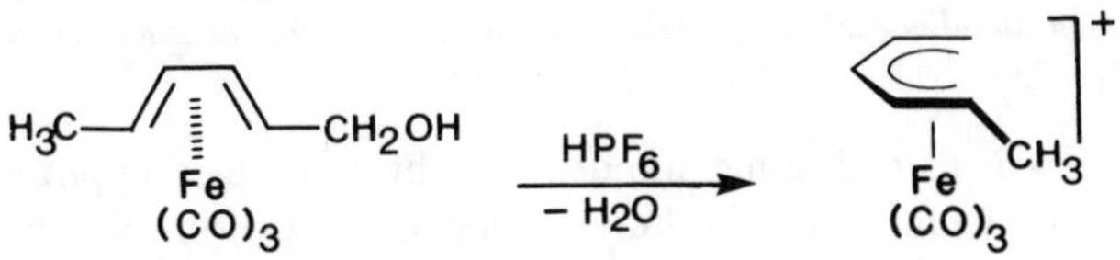

3. Metal Salt + Olefin + Reducing Agent

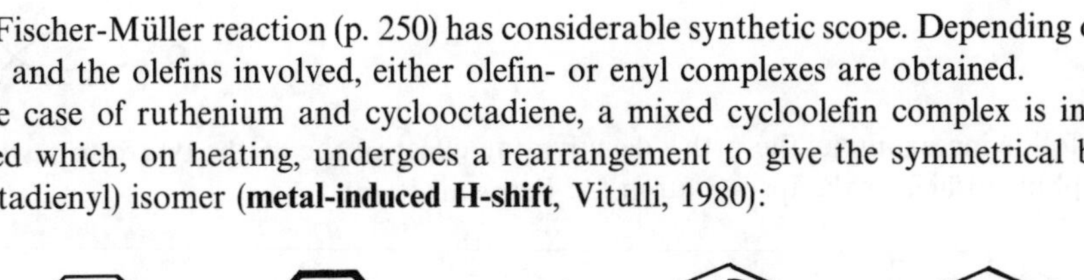

This Fischer-Müller reaction (p. 250) has considerable synthetic scope. Depending on the metal and the olefins involved, either olefin- or enyl complexes are obtained.
In the case of ruthenium and cyclooctadiene, a mixed cycloolefin complex is initially formed which, on heating, undergoes a rearrangement to give the symmetrical bis(cyclooctadienyl) isomer (**metal-induced H-shift**, Vitulli, 1980):

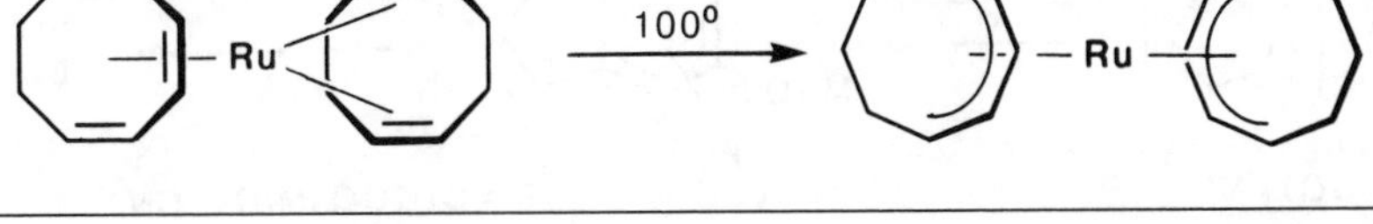

STRUCTURE AND PROPERTIES

The bonding in these complexes, involving synergetic bonding and backbonding contributions, is analogous to that of other π ligands. Structural studies have established planarity and uniform bond orders within the metal-coordinated enyl moiety.

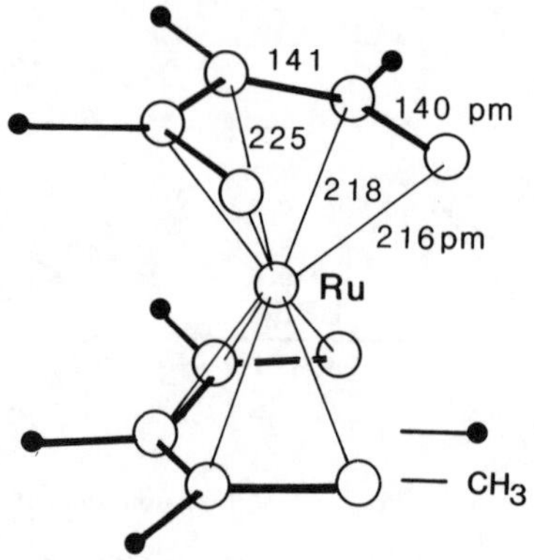

Structure of ***bis(2,3,4-trimethylpentadienyl)ruthenium.*** *The* M – C *bond lengths differ from each other only slightly and they are very similar to those in the corresponding cyclopentadienyl complex, justifying the designation* **"open metallocene"**. *The angle of twist in the ecliptical conformation shown above is* 52.5° (*Ernst*, 1983).

Enyl complexes have gained some importance through their application in **organic synthesis**. This is particularly true for $[(\text{dienyl})Fe(CO)_3]^+$ cations, to which nucleophiles add regioselectively (at the terminal carbon) and stereospecifically (antifacial to iron) (A. Pearson, 1980 R). The $Fe(CO)_3$ group can be subsequently removed by oxidation:

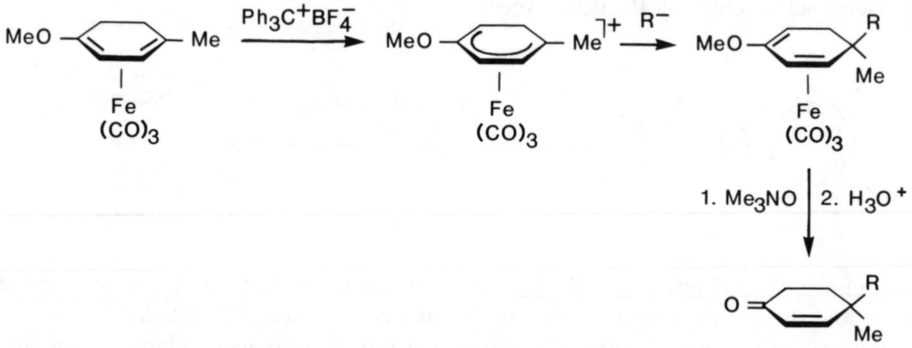

In the following example, the ortho-deactivating effect of the methoxy group results in nucleophilic attack at the ipso-C atom (A. Pearson, 1978):

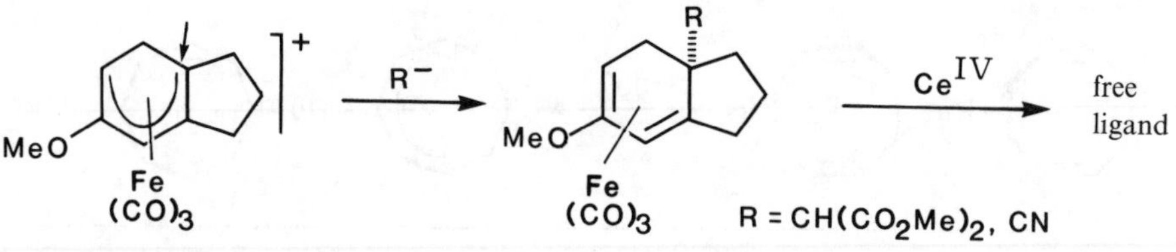

Nucleophilic addition to cationic π-complexes is one of the most important methods for the conversion of η^n-bonded ligands into the corresponding η^{n-1} ligands. Allyl and enyl ligands as well as unsaturated hydrocarbons like ethylene (p. 267), butadiene (p. 283) or benzene (p. 351f), which do not normally undergo nucleophilic addition reactions, are readily attacked by nucleophiles like H^-, CN^- or MeO^- when coordinated to a transition-metal cation. This enhanced reactivity of coordinated polyenes is a consequence of metal-ligand bonding, which results in a transfer of electron density from the hydrocarbon to the positively charged metal atom. Based on perturbation-theoretical considerations, Davies, Green and Mingos (1978) have proposed three general rules which allow predictions of the **direction of kinetically controlled nucleophilic attack** at 18 VE cationic complexes containing unsaturated hydrocarbon ligands. Applications of these rules require the ligands to be classified as **even** (η^2, η^4, η^6) or **odd** (η^3, η^5). Furthermore, a distinction between **closed** (cyclic conjugated) and **open** ligands is necessary.
The propensity for kinetically controlled nucleophilic addition to 18 VE cationic complexes decreases in the following order:

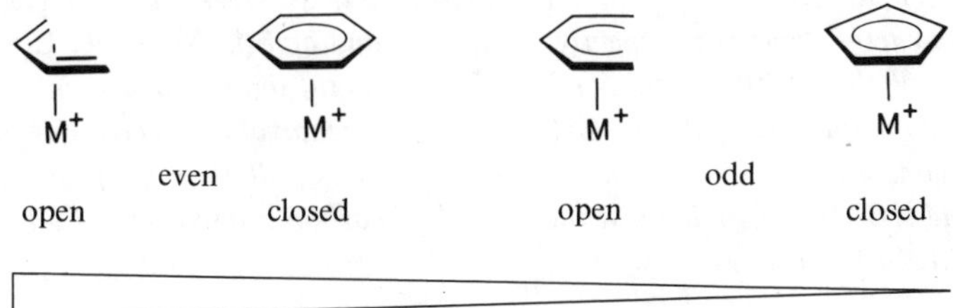

① *Nucleophilic attack occurs preferentially at* ***even*** *coordinated polyenes.*
② *Nucleophilic addition to* ***open*** *coordinated polyenes is preferred to addition to closed polyene ligands.*
③ *In the case of even open polyenes, nucleophilic attack always occurs at the* ***terminal carbon atom****; for odd open polyenyls, attack at the terminal carbon occurs only if* L_nM^+ *is a strongly electron-withdrawing fragment.*

The **Davies-Green-Mingos (DGM) rules** have to be applied in the sequence ① ② ③.
Examples:

Rule

① ③ — Ru —]+ —H⁻→ — Ru —

① ② ③ Mo]+ —H⁻→ Mo CH₃

②
③

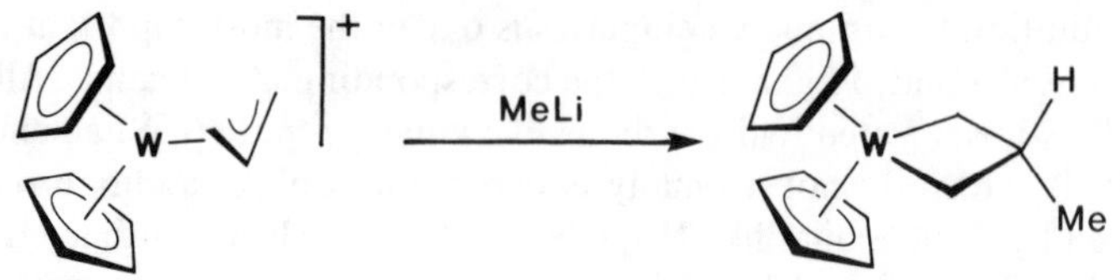

Understanding the DGM Rules

Let us start from the premise that the readiness of a ligand carbon atom to accept a nucleophile is governed by its partial positive charge. Chemo- and regioselectivity of nucleophilic attack then reflects unequal charges at the ligand carbon atoms. These charge differences must be a result of metal coordination, since for free alternating polyenes C_nH_{n+2} as well as for cyclic π-perimeters C_nH_n the charge distribution along the C_n chain is uniform. The occurrence of charge differentiation in the complex cations may be explained using the prototypes $C_3H_3^{\bullet}$ (allyl) and C_4H_6 (butadiene) as examples. Herein the complex cations are mentally constructed from the respective neutral ligand and the fragment L_nM^+, realizing that – because of the positive charge on L_nM^+ – the metal-ligand bond should be dominated by the donor contribution $L_nM^+ \leftarrow L(HOMO)$. The removal of electron density from the ligand with concomitant generation of partial charges will then be proportional to the squares of the HOMO coefficients at the individual carbon atoms, depicted qualitatively by the sizes of the orbital lobes.

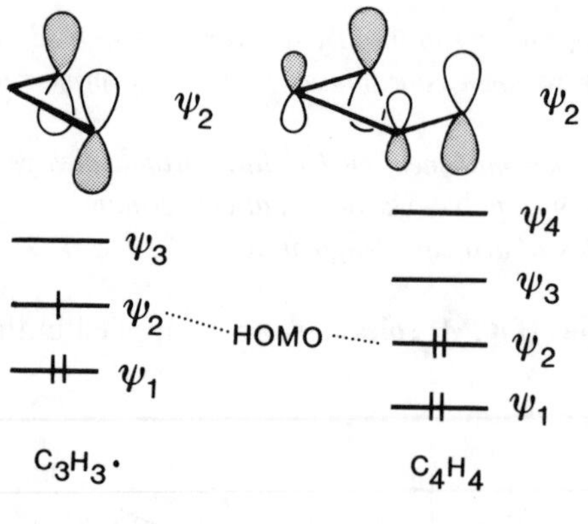

As suggested by the diagram, removal of charge dominates at the ***terminal*** *carbon atoms, which consequently attain higher partial charges δ^+ and which therefore are the preferred sites for nucleophilic attack (Rule ③).*

The limitation to rule ③ is apparent as well. In the case of even polyenes, the ligand HOMO is occupied by two electrons which in the complex populate a bonding MO. *Thus, irrespective of the acceptor strength of L_nM^+,* MO *formation will always generate partial positive charges ($0 < q < +2$) on the polyene. Odd polyenes, on the other hand, contribute with a singly occupied ligand HOMO to metal-ligand bonding with the effect that, depending on the acceptor strength of L_nM^+, negative or positive partial charges ($-1 < q < +1$) can accrue* *on the ligand. In the prototypical example η^3-allyl,* $C_3H_5^-$- *or* $C_3H_5^+$*-like behavior can therefore arise:*

L_nM^+ *is a strong electron acceptor*
(η^3-C_3H_5) *reacts like* $C_3H_5^+$

L_nM^+ *is a weak electron acceptor*
(η^3-C_3H_5) *reacts like* $C_3H_5^-$

Since charge transfer is more extensive for even polyenes ($0 < q < +2$) *relative to odd polyenes* ($-1 < q < +1$), *higher reactivity of the complexes is expected for the former, as compared to the latter (Rule ①).*
The preferred nucleophilic attack at complexes of open polyenes probably results from the fact that in open polyene ligands, charge distribution is less balanced than in the cyclic polyene counterparts (Rule ②). Refined discussion must, of course, include substituent effects and possible competing reactions like nucleophilic attack at CO *ligands.*

Excursion

^{13}C- and ^{1}H-NMR Spectroscopy of Organometallics

^{13}C-NMR

Despite the low natural abundance (1.1 %) and the relatively low receptivity of the ^{13}C nucleus (p. 25), ^{13}C-NMR has become an indispensable tool in the study of organometallic compounds. Essential features of ^{13}C-NMR as applied to organometallics are the following:

1. Proton decoupled $^{13}C\{^1H\}$ spectra, in the absence of other magnetic nuclei, display sharp singlets. Since signal dispersion is far superior in ^{13}C-NMR as compared to ^{1}H-NMR, ^{13}C spectra lend themselves to the analysis of elaborate molecules, of mixtures of isomers as well as to the recognition of fine structural details.
If the central atom possesses a magnetic moment, high isotopic abundance and relaxation times T_1 and T_2 which are not too short (e.g. ^{103}Rh, ^{195}Pt), scalar couplings $J(M, {}^{13}C)$ are observed. The latter provide insight into the nature of the metal-ligand bond.
2. The analysis of coupled ^{13}C-NMR spectra furnishes the coupling constants $J({}^{13}C, {}^{1}H)$. Together with the values $J({}^{1}H, {}^{1}H)$ from ^{1}H-NMR, they reflect ligand structure and the hybridization states of the carbon atoms. Particularly revealing are geminal and vicinal coupling constants which offer hints to coordination induced changes in the intra-ligand π-bond order.
3. Special techniques allow isotopomeric species to be studied (even at natural abundance) which contain two neighboring ^{13}C nuclei so that coupling constants $^{1}J({}^{13}C, {}^{13}C)$ also become available. This parameter again reveals the hybridization states of the respective carbon atoms as well as C, C bond orders in the ligand.

Furthermore, for complexes of a certain metal, linear correlation usually exists between the coupling constants $^1J(^{13}C, ^{13}C)$ and the bond lengths $d(C-C)$.

4. ^{13}C-NMR is particularly well-suited for the study of exchange processes. Dispersion of the chemical shifts in ^{13}C-NMR spectra being about five times as large as in ^{1}H-NMR, considerably faster processes may be studied by ^{13}C-NMR over the total dynamic range. Thus, ligand reorientation in (cyclooctatetraene)-$Fe(CO)_3$ (p. 366) by means of ^{1}H-NMR is amenable to study in the fast exchange region only, whereas using ^{13}C-NMR, the total range from slow to fast exchange can be investigated. Additionally, in ^{13}C-NMR spectra simpler splitting patterns are often encountered, rendering line-shape analyses less cumbersome than for the respective ^{1}H-NMR spectra.

I. Chemical Shifts for ^{13}C Nuclei in Diamagnetic Organometallics

In attempts at a quantitative interpretation, the shielding terminology (σ) introduced by Ramsey is usually employed instead of the chemical shift (δ). Therein, the chemical shifts are governed by diamagnetic and paramagnetic shielding terms, Eq. (1). Note that, by convention, the shift and the shielding possess opposite signs.

$$\sigma_i = \sigma_i^{\text{dia}} + \sigma_i^{\text{para}} + \sigma_{ij} \qquad (1)$$

The **diamagnetic shielding term σ_i^{dia}** accounts for the unperturbed spherical motion of the electrons; it dominates in ^{1}H-NMR.

The **paramagnetic shielding term σ_i^{para}**, which corrects for eventual perturbations of electronic motion and for non-spherical charge distribution, is opposed to σ_i^{dia}. According to Karplus and Pople, σ_i^{para} may be evaluated if a mean electronic excitation energy ΔE is adopted:

$$\sigma_i^{\text{para}} = -\frac{\mu_0 \mu_B}{2\pi} \langle r^{-3} \rangle_{np} [Q_i + \sum_{i \neq j} Q_j]/\Delta E \qquad (2)$$

μ_0 = *permeability (vacuum)*
μ_B = *Bohr magneton*

Besides the mean radius r of the $2p$ orbitals of the respective carbon atom, Eq. (2) contains the parameters Q_i (electron density at atom i) and Q_j (bond orders between atoms i and j). Like the mean excitation energy ΔE, which is difficult to infer from experiment, Q_i and Q_j are often derived from MO calculations. ΔE is usually approximated by the energy gap between the HOMO and the LUMO. The third, **nonlocal term σ_{ij}** describes the influence of the remote electrons (e.g. ring current effects); for the ^{13}C nucleus in rudimentary discussions, this term is generally neglected. Apart from hydrogen, for all other magnetic nuclei the paramagnetic shielding term σ_i^{para} dominates. In a number of cases Eq. (2) provides a link between the chemical shift $\delta\,^{13}$C and the electronic structure of the respective molecule. As for other nuclei, however, a quantitative interpretation of $\delta\,^{13}$C is not yet feasible, mainly because various factors which in addition to ΔE influence the paramagnetic term, cannot be assessed precisely. Coordination of an organic ligand to a metal fragment affects the resonance frequencies of the ligand nuclei in a characteristic way. This **coordination shift $\Delta\delta$** is defined as the chemical shift difference between the metal-bonded and the free ligand:

$$\Delta\delta = \delta^{(\text{complex})} - \delta^{(\text{ligand})} \qquad (3)$$

In Figure 15-1 representative examples for various classes of organometallic compounds are depicted, together with values $\delta\,^{13}$C for the metal-coordinated carbon atoms.

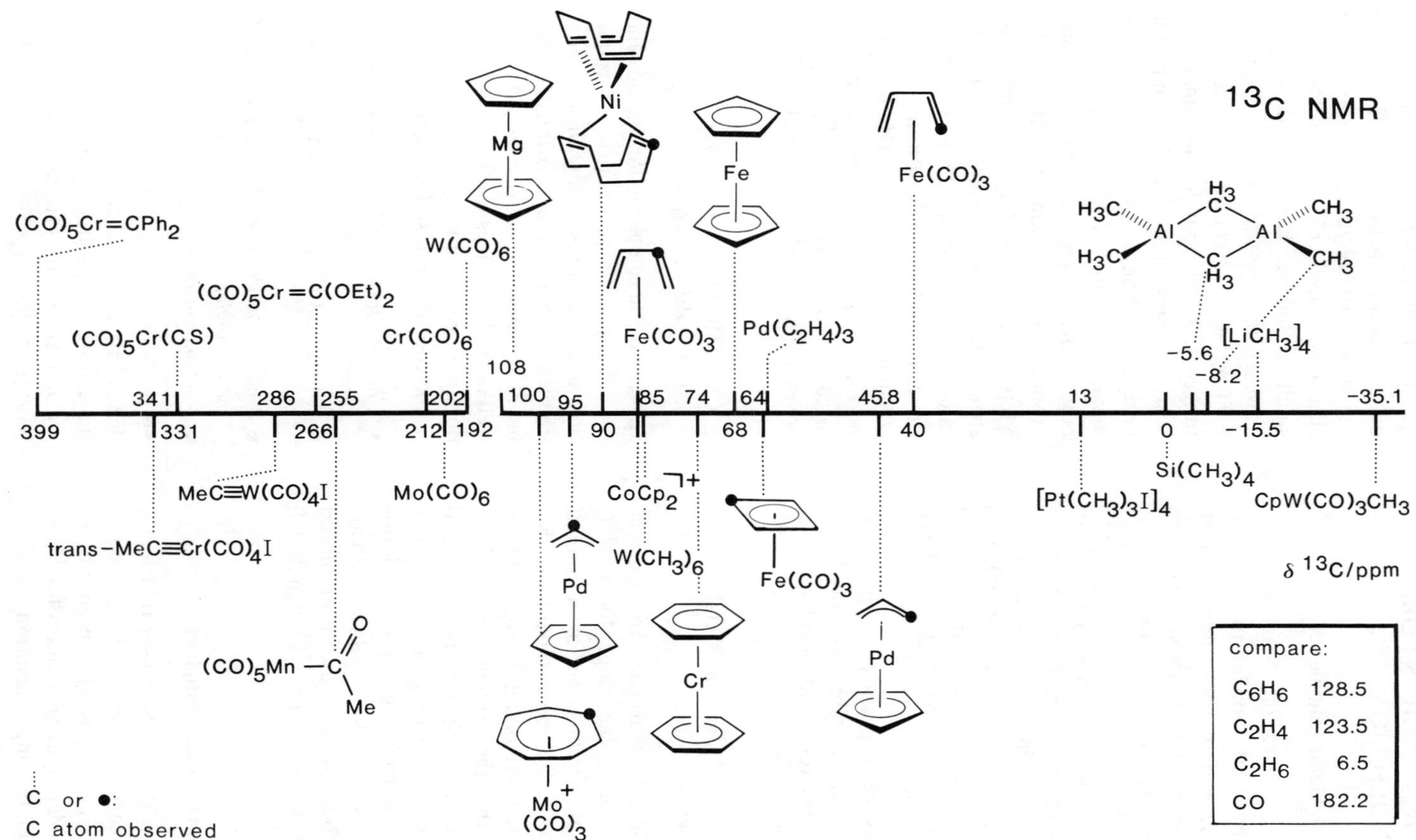

Figure 15-1: *Survey of* δ ^{13}C *values for metal-coordinated carbon atoms of various classes of organometallic compounds.*

1. CHEMICAL SHIFTS FOR σ-BONDED LIGANDS

a) σ-Alkyl Metal Complexes

The $\delta\,^{13}C$ values of α-carbon atoms which are directly bonded to the metal, strongly depend on the nature of the metal and the ancillary ligands. Positive as well as negative coordination shifts are observed. Positive values $\Delta\delta\,^{13}C$ (shifts to lower field) are primarily found for Zr, Hf, Nb and Ta. Note, however, that the coordination sphere of the metal fragment, which is σ-bonded to an alkyl group, has a dramatic effect on $\delta\,^{13}C$ (compare: $W(\mathbf{CH_3})_6$, $H_3C\mathbf{CH_3}$, $Cp(CO)_3W\mathbf{CH_3}$, p. 297). For the β-carbon atoms, small positive values $\Delta\delta\,^{13}C$ are usually encountered whereas for γ-carbon atoms, very small negative values $\Delta\delta\,^{13}C$ arise.

b) Carbene- and Carbyne Metal Complexes

$\delta\,^{13}C$ values for carbene- and carbyne complexes lie in the range $400 > \delta\,^{13}C > 200$ ppm. Deshielding of a similar magnitude is also encountered for carbenium ions. Therefore, it is reasonable to conclude that the carbene- and carbyne carbon atoms in their respective metal complexes bear a partial positive charge. Similarly to carbenium ions, π-donor substituents like OR or NR_2 at the carbene carbon atom cause pronounced high-field shifts of the ^{13}C signals (cf. p. 214).

c) Carbonyl Metal Complexes

Let it be stated at the outset that metal carbonyl carbon atoms can be difficult to detect. This is because the absence of directly bonded hydrogen atoms for ^{13}CO may result in long relaxation times T_1 (p. 24). Furthermore, in double-resonance experiments like $^{13}C\{^1H\}$, due to the large carbon-hydrogen distance, nuclear Overhauser enhancement (NOE) is small. Techniques have been developed, however, to cope with this situation.

$\delta\,^{13}C$ values for **metal carbonyls** lie in the ranges 150–220 ppm for terminal carbonyl groups and 230–280 ppm for bridging carbonyls. Within a group of metals, shielding of the carbonyl carbon nucleus increases with increasing atomic number [example: $Cr(CO)_6$ 212, $Mo(CO)_6$ 202, $W(CO)_6$ 192 ppm]. For structurally related complexes of the same central metal, correlations between the chemical shift $\delta\,^{13}CO$ and the force constant $k(CO)$ of the stretching vibration have been established: increasing force constants are paralleled by increased shielding (I. S. Butler, 1979 R).

2. CHEMICAL SHIFTS FOR π-BONDED LIGANDS

In **olefin metal complexes** for protons and for ^{13}C nuclei large negative coordination shifts $\Delta\delta$ (shifts to high field or low frequency) are observed. Various authors have attempted to rationalize $\Delta\delta\,^1H$ as well as $\Delta\delta\,^{13}C$ values on the basis of **one** dominant contribution. Possible sources of the coordination shifts include:

- *Changes in charge density on the ligand atoms*
- *Changes in the intra-ligand π-bond orders*
- *Changes in the hybridization states of metal-bonded carbon atoms*
- *Various anisotropy effects*

Since these effects are in part mutually dependent, controversial statements as to the nature of the dominating influence are not surprising (Trahanovsky, 1974; Norton, 1974).

Characteristic ^{13}C-NMR spectra are also displayed by **η^3-allyl- and η^4-diolefin complexes**. In allyl complexes, the terminal carbon nuclei generally are shielded more strongly ($80 > \delta\,^{13}C > 40$ ppm) as compared to the central carbons ($110 > \delta\,^{13}C > 70$ ppm). For allyl complexes of the d^{10} metals, shielding of terminal as well as of central carbon nuclei usually increases in the order Pd < Ni < Pt.
For olefin complexes as well, $\Delta\delta\,^{13}C$ values for the terminal carbon atoms C_1 exceed those of the central carbons C_2 up to three-fold. The ratio $\Delta\delta(C_1)/\Delta\delta(C_2)$ within a group of metals increases with increasing atomic number.
Example:

(η^4-butadiene)M(CO)$_3$	M = Fe	Ru	Os
$\Delta\delta(C_1)/\Delta\delta(C_2)$	1.47	1.65	1.68

A possible rationale for this trend is the more extensive rehybridization $sp^2 \rightarrow sp^3$ of the terminal carbon atoms in complexes of the heavier metals. The $^{13}C\{^1H\}$-NMR spectrum of (η^4-cycloheptatriene)Fe(CO)$_3$ may serve to illustrate the shielding difference between coordinated and non-coordinated olefinic carbon atoms as well as between terminal and central carbon atoms in the η^4-butadiene segment of a coordinated oligoolefin (Figure 15-2).
Free C_7H_8: $\delta\,^{13}C = 120(C_1),\ 127(C_2),\ 131(C_3),\ 28(C_7)$. Note the low intensity of the signal for the CO groups relative to C_7 (p. 298).

II. Coupling Constants $J(^{13}C, X)$

In fluid solution, by virtue of thermal molecular motion, through-space dipolar

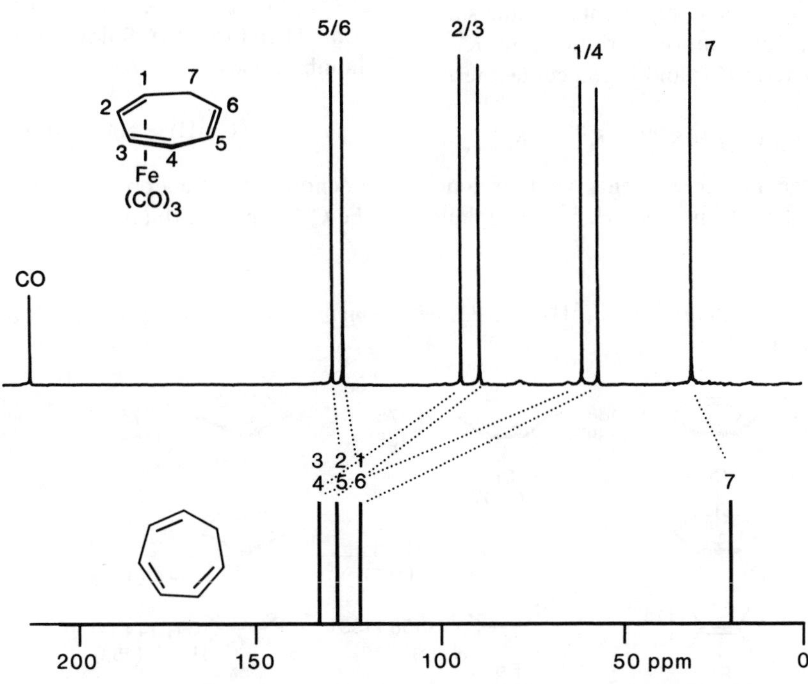

Figure 15-2: *$^{13}C\{^1H\}$-NMR spectra of* (η^4*-cycloheptatriene*)Fe(CO)$_3$ *and of free cycloheptatriene.*

interactions are averaged to zero and only the scalar spin-spin interactions, which are transmitted by the electrons, are observed.

Scalar coupling ^nJ (A, X) is independent of the external magnetic field; it is usually expressed in frequency units (Hz), n designating the number of bonds between the atoms A and X.

If scalar couplings between different kinds of nuclei are to be compared, rather than using the coupling constants ^nJ (A, X) themselves, it is more convenient to employ the reduced coupling constants ^nK (A, X). The latter reflect the influence of the electronic environment in a more direct way, being independent of the magnitudes and signs of the magnetogyric ratios γ. Reduced coupling constants are defined according to:

$$^nJ\,(\mathrm{A, X}) = h\,\frac{\gamma_\mathrm{A}}{2\pi}\,\frac{\gamma_\mathrm{X}}{2\pi}\,{}^nK\,(\mathrm{A, X}) \qquad (4)$$

They are usually represented as the sum of the contributions orbital term K^{OD}, dipole term K^{DD} and Fermi contact term K^{FC}:

$$^nK\,(\mathrm{A, X}) = K^{\mathrm{OD}} + K^{\mathrm{DD}} + K^{\mathrm{FC}} \qquad (5)$$

For the lighter elements void of lone pairs of electrons (e.g. ^{1}H, ^{13}C) coupling across **one** bond is dominated by the Fermi contact term K^{FC} and the following grossly simplified relation has been proposed (Pople, 1964):

$$^nK\,(\mathrm{A, X}) \approx K^{\mathrm{FC}} = 16\pi/9\;\mu_0\mu_\mathrm{B}^2(^3\Delta E)^{-1}\cdot[S_\mathrm{A}^2(0)\,S_\mathrm{X}^2(0)\,\mathrm{P}^2 s_\mathrm{A} s_\mathrm{X}] \qquad (6)$$

$^3\Delta E$ is the mean triplet excitation energy, $S_\mathrm{A}^2(0)$ and $S_\mathrm{X}^2(0)$ represent the electron densities in valence s orbitals at the nuclei A and X, respectively, and $\mathrm{P}^2 s_\mathrm{A} s_\mathrm{X}$ is the s bond order between A and X. In discussions of spin-spin coupling across several bonds and for atoms bearing lone pairs (e.g. ^{19}F) the terms K^{OD} and K^{DD} must also be considered.

Hence, the **coupling constants $^1J(^{13}\mathrm{C}, {}^1\mathrm{H})$** in principle provide hints as to the hybridization state of a certain carbon atom. Typical values of $^1J(^{13}\mathrm{C}, {}^1\mathrm{H})$ in free hydrocarbons are 125, 157, and 250 Hz for sp^3-, sp^2-, and sp hybridization. Therefore, the following rule of thumb applies:

$$^1J(^{13}\mathrm{C}, {}^1\mathrm{H}) \approx 500\,\mathrm{s} \quad (\mathrm{Hz})$$

s denoting the fraction of s character in the C–H bond orbital.

Coupling constants $^1J(^{13}\mathrm{C}, {}^1\mathrm{H})$ *for selected* π*-complexes and the respective free ligands* (*in brackets*):

H 168 (158)
Cr

H 175 (158)
Cr
$(CO)_3$

H 175 (159)
Fe

H 146 (157)
Pt
$(Ph_3P)_2$

H 161 (155)
H 158 (159)
H 169 (153)
Fe
$(CO)_3$

H 144 (155)
H 144 (159)
H 156 (153)
Zr
$(C_5H_5)_2$

Obviously, π-coordination may increase the coupling constant $^{1}J(^{13}C, ^{1}H)$, leave it unaffected or decrease it. Whereas a simple explanation for the increase of $^{1}J(^{13}C, ^{1}H)$, encountered for η^6-arene- and η^5-cyclopentadienyl complexes, is not available at present (Günther, 1980), *the coupling constants for olefin complexes may be discussed with reference to structural data. Thus, the decrease of $^{1}J(^{13}C, ^{1}H)$ upon complexation for ethylene in* $(Ph_3P)_2Pt(\eta^2\text{-}C_2H_4)$ *can be rationalized by a partial rehybridization from sp^2 to sp^3 which is in line with structural data (p. 259). Similarly, the decrease of $^{1}J(^{13}C, ^{1}H)$ for the terminal carbon atoms in* $(C_5H_5)_2Zr(\eta^4\text{-}C_4H_6)$ *(p. 263) is a result of rehybridization which diminishes the s orbital contribution to the* C–H *bond. The central carbon atoms apparently remain sp^2-hybridized. Note that butadiene complexes of the early transition metals – as judged from structural data – are better described as metallacyclopentenes (p. 263). For complexes of the type* $(CO)_3Fe(\eta^4$-*diene) the discussion is more involved since here, in addition to rehybridization, twisting along the terminal* C–C *bond arises (v. Philipsborn, 1976).*

While it is often possible to find *ex post facto* explanations for an observed trend in the coupling constants $^{1}J(^{13}C, ^{1}H)$, this parameter may not yet be regarded as an unambiguous diagnostic tool.

The **coupling constants $^{1}J(M, ^{13}C)$** contain information concerning the nature of the metal-carbon bond: for $M \overset{\sigma}{-} C$ bonds, considerably higher values $^{1}J(M, ^{13}C)$ are observed as compared to $M \overset{\pi}{-} C$ bonds. Again, the higher *s* character in the former bond type is responsible.

As a consequence of pronounced differences in the magnetogyric ratios γ_M, the parameters $^{1}J(M, ^{13}C)$ are spread over a wide range. Thus, in contrast to the small

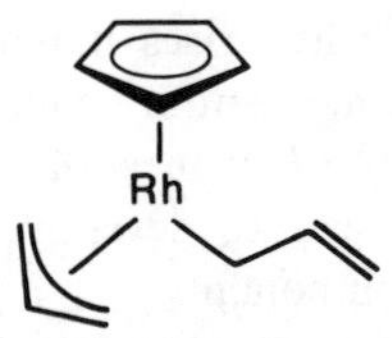

In the complex **(Cp)Rh(σ-allyl)(π-allyl)**, *the* $Rh \overset{\sigma}{-} C$ *bond is characterized by a value* $^{1}J(^{103}Rh, ^{13}C)$ of 26 Hz *whereas for the π-bonded allyl ligand* $^{1}J(^{103}Rh, ^{13}C)$ *amounts to* 14–16 Hz.

The ***α-ferrocenylcarbenium ion,*** *enriched in* ^{57}Fe, *features the coupling constant* $^{1}J(^{57}Fe, ^{13}C) = 1.5$ Hz *at the exocyclic carbon atom. Therefore, this carbon atom is probably not σ-bonded to the central metal since for* $Fe \overset{\sigma}{-} C$ *bonds, coupling constants of about* 9 Hz *are typical, whereas for* $Fe \overset{\pi}{-} C$ *bonds, values of* 1.5–4.5 Hz *are observed (Koridze, 1983).*

values $^{1}J(^{103}Rh, ^{13}C)$ and $^{1}J(^{57}Fe, ^{13}C)$ mentioned above, considerably larger coupling constants are encountered for other metals. Examples:
$^{1}J(^{183}W, ^{13}C) = 43$ Hz in WMe_6;
$^{1}J(^{183}W, ^{13}C) = 126$ Hz in $W(CO)_6$;
$^{1}J(^{195}Pt, ^{13}C_{Me}) = 568$ Hz,
$^{1}J(^{195}Pt, ^{13}C_{CO}) = 2013$ Hz
in *cis* $[Cl_2Pt(CO)Me]^-$.

Even in the case of high isotopic abundance, a coupling pattern due to $^{1}J(M, C)$ is only observable if the relaxation times T_1 of both nuclei engaged in spin-spin coupling are long relative to $1/2\pi J$ ($T_1 > 10 \cdot 1/2\pi J$). This conditon is usually fulfilled for interactions between $I = 1/2$ nuclei. If, on the other hand, T_1 for one of the coupling nuclei is very short ($T_1 \ll 1/2\pi J$), a sharp singlet appears instead of the multiplet. This situation typically arises if nuclei with a quadrupole moment are coupled to ^{13}C. T_1- and $1/2\pi J$ values of comparable magnitude lead to severely broadened multiplet components.

Geminal coupling constants $^2J(X, {}^{13}C)$ provide stereochemical information:

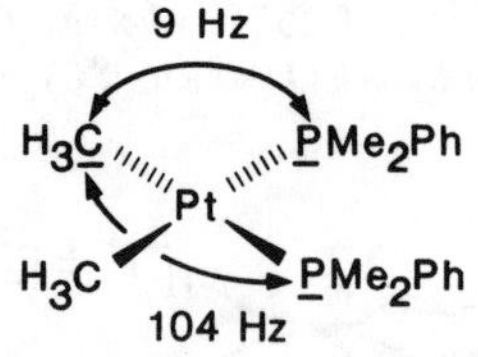

For complexes of heavy transition metals like Ru, Os, Rh, Ir *and* Pt, ***cis- and trans coupling*** *constants $^2J(X, {}^{13}C)$ strongly differ in magnitude. Example: in cis*-$Pt(Me)_2(PMe_2Ph)_2$ *the coupling $^2J_{trans}({}^{31}P, {}^{13}C)$ amounts to* 104 Hz, *the value $^2J_{cis}({}^{31}P, {}^{13}C)$ only to* 9 Hz.

Like the coupling constants $^3J({}^1H, {}^1H)$ in transition-metal complexes, **vicinal coupling $^3J(M, {}^{13}C)$ and $^3J({}^{13}C, {}^1H)$** is governed by the length of the central single- or double bond, the torsional angle φ and the electronegativity of substituents. The dependence on the torsional angle is cast in form of an equation of the Karplus type (see textbooks on NMR spectroscopy):

$$^3J(X, C) = A\cos^2\varphi + B\cos\varphi + C$$

A, B, and C are empirical parameters.

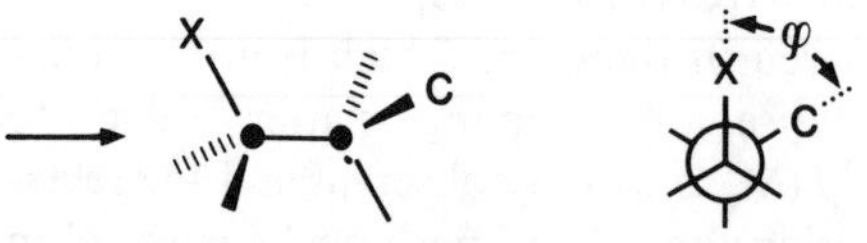

Thus ^{13}C-NMR studies of platinacycles and of ***exo-*** *and* ***endo-***norbornyl *trimethylstannanes have demonstrated that the coupling constants $^3J(M, C)$ in the* M–C–C–C *segments adhere to a Karplus equation.*

Example:

Coupling constants $^3J({}^{119}Sn, {}^{13}C)$ in Hz

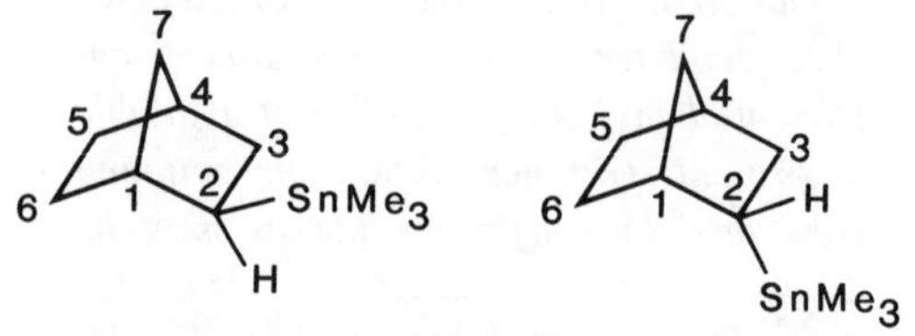

Sn–C(4) = 23.0	Sn–C(4) = 22.0	A = 25.2
Sn–C(6) = 69.0	Sn–C(6) = 34–38	B = –7.6
Sn–C(7) = 0.0	Sn–C(7) = 59.0	C = 30.4

The parameters A, B and C, derived from experimental data, provide a **Karplus relation** *for $^3J({}^{119}Sn, {}^{13}C)$ which can be applied in the study of structure and conformation of organotin compounds (Kuivila,* 1974).

In (η^4-1,3-diene)metal complexes, where the ligand is π- rather than σ-bonded, vicinal couplings $^3J(X, H)$, X = C or H, furnish information concerning ligand geometry. Upon coordination to a metal fragment ML_n, the vicinal ***cis-*** *and* ***trans*** *couplings $^3J({}^{13}C, {}^1H)$ and $^3J({}^1H, {}^1H)$ are reduced to different extents. This effect may be explained by a twisting of the terminal* C–C *bond such that the endo substituent moves away from the metal and the exo substituent towards the metal* (*cf.* p. 264). *In the case of* (4-*methyl*-1,3-*pentadiene*)$Fe(CO)_3$ *the vicinal coupling $^3J_{trans}({}^{13}C_{Me}, {}^1H)$ is reduced to such a degree that it becomes smaller than the coupling $^3J_{cis}({}^{13}C_{Me}, {}^1H)$ – a situation never encountered for free olefins (v. Philipsborn,* 1988).

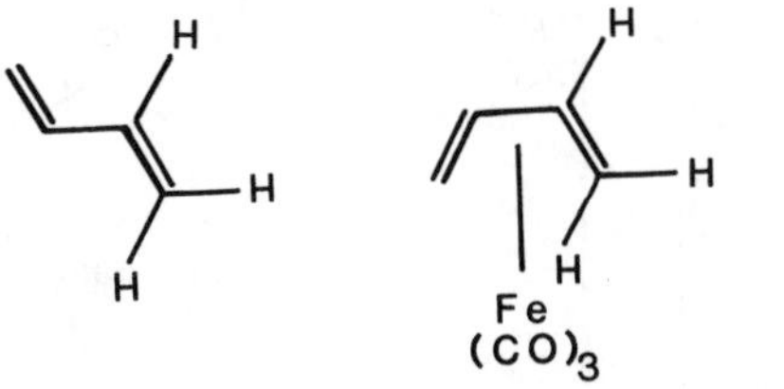

		Reduction
J_{cis}(H, H) = 10.17	$^3J_{cis}$(H, H) = 6.93 Hz	32%
J_{trans}(H, H) = 17.05	$^3J_{trans}$(H, H) = 9.33 Hz	45%

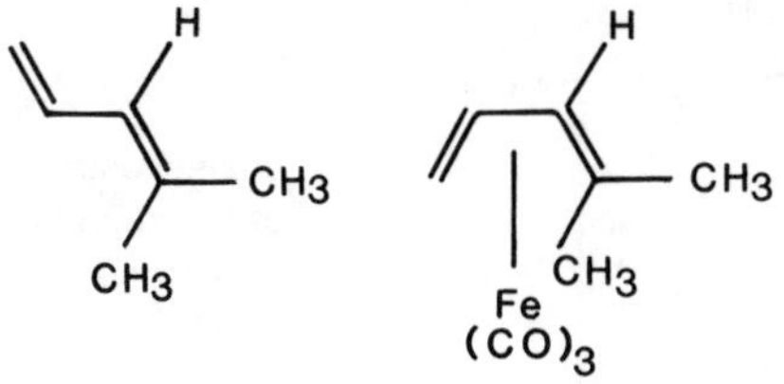

		Reduction
J_{cis}(C, H) = 7.0	$^3J_{cis}$(C, H) = 5.0 Hz	28%
J_{trans}(C, H) = 8.2	$^3J_{trans}$(C, H) = 3.5 Hz	57%

III. Influence of Dynamic Processes

^{13}C-NMR is not limited to the analysis of static molecular structures; it is well-suited for the study of reversible intra- and intermolecular processes and the determination of the respective activation energies, since the line-shapes of the resonance signals in these cases show temperature dependence. For dynamic processes in which an observed nucleus is transferred from the magnetic environment A to the environment B with a rate constant k, the following cases may be distinguished:

- *limit of slow exchange* ($k \ll \Delta\omega$)
- *coalescence region* ($k \approx \Delta\omega$)
- *limit of fast exchange* ($k \gg \Delta\omega$)

wherein $\Delta\omega$ is the difference in the resonance frequencies of the observed nuclei in the two magnetic environments. The quantitative interpretation of these spectra is possible without computer simulation only for the simplest case which involves a periodic change between two magnetic environments ("two-site exchange"). For this type of exchange process, which frequently occurs in practice, temperature-dependent rate constants are readily calulated with the aid of approximate formulae. Assuming equal populations of the sites A and B, the following expressions apply:

slow-exchange region

$$k = \pi(W_{1/2} - Wo_{1/2})$$

coalescence

$$k = \pi\,\Delta\omega/\sqrt{2} = 2.22\,\Delta\omega$$

fast-exchange region

$$k = (\pi\,\Delta\omega^2)/2\,(W_{1/2} - Wo_{1/2})$$

$Wo_{1/2}$ represents the half-width in the absence of exchange (e.g. width of the solvent signal in Hz), and $W_{1/2}$ the half-width of the signal broadened by exchange. If the spectrum is recorded at different temperatures, the free enthalpy of activation ΔG^{+} as well as ΔH^{+} and ΔS^{+} can be calculated by means of the Eyring formula. The activation enthalpy ΔH^{+} is a measure of the height of the energy barrier, the activation entropy ΔS^{+} reflects the change in geometry upon approaching the transition state.

In many exchange processes, several magnetically active nuclei (e.g. ^{1}H, ^{13}C, ^{31}P) are affected simultaneously. The appropriate choice of the nucleus to be observed by NMR then simplifies the analysis of the process considerably. Alternatively, in order to check for internal consistency, an investigation of the exchange process using all NMR active nuclei is advisable.

Example: The cation $[(C_5Me_5)Rh(2,3$-*dimethylbutenyl*$)]^{+}$ *undergoes an interconversion between the enantiomeric forms* A

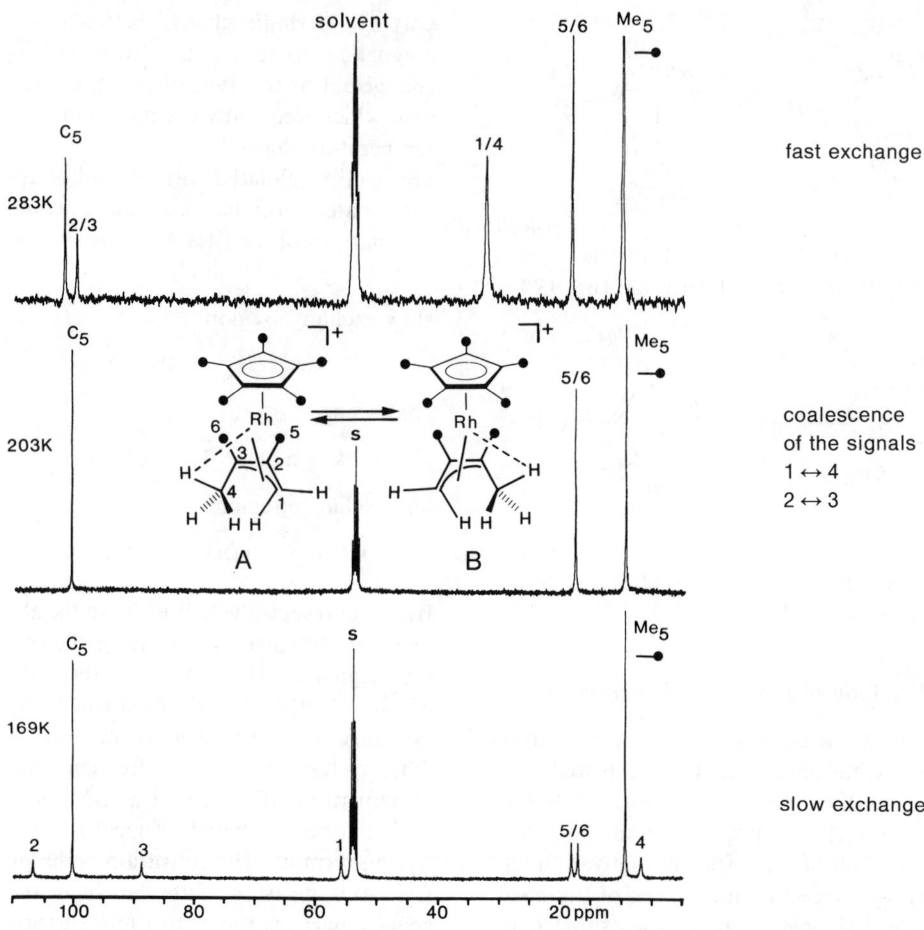

Figure 15-3: $^{13}C\{^1H\}$-*NMR spectrum of* $[(C_5Me_5)Rh(2,3\text{-}dimethylbutenyl)]BF_4$ in CD_2Cl_2 at 100.8 MHz.

and B (*p.* 264). *In* **^{1}H-*NMR*** *this fluxionality represents a "five-site exchange" because the rotation of the terminal methyl group, which is fixed to the metal by means of an agostic interaction* C−H⇀M, *must be taken into account. In* **^{13}C-*NMR***, *on the other hand, the simpler "two-site exchange" case applies* (1↔4, 2↔3, 5↔6) (*Figure* 15-3). *For the exchange* A↔B *the following parameters where derived:* $\Delta G^{\ddagger}_{300\,K} = 38.9$ kJ/mol, $\Delta H^{\ddagger} \approx 29$ kJ/mol, *and* $\Delta S^{\ddagger} \approx -35$ J/mol K (*Salzer, v. Philipsborn*, 1987).

^{1}H-NMR

Proton magnetic resonance for diamagnetic organotransition metal compounds is observed in the chemical shift range $25 > \delta\,^1H > -40$ ppm. Paramagnetic organometallics may give rise to chemical shifts of several hundred ppm [Example: $(C_6H_6)_2V^{\cdot}$, $\delta\,^1H = 290$ ppm]. In the

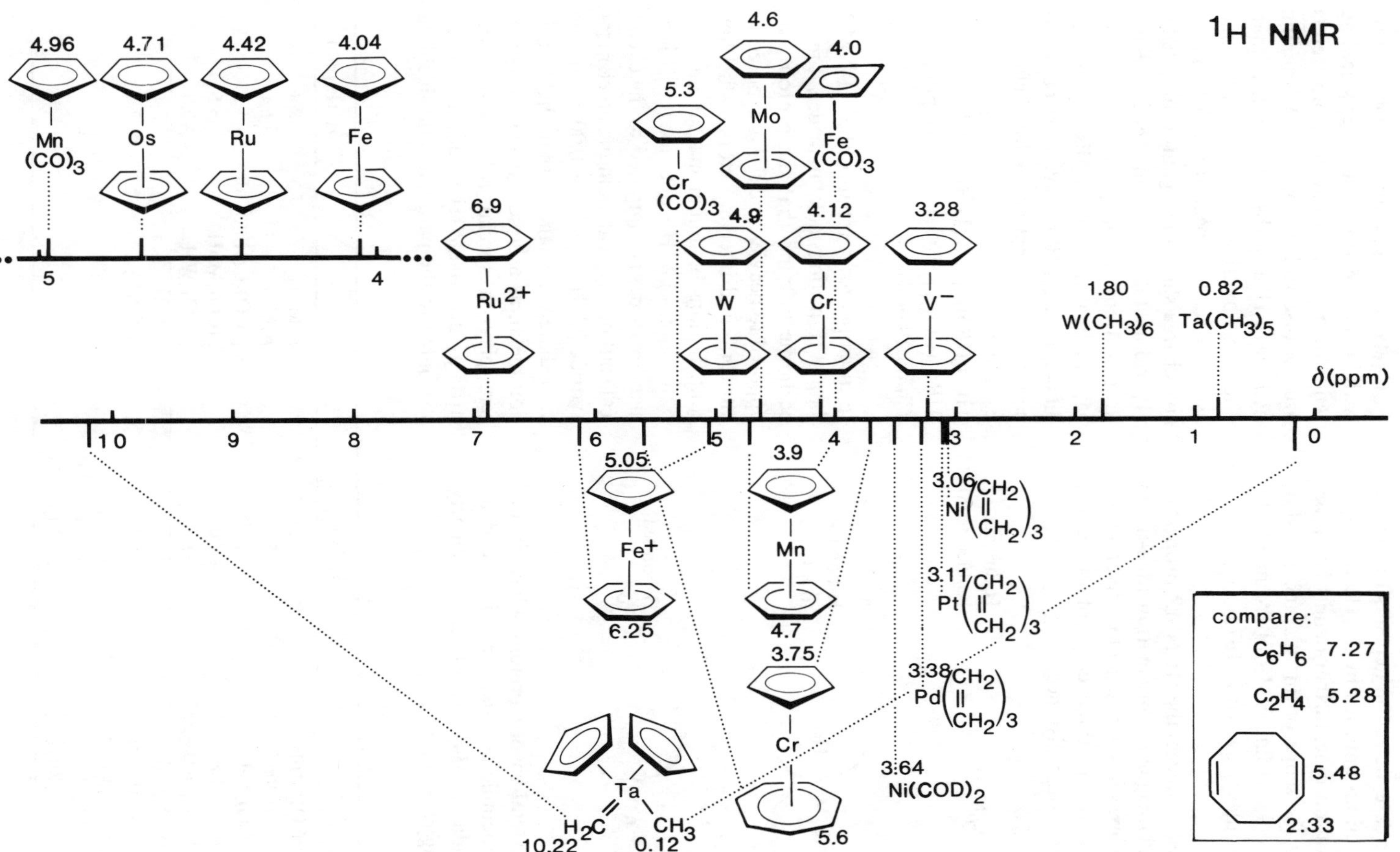

Figure 15-4: $\delta\,^{1}H$ *values for diamagnetic organotransition-metal compounds.*

latter case, detection depends on certain restrictions imposed by the electron spin relaxation time, further details being beyond the scope of this text (cf. La Mar, Horrocks, Holm, NMR of Paramagnetic Molecules, New York, 1973).

In discussions of the ^{1}H-NMR properties of **diamagnetic organotransition-metal complexes** it is convenient to differentiate the following three cases of increasing metal-proton distance:

Ⓐ $L_nM{-}H$ — H directly bonded to M

Ⓑ $L_nM{-}C{-}H$ — H bonded to η-C

Ⓒ $L_nM{-}C$ … $C{-}H$ — H bonded to C

Ⓐ **Metal-bonded protons** which in organometallic chemistry are regarded as hydridic (p. 234), experience particularly strong shielding:

For related compounds, across a row with increasing atomic number shielding tends to increase; smooth group trends are not discernible, however. A bridging hydride (μ-H) absorbs at higher field than a terminal hydride. Positive charge causes resonance to appear at low field, negative charge shifts the signal to high field. Hydridometal clusters in which the hydrogen atom is buried in the cavity of M_n-polyhedron (p. 406), display totally different ^{1}H-NMR behavior, in that resonance occurs at extremely low field.
Examples:
$[HCo_6(CO)_{15}]^-$, $\delta\,^1H = +23.2$;
$[HRu_6(CO)_{18}]^-$, $\delta\,^1H = +16.4$ ppm
(Chini, 1979 R).

In attempts to rationalize the ^{1}H-NMR data of transition-metal hydrides, it must be borne in mind that the proton – as opposed to heavier nuclei – is exposed to intra (σ_i)- and inter (σ_{ij}) atomic shielding contributions of similar magnitude.
If the range of $\delta\,^1H$ values for transition-metal hydrides is compared to the chemical shift of a typical main-group element hydride ($[AlH_4]^-$, $\delta\,^1H = 0.00$ ppm), one immediately recognizes that the transition metal, as a bond partner to hydrogen, plays a special role. Hence, the strongly negative $\delta\,^1H$ values have been attributed to the paramagnetic shielding

		$\delta(^1H)$/ppm		$\delta(^1H)$/ppm
	$(CO)_5MnH$	−7.5	Cp_2MoH_2	−8.8
	$(CO)_5ReH$	−5.7	$Cp_2MoH_3^+$	−6.1
	$(CO)_4FeH_2$	−11.1	$Cp(CO)_3CrH$	−5.5
	$(CO)_4RuH_2$	−7.6	$Cp(CO)_3MoH$	−5.5
	$(CO)_4OsH_2$	−8.7	$Cp(CO)_3WH$	−7.3
	$(CO)_4CoH$	−10.7	Cp_2FeH^+	−2.1
	trans-$(Et_3P)_2ClPtH$	−16.8	Cp_2RuH^+	−7.2
μ-H	$(CO)_{10}Cr_2H^-$	−19.5 (p. 235)		
μ-H	$(CO)_{11}Fe_3H^-$	−15.0 (p. 234)		
μ-H	$(CO)_{24}Rh_{13}H_3^{2-}$	−29.3 (p. 410)		

term σ_i^{para} of the neighboring transition metal (Buckingham, 1964). A characteristic feature of transition-metal complexes is the existence of low-lying, electronically excited states which are admixed to the ground state under the influence of the external magnetic field. In this way, the field induces a magnetic moment at the metal which exerts a deshielding effect at the central atom and – due to the distance relationship – a shielding effect at the hydride ligand (σ_{ij}, non-local effect).

Ⓑ **Protons** which are **fixed to metal-bonded carbon atoms** display a much smaller span of chemical shifts. The coordination shifts usually lie in the range $1 < \Delta\delta\,^{1}H < 4$ ppm; they are always negative.

The gradation of the coordination shifts for a ligand of low symmetry can be read off the ^{1}H-NMR spectrum of (η^4-cis-1,3-pentadiene) Fe(CO)$_3$:

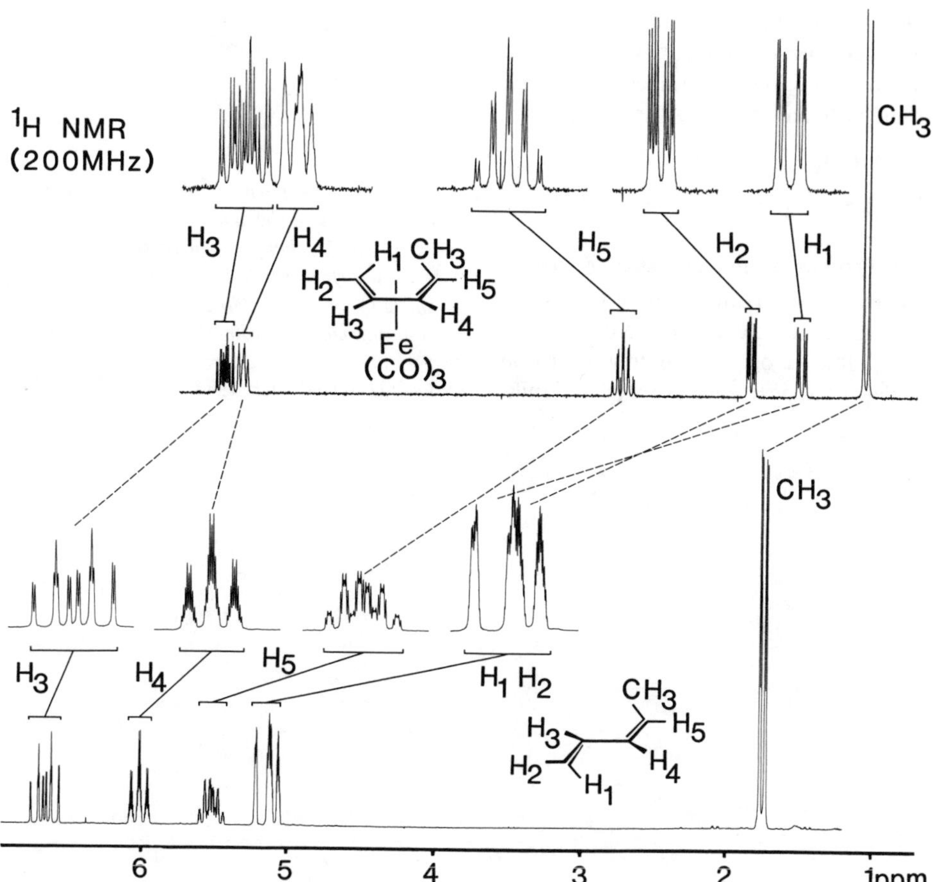

^{1}H-*NMR spectra of* (η^4-*cis*-1,3-*pentadiene*)Fe(CO)$_3$ *and of free* 1,3-*pentadiene in* CDCl$_3$ *at* 200 MHz.

$\Delta\delta\,^{1}$H *decreases in the order terminal, endo* (H$_1$) > *terminal, exo* (H$_{2,5}$) > *central* (H$_{3,4}$)

The coordination shifts $\Delta\delta\,^1H$ of η-CH units are ruled by local effects (partial charge, rehybridization) and by non-local effects (neighboring group anisotropy of the metal, coordination-induced perturbation of diamagnetic ring currents). A quantitative assessment of these contributions is not yet possible. Notwithstanding, the coordination shift $\Delta\delta\,^1H$ is of considerable heuristic value.

The **coupling constants $J(^1H, ^1H)$** – like the values $J(^{13}C, ^1H)$ – yield information with regard to intra-ligand bond orders, hybridization states of the carbon atoms and therefore ligand geometry. Occasionally, **coupling constants $^2J(M, ^1H)$** are also resolved; they characterize the type of metal-ligand bonding. Examples:

$$W(CH_3)_6,\ ^2J(^{183}W, ^1H) = 3\ Hz;$$
$$(\eta^2\text{-}C_2H_4)_3Pt,\ ^2J(^{195}Pt, ^1H) = 57\ Hz.$$

(c) **Protons at non-coordinated carbon atoms**, not surprisingly, show very small coordination shifts $\Delta\delta\,^1H$. In contrast to their influence on the shielding of carbon nuclei, non-local effects make a significant contribution to the shielding of protons. Specifically, non-local contributions to $\Delta\delta\,^1H$ are expected in cases where metal coordination affects the diamagnetic ring current which governs the shielding in the periphery of arenes.

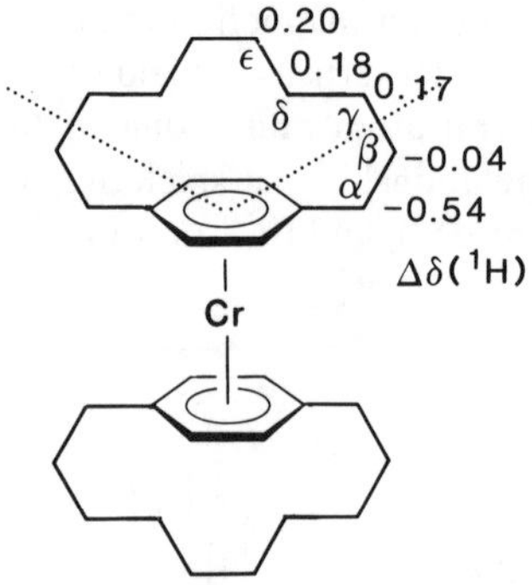

Bis(1,4-decamethylene-η^6-benzene)chromium may serve as an example. The area in space, where $\Delta\delta\,^1H$ changes sign corresponds to the conical surface which, for free arenes, separates shielding and deshielding regions. The conclusion that metal coordination partially quenches the diamagnetic ring current therefore suggests itself (Elschenbroich, 1984).

15.4 Complexes of the Cyclic π-Perimeters C_nH_n

Cyclic conjugated ligands $C_nH_n^{+,0,-}$ are known to occur in five different classes of compounds:

I Sandwich complexes

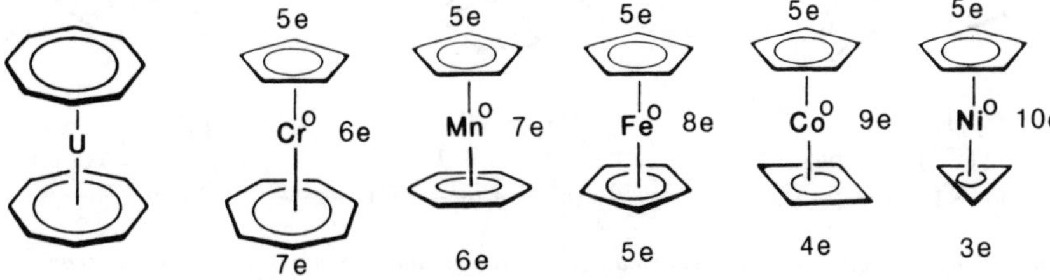

If an oxidation state of zero is assigned to the central metal atom, the rings in this series become 7- to 3-electron ligands. The combinations shown above result in 18 VE complexes. They have all been synthesized. The 18 VE rule cannot, however, be applied to lanthanoid and actinoid sandwich complexes (*cf.* p. 364).

η^5-$C_5H_5^-$ 6e

Mn^+ 6e Mn^I

η^6-C_6H_6 6e

An alternative approach is to assign the complexed ligand the same charge as in its most stable uncoordinated form. The formal oxidation state of the metal is then derived by balancing the charge.

II Half-sandwich complexes

III Multidecker sandwich complexes

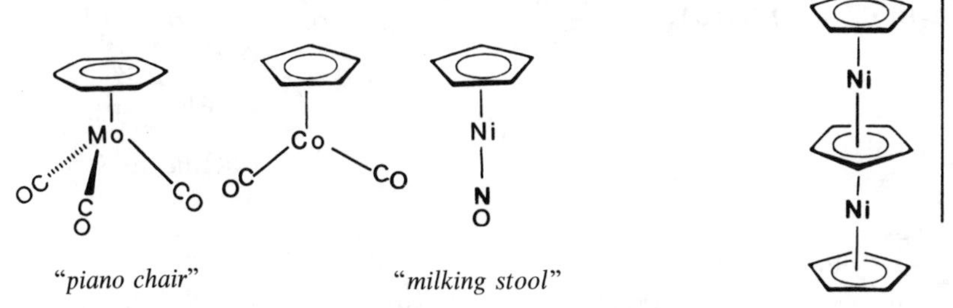

"*piano chair*" "*milking stool*"

IV Complexes with tilted sandwich structure

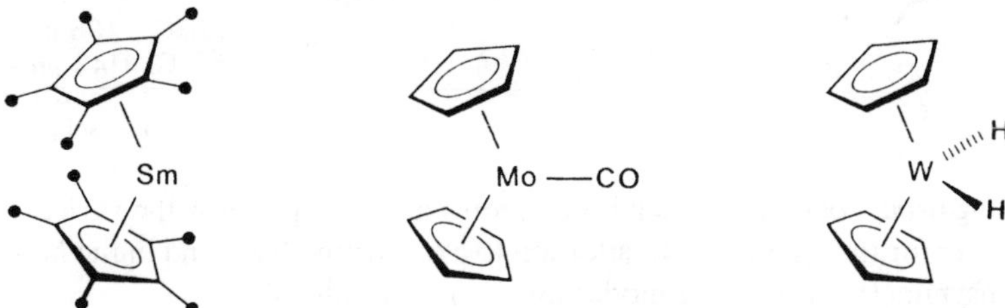

V Complexes with more than two C_nH_n ligands

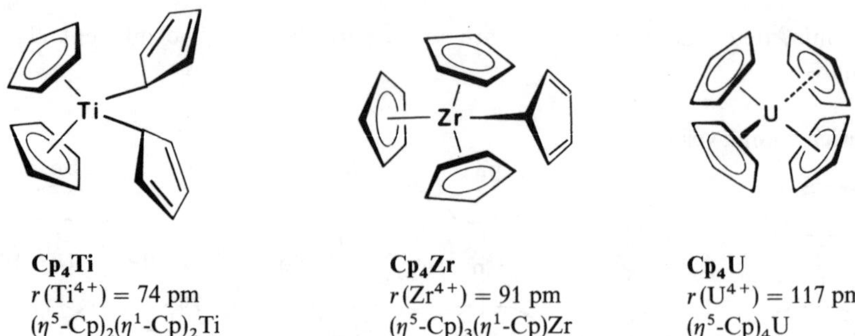

Cp_4Ti
$r(Ti^{4+}) = 74$ pm
$(\eta^5\text{-}Cp)_2(\eta^1\text{-}Cp)_2Ti$

Cp_4Zr
$r(Zr^{4+}) = 91$ pm
$(\eta^5\text{-}Cp)_3(\eta^1\text{-}Cp)Zr$

Cp_4U
$r(U^{4+}) = 117$ pm
$(\eta^5\text{-}Cp)_4U$

Four η^5-Cp ligands can only be accomodated by the actinoid metals. This is an outcome of size considerations, the availability of 5 f orbitals and the readiness of the actinoids to engage in covalent bonding (p. 422).

15.4.1 $C_3R_3^+$ as a Ligand

A cyclopropenyl cation $Ph_3C_3^+$, the most simple aromatic system obeying the Hückel rule [$(4n + 2)$ π-electrons, $n = 0$], was first synthesized by Breslow in 1957, the parent ion $C_3H_3^+$ appearing in 1967. The number of η^3-cyclopropenyl complexes known to date is still relatively small. Furthermore, since the overlap of the small π-perimeter with the M(d) orbitals is unfavorable, only M(3d) – η^3-C_3R_3 complexes are known.

Ph_3C_3Br + $Ni(CO)_4$ $\xrightarrow[\text{OA}]{\text{reflux}}$ $[(\eta^3\text{-}C_3Ph_3)Ni(CO)(\mu\text{-}Br)]_2$

(Kettle, 1965)

$(\eta^3\text{-}C_3Ph_3)NiBr(py)_2$ $\xrightarrow[C_6H_6,\ RT]{\text{TlCp}}$ $(\eta^5\text{-}Cp)Ni(\eta^3\text{-}C_3Ph_3)$

d(Ni-C) 196 pm
d(C-C) 143 pm

(Rausch, 1970)

An interesting aspect of cyclopropenyl transition-metal complexes is the variety of coordination modes for the C_3 moiety. In addition to axial symmetry found in the Ni–C_3Ph_3 complex, unsymmetrical bonding modes are also encountered.

One short, two long M – C bonds (Sacconi, 1980):

$$(C_2H_4)Ni(PPh_3)_2 + C_3Ph_3PF_6 \xrightarrow{MeOH}$$

One long, two short M – C bonds (Weaver, 1973):

$$(C_2H_4)Pt(PPh_3)_2 + C_3Ph_3PF_6 \xrightarrow{CH_2Cl_2}$$

The difference in bond lengths, established for $(Me_3P)_2Cl(CO)Ir(Ph_3C_3)$, may be regarded as a continuation of the trend discernible for the Ni- and Pt complexes above (Weaver, 1972):

$$trans\text{-}Ir^{I}(PMe_3)_2(Cl)CO + Ph_3C_3PF_6 \longrightarrow$$

This structural gradation from the (η^3-cyclopropenyl)Ni complex to the iridacyclobutane structure is interesting since it mimics the various stages of strained three-membered ring opening (cf. Chapter 17.2).

15.4.2 C_4H_4 as a Ligand

The preparation of free cyclobutadiene by photolytic methods in low-temperature matrices (8 – 20 K) is a relatively recent accomplishment. The stabilization of C_4H_4 by coordination to a transition metal had, however, been predicted on theoretical grounds by Longuet-Higgins and Orgel in 1956, the first synthesis of a cyclobutadiene transition-metal complex being carried out by Criegee in 1959.

PREPARATION

1. Dehalogenation of Cyclobutene Dihalides

$$\text{Me}_4\text{C}_4\text{Cl}_2 + \text{Ni(CO)}_4 \xrightarrow[-\text{NiCl}_2]{} [(\text{C}_4\text{Me}_4)\text{NiCl}(\mu\text{-Cl})]_2$$

—● = —CH_3

d(Ni–C) = 199–205 pm, d(C–C) = 140–145 pm
air-stable, red-violet crystals (*Criegee*, 1959)

$$\text{C}_4\text{H}_4\text{Cl}_2 + \text{Fe}_2(\text{CO})_9 \xrightarrow[-\text{FeCl}_2]{} (\text{C}_4\text{H}_4)\text{Fe(CO)}_3 + \text{FeCl}_2$$

planar C_4 *ring*, d(C–C) = 146 pm, *diamagnetic*,
mp 26 °C (*Pettit*, 1965)

2. Dimerization of Acetylenes

$$\text{CpCo(CO)}_2 \xrightarrow[\Delta T]{\text{RC}\equiv\text{CR}} \text{CpCo}(\text{C}_4\text{R}_4) + \text{C}_6\text{R}_6$$

$$\text{Na}_2\text{PdCl}_4 + 2\ \text{PhC}\equiv\text{CPh} \xrightarrow{\text{EtOH}} [(\text{C}_4\text{Ph}_4)\text{PdCl}_2]_2$$

3. Ligand Transfer

$$[(\text{C}_4\text{Ph}_4)\text{PdBr}_2]_2 + \text{Fe(CO)}_5 \longrightarrow (\text{C}_4\text{Ph}_4)\text{Fe(CO)}_3$$

4. Ring Contraction of Metallacyclopentadienes

$\xrightarrow{NiBr_2}$ $[(C_4Ph_4)NiBr_2]_2$ + 1/n $(Me_2Sn)_n$

1. Li
2. $NiBr_2$

first binary cyclobutadiene complex (*Hoberg*, 1978)

5. Decarboxylation of photo-α-Pyrone

hν

+CpCo(CO)$_2$ − CO_2

− CO_2 + $Fe(CO)_5$

d(Co-C) 204 139

d(Co-C) 197 143

$Fe(CO)_3$

(Davis, 1976)

STRUCTURE AND BONDING

While free cyclobutadiene, according to spectroscopic evidence, is rectangular (Schweig, 1986 R), η^4-cyclobutadiene in transition-metal complexes clearly adopts a square structure. Square cyclobutadiene should possess two unpaired electrons (single occupancy of the two degenerate MO's ψ_2 and ψ_3). The diamagnetism of η^4-C_4H_4 complexes is explained by the interaction of these singly-occupied π MO's with singly-occupied metal orbitals of matching symmetry (formation of two covalent bonds).

Metal-ligand interactions in cyclobutadiene complexes:

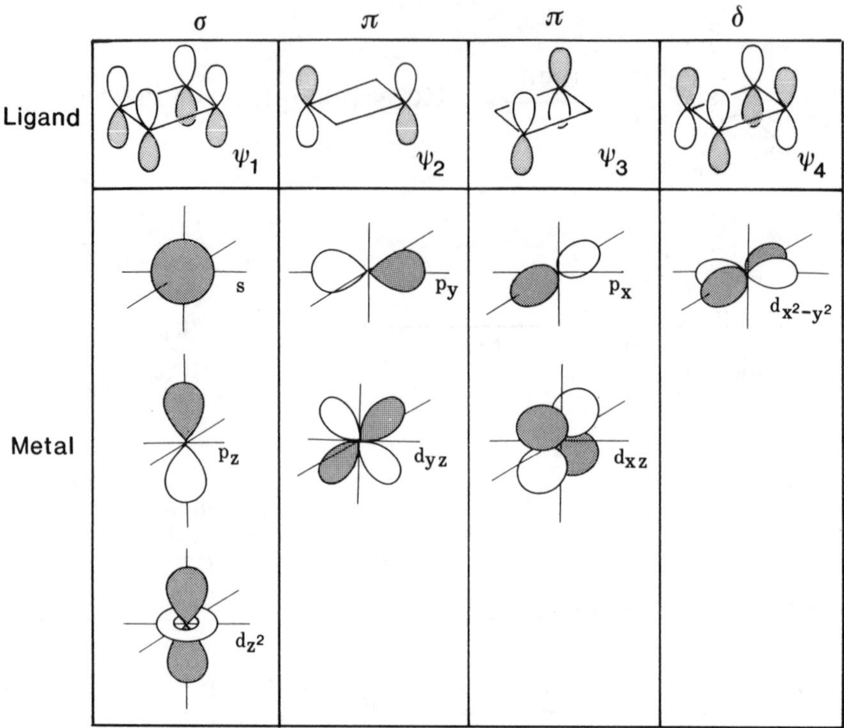

REACTIVITY OF CYCLOBUTADIENE COMPLEXES

The most important feature of $(\eta^4\text{-}C_4H_4)Fe(CO)_3$ is its similarity to aromatic compounds; this is exemplified by the ease with which it undergoes electrophilic substitution at the ring to form $C_4H_3RFe(CO)_3$ (R = COMe, COPh, CHO, CH_2Cl etc). The following mechanism has been proposed for this reaction:

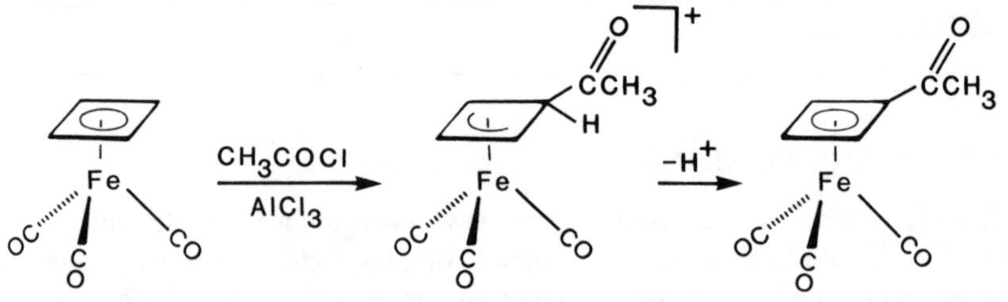

α-CH_2X-groups attached to $C_4H_4Fe(CO)_3$ are easily solvolyzed. Apparently, coordination to iron results in a **stabilization of the α-carbenium ion**; this effect is also attributed to the ferrocenyl group (Watts, 1979 R). The intermediate cation may be isolated as a PF_6^- salt:

CH_2Cl Fe $(CO)_3$ $\xrightarrow{SbCl_5}$ { $\overset{+}{C}H_2$ Fe $(CO)_3$ ⟷ CH_2 Fe $(CO)_3$ } $SbCl_6^-$ $\underset{-H_2O}{\overset{+H_2O}{\rightleftharpoons}}$ OH CH_2 Fe $(CO)_3$ + $HSbCl_6$

In organic synthesis, $C_4H_4Fe(CO)_3$ serves as a **source of free cyclobutadiene.** After low-temperature oxidation, the liberated C_4H_4 reacts with alkynes to give Dewar-benzene derivatives, while its treatment with *p*-quinones opens up an elegant route for the synthesis of cubane (Pettit, 1966):

HC≡CR R OC Fe CO C O Ce^{IV} Br O O Br Br O O Br hν Br O O Br KOH H_2O COOH COOH t-Bu-perester thermolysis

15.4.3 $C_5H_5^-$ as a Ligand

HISTORICAL BACKGROUND

1901 Thiele: Synthesis of KC_5H_5 from potassium and C_5H_6 in benzene.

1951 Miller, Tebboth, Tremaine: Formation of $Fe(C_5H_5)_2$ in the reaction of C_5H_6 vapor with freshly reduced iron at 300 °C.

$$2\,C_5H_6 + Fe \longrightarrow Fe(C_5H_5)_2 + H_2$$

1951 Kealy, Pauson:

$$3\,C_5H_5MgBr + FeCl_3 \longrightarrow Fe(C_5H_5)_2 + 1/2\,C_{10}H_{10} + 3\,MgBrCl$$

Their actual goal was the synthesis of fulvalene:

$$2\ C_5H_5MgBr \xrightarrow{FeCl_3} 2\ C_5H_5^{\bullet} \longrightarrow C_{10}H_{10} \xrightarrow[-H_2]{FeCl_3} C_{10}H_8$$

Fulvalene $\equiv C_{10}H_8$

Structures which were originally proposed: (C_5H_5)–Fe–(C_5H_5) (σ-bonded, H on each ring carbon), $[C_5H_5^- Fe^{2+} C_5H_5^-]$

1952 E. O. Fischer: "**Double-cone structure**"

based on:

- *X-ray structural analysis (compare the electron density contour lines)*
- *diamagnetism*
- *chemical behavior*

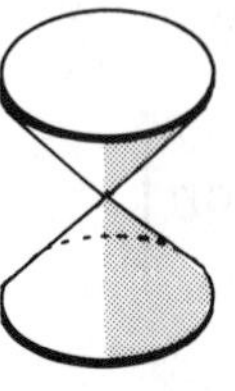

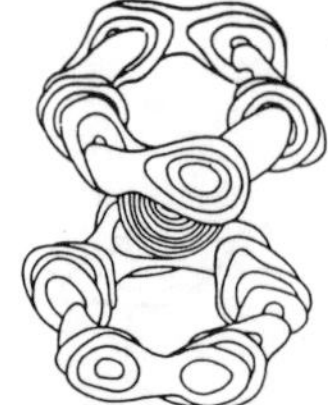

1952 G. Wilkinson, R. B. Woodward: "**Sandwich structure**"

based on:

- *IR spectroscopy*
- *diamagnetism*
- *dipole moment* = 0

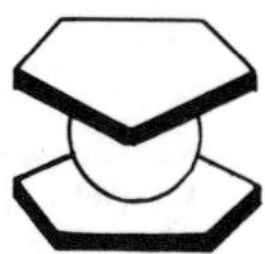

Woodward: The cyclopentadienyl rings of $Fe(C_5H_5)_2$ *are amenable to electrophilic substitutions. This similarity to the aromatic behavior of benzene led to the trivial name* ***ferrocene****, later being extended to the designation* ***metallocenes*** *for compounds* $M(\eta^5\text{-}C_5H_5)_2$ *in general.*

15.4.3.1 Binary Cyclopentadienyl Transition-Metal Complexes

PREPARATION

1. Metal Salt + Cyclopentadienyl Reagent

Dicyclopentadiene has first to be "cracked" in a retro-Diels-Alder reaction to give monomeric C_5H_6. Cyclopentadiene is a weak acid, $pK_a \approx 15$; it can be deprotonated by strong bases or by alkali metals. Cyclopentadienylsodium (NaCp) is the most common reagent for the introduction of cyclopentadienyl ligands.

$$\text{dicyclopentadiene} \xrightarrow{180°} 2\ C_5H_6 \xrightarrow{Na} 2\ NaC_5H_5 + H_2$$

$$MCl_2 + 2\,NaC_5H_5 \longrightarrow (C_5H_5)_2M \quad M = V, Cr, Mn, Fe, Co$$
Solvent = THF, DME, NH_3 (l)

$$Ni(acac)_2 + 2\,C_5H_5MgBr \longrightarrow (C_5H_5)_2Ni + 2\,acacMgBr$$

$$MCl_2 + (C_5H_5)_2Mg \longrightarrow (C_5H_5)_2M + MgCl_2$$

Apart from being a ligand source, NaC_5H_5 can also act as a reducing agent:

$$CrCl_3 + 3\,NaC_5H_5 \longrightarrow (C_5H_5)_2Cr + 1/2\,C_{10}H_{10} + 3\,NaCl$$

In many cases, treatment of NaC_5H_5 with metal (4*d*, 5*d*) salts does not give MCp_2 complexes, but Cp-metal hydrides or complexes containing σ-bonded cyclopentadienyl rings:

$$TaCl_5 \xrightarrow{NaCp} (\eta^5\text{-}C_5H_5)_2TaCl_3 \xrightarrow{NaCp} (\eta^5\text{-}C_5H_5)_2Ta(\eta^1\text{-}C_5H_5)_2$$

$$ReCl_5 \xrightarrow{NaCp} (\eta^5\text{-}C_5H_5)_2ReH$$

2. Metal + Cyclopentadiene

$$M + C_5H_6 \longrightarrow MC_5H_5 + 1/2\,H_2 \quad M = Li, Na, K$$

$$M + 2\,C_5H_6 \xrightarrow{500\,°C} (C_5H_5)_2M + H_2 \quad M = Mg, Fe$$

The less electropositive metals react only at elevated temperatures or as metal atoms in the cocondensation method (cf. p. 255).

3. Metal Salt + Cyclopentadiene

$$(Et_3P)Cu(t\text{-}BuO) + C_5H_6 \longrightarrow (\eta^5\text{-}C_5H_5)Cu(Et_3P) + t\text{-}BuOH$$

In this process, an auxiliary base is required if the basicity of the salt anion is insufficient to deprotonate cyclopentadiene:

$$Tl_2SO_4 + 2\,C_5H_6 + 2\,OH^- \xrightarrow{H_2O} 2\,TlC_5H_5 + 2\,H_2O + SO_4^{2-}$$

$$FeCl_2 + 2\,C_5H_6 + 2\,Et_2NH \longrightarrow (C_5H_5)_2Fe + 2\,[Et_2NH_2]Cl$$

In other cases, a reducing agent is needed:

$$RuCl_3(H_2O)_x + 3\,C_5H_6 + 3/2\,Zn \xrightarrow{EtOH} (C_5H_5)_2Ru + C_5H_8 + 3/2\,Zn^{2+}$$

ELECTRONIC STRUCTURE AND BONDING IN THE COMPLEXES $(C_5H_5)_nM$

The properties and bonding types of (cyclopentadienyl)metal compounds can vary over a wide range. In the following table, main-group compounds are also included.

Character	Bonding	Properties	Examples
Ionic	Ionic lattice $M^{n+}[C_5H_5^-]_n$ similar to the halides MX_n	Highly reactive towards air, water and other compounds with active hydrogen, not sublimable	$n = 1$: alkali metals $n = 2$: heavy alkaline earth metals $n = 2, 3$: lanthanoids
Intermediate		Partially sensitive to hydrolysis (exception: TlCp), sublimable	$n = 1$: In, Tl $n = 2$: Be, Mg, Sn, Pb, Mn, Zn, Cd, Hg
Covalent	molecular lattice π-MO(C_5H_5) $\downarrow$ M (*s*, *p*, *d*) and π^*-MO(C_5H_5) $\uparrow$ M (*d*)	Only partially air-sensitive, in general stable to hydrolysis, sublimable	$n = 2$: (Ti), V, Cr, (Re), Fe, Co, Ni, Ru, Os, (Rh), (Ir) $n = 3$: Ti $n = 4$: Ti, Zr, Nb, Ta, Mo, U, Th

The sequence of the chemically relevant **frontier orbitals** which follows from MO-calculations on metallocenes with axial symmetry is already apparent from a simple electrostatic model. This is based on the presence of a crystal field, generated by two negatively charged rings, which splits the energy levels of the metal *d* orbitals as a result of disparate electrostatic "repulsions" between the metal *d* electrons and the charge loops, intended to represent the ligands.

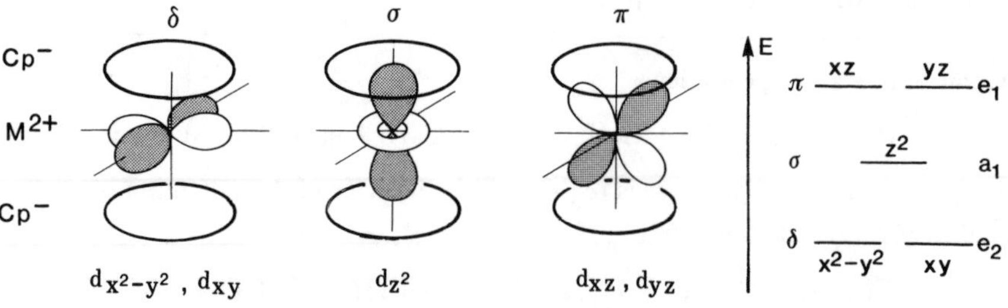

For ferrocene (Fe^{II}, d^6), the (correct) electronic configuration $\ldots(e_2)^4(a_1)^2(e_1)^0$ in the frontier region is thus obtained. This crystal-field picture does not, however, do justice to the nature of chemical bonding in metallocenes which, in the case of transition metals, is essentially covalent.

The MO treatment of the cyclopentadienyl-transition-metal bond is based on the principle outlined for olefin complexes (Chapter 15.1). The π MO's of C_5H_5 may be classified in order of increasing energy as follows:

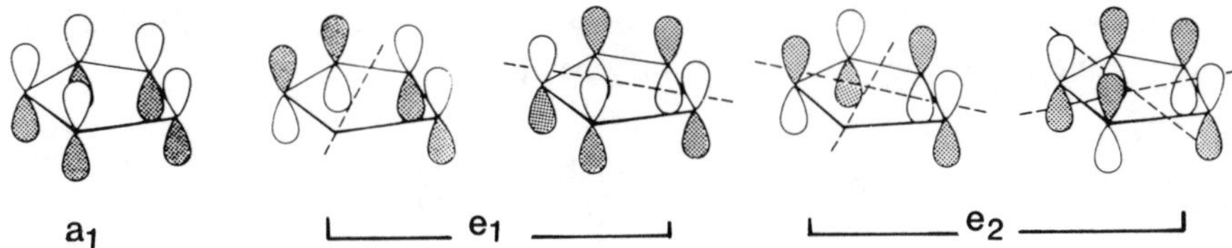

These ligand π orbitals are united pairwise, using plus and minus signs, to form symmetry-adapted linear combinations (SALC) which then overlap with metal orbitals of appropriate symmetry. For a metallocene in its staggered conformation (D_{5d}) the following interactions result:

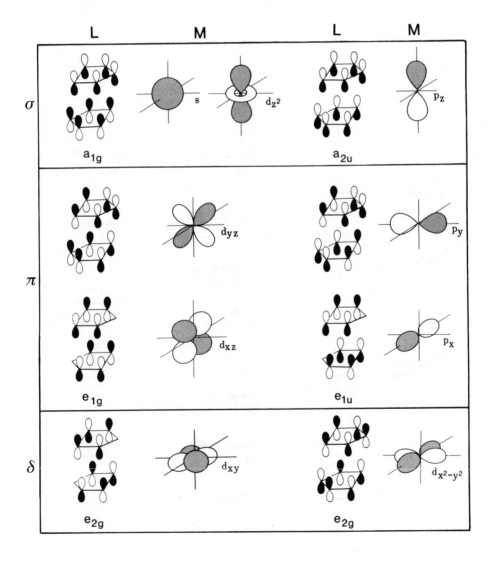

	L	M		L	M
σ	a_{1g}	s	d_{z^2}	a_{2u}	p_z
π	e_{1g}	d_{yz}		e_{1u}	p_y
	e_{1g}	d_{xz}		e_{1u}	p_x
δ	e_{2g}	d_{xy}		e_{2g}	$d_{x^2-y^2}$

If the metallocenes Cp_2M are visualized as being comprised of the components M^{2+} and $2\,Cp^-$, the concept of synergism (p. 185) may be applied:

$$\text{Donor bond} \begin{cases} \sigma & Cp(a_1) \longrightarrow M(4s, 3p_z) \\ \pi & Cp(e_1) \longrightarrow M(d_{xz,yz}; p_{x,y}) \end{cases}$$

$$\text{Acceptor bond (Backbonding)} \quad \delta \quad Cp(e_2) \longleftarrow M(d_{x^2-y^2,xy})$$

A generally accepted qualitative MO diagram for ferrocene in its staggered conformation (D_{5d}) has the following appearance (intra-ligand σ orbitals are not included):

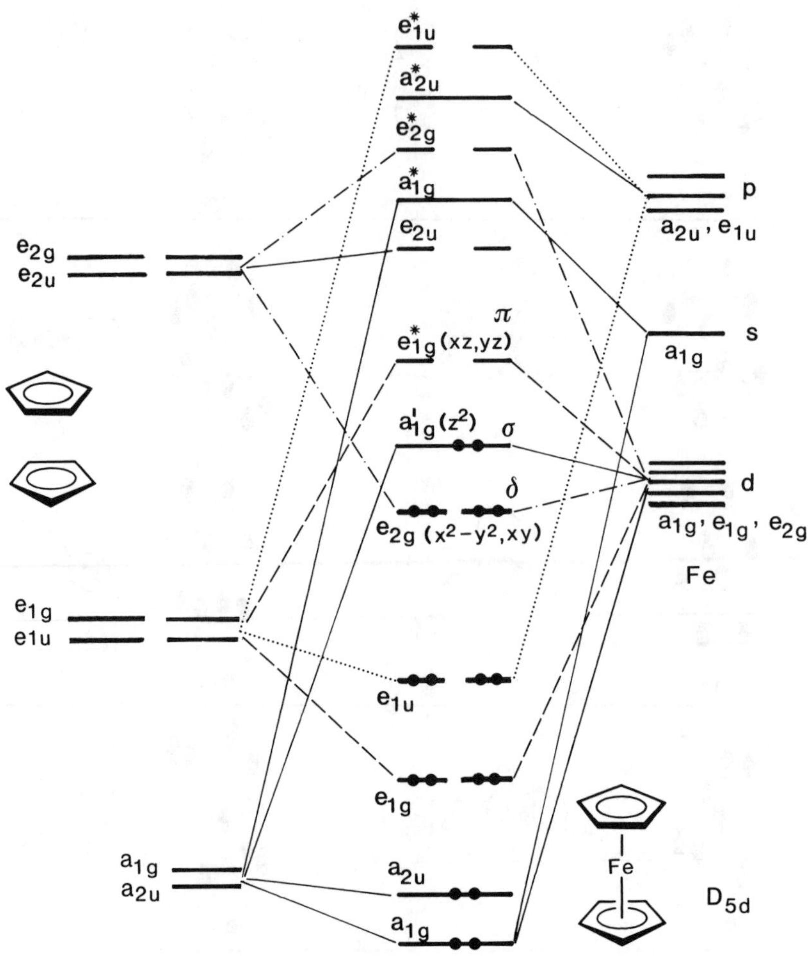

Analogous metal-ligand interactions, albeit of different group-theoretical designation, pertain to eclipsed ferrocene (D_{5h}) or to any other rotamer (D_5), the bond energy being nearly invariant to rotation. The preference for one or the other conformation in specific cases is governed by packing forces or by repulsive effects of peripheral substituents. A theoretical interpretation for the preference "eclipsed" over "staggered" is given by Murrell (1980).

Details in the MO diagram such as the energetic separation or even the sequence of the molecular orbitals can vary depending on the quantum-chemical method used. Furthermore, upon ionization, the sequence of certain MO's may invert, which means that the MO scheme should not be regarded as "frozen", but subject to modification with changing complex charge. Koopmans' theorem (Ionization energy = − orbital energy) therefore occasionally fails in organometallic chemistry, the most prominent example being the apparent inconsistency in the electron configurations of Cp_2Fe and Cp_2Fe^+ (Veillard, 1972). This "failure" is, of course, a result of the oversimplification inherent in equating electron configurations with electronic states whereby electron correlation effects are neglected.

In the chemically relevant frontier-orbital region, the order $e_2 < a_1 < e_1$, already deduced from the crystal-field model, reappears. Judging from their energy relative to the basis orbitals, the frontier orbitals must be regarded as bonding (e_2), nonbonding (a_1) and antibonding (e_1^*).

For some metallocenes, electron configurations can be inferred from the general MO scheme (p. 320) which correctly predicts magnetic properties. In other cases, an *a priori* decision between a high-spin or low-spin form cannot be made (example: Cp_2V), and a more refined approach is called for. Thus, the magnetic moment of the Cp_2Fe^+ cation, which considerably exceeds the 'spin-only' value, is not directly deduced from a rigid MO scheme, which would suggest a ground state $^2A_{1g}$ at variance with the experimentally found ground state $^2E_{1g}$.

Electron Configurations and Magnetism of Metallocenes

	Electron configuration $\begin{Bmatrix}(a_{1g})^2(a_{2u})^2\\(e_{1g})^4(e_{1u})^4\end{Bmatrix}$ +	Number of unpaired electrons n	Spin only value $\sqrt{n(n+2)}$	Magnetic moment (in Bohr magnetons) expected	found
Cp_2Ti^+	$(e_{2g})^1$	1	1.73	>1.73	2.29 ± 0.05
Cp_2V^{2+}	$(e_{2g})^1$	1	1.73	>1.73	1.90 ± 0.05
Cp_2V^+	$(e_{2g})^2$	2	2.83	2.83	2.86 ± 0.06
Cp_2V	$(e_{2g})^2(a'_{1g})^1$	3	3.87	3.87	3.84 ± 0.04
Cp_2Cr^+	$(e_{2g})^2(a'_{1g})^1$	3	3.87	3.87	3.73 ± 0.08
Cp_2Cr	$(e_{2g})^3(a'_{1g})^1$	2	2.83	>2.83	3.20 ± 0.16
Cp_2Fe^+	$(e_{2g})^3(a'_{1g})^2$	1	1.73	>1.73	2.34 ± 0.12
Cp_2Fe	$(e_{2g})^4(a'_{1g})^2$	0	0	0	0
Cp_2Mn	$(e_{2g})^2(a'_{1g})^1(e^*_{1g})^2$	5	5.92	5.92	5.81
Cp_2Co^+	$(e_{2g})^4(a'_{1g})^2$	0	0	0	0
Cp_2Co	$(e_{2g})^4(a'_{1g})^2(e^*_{1g})^1$	1	1.73	>1.73	1.76 ± 0.07
Cp_2Ni^+	$(e_{2g})^4(a'_{1g})^2(e^*_{1g})^1$	1	1.73	>1.73	1.82 ± 0.09
Cp_2Ni	$(e_{2g})^4(a'_{1g})^2(e^*_{1g})^2$	2	2.83	2.83	2.86 ± 0.11

Some Further Properties of Metallocenes

Complex	Color	Melting point °C	Miscellaneous
"$(C_5H_5)_2Ti$"	green	200 (decomp.)	Dimeric with two μ-H bridges and a fulvalenediyl bridging ligand
$(C_5H_5)_2V$	purple	167	very air-sensitive
"$(C_5H_5)_2Nb$"	yellow		Dimeric with $\eta^1{:}\eta^5$-C_5H_4 bridges and terminal hydride ligands
$(C_5H_5)_2Cr$	scarlet	173	very air-sensitive
"$(C_5H_5)_2Mo$"	black		several dinuclear isomers with
"$(C_5H_5)_2W$"	yellow green		fulvalenediyl or $\eta^1 - \eta^5$ bridges and terminal hydride ligands, diamagnetic
$(C_5H_5)_2Mn$	brown	173	air-sensitive and readily hydrolyzed; at 158 °C converted into a pink form
$(C_5H_5)_2Fe$	orange	173	air-stable, can be oxidized to blue-green $(C_5H_5)_2Fe^{+\cdot}$
$(C_5H_5)_2Co$	purple-black	174	air-sensitive, oxidation gives the air-stable, yellow cation $(C_5H_5)_2Co^{+\cdot}$
$(C_5H_5)_2Ni$	green	173	slow oxidation by air to the labile, orange cation $(C_5H_5)_2Ni^{+\cdot}$

SPECIAL STRUCTURAL FEATURES

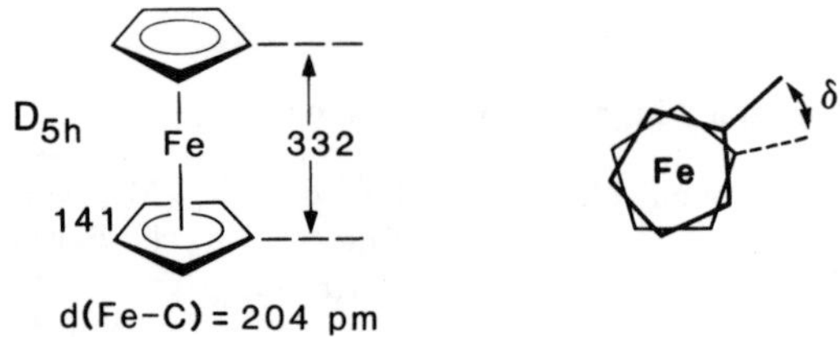

Ferrocene *at room temperature crystallizes in a monoclinic, at $T < 164$ K in a triclinic, and at $T < 110$ K in an orthorhombic modification. In the monoclinic form, disorder phenomena feign a staggered conformation* (D_{5d}) *of individual sandwich molecules. In the triclinic form, the molecules deviate from the eclipsed conformation* (D_{5d}) *by $\delta = 9°$, while in the orthorhombic form, the rings are fully eclipsed* (D_{5h}) *(Dunitz, 1982). In the gas phase, ferrocene also adopts an eclipsed conformation, the rotational barrier being small* ($\approx$ 4 kJ/mol, *Haaland,* 1966). *Decamethylferrocene* $(Cp^*)_2Fe$, *however, has been shown to exist in a staggered conformation* (D_{5d}) *in both the solid state* (*Raymond,* 1979) *as well as in the gas phase* (*Haaland,* 1979).

Interestingly, for the salt $[Cp_2^*Fe^{III}]$ [TCNE] (TCNE = tetracyanoethylene), which crystallizes in an alternating donor acceptor ... $D^{\dot{+}} A^{\dot{-}} D^{\dot{+}} A^{\dot{-}}$ ... linear-chain structure, bulk ferromagnetic behavior (spontaneous magnetization) has been established (Miller, 1988 R).

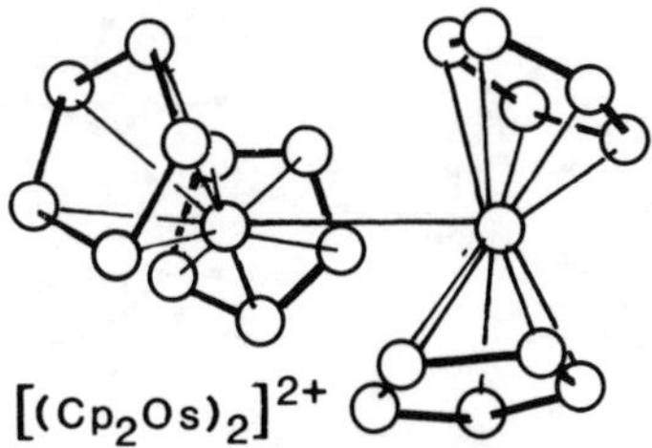

*The neutral complexes **ruthenocene and osmocene** in the solid state exhibit the eclipsed conformation of orthorhombic ferrocene, with* d(Ru–C) = 221 pm, d(Os–C) = 222 pm. *The **osmocinium cation*** $[Cp_2Os]^+$ *forms a dimer with a long metal-metal bond,* d(Os–Os) = 304 pm (*Taube*, 1987).
The permethylated cation $[Cp_2^*Os]^+$, *however, possesses a monomeric sandwich structure, the ligands assuming a staggered conformation* (D_{5d}) (*J. S. Miller*, 1988).

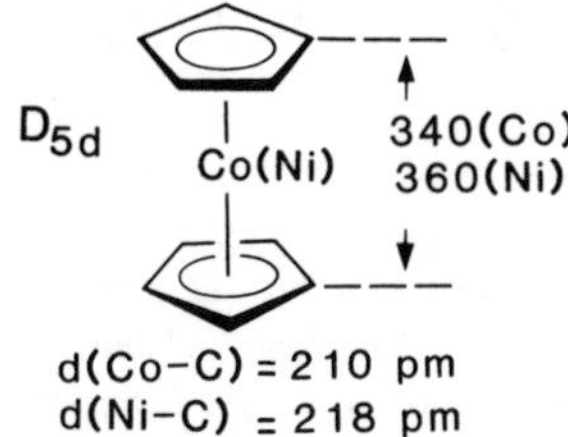

*Structure of **cobaltocene and nickelocene**: The five-membered rings are reported to be staggered (E. Weiss, 1975). A comparison of* M–C *distances in various metallocenes shows a distinct minimum for ferrocene, in accordance with the occupation of all bonding orbitals in this molecule. Placing one or two electrons, respectively, into the antibonding orbitals of* $CoCp_2$ *und* $NiCp_2$ *weakens the metal-ring bond, thereby increasing* d(M–C).

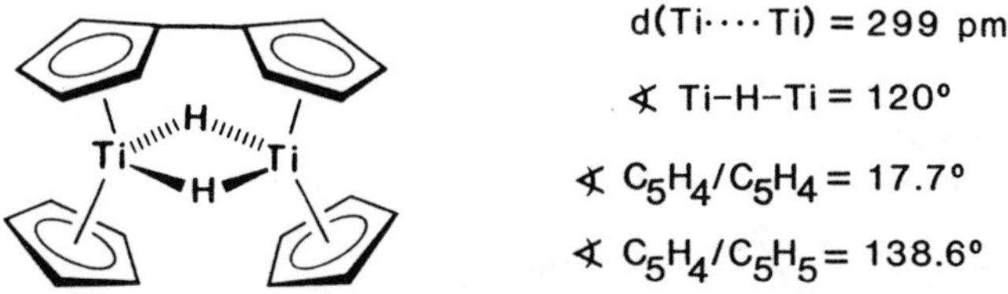

"Titanocene". *At least two isomers exist, possibly more, depending on the method used for their synthesis. Reduction of* Cp_2TiCl_2 *or treatment of* $TiCl_2$ *with* NaCp *leads to the formation of a dimer with a fulvalenediyl bridge and two hydride bridges. Under the assumption of* 2e 3c M–H–M *bridges and a* Ti–Ti *bond, both titanium atoms acquire a* 16 VE *configuration. The X-ray structural analysis of "titanocene", as depicted above, is of very recent origin* (*Troyanov*, 1992).

Decamethyltitanocene at room temperature exists as an equilibrium mixture of two isomers: a yellow paramagnetic 14 VE complex and a green diamagnetic 16 VE compound formed by insertion of titanium into the C–H bond of a methyl group of the pentamethylcyclopentadienyl ligand (Bercaw, 1974):

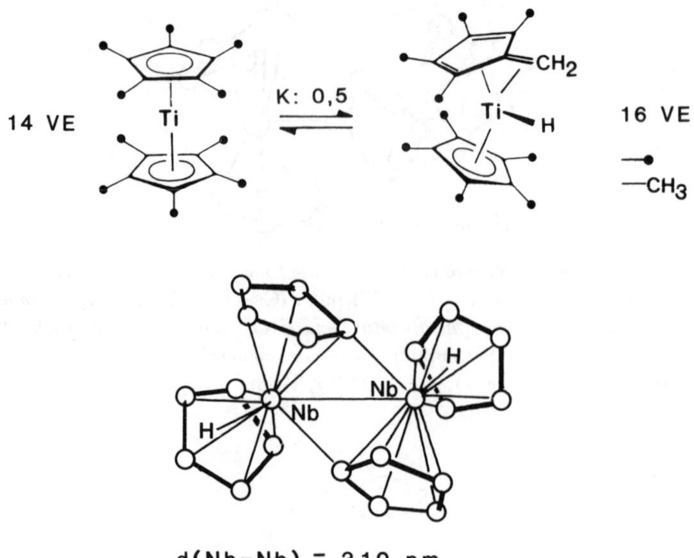

A complex of the stoichiometric composition **"$(C_5H_5)_2Nb$"** *also represents a dimeric species* $[C_5H_5)(C_5H_4)NbH]_2$. *It contains two η^1:η^5-cyclopentadienediyl bridges and two terminal hydride ligands. This complex, in contrast to fictious "niobocene", has an* 18 VE *shell.*

Monomeric metallocenes which would be analogous to chromocene are unknown for molybdenum and tungsten. So far, four dimeric isomers of composition $\mathbf{[C_{10}H_{10}M]_2}$ (M = Mo, W) have been characterized, some of which appear to be interconvertible. Like the titanium and niobium dimers, they possess η^1:η^5-cyclopentadienediyl bridges or η^5:η^5-fulvalenediyl bridges as well as terminal hydride ligands (M. L. H. Green, 1982):

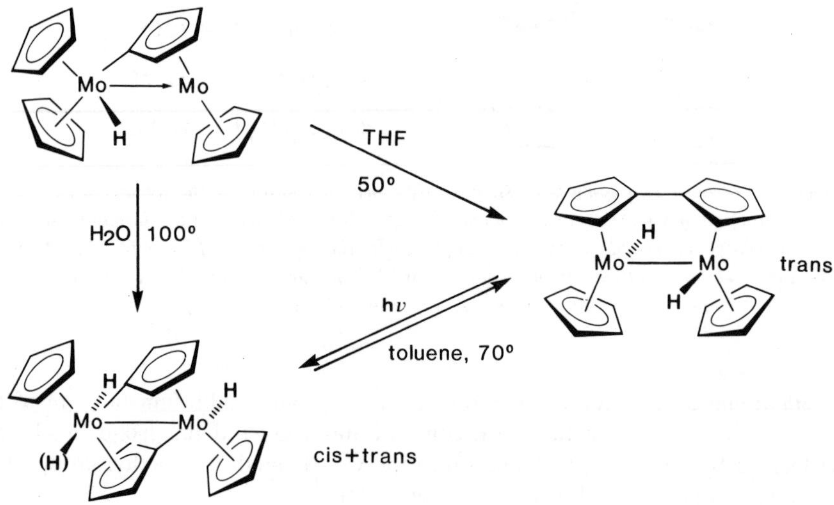

Manganocene Cp_2Mn in the solid state has a chain structure without individual sandwich entities (E. Weiss, 1978):

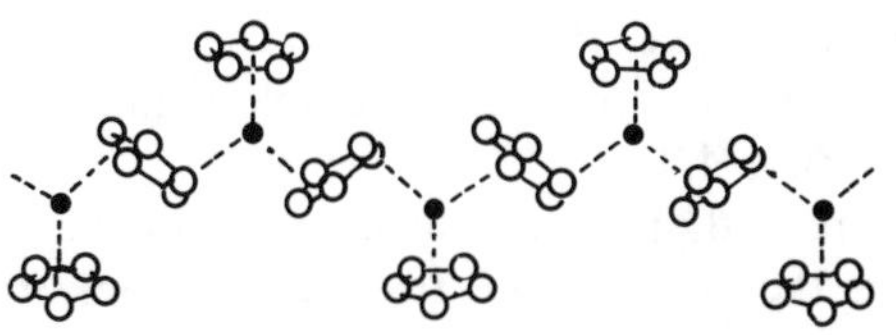

$d(\text{Mn}-\text{C})_{terminal} = 242$ pm
$d(\text{Mn}-\text{C})_{bridge} = 240-330$ pm

The magnetic moment of $(C_5H_5)_2Mn$ *shows an unusual temperature dependence: as a solid at* $T > 432$ K, *in the melt and in solutions at room-temperature, the complex has a magnetic moment of* $\mu = 5.9$ B.M. *as expected from an electron configuration* $(e_{2g})^2\,(a'_{1g})^1\,(e^*_{1g})^2$ [*ground state* $^6A_{1g}$]. *Between* 432 *and* 67 K, *with decreasing temperature increasing antiferromagnetic behavior is observed, corresponding to an increase of cooperative interactions. A dilute solid solution of* $(C_5H_5)_2Mn$ (8%) *in* $(C_5H_5)_2Mg$ *has a magnetic moment of* $\mu = 5.94$ B.M. *over the whole temperature range* (*Wilkinson*, 1956).

Decamethylmanganocene Cp^*_2Mn displays a "normal" sandwich structure (Raymond, 1979), and the magnetic moment $\mu = 2.18$ B.M. points to a low-spin Mn(d^5) configuration [ground state $^2E_{2g}$]. 1,1′-Dimethylmanganocene, on the other hand, in the gas phase consists of an equilibrium mixture of high-spin ($^6A_{1g}$) and low-spin ($^2E_{2g}$) states according to He-photoelectron spectroscopic evidence.

Other derivatives of manganocene also show a puzzling dependence of the magnetic behavior on the molecular environment. Today, it is known that the free molecules $(C_5H_5)_2Mn$ and $(CH_3C_5H_4)_2Mn$ are so close to the **high-spin/low-spin cross-over** point, that the small intermolecular forces existing in frozen solutions or in molecular crystals are sufficient to incite the observed changes of the electronic ground-state (Ammeter, 1974). As expected from ligand-field theory, environments which tolerate a large Mn−Cp distance favor the high-spin state ($^6A_{1g}$), while matrices enforcing a small Mn−Cp distance promote a low-spin state ($^2E_{2g}$). The energy difference is only 2 kJ/mol. (In this context, compare the pressure dependence of the electronic spectra of metal complexes.)

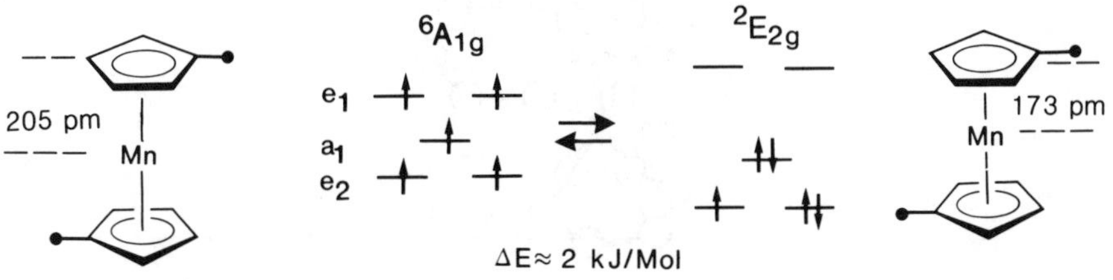

205 pm Mn — $^6A_{1g}$: e_1, a_1, e_2 ⇄ $^2E_{2g}$ — Mn 173 pm

$\Delta E \approx 2$ kJ/Mol

Rhenocene Cp_2Re, isoelectronic with the osmocinium cation, is thought to have a dimeric structure with a metal−metal bond (Pasman, 1984).

Decamethylrhenocene Cp^*_2Re, however, like decamethylmanganocene is monomeric. The magnetic behaviour of decamethylrhenocene is as complicated as that of manganocene. The application of an arsenal of spectroscopic and magnetochemical methods has led to the conclusion that Cp^*_2Re, depending on the state of aggregation and the medium, exists as an equilibrium of $^2A_{1g}$ and $^2E_{5/2}$ species (Cloke, 1988).
These examples show that **ring methylation** can have a considerable influence on the structure and properties of metallocenes (cf. p. 47).
A special type of alkylmetallocene is the recently prepared superferrocenophane (Hisatome, 1986). In this sandwich complex, the shielding of the central atom from its environment is perfect:

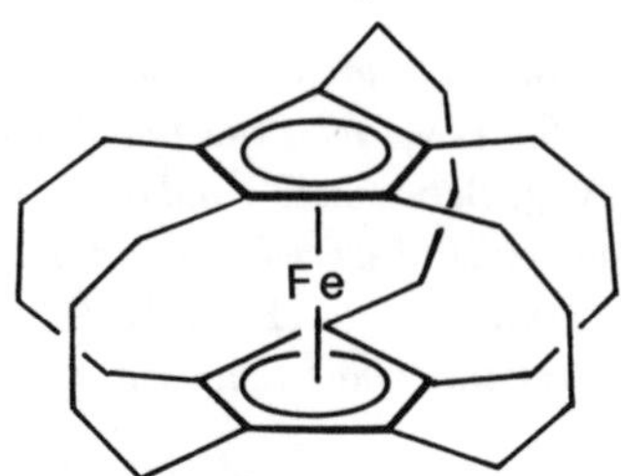

Several ring-anellated derivatives of the cyclopentadienyl anion can also form metal π-complexes.

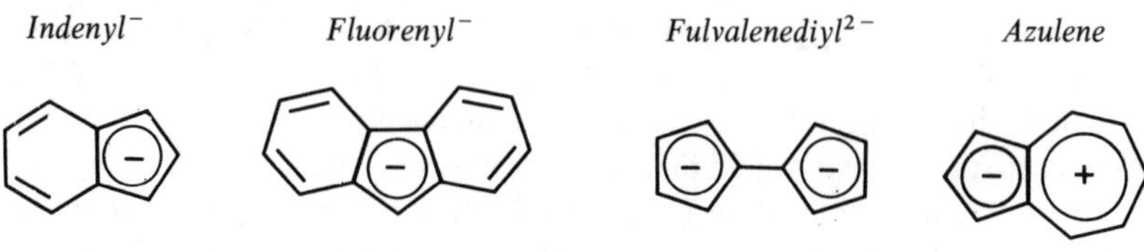

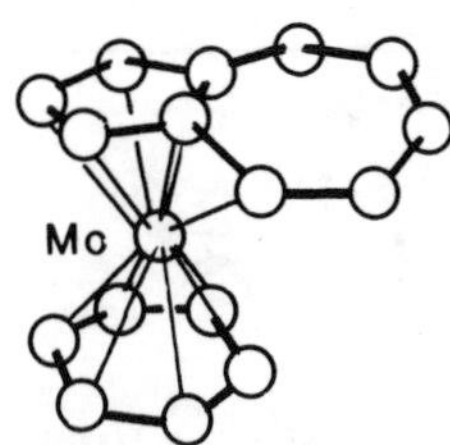

Azulene *prefers η^5-coordination via the 5-membered ring. An interesting variant is revealed in the structure of* ***azulene(benzene)Mo****: the azulene ring coordinates in an η^6-fashion via a fulvene-like unit (Behrens, 1987). The deviation from ligand planarity is reminiscent of the structure of ferrocenyl carbenium ions (p. 329).*

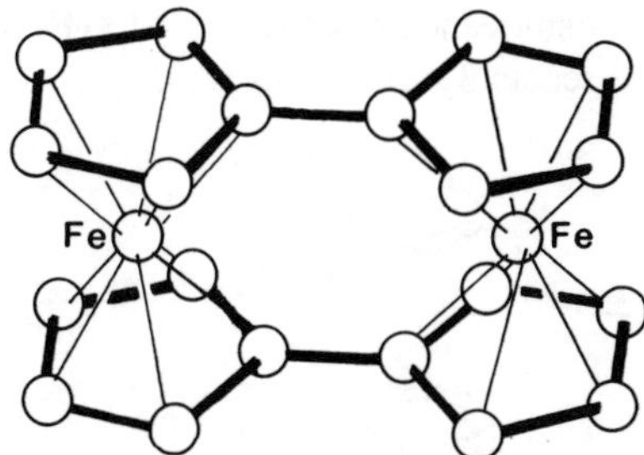

Bis-μ(fulvalenediyl)diiron *(Biferrocenylene) interestingly forms a dication* (Fe^{3+}, Fe^{3+}) *which is diamagnetic. It has been suggested that despite the large* Fe—Fe *distance of* 398 pm, *direct exchange interaction* (*p.* 388) *between the two central metal ions M* (d^5) *accounts for spin pairing here* (*Hendrickson*, 1975).

Intersandwich compounds with fulvalenediyl bridges have been extensively studied as model compounds in the context of intramolecular electron transfer and the mixed-valence phenomenon. The following example also illustrates an application of ^{57}Fe-Mössbauer spectroscopy (Hendrickson, 1985):

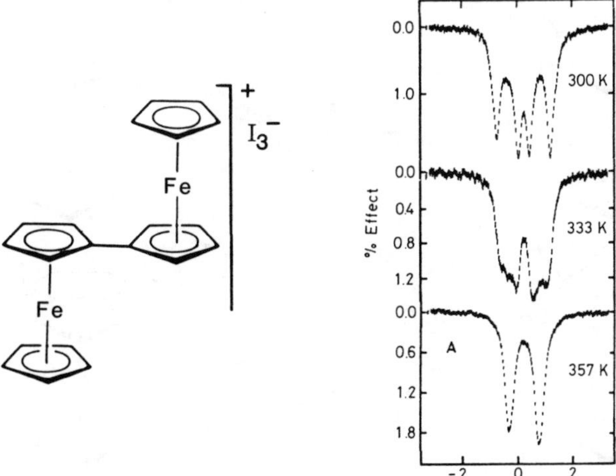

The observation of ***two*** *doublets at room temperature points to the presence of both* Fe^{2+} *and* Fe^{3+}; *it demonstrates that intramolecular electron transfer is slow on the time scale* ($\approx 10^{-8}$ s) *of the Mössbauer experiment,* ***mixed valence*** *case.*
The merging into ***one*** *doublet at temperatures* > 350 K *indicates but one type of iron to which an oxidation state of* $Fe^{2.5+}$ *must be assigned; electron transfer is fast at this temperature,* ***intermediate valence*** *case.*

SELECTED REACTIONS OF METALLOCENES

The cyclopentadienyl ring in ferrocene carries a partial negative charge. This is reflected in the following properties:

$C_5H_5FeC_5H_4NH_2$ *is a stronger base than aniline*
$C_5H_5FeC_5H_4COOH$ *is a weaker acid than benzoic acid*
} *The ferrocenyl group therefore acts as an electron donor*

Ferrocene, ruthenocene and osmocene are susceptible to **electrophilic substitution.** Compared to benzene, ferrocene reacts 3×10^6 times faster.

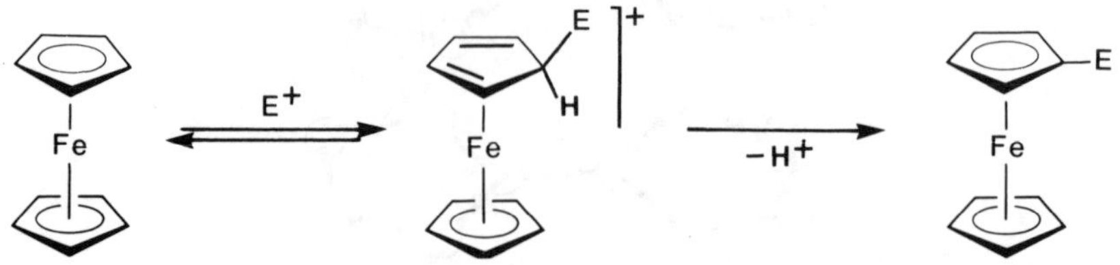

The electrophile should not be an oxidizing agent, as substitution would then be surpressed by oxidation to the ferricinium ion $[FeCp_2]^+$ which is inert to attack by electrophiles. That electrophilic substitution of ferrocene does not involve direct participation of the metal has been established by the intramolecular acylation of two isomeric ferrocene carboxylic acids. It was found that the ***exo***-isomer, in which the acylium ion cannot directly interact with the metal, cyclizes faster than the ***endo***-isomer (Rosenblum, 1966):

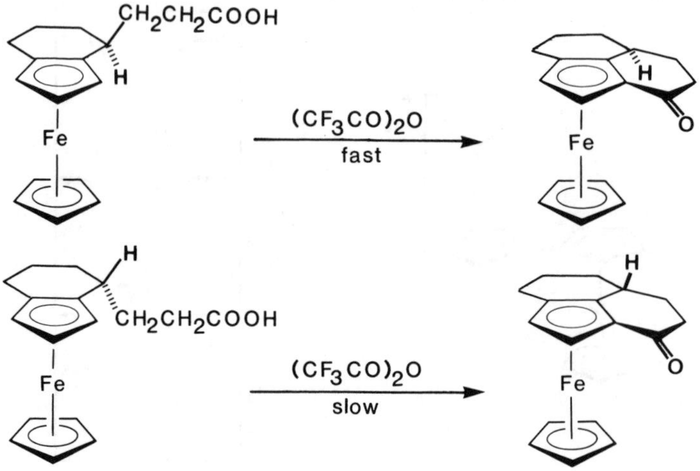

This result suggests that electrophilic attack occurs from the ***exo***-direction.

Using strong acids, ferrocene can, however, be protonated at the metal; the cation $[(C_5H_5)_2FeH]^+$ is detectable by ^{1}H-NMR spectroscopy, ($\delta = -2.1$ ppm) (Rosenblum, 1960):

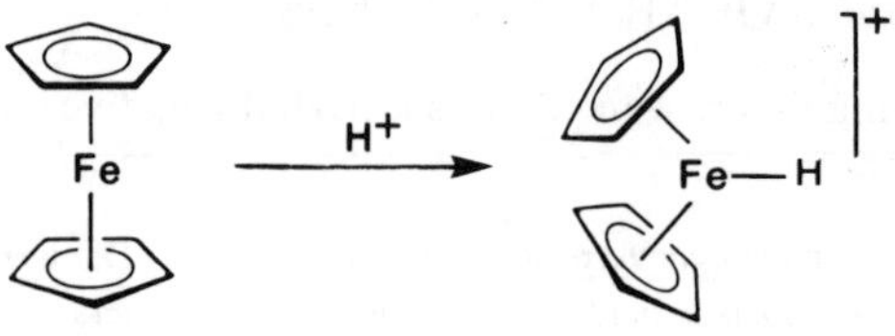

Friedel-Crafts acylation:

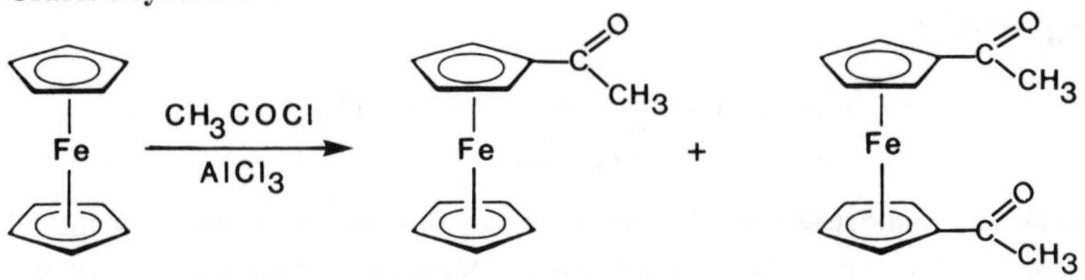
CH3COCl
AlCl3
Fe
CH3
+

Mannich reaction (Aminomethylation):

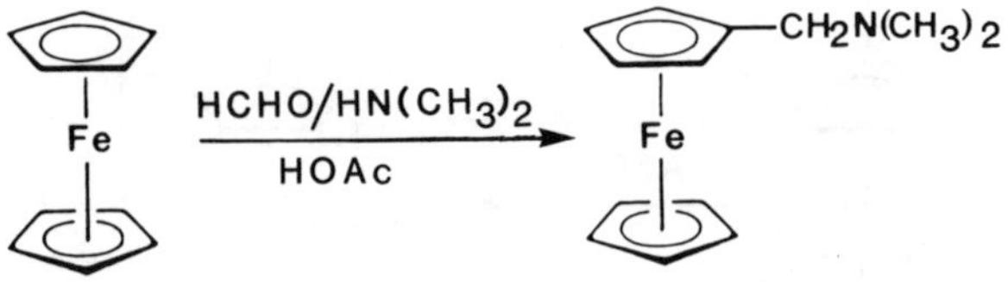
HCHO/HN(CH3)2
HOAc
Fe
CH2N(CH3)2

Metallation:

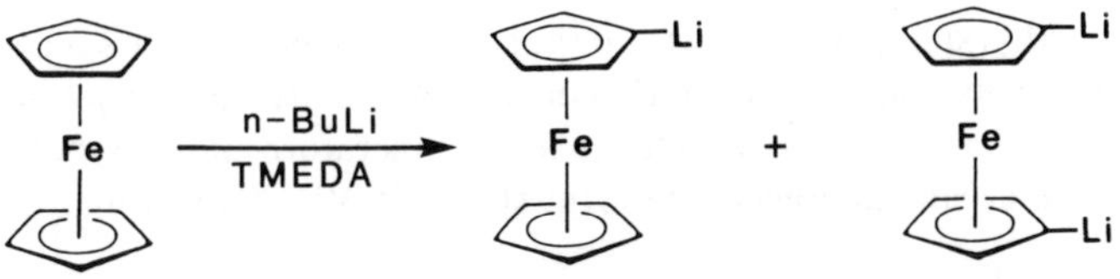
n-BuLi
TMEDA
Fe
Li
+

Ferrocene exhibits an extensive organic chemistry. **α-Ferrocenyl carbenium ions** are particularly stable and even isolable as salts with BF_4^-, PF_6^- as counterions (Watts, 1979 R):

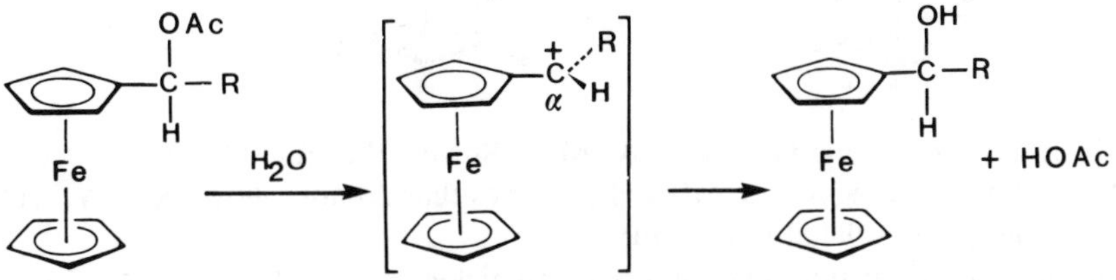
OAc
C — R
H
Fe
H2O
α
OH
+ HOAc

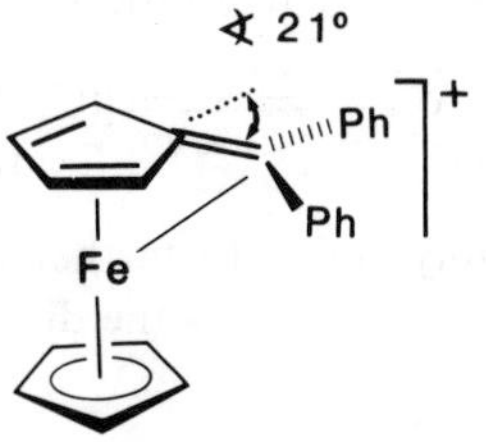
∢ 21°
Ph
Ph
Fe

d(Fe-C_{exo}) = 271 pm

These species may also be regarded as ***pentafulvene complexes.*** *X-ray structural data reveal that the α-carbon atom is involved in bonding to the metal, since the iron atom is displaced from its position below the center of the substituted ring towards the substituent* (Behrens, 1979). *As to the nature of the* Fe$-$$C_\alpha$ *bond, compare* *p.* 301.

Ferrocene can be oxidized both electrochemically and by many chemical oxidizing agents, *e.g.* HNO_3:

$$\underset{18\text{ VE}}{(C_5H_5)_2Fe} \underset{\substack{E^0 = +0.31\text{ V in} \\ \text{AN/LiClO}_4\text{ versus SCE}}}{\overset{-e}{\rightleftharpoons}} \underset{17\text{ VE}}{[(C_5H_5)_2Fe]^+}$$

Ruthenocene and osmocene can also be oxidized, but the corresponding metallicinium ions are not stable as monomers. Dimerization occurs instead (see p. 323) or disproportionation to give MCp_2 und $[MCp_2]^{2+}$. In contrast, the permethylated complex $(C_5Me_5)_2Ru$ forms a cation $[(C_5Me_5)_2Ru]^+$, which is stable at $-30\,°C$ (Kölle, 1983):

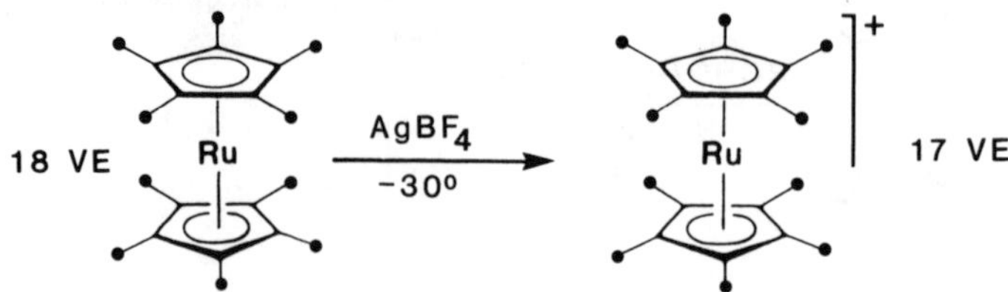

The influence methyl substitution exerts on the stability of metallocenes is also manifested in the properties of manganocenes and rhenocenes. While $MnCp_2$ cannot be reduced and monomeric $ReCp_2$ has only been detected in low-temperature matrices, both **decamethylmanganocene and decamethylrhenocene** form stable 18 VE anions (Smart, 1979; Cloke, 1985):

17 VE M —K→ M 18 VE (−) M= Mn, Re

$[(C_5Me_5)_2Ru]^+$, $[(C_5Me_5)_2Os]^+$ and $(C_5Me_5)_2Re$ are therefore the only documented examples of metallocenes of the 4 *d*- and 5 *d* series that do not follow the 18 VE rule but can nevertheless be isolated in quantity.

The chemistry of the metallocenes of cobalt and nickel and their homologues is influenced by their tendency to attain an 18 VE configuration:

$$\underset{19\text{ VE}}{(C_5H_5)_2Co} \underset{\substack{E^0 = -0.90\text{ V} \\ \text{versus SCE}}}{\overset{-e^-}{\rightleftharpoons}} \underset{18\text{ VE}}{[(C_5H_5)_2Co]^+}$$

Cobalticinium salts are remarkably stable to further oxidation; with strong oxidizing agents, the dimethylcobalticinium cation gives the dicarboxylic acid without rupture of the Co–Cp bonds:

$[(CH_3C_5H_4)_2Co]^+ \xrightarrow{HNO_3} [(HOOCC_5H_4)_2Co]^+$

Towards alkyl halides, cobaltocene acts as a reducing agent:

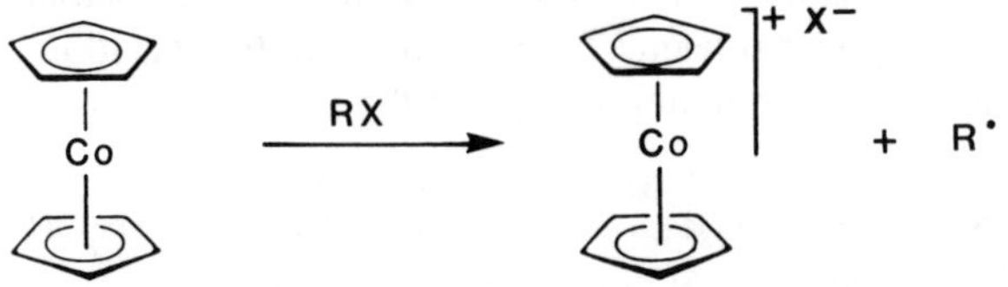

Organic radicals add to cobaltocene. The η^5-cyclopentadienyl ligand is thereby converted to η^4-cyclopentadiene with attendant change from a 19 VE- to an 18 VE configuration:

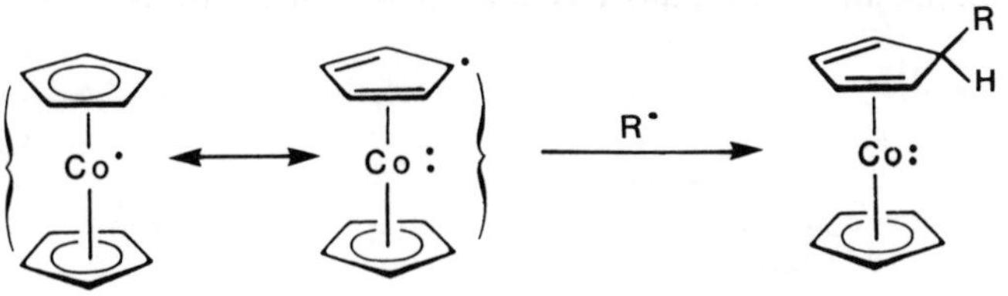

The biradical O_2 as well adds to cobaltocene; the resulting bis(cobaltocenyl)peroxide may only be isolated at low temperatures:

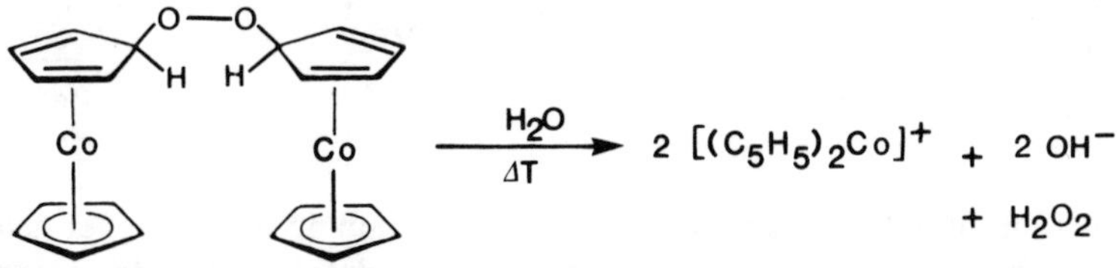

Nucleophilic addition to cobalticinium ions occurs from an ***exo***-direction:

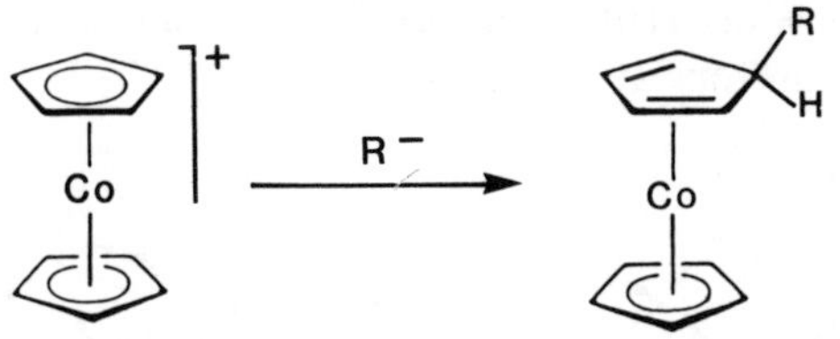

The 18 VE ions $[RhCp_2]^+$ and $[IrCp_2]^+$, like $[CoCp_2]^+$, are very stable, as opposed to the neutral monomers $RhCp_2$ und $IrCp_2$ which as yet have only been observed in matrix isolation. At room temperature, they immediately dimerize. ESR measurements and EH-MO calculations suggest that the unpaired electron in these complexes is located mainly on the C_5H_5 rings.

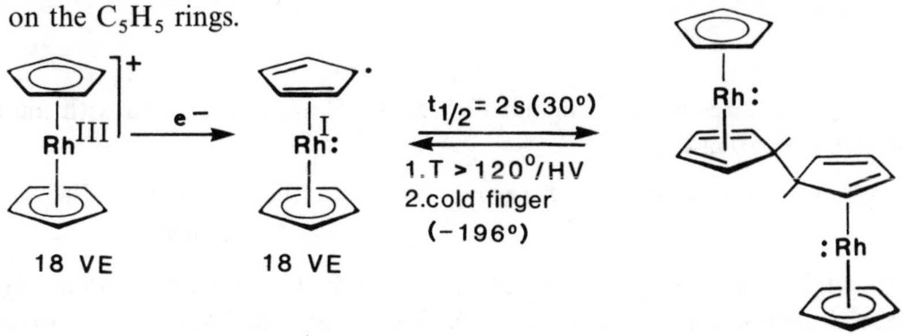

Nickelocene is the only metallocene with 20 valence electrons. It is paramagnetic and easily oxidized to the nickelicinium ion (19 VE), which is of low stability. At low temperatures and in extremely pure solvents, further oxidation to a dication has been observed as well as reduction to a monoanion (Geiger, 1984 R):

$$\underset{21\ \text{VE}}{[(C_5H_5)_2Ni]^-} \underset{-1.6\ \text{V},\ -60\,^\circ\text{C}}{\overset{-e^-}{\rightleftharpoons}} \underset{20\ \text{VE}}{(C_5H_5)_2Ni} \underset{0.1\ \text{V}}{\overset{-e^-}{\rightleftharpoons}} \underset{19\ \text{VE}}{[(C_5H_5)_2Ni]^+} \underset{0.74\ \text{V}}{\overset{-e^-}{\rightleftharpoons}} \underset{18\ \text{VE}}{[(C_5H_5)_2Ni]^{2+}}$$

Quite unexpectedly, upon treating nickelocene with HBF_4 in propionic anhydride, the prototype of the **multidecker-sandwich complexes** is formed (Werner, 1972):

While numerous multidecker-sandwich complexes containing larger rings or heterocycles have been synthesized since, the cation $[(C_5H_5)_3Ni_2]^+$ (**tripledecker sandwich**) remains the only complex of its type which features unsubstituted cyclopentadienyl rings exclusively. Intermediates can only be detected under special conditions. On protonating **decamethylnickelocene,** the reaction stops at the first stage:

Tripledecker complexes with the ligand $C_5Me_5^-$ (Cp*) are also formed with the heavier group 8 metals (Rybinskaya, 1987):

$$Ru(C_5Me_5)_2 + [(C_5Me_5)Ru(AN)_3]^+PF_6^- \longrightarrow \underset{30\ \text{VE}}{[(C_5Me_5)_3Ru_2]^+PF_6^-}$$

These two tripledecker complexes are representative for the two series containing **34** and **30 valence electrons,** which, according to MO calculations should display exceptional stability (R. Hoffmann, 1976). A paramagnetic cobalt tripledecker with 33 valence electrons has, however, also recently been prepared by metal-vapor techniques (Schneider, 1991):

Co(g) + [C_5Me_5H] $\xrightarrow[\text{cyclo-}C_6H_{11}CH_3]{-120°/10^{-3}\text{torr}}$ [Cp*Co(C_5Me_5H)] + [Cp*Co(μ-C_5Me_5)CoCp*] (33 VE) + Cp*Co=CoCp*

Of particular interest is the molecule Cp*Co=CoCp* which features an unsupported Co=Co double bond. For Cp^*Ni^+ dimerization is frustrated by the positive charge. Further examples of tripledecker complexes with 31–33 valence electrons will be described later (p. 383), as well as examples in which the number of valence electrons falls considerably short of the "magic number" (p. 357).

15.4.3.2 Cyclopentadienyl Metal Carbonyls

PREPARATION

1. Metal Carbonyl + Cyclopentadiene

$Fe(CO)_5$ or $Ru_3(CO)_{12}$ + C_5H_6 ⟶ (η^4-C_5H_6)M(CO)$_3$ $\xrightarrow{-CO}$ CpM(CO)$_2$H $\xrightarrow{-H_2}$ $[CpM(CO)_2]_2$

M= Fe, Ru

The cyclopentadiene- and cyclopentadienyl hydride complexes which appear as intermediates here can both be independently synthesized under mild conditions. They are sensitive to heat and sunlight, readily dimerizing under elimination of H_2 *most likely via a radical pathway* (*Baird*, 1990).

$Re_2(CO)_{10}$ $\xrightarrow[210°,\ -H_2]{C_5Me_5H}$ 2 Cp*Re(CO)$_3$ $\xrightarrow[-3\ CO]{h\nu}$ Cp*Re(μ-CO)$_3$ReCp* (Re≡Re)

2. Metal Carbonyl + Cyclopentadienyl Reagent

$$Na[V(CO)_6] + C_5H_5HgCl \longrightarrow C_5H_5V(CO)_4 + Hg + NaCl + 2\,CO$$

This is an unusual method for introducing a cyclopentadienyl group; Hg(II) acts as the oxidizing agent here.

$$W(CO)_6 + NaC_5H_5 \longrightarrow Na[C_5H_5W(CO)_3] \xrightarrow{H^+} C_5H_5W(CO)_3H$$

$$Na[C_5H_5W(CO)_3] \xrightarrow{Fe^{3+}} [C_5H_5W(CO)_3]_2 + H_2 \qquad C_5H_5W(CO)_3H \xrightarrow{h\nu} [C_5H_5W(CO)_3]_2 + H_2$$

As in the case of $CpM(CO)_2H$ (M = Fe, Ru) the corresponding Mo- and W analogues tend to dimerize, H_2 being lost. An alternative method is oxidative coupling of the cyclopentadienyl carbonyl metallates.

3. Metal Carbonyl Halide + Cyclopentadienyl Reagent

$$[\mu\text{-ClRh(CO)}_2]_2 + 2\,TlCp \xrightarrow[RT]{hexane} 2\,CpRh(CO)_2 + 2\,TlCl$$

4. Cyclopentadienyl Metal Halide + Reducing Agent + CO

$$Cp_2TiCl_2 \xrightarrow{1.\ Zn,\ THF;\ 2.\ CO;\ 3.\ NH_3} Cp_2Ti(CO)_2 + [Zn(NH_3)_4]Cl_2$$

5. Metallocene + CO

$$MnCp_2 + 3\,CO \xrightarrow[15h,\ 90-150^0]{200\ bar\ CO} CpMn(CO)_3$$

"cymantrene"

6. Metallocene + Metal Carbonyl

$$Ni(C_5H_5)_2 + Ni(CO)_4 \longrightarrow [C_5H_5Ni(CO)]_2 + 2\,CO$$

STRUCTURES

The monomeric complexes $CpM(CO)_n$ are called half-sandwich complexes.

The previously mentioned capability of the carbonyl group to act as a terminal or as a bridging ligand is also exhibited by the dinuclear (cyclopentadienyl)metal carbonyls:

Note that carbonyl bridging is avoided by the heavier metals.

$[CpFe(CO)_2]_2$ is a **fluxional molecule.** In solution, the following processes occur:

- *cis/trans isomerization*
- *"scrambling" (redistribution of terminal and bridging* CO *ligands).*

By IR and NMR spectroscopy, four different species can be detected:

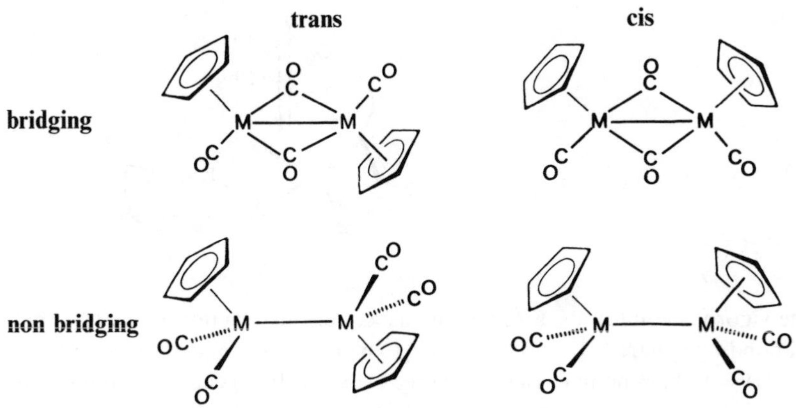

The structural dynamics have been unravelled by Cotton (1973).

REACTIONS OF CYCLOPENTADIENYL METAL CARBONYLS

(a) Reduction

$$CpV(CO)_4 \xrightarrow[-CO]{Na/Hg} Na_2[CpV(CO)_3]$$

$$[CpFe(CO)_2]_2 \xrightarrow{Na/Hg} 2\,Na[CpFe(CO)_2] \xrightarrow{CH_3I} CpFe(CO)_2CH_3 + NaI$$

(b) Reaction with Halogens

$$[CpFe(CO)_2]_2 + X_2 \longrightarrow 2\,CpFe(CO)_2X \xrightarrow{AlX_3/CO} [CpFe(CO)_3]^+$$

$(X = Br, I)$

Under photochemical conditions, CCl_4 or $PhCH_2Cl$ serve as the halide source.

(c) Ring Substitution

$CpV(CO)_4$, $CpMn(CO)_3$ and $CpRe(CO)_3$ readily undergo electrophilic substitution, similar to ferrocene.

(d) Substitution of CO

$CpMn(CO)_3 + L \longrightarrow CpMn(CO)_2L + CO$ L = *phosphane, olefin, etc.*

$CpCo(CO)_2 + L_2 \longrightarrow CpCoL_2 + 2\,CO$ L = *phosphane or another Lewis base*

L_2 = *diolefin*

Thermal as well as photochemical reaction conditions are suitable.

(e) Oxidative Decarbonylation

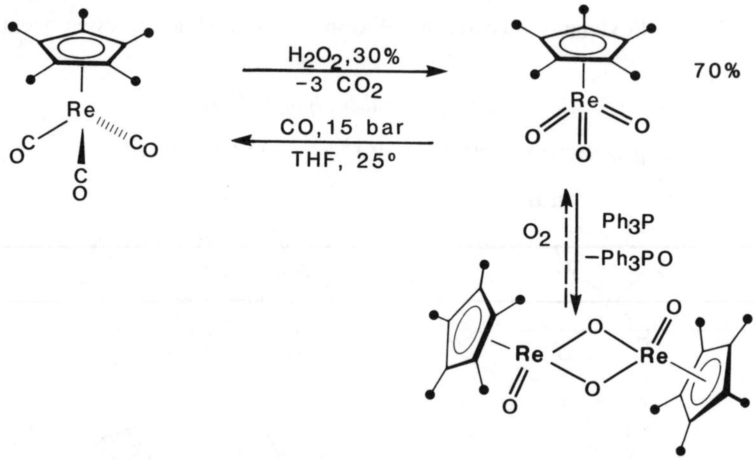

Apart from the vigorous conditions, a surprising aspect of this reaction is the fact that $(\eta^5\text{-}C_5R_5)-M$ bonding apparently is not restricted to metals in low oxidation states. The distance $d(Re-O) = 170$ pm points to a high degree of double bond character and shows that the oxo ligand acts as a strong π-donor. Partial deoxygenation leads to organometallic oxides with terminal as well as bridging oxygen atoms (Herrmann, 1987 R).

15.4.3.3 Cyclopentadienyl Metal Nitrosyls

Nitric oxide is a very versatile ligand; it can function in two ways:

as a coordinated NO^+ *ion* (*isoelectronic with* CO) — *as a coordinated* NO^- *ion* (*isoelectronic with* O_2)

$$L_{n-1}\overset{-I}{M}\cdots|N{\equiv}O|^{\oplus} \xleftarrow[-L]{\;\dot{N}=O\;} L_nM^0 \xrightarrow{\;\dot{N}=O\;} L_n\overset{+I}{M}\cdots\overset{\ominus}{N}=O$$

18 VE — 17 VE — 18 VE

NO *as* 3*e ligand*
linear structure MNO

NO *as* 1*e ligand*
bent structure MNO

The conversion MNO(linear) to MNO(bent) is triggered by intramolecular transfer of two electrons from the metal to the ligand; during this process, a free coordination site is generated and associative mechanisms of substitution become possible (cf. p. 232).
In organometallic chemistry, NO is generally regarded as a 3*e* ligand; isoelectronic complexes are obtained by substituting three CO groups for two NO ligands, or by compensating the substitution of one CO for one NO by the introduction of a positive charge. Examples:

$Cr(NO)_4$	$Mn(NO)_3CO$	$Fe(NO)_2(CO)_2$	$Co(NO)(CO)_3$	$Ni(CO)_4$
	$C_5H_5V(CO)(NO)_2$,	$C_5H_5Cr(CO)_2NO$,	$[C_5H_5Mn(CO)_2NO]^+$	

PREPARATION

$$Ni(C_5H_5)_2 + NO \xrightarrow[-1/2\,C_{10}H_{10}]{} C_5H_5NiNO$$

$$2\,C_5H_5Co(CO)_2 + 2\,NO \xrightarrow[-4\,CO]{} [C_5H_5CoNO]_2$$

$$C_5H_5Re(CO)_3 + NO^+HSO_4^- \xrightarrow[-CO]{CH_2Cl_2} [C_5H_5Re(CO)_2NO]^+ + HSO_4^-$$

STRUCTURE AND PROPERTIES

Like CO, NO can serve as a terminal as well as a bridging ligand; an additional variable is the MNO angle of terminal nitrosyls. In certain cases both linear and bent NO configurations coexist in the same molecule:

ν(NO) [cm^{-1}] $\quad \nu(NO)_{linear} = 1677$
$\nu(NO)_{bridge} = 1518$
(*Fontana*, 1974)

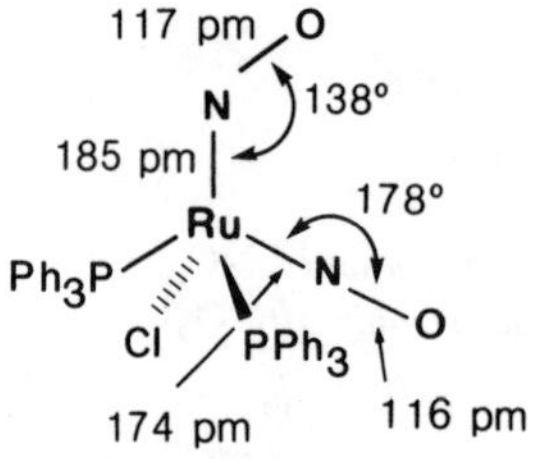

117 pm O; N 138°; 185 pm; Ru 178°; Ph_3P; N; Cl; PPh_3; O; 174 pm; 116 pm

$\nu(NO)_{linear} = 1845$
$\nu(NO)_{bent} = 1687$
(Eisenberg, 1975 R)

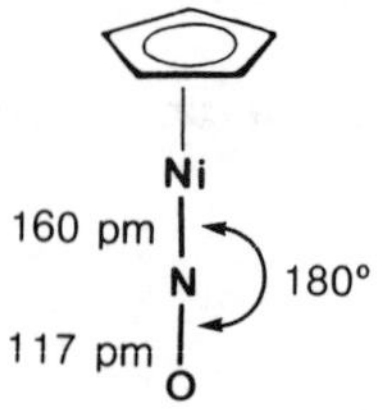

Ni; 160 pm; N 180°; 117 pm O

$\nu(NO)_{linear} = 1830$
(Cox, 1970)

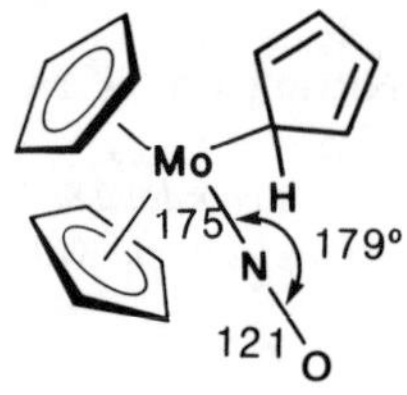

Mo; H; 175; 179°; N; 121; O

$\nu(NO)_{linear} = 1610$
fluxional at room temperature
(Cotton, 1969)

Compare: $\nu(NO)_{free} = 1876\ cm^{-1}$, $\nu(NO^+)_{free} = 2250\ cm^{-1}$

The ν_{CO} frequency is not a reliable criterion to distinguish between linear or bent MNO units, because the two absorption ranges overlap (Ibers, 1975). Bridging NO groups, however, exhibit considerably lower ν_{NO} frequencies and therefore are easily identified. Starting from $CpMn(CO)_3$, optically active compounds have been synthesized with **manganese as a centre of chirality:**

$$C_5H_5Mn(CO)_3 \xrightarrow[-CO]{NO^+PF_6^-,\ AN} [C_5H_5Mn(CO)_2NO]^+$$

$$\downarrow +PPh_3,\ -CO$$

$$[C_5H_5Mn^*(CO)(NO)PPh_3]^+$$

$$\downarrow +OR^{*-}$$

$$C_5H_5Mn^*(COOR^*)(NO)PPh_3$$

The preferential substitution of CO *indicates that its bond to the metal is weaker, as compared to* NO.

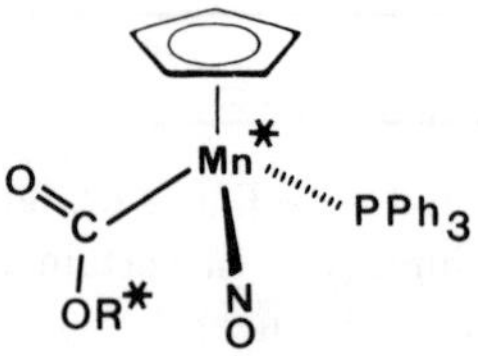

Mn*; O; C; OR^*; PPh_3; N; O

For $OR^- = OC_{10}H_{19}^-$ (L-*mentholate*), ***diastereomers*** *can be separated based on their different solubilities. In the solid state, the diastereomers are configurationally stable.*

The significance of these chiral complexes derives from their use in the study of organometallic reaction mechanisms, the observation of retention, inversion or racemisation serving as an experimental probe (Brunner, 1971 R).

15.4.3.4 Cyclopentadienyl Metal Hydrides

This type of complex is mainly encountered with $4d$- and $5d$ metals:

Cp_2MH_3 (M = Nb, Ta)	Cp_2MH_2 (M = Mo, W)	Cp_2MH (M = Tc, Re)

Corresponding complexes of the $3d$ metals tend to dimerize with attendant loss of dihydrogen.

PREPARATION

$$MoCl_5 + 2\,NaC_5H_5 + NaBH_4 \xrightarrow[-78\,°C/65\,°C]{THF} (C_5H_5)_2MoH_2$$

$$\underset{1\ :\ 10}{ReCl_5 + NaC_5H_5} \xrightarrow{THF} \underset{30\text{–}40\%\ \text{yield}}{(C_5H_5)_2ReH} \qquad \delta^1H = \begin{array}{l} 3.64\ \text{ppm (10 H)} \\ -23.5\ \text{ppm (1 H)} \end{array}$$

Cp_2ReH (Wilkinson, 1955) in contrast to the expected product Cp_2Re (cf. p. 325) is air-stable and diamagnetic. The investigation of Cp_2ReH was one of the first applications of ^1H-NMR spectroscopy in organometallic chemistry.

Re—H

Mo (H)(H)

Structure: In all complexes of composition $(C_5H_5)_mMH_n$ *known to date, the planes of the* C_5H_5 *rings are tilted in a fashion similar to* $Sn(C_5H_5)_2$ *(see p. 134).*

The ability of Re^{VII} to accomodate a large number of hydride ligands in its coordination shell [cf. $(R_3P)_2ReH_7$, $K_2(ReH_9)$] also manifests itself in the (cyclopentadienyl)metal hydrides (Herrmann, 1986):

$$Cp^*Re(O)Cl_2 \xrightarrow[2.\ CH_3OH]{1.\ LiAlH_4} Cp^*ReH_6 \qquad \nu_{(ReH)} = 2018\ cm^{-1}$$

At −90 °C, the ^1H-*NMR spectrum shows separate signals for the axial and the equatorial hydride ligands, which at 60 °C coalesce due to rapid exchange between the two positions. Proton chemical shifts of transition-metal hydrides are listed on p. 306.*

The **reactivity** of cyclopentadienyl metal hydrides is largely governed by the **Lewis-basicity of the central metal.** These complexes are readily protonated:

$$(C_5H_5)_2MH_2 \underset{OH^-}{\overset{H^+}{\rightleftharpoons}} [(C_5H_5)_2MH_3]^+ \quad (M = Mo, W)$$

$$(C_5H_5)_2ReH \underset{OH^-}{\overset{H^+}{\rightleftharpoons}} [(C_5H_5)_2ReH_2]^+ \quad \text{(base strength comparable to } NH_3)$$

Other Lewis acids like BF_3 can also add to cyclopentadienyl metal hydrides:

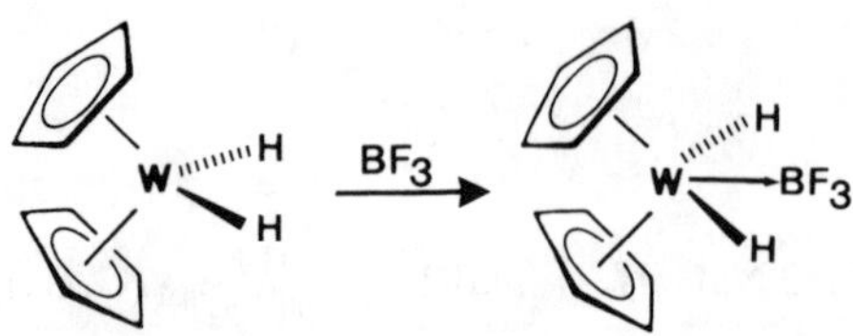

Adduct formation of a different type is realized in titanocenyl boranate where the moieties Cp_2TiH and BH_3 – unknown in free form – associate through $2e\,3c$ hydride bridges:

$$Cp_2TiCl_2 \xrightarrow{NaBH_4} Cp_2Ti(\mu\text{-}H)_2BH_2$$

The formation of Cp_2MH structural units, which then dimerize in a variety of ways, is a characteristic feature of the electron-poor metallocenes, Σ VE < 16 (cf. p. 323).

15.4.3.5 Cyclopentadienyl Metal Halides

Halide ions and hydride ions share the property of both being anionic $2e$ ligands. The variety of Cp-metal halides considerably outweighs that of the Cp-metal hydrides.

$CpMCl_2$ (Ti, V, Cr)	Cp_2MCl_2 (Ti, Zr, Hf, V, Mo, W)	Cp_3MCl (Th, U)
$CpMCl_3$ (Ti, Zr, V)	Cp_2MBr_2 (Nb, Ta)	
$CpMCl_4$ (Nb, Ta, Mo)	Cp_2MCl (Ti, V)	

PREPARATION

$$TiCl_4 + 2\,NaC_5H_5 \xrightarrow[25\,°C]{THF,} (C_5H_5)_2TiCl_2 + 2\,NaCl$$

$$TiCl_4 + (C_5H_5)_2TiCl_2 \xrightarrow[130\,°C]{Xylene} 2\,(C_5H_5)TiCl_3$$

$$(C_5H_5)_2V + (C_5H_5)_2VCl_2 \longrightarrow 2\,(C_5H_5)_2VCl$$

$$(C_5H_5)_2Cr + CCl_4 \xrightarrow{0\,°C} (C_5H_5)CrCl_2$$

$$(C_5H_5)_2MoH_2 + 2\,CHCl_3 \xrightarrow{60\,°C} (C_5H_5)_2MoCl_2 + 2\,CH_2Cl_2$$

129° M Cl 95.2° Cl

M = Ti,Zr,Hf

Complexes of the type **Cp_2MCl_2** (M = Ti, Zr, Hf) *have a "pseudotetrahedral" structure. They are air-stable and sublimable. The bond parameters quoted here refer to the titanium derivative.*

Cp-metal halides are starting materials for the synthesis of Cp-metal alkyls and Cp-metal aryls:

$$Cp_2ZrCl_2 \xrightarrow{2\ CH_3Li} Cp_2Zr(CH_3)_2 + 2\ LiCl$$

The Cp_2-metal chloride of Ti(III) is dimeric:

$$2\ TiCl_3 + 2\ MgCp_2 \longrightarrow [Cp_2TiCl]_2$$

Measurements of the magnetic susceptibility of $[Cp_2TiCl]_2$ *show that the singlet and triplet states coexist in thermal equilibrium thereby signalizing a weak intramolecular* Ti–Ti *interaction* (*R. L. Martin*, 1965).

The mixed Cp-metal chloride hydride $Cp_2ZrCl(H)$, presumably of polymeric structure, was introduced into organic synthesis as **Schwartz's reagent: By hydrozirconation** and subsequent oxidative decomplexation alkenes are converted into substituted alkanes (Schwartz, 1976 R).

Zr Cl H + (cyclohexene) → Zr Cl (cyclohexyl)

LiAlH$_4$ ↑ $Cp_2ZrClBr$ +

Br_2 → Br–(cyclohexyl); O_2 → HO–(cyclohexyl); D_2O → D–(cyclohexyl)

Mixed **Cp-metal carbonyl halides** are important reagents in organometallic synthesis:

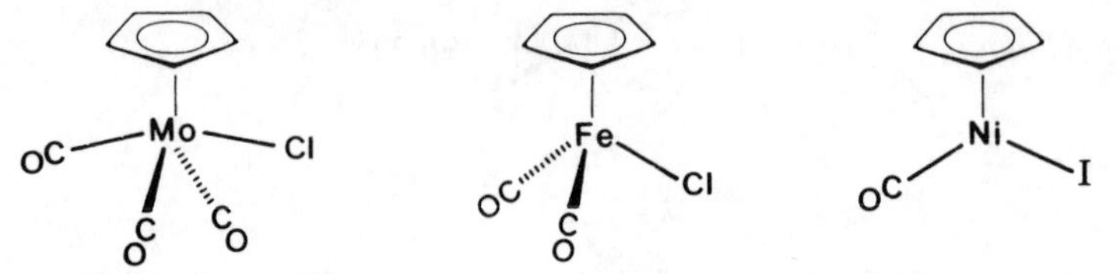

These compounds are prepared through the oxidative cleavage of dinuclear precursors by dihalogens:

$$[CpFe(CO)_2]_2 + I_2 \longrightarrow 2\,CpFe(CO)_2I$$

They serve as starting materials in the formation of new transition-metal carbon bonds:

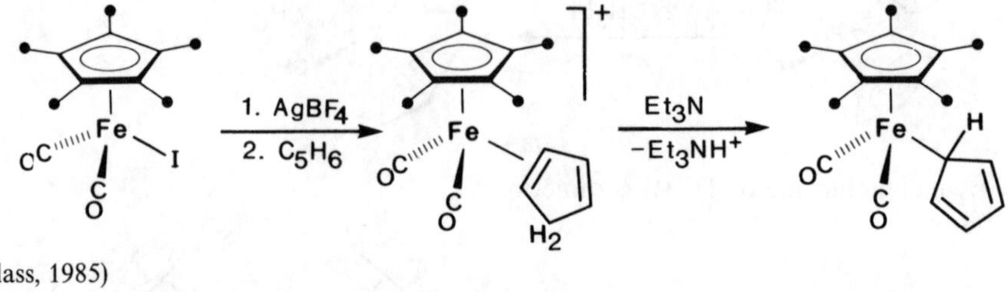

(Glass, 1985)

$(\eta^1\text{-Cp})(\eta^5\text{-Cp})Fe(CO)_2$ was the first transition-metal complex for which the phenomenon of fluxionality (non-rigidity) was studied (Wilkinson, 1956; Cotton, 1966). The two Cp-rings display the different bonding modes $\pi(\eta^5\text{-})$ and $\sigma(\eta^1\text{-})$ respectively, the latter being subject to haptotropic shifts, which cause the position 1–5 to become equivalent at room temperature (^{1}H-NMR, p. 343):

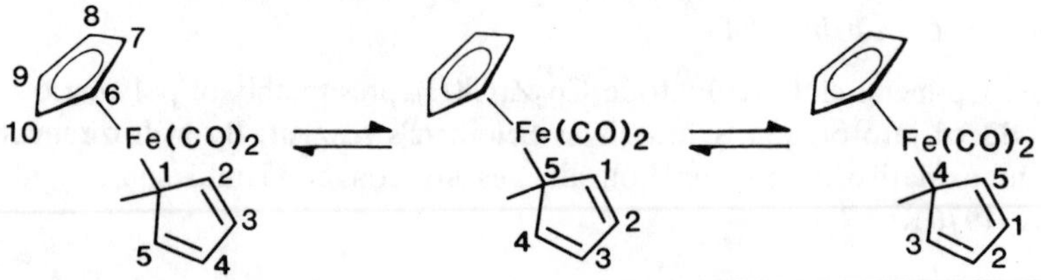

15.4.4 C_6H_6 as a Ligand

The coordination of neutral arenes to metals has been previously discussed in the context of main-group organometallics (p. 90). However, whereas for main-group elements metal-arene coordination must be regarded as a weak interaction, more or less confined to the solid state, the transition elements form the basis of a large variety of molecules containing (η^6-arene)metal segments. As early as 1919, F. Hein isolated "phenylchromium compounds" by treating $CrCl_3$ *with* PhMgBr *in diethylether. Although unrecognized at that time, these materials were bis(η^6-arene)chromium(I) cations. The first rational synthesis of bis(arene)metal complexes was developed by E. O. Fischer and W. Hafner in 1955.*

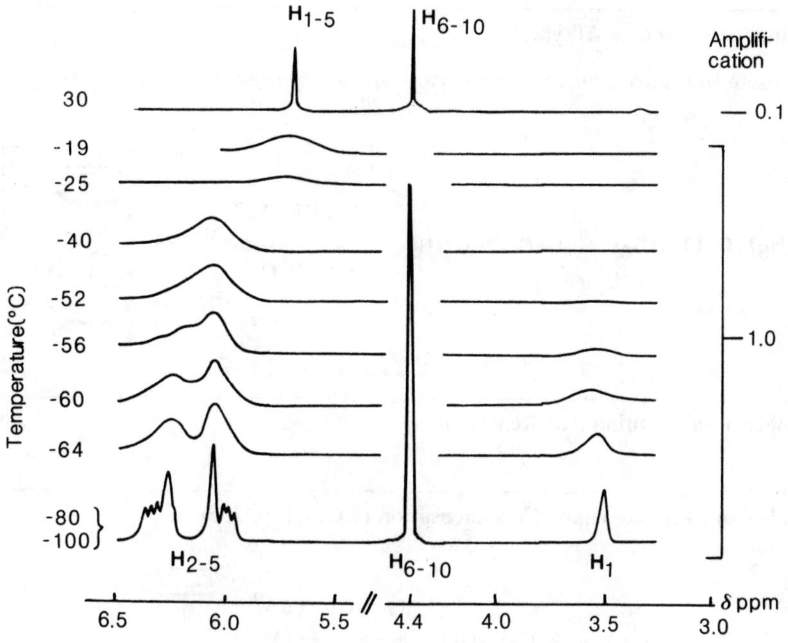

^{1}H-*NMR spectrum of* (η^1-Cp) (η^5-Cp)Fe(CO)$_2$ (60 MHz)

15.4.4.1 Bis(arene)metal Complexes

PREPARATION

---

1. Fischer-Hafner Synthesis

$$3\,CrCl_3 + 2\,Al + 6\,ArH \xrightarrow[2.\ H_2O]{1.\ AlCl_3} 3\,[(ArH)_2Cr]^+ + 2\,Al^{III}$$

$$\downarrow Na_2S_2O_4,\ KOH$$

$$(\eta^6\text{-}ArH)_2Cr$$

Scope:

V	Cr	–	Fe	Co	Ni
–	Mo	Tc	Ru	Rh	–
–	W	Re	Os	Ir	–

Limitations:

The arene must be inert towards $AlCl_3$. Alkylated arenes are isomerized by $AlCl_3$. Aromatic ligands with substituents having lone pairs like halobenzenes, dimethylaniline or phenols cannot be used since they form Lewis adducts with $AlCl_3$.

---

2. Cyclotrimerization of Alkynes

This route to bis(arene)metal complexes is of minor preparative importance (cf. p. 277).

$$(C_6H_5)_3Cr(THF)_3 + H_3CC{\equiv}CCH_3 \xrightarrow[2)\ H_2O,\,O_2]{1)\ THF,\,20^{\circ}} [(\eta^6\text{-}C_6Me_6)_2Cr]^+$$

—• = —CH_3

3. Peripheral Substitution (see Reactions)

4. Metal-Atom Ligand-Vapor Cocondensation (CC) (cf. p. 255)

$$Ti(g) + 2\ C_6H_6(g) \xrightarrow[2)\ 25^{\circ}]{1)\ CC,\,-196^{\circ}} (C_6H_6)_2Ti$$

The following simple bis(η^6-arene)metal complexes as yet can only be prepared by means of cocondensation techniques:

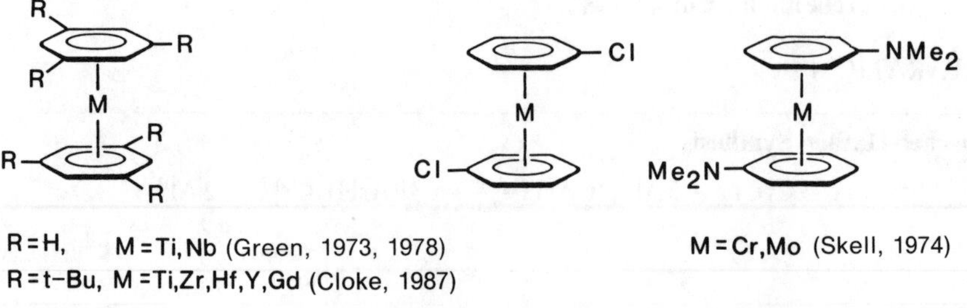

R = H, M = Ti, Nb (Green, 1973, 1978)
R = t-Bu, M = Ti, Zr, Hf, Y, Gd (Cloke, 1987)

M = Cr, Mo (Skell, 1974)

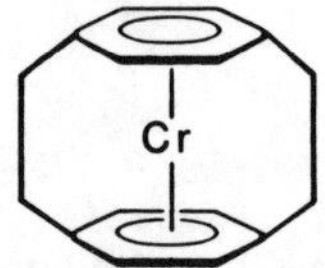

(η^{12}-[2.2]Paracyclophane)chromium(0) *is a compressed sandwich complex; the average distance between the two rings in the free ligand (ca.* 290 pm) *falls considerably short of the corresponding distance in* (η^6-C_6H_6)$_2$Cr (322 pm) (*Elschenbroich,* 1978).

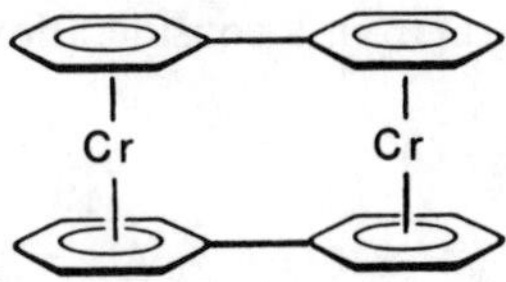

Bis[μ-(η^6 : η^6-biphenyl)]dichromium. *In solution, the monocation [ ]$^+$ represents the* ***intermediate valence*** *case* $Cr^{+1/2}$, $Cr^{+1/2}$. *The paramagnetic dication [ ]$^{2+}$ is an organometallic triplet species* (*Elschenbroich,* 1979). *In contrast, the structurally related cation* [*bis-μ-(η^5 : η^5-fulvalenediyl)diiron*]$^{2+}$ *is diamagnetic* (*p.* 327).

Neutral sandwich complexes of anellated arenes are also accessible by the CC method only (Elschenbroich, Kündig, Timms, 1977f). Examples:

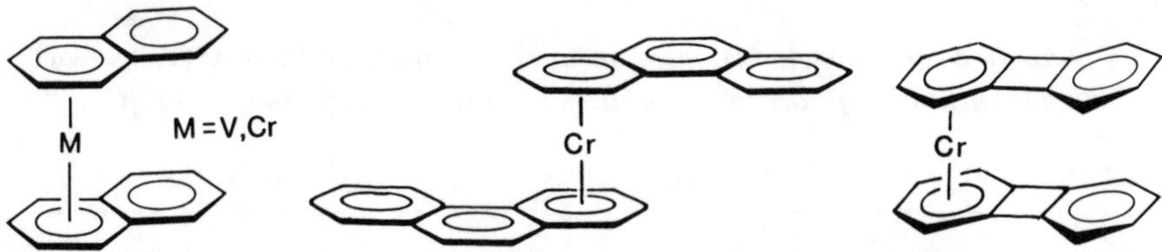

In the case of higher anellated arenes like phenanthrene, triphenylene and coronene, the metal atom preferentially coordinates to the ring with the highest index of local aromaticity (ILA), i.e. the ring which among the canonical forms most frequently appears with a Kekulé structure. This invariably is the terminal, least anellated ring:

ELECTRONIC STRUCTURE AND BONDING IN BIS(ARENE)METAL COMPLEXES

The situation largely resembles that of the metallocenes. Shifts of the energy levels in the MO diagram of bis(benzene)chromium, as compared to ferrocene, result from differences in the energies of the respective basis orbitals. These are lower for C_6H_6 than for $C_5H_5^-$ and higher for Cr^0 than for Fe^{2+}. As suggested by the X-ray photoelectron spectrum (ESCA), the central metal atom in bis(benzene)chromium(0) bears a partial positive charge (ca. + 0.7) and the ligands partial negative charges (ca. − 0.35).
The **bond energy** per ring $\boldsymbol{E(Cr-C_6H_6)}$, as determined by thermochemical measurements, amounts to about **170 kJ/mol;** it is lower than the corresponding value **(220 kJ/mol)** found for **ferrocene.**

As usual, the MO treatment starts from a consideration of the ligand π MO's:

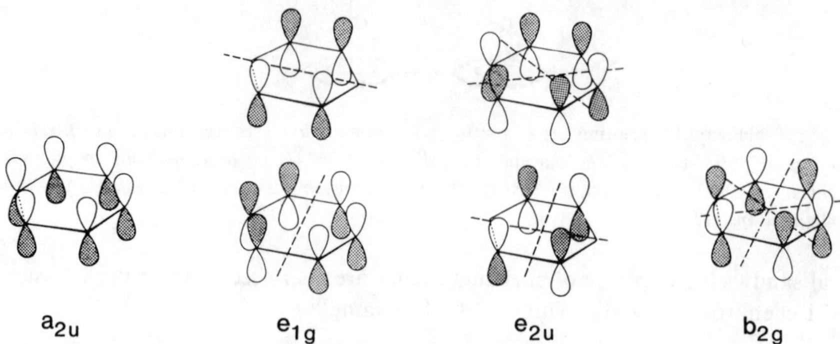

Interactions of symmetry-adapted linear combinations of the π MO's of two C_6H_6 ligands with appropriate metal orbitals in bis(benzene)chromium (D_{6h}):

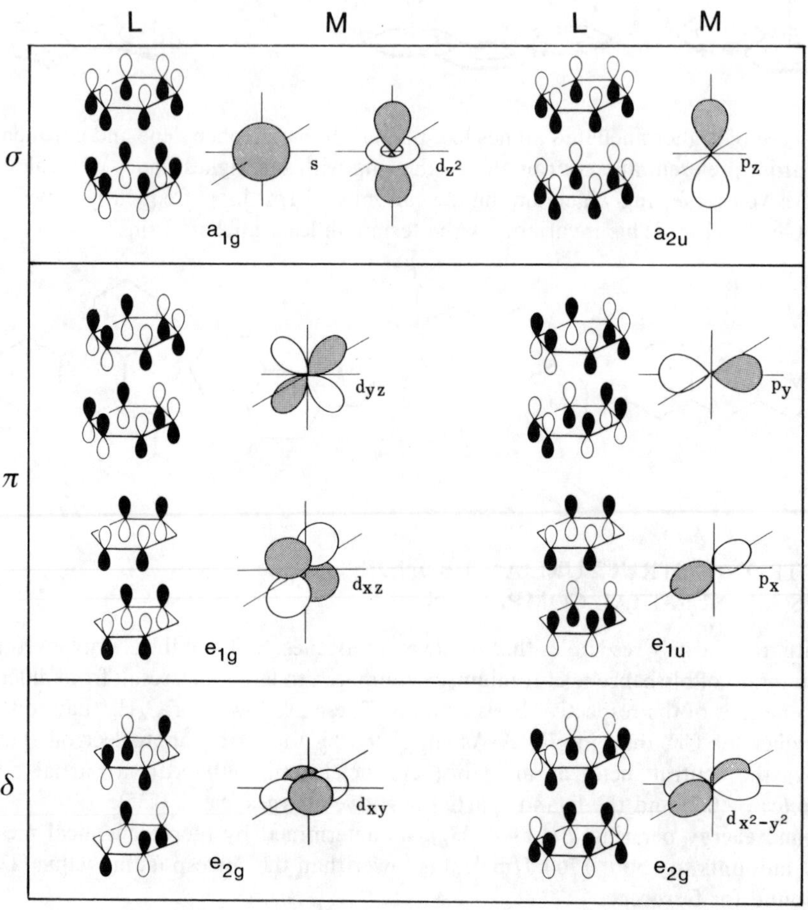

A qualitative MO diagram for bis(benzene)chromium (D_{6h}, intra-ligand σ orbitals are not included) is shown below:

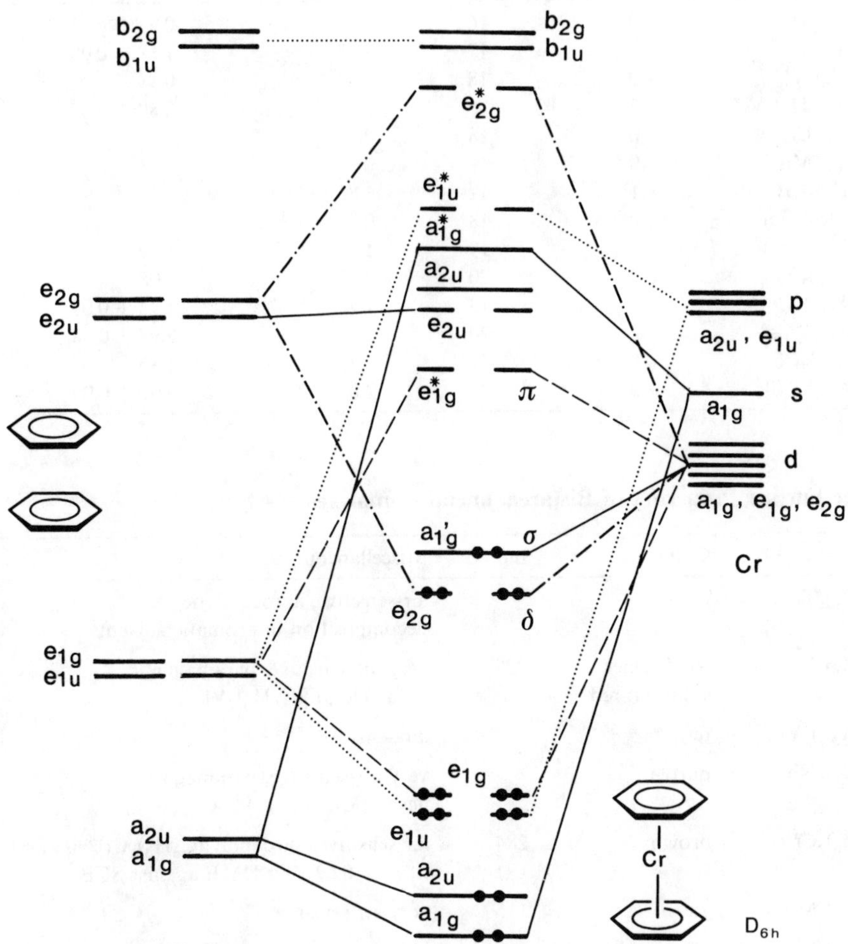

This MO diagram for $(C_6H_6)_2Cr$ is similar to that of $(C_5H_5)_2Fe$. As the basis orbitals of symmetry e_{2g} are closer in energy for the combination Cr^0/C_6H_6 than for $Fe^{2+}/C_5H_5^-$, the δ bond (backbonding, $C_6H_6 \leftarrow M$) contributes more significantly in bis(benzene)-chromium as compared to ferrocene.

Magnetic Properties of Bis(arene)metal Complexes

	Number of unpaired electrons	Σ VE	Magnetic moment	
			Spin-only value in B.M.	Experimental value in B.M.
$(C_6H_6)_2Ti$	0	16	0	0
$(C_6H_6)_2V$	1	17	1.73	1.68 ± 0.08
$[(C_6H_6)_2V]^-$	0	18	0	0
$[(C_6Me_3H_3)_2V]^+$	2	16	2.83	2.80 ± 0.07
$(C_6H_6)_2Cr$	0	18	0	0
$(C_6H_6)_2Mo$	0	18	0	0
$[(C_6H_6)_2Cr]^+$	1	17	1.73	1.77
$[(C_6Me_6)_2Fe]^{2+}$	0	18	0	0
$[(C_6Me_6)_2Fe]^+$	1	19	1.73	1.89
$(C_6Me_6)_2Fe$	2	20	2.83	3.08
$[(C_6Me_6)_2Co]^{2+}$	1	19	1.73	1.73 ± 0.05
$[(C_6Me_6)_2Co]^+$	2	20	2.83	2.95 ± 0.08
$(C_6Me_6)_2Co$	1	21	1.73	1.86
$[(C_6Me_6)_2Ni]^{2+}$	2	20	2.83	3.00 ± 0.09

Some Further Properties of Bis(arene)metal Complexes

Complex	Color	mp/°C	Miscellaneous
$(C_6H_6)_2Ti$	red	–	air-sensitive, autocatalytic decomposition in aromatic solvents
$(C_6H_6)_2V$	solid: black solution: red	227	very air-sensitive, paramagnetic, reducible to $[(C_6H_6)_2V]^-$
$(C_6H_5F)_2V$	red	–	air-sensitive
$(C_6H_6)_2Nb$	purple		very air-sensitive, paramagnetic, decomposes at ca. 90 °C
$(C_6H_6)_2Cr$	brown	284	air-sensitive, the cation $[(C_6H_6)_2Cr]^+$ is air-stable, $E^0 = -0.69$ V in DME against SCE
$(C_6H_6)_2Mo$	green	115	very air-sensitive
$(C_6H_6)_2W$	yellow-green	160	less air-sensitive than $(C_6H_6)_2Mo$
$[(C_6Me_6)_2Mn]^+$	pale pink	–	diamagnetic
$[(C_6Me_6)_2Fe]^{2+}$	orange	–	reducible to $[(C_6Me_6)_2Fe]^+$, violet, and to $(C_6Me_6)_2Fe^0$, black, paramagnetic, extremely air-sensitive
$[(C_6Me_6)_2Ru]^{2+}$	colorless	–	air-stable, diamagnetic, reducible to $(C_6Me_6)_2Ru$, orange, diamagnetic
$[(C_6Me_6)_2Co]^+$	yellow		paramagnetic, reducible to $(C_6Me_6)_2Co^0$, very air-sensitive

STRUCTURAL FEATURES

In the years immediately following the synthesis of **bis(benzene)chromium,** the structure and symmetry of this complex (D_{3d} or D_{6h}?) was the subject of considerable debate. While IR spectra and neutron-diffraction studies spoke for "localized" electron pairs and alternating C–C bond lengths as in a fictitious cyclohexatriene, electron diffraction and low-temperature X-ray diffraction studies confirmed a geometry with equal bond lengths (symmetry D_{6h}). This is the structure generally accepted today.

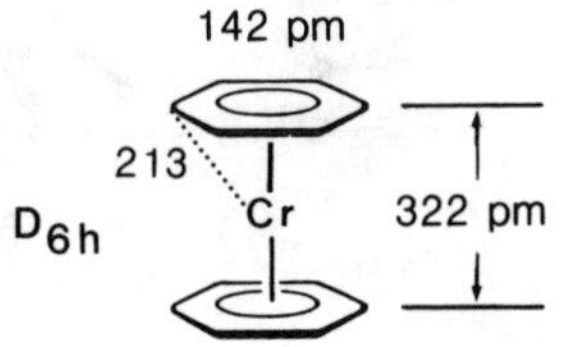

142 pm
213
Cr
322 pm
D_{6h}

(Jellinek, 1966)

The inter-ring distance is approximately equal to the van der Waals separation of two π-systems. Compared to free benzene, d(C–C) is somewhat elongated in the complex as would be expected from the participation of antibonding benzene orbitals (π^*, e_{2u}) in metal-ligand bonding. The rotational barrier is very low, probably $\lesssim$ 4 kJ/mol *in the gas phase. According to ENDOR (electron nuclear double resonance) experiments on **bis(benzene)vanadium**, ring rotation prevails even in the solid state (Schweiger,* 1984).

Paramagnetic **bis(hexamethylbenzene)rhenium** is unstable as a monomer at room temperature. The monomeric form can only be detected by means of ESR spectroscopy at – 196 °C. At room temperature, the dimer is irreversibly formed (E. O. Fischer, 1966):

$$\underset{18\ \text{VE}}{[(C_6Me_6)_2Re]PF_6} \xrightarrow[200\,^\circ C]{Li(\ell)} \underset{\substack{19\ \text{VE}\\ \text{black}}}{(C_6Me_6)_2Re^{\bullet}} \xrightarrow[RT]{} \underset{\substack{18\ \text{VE}\\ \text{yellow}}}{[(C_6Me_6)_2Re]_2}$$

This process can be formally ascribed to a transition Re(0) → Re(I) and the formation of a cyclohexadienyl ligand bearing an unpaired electron which is then used in inter-ligand bonding. The transition metal thus retains its 18 VE configuration. (Compare the analogous dimerization of rhodocene!) The most likely structure of the dimer is:

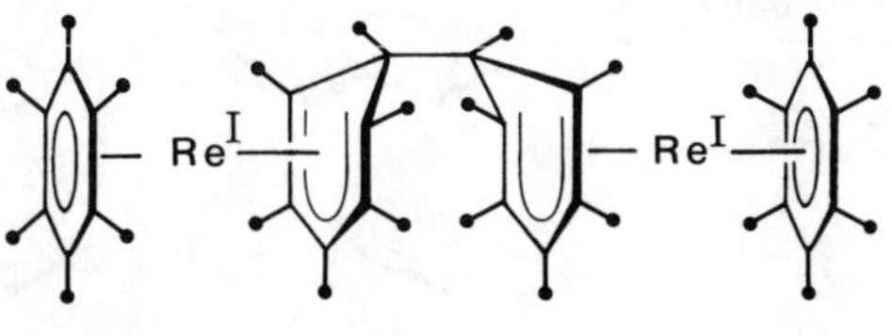

Re^I Re^I

Bis(hexamethylbenzene)ruthenium(0) as an axially symmetrical sandwich complex would be a 20 VE system. However, since one of the rings is only η^4-coordinated, an 18 VE configuration is attained. The complex is **fluxional** in solution (E.O. Fischer, 1970):

$(\eta^6\text{-}Me_6C_6)_2Ru^{2+}$ → (pentane) Ru^0 (η^6, η^4, 43°)

—● = —CH_3

18 VE 18 VE

The ^{1}H *NMR spectrum displays coalescence to a single methyl proton signal at* $T = 35\,°C$. *An activation enthalpy of* 65.3 kJ/mol *has been determined for this fluxional process* (*Muetterties*, 1978).

REACTIONS

(a) Oxidation

All neutral bis(arene)metal complexes with hydrocarbon ligands are **air-sensitive.** This sensitivity to oxidation is diminished on replacement of the ring hydrogen atoms by electron-withdrawing groups. Example: $(\eta^6\text{-}C_6H_5Cl)_2Cr$ is air-stable.

(b) Ligand Exchange

Bis(benzene)chromium(0) is **kinetically inert.** Ring exchange occurs only in the presence of a Lewis acid like $AlCl_3$. This reaction is, however, unsuitable for the synthesis of other derivatives (compare the limitations to the Fischer-Hafner synthesis, p. 343). In contrast, binary complexes of condensed arenes are **labile**. Example:

$$(\eta^6\text{-}C_{10}H_8)_2Cr + 3\,\text{bipy} \xrightarrow{25\,°C} (\text{bipy})_3Cr + 2\,C_{10}H_8$$

This lability of $(C_{10}H_8)_2Cr$ renders possible the synthesis of a mixed benzene-naphthalene dinuclear complex (Lagowski, 1987):

$(C_{10}H_8)_2Cr$ → (THF, C_6H_6, 100°, autoclave) Cr + Cr

(c) Electrophilic Aromatic Substitution

In contrast to ferrocene, this reaction is not applicable to bis(benzene)chromium, since the attacking electrophile oxidizes Cr(0) to Cr(I), yielding the cation $[(C_6H_6)_2Cr]^+$ which does not undergo peripheral electrophilic substitution.

(d) Metallation

H/metal exchange is effected by means of *iso*-amylsodium or, better, by *n*-butyllithium in the presence of *N,N,N',N'*-tetramethylethylenediamine (TMEDA). π-Bonded benzene is metallated more readily than free benzene, i.e. the kinetic acidity of benzene is increased upon coordination to chromium. A mixture of lithiation products is obtained, however. This reaction paves the way for the preparation of substituted bis(arene)metal complexes. Examples:

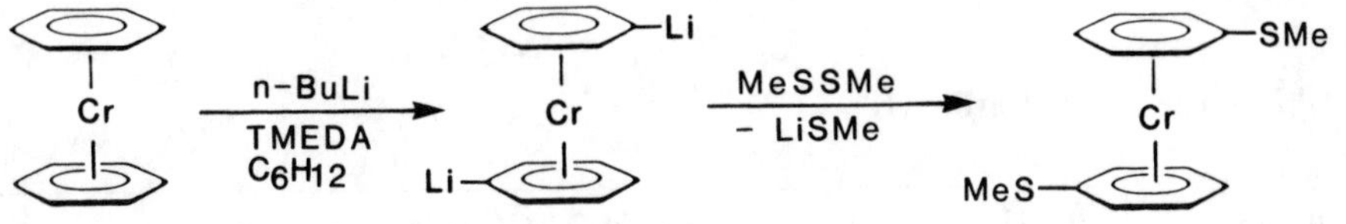

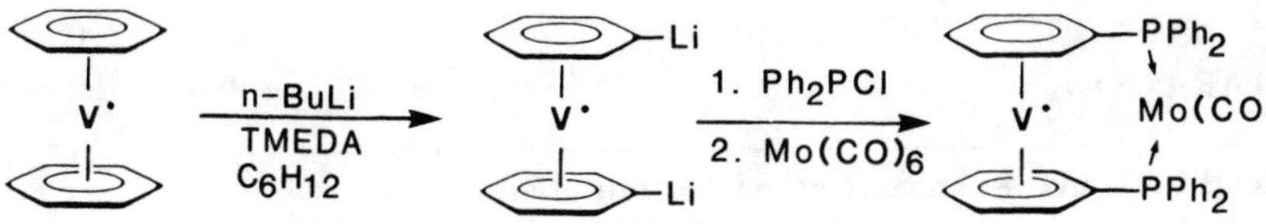

The vanadium complex with π-bonded PPh_3 may function as a paramagnetic bidentate ligand.

(e) Addition of nucleophiles

Cationic bis(arene)metal complexes readily react with nucleophiles by ring addition:

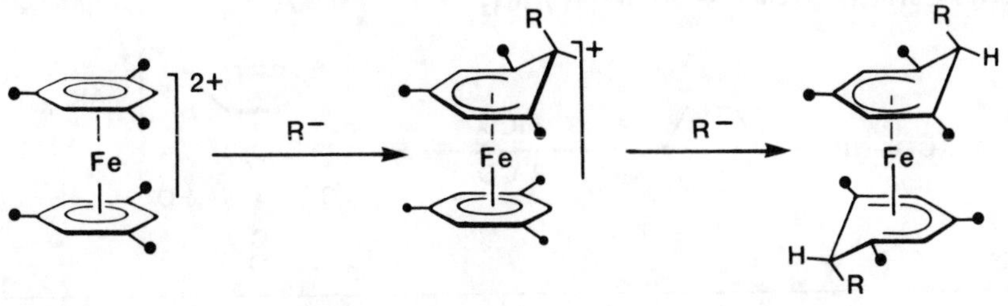

The second addition of R^- to the iron complex is in accordance with the DGM rule ① (p. 293). The analogous ruthenium complex, on the other hand, also undergoes homoannular secondary addition, two products being obtained:

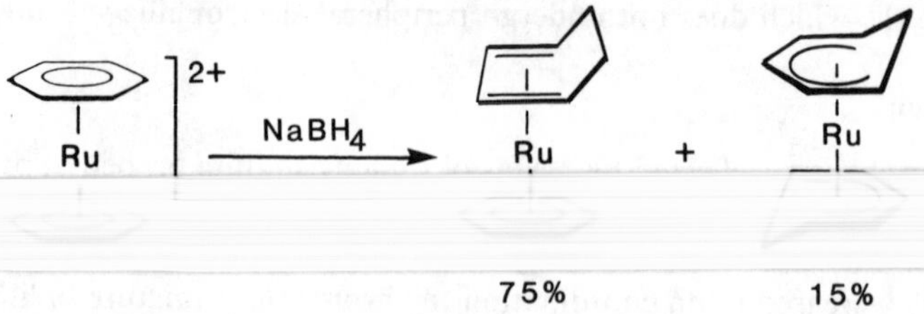

75% 15%

This difference could be an outcome of thermodynamic control, in that the more noble metal ruthenium prefers the oxidation state M^0. (Note that the DGM-rules only apply to kinetically controlled reactions).

15.4.4.2 Arene Metal Carbonyls

Examples:

$[C_6H_6V(CO)_4]^+$	$C_6H_6M(CO)_3$ (M = Cr, Mo, W)
$[C_6H_6Mn(CO)_3]^+$	$C_6H_6Fe(CO)_2$
$[(C_6H_6)_3Co_3(CO)_2]^+$	

PREPARATION

1. Metal Carbonyl + Arene (Carbonyl Substitution)

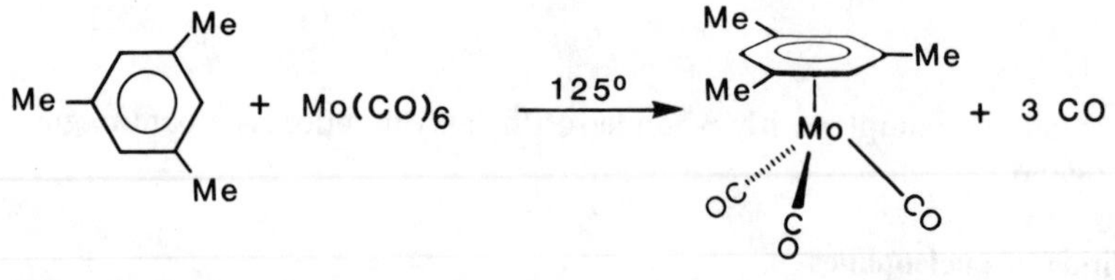

This CO substitution does not proceed beyond the stage $(ArH)M(CO)_3$.

2. Metal-Carbonyl Halide + Arene + $AlCl_3$

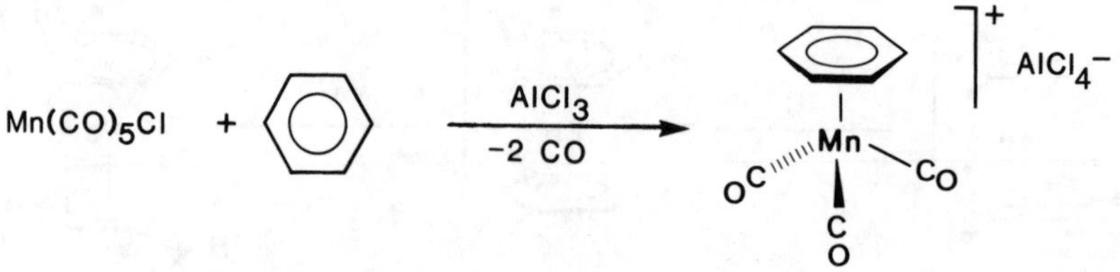

3. Ligand Exchange

$$(ArH)Cr(CO)_3 + ArH' \xrightarrow{140-160\,°C} (ArH')Cr(CO)_3 + ArH$$
$$(CH_3CN)_3Cr(CO)_3 + ArH \xrightarrow{25\,°C} (ArH)Cr(CO)_3 + 3\,CH_3CN$$
$$py_3Mo(CO)_3 + ArH \xrightarrow{Et_2OBF_3} (ArH)Mo(CO)_3 + 3\,pyBF_3$$

STRUCTURAL FEATURES

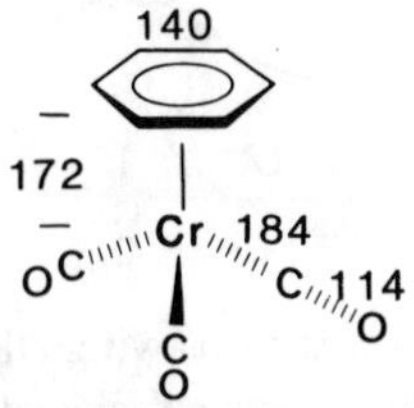

The prototype "piano-chair" complex, **η^6-benzene(tricarbonyl)chromium,** *does not present any structural surprises. The* $M(CO)_3$ *unit and the arene are staggered and the six-fold symmetry of η^6-benzene is maintained* (*Dahl*, 1965).
Compare $d(C{\equiv}O)_{free} = 113$ pm.

(Carbonyl)metal units also furnish examples for bonding modes of benzene which diverge from the usual η^6-type:

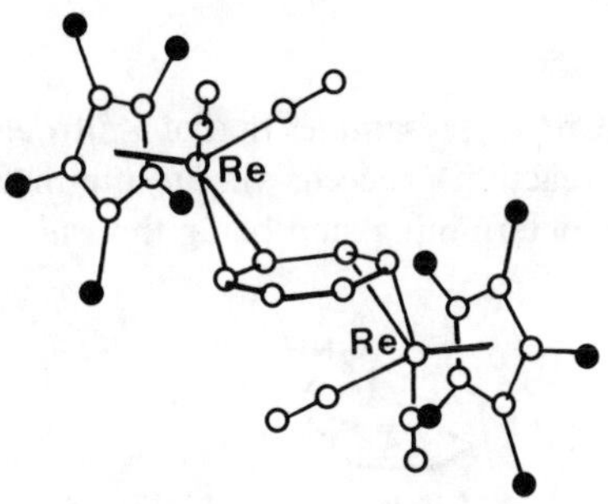

μ_2-($\eta^2:\eta^2$-C_6H_6) [(η^5-C_5Me_5)Re(CO)$_2$]$_2$. *This complex forms on irradiation of* $(C_5Me_5)Re(CO)_3$ *in benzene solution* (*Orpen*, 1985). *The benzene ring shows pronounced* C–C *bond length alternation.*

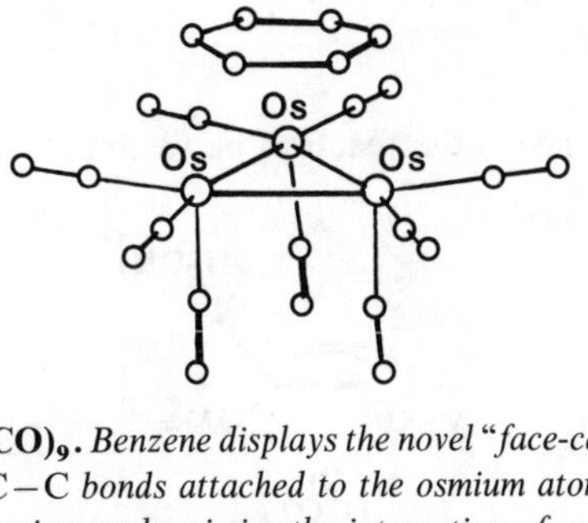

μ_3-($\eta^2:\eta^2:\eta^2$-C_6H_6)$Os_3(CO)_9$. *Benzene displays the novel "face-capping" ligation here* (*B. F. G. Johnson, J. Lewis*, 1985); *the* C–C *bonds attached to the osmium atoms are shorter than the unattached* C–C *bonds. This coordination mode mimics the interaction of arenes with metal surfaces.*

REACTIONS OF ARENE METAL CARBONYLS

Complexation of an arene to the $Cr(CO)_3$ fragment alters its reactivity in several ways, as summarized below (adapted from Semmelhack, 1976):

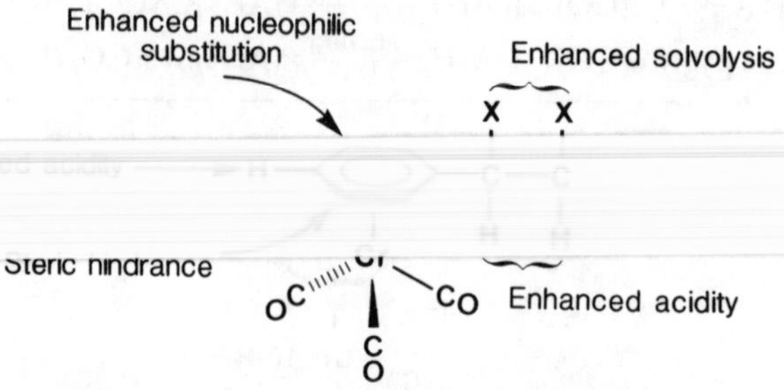

Enhanced nucleophilic substitution

Enhanced solvolysis

Enhanced acidity

Steric hindrance

Enhanced acidity

The $Cr(CO)_3$ unit exerts an electron-withdrawing effect. Compared to the free arene, the complexed arene is thus deactivated toward **electrophilic substitution,** but particularly susceptible to **nucleophilic substitution:**

NaOMe / MeOH

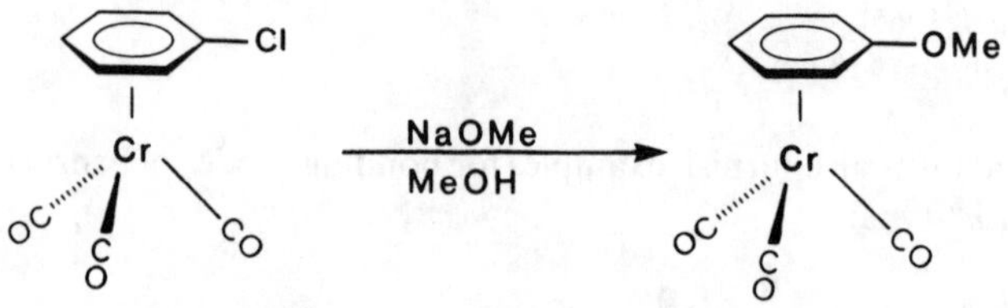

The reaction rate of $(C_6H_5Cl)Cr(CO)_3$ resembles that of *p*-nitrochlorobenzene. According to kinetic measurements, the reaction proceeds via an intermediate with an η^5-bonded cyclohexadienyl ring, the rate-determining step being the *endo*-loss of the halide ion.

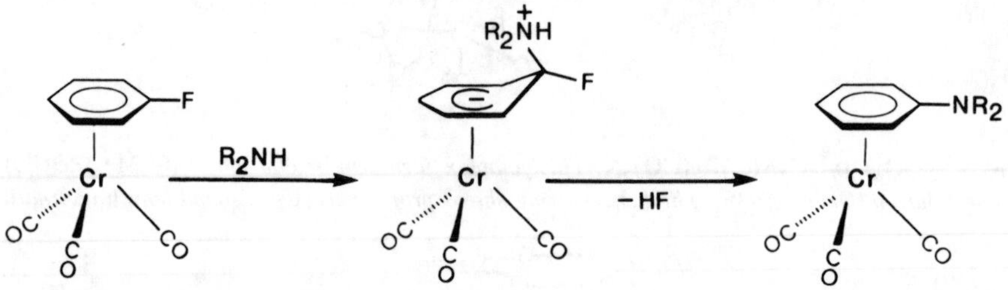

R_2NH ; $-HF$

Of greater preparative importance is the reaction between substituted arene chromium tricarbonyls and carbanions:

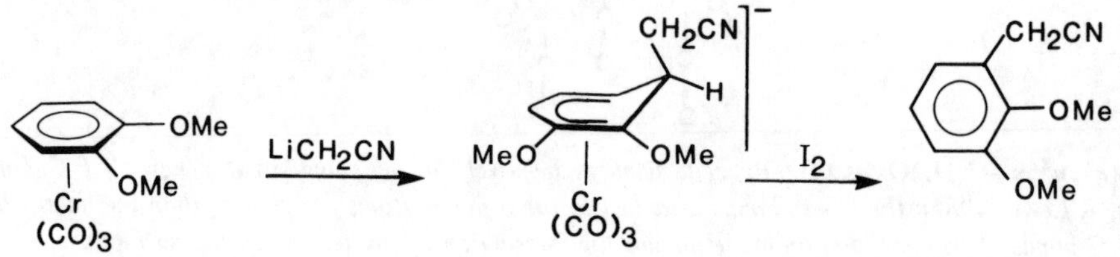

$LiCH_2CN$; I_2

Nucleophilic attack occurs at the uncomplexed face of the arene ligand (*exo*). The intermediary anionic cyclohexadienyl complex has been isolated in some cases (Semmelhack, 1981 R).

Attack of σ-donor/π-acceptor ligands can cause a **displacement of the η^6-arene ligand:**

$$C_6H_3Me_3Cr(CO)_3 + 3\,PF_3 \xrightarrow[140^\circ-150^\circ]{} fac\text{-}(PF_3)_3Cr(CO)_3$$

Oxidation of 18 VE (arene)metal carbonyls in contrast to bis(arene)metal complexes does not lead to stable 17 VE radical cations. On oxidation of (arene)molybdenum tricarbonyl with iodine, molybdenum is oxidized to Mo^{II} and an unusual counterion is formed (Calderazzo, 1986):

$$\underset{18\ VE}{3\,(\eta^6\text{-}ArH)Mo(CO)_3} \xrightarrow{I_2} \underset{18\ VE}{[(\eta^6\text{-}ArH)Mo(CO)_3I]^+} + [Mo_2I_5(CO)_6]^-$$

15.4.4.3 Other Complexes of the General Type (η^6-ArH)ML_n

Only two reactions of fairly extensive scope are presented here:

$$2M + 2\,C_6F_5Br + C_6H_5Me \xrightarrow[-\,MBr_2]{CC} (\eta^6\text{-}C_6H_5Me)M(C_6F_5)_2 \xrightarrow{Arene} (\eta^6\text{-arene})M(C_6F_5)_2$$

M = Co, Ni

In this way, $(C_6F_5)_2M$ fragments can be transferred to a large number of arenes (Klabunde, 1978).

Cationic complexes of the type $[(\eta^6\text{-arene})Ru(solv)_3]^{2+}$ are useful reagents for transferring $(arene)Ru^{2+}$ half-sandwich units to other arenes:

Cyclohexadiene + $RuCl_3$ $\xrightarrow{C_2H_5OH}$ $[(C_6H_6)RuCl_2]_2$ (Cl-bridged dimer)

$[(C_6H_6)RuCl_2]_2$ $\xrightarrow{L}$ $(C_6H_6)RuCl_2L$; L = PR_3, CO etc.

$[(C_6H_6)RuCl_2]_2$ $\xrightarrow[-\,2\,AgCl]{AgBF_4,\ acetone}$ $[(C_6H_6)Ru(acetone)_3]^{2+}$ 2 BF_4^-

Bennett (1979)

$[(C_6H_6)Ru(acetone)_3]^{2+}$ + [2.2]paracyclophane → $[(C_6H_6)Ru(paracyclophane)Ru(C_6H_6)]^{4+}$

Boekelheide (1980)

15.4.4.4 Benzene Cyclopentadienyl Metal Complexes

Complexes containing both six- and five-membered rings are accessible by several routes (E.O. Fischer, 1966):

$$MnCl_2 + NaC_5H_5 + C_6H_5MgBr \xrightarrow[2.\ H_2O]{1.\ THF} (C_5H_5)Mn(C_6H_6) \quad 2\%$$

(Biphenyl)MnC_5H_5 is formed as the main product; a further byproduct is the dinuclear complex µ-(η^6:η^6-biphenyl)bis[(cyclopentadienyl)manganese]. $(C_6H_6)CrC_5H_5$ (17 VE) can be prepared in an analogous manner.

One cyclopentadienyl ring of ferrocene can be exchanged for an arene (Nesmeyanow, 1963):

$$(C_5H_5)_2Fe + C_6H_5R \xrightarrow[3\ :\ 1\ :\ 1]{AlCl_3/Al/H_2O} [(C_5H_5)Fe(C_6H_5R)]^+ + C_5H_5$$

Reduction of these 18 VE (arene) (cyclopentadienyl)iron cations affords stable 19 VE radicals (Astruc, 1983 R):

$$(C_5H_5)Fe(C_6Me_6H) \xleftarrow{NaBH_4} [(C_5H_5)Fe^{II}(C_6Me_6)]^+ \xrightarrow[DME]{Na/Hg} (C_5H_5)Fe^{I}(C_6Me_6)$$

18 VE (d^6) 19 VE (d^7)

In concordance with the DGM rules (p. 293) nucleophilic addition occurs exclusively at the six-membered ring, neutral cyclohexadienyl complexes being formed.

(Arene) (cyclopentadienyl)ruthenium cations are only obtained in low yields from ruthenocene; a better synthetic route proceeds from dimeric (benzene)ruthenium dichloride (Baird, 1972):

$$[C_6H_6RuCl_2]_2 + 2\ TlCp \xrightarrow[2.\ H_2O,NH_4PF_6]{1.\ EtOH} 2\ [(C_5H_5)Ru(C_6H_6)]^+ \ PF_6^-$$

The cation $[(C_5H_5)Ru(C_6H_6)]^+$ is photolabile, forming solvates $[(C_5H_5)Ru(solv)_3]^+$ which transfer **$(C_5H_5)Ru^+$ fragments** to other arenes (cf. p. 333, 368):

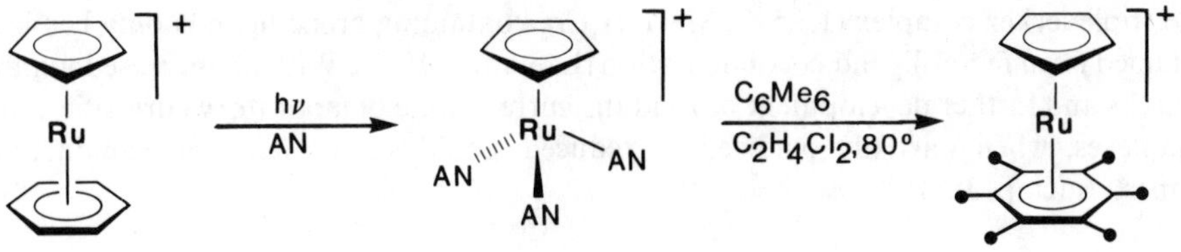

(K. R. Mann, 1982)

Successive elimination of two H^- ions from a cyclohexadiene complex can also lead to the formation of an $(\eta^5\text{-}C_5H_5)M(\eta^6\text{-}C_6H_6)$ species:

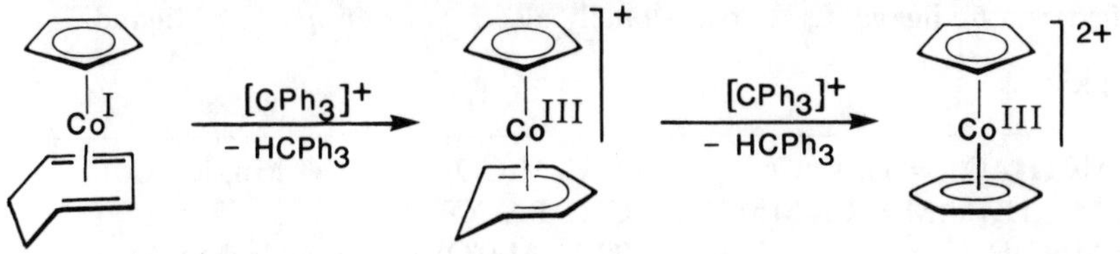

The class of mixed ligand complexes features examples in which arenes act as bridging ligands, displaying *syn-* as well as *anti*-coordination. In the following *syn*-complex of iron, the bridging ligand is generated by trimerization of acetylenes (Jonas, 1983):

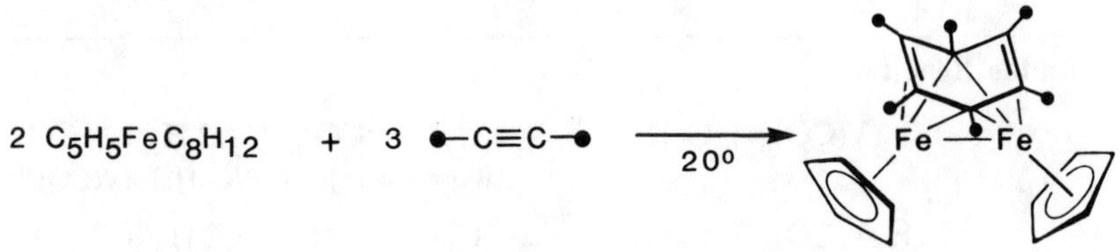

Noteworthy is also the first **tripledecker complex** with a bridging benzene ligand $[C_5H_5V]_2C_6H_6$ (Jonas, 1983):

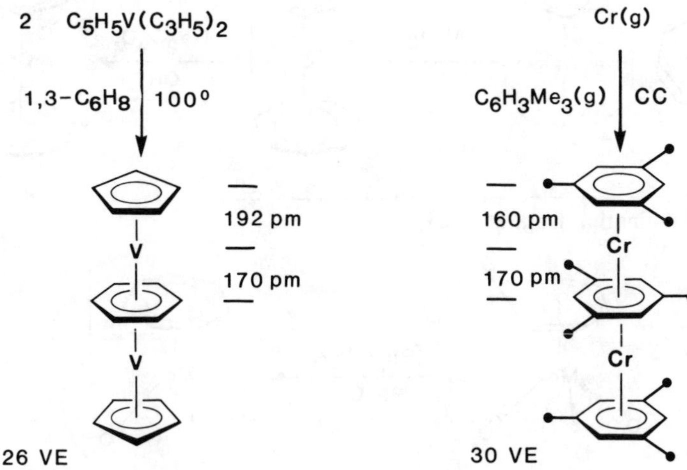

The tripledecker complex $(1,3,5\text{-}C_6Me_3H_3)_3Cr_2$ containing arene ligands only has been obtained from metal-ligand cocondensation (Lamanna, 1987). With an increased number of decks and further development beyond the current state of laboratory curiosities, such complexes, when partially oxidized or reduced, could show interesting conductivity properties (cf. p. 383).

15.4.5 $C_7H_7^+$ as a Ligand

η^7-C_7H_7 metal complexes are almost exclusively of the mixed-ligand or half-sandwich type, other ligands being present in addition to the cycloheptatrienyl ring. The ligand C_7H_7 is generally regarded as a coordinated aromatic tropylium ion η^7-$C_7H_7^+$ and is thus classified as a **6*e* ligand.** C_7H_7 occasionally also acts as an η^3- or η^5-ligand.

Examples:

$C_7H_7MC_5H_5$ (M = Ti, V, Cr)	$C_7H_7M(CO)_3$ (M = V, Mn, Re, Co)
$[C_7H_7MC_5H_5]^+$ (M = Cr, Mn)	$C_7H_7Re(CO)_4$
$[C_7H_7MoC_6H_6]^+$	$[C_7H_7M(CO)_3]^+$ (M = Cr, Mo, W, Fe)
$[(C_7H_7)_2V]^{2+}$	$[C_7H_7Fe(CO)_3]^-$

PREPARATION

1. Substitution Reactions

$$V(CO)_6 + C_7H_8 \xrightarrow{65\,°C} C_7H_7V(CO)_3 + 1/2\,H_2 + 3\,CO$$

Byproduct: $[C_7H_7VC_7H_8]^+[V(CO)_6]^-$

$$C_5H_5V(CO)_4 + C_7H_8 \xrightarrow{120\,°C} C_5H_5VC_7H_7 + 1/2\,H_2 + 4\,CO$$

$$[C_7H_7Mo(CO)_3]^+ + C_6H_5CH_3 \xrightarrow{115\,°C/66\,h} [C_7H_7MoC_6H_5CH_3]^+ + 3\,CO$$

C_7H_8 + Cr (sandwich) $\xrightarrow{AlCl_3}$ [Cr]$^+$ $\xrightarrow[OH^-]{Na_2S_2O_4}$ Cr

2. Hydride Elimination from η^6-C_7H_8

η^6-$C_7H_8Mo(CO)_3$ $\xrightarrow[-Ph_3CH]{[Ph_3C]BF_4}$ $[\eta^7\text{-}C_7H_7Mo(CO)_3]^+$

$(C_7H_8)_2V \xrightarrow[-2\ Ph_3CH]{2\ [Ph_3C]\ BF_4}$ $[(C_7H_7)_2V]^{2+}$

3. Ring Expansion

$C_5H_5MnC_6H_6 \xrightarrow[-40^0]{RCOCl/AlCl_3}$ $[C_5H_5MnC_7H_6R]^+$

Formally, this reaction is an insertion of RC^+ into the six-membered ring. This remarkable transformation is only feasible for the combination $(\eta^5\text{-}C_5H_5)/(\eta^6\text{-}C_6H_6)$; it is also observed for $C_5H_5CrC_6H_6$, but not for $C_6H_6Cr(CO)_3$ or for free C_6H_6.

4. Photochemically or Chemically Induced σ/π-Rearrangement

$C_7H_7COCl \xrightarrow[-Cl^-]{[Re(CO)_5]^-} C_7H_7C(O)Re(CO)_5 \xrightarrow[-CO]{h\nu} C_7H_7Re(CO)_5 \longrightarrow$

$\xrightarrow{Me_3NO} C_7H_7Re(CO)_4 \xrightarrow{Me_3NO} C_7H_7Re(CO)_3$

The analogous irradiation of $C_7H_7COMn(CO)_5$ at $-68\,°C$ yields $\eta^5\text{-}C_7H_7Mn(CO)_3$ directly, intermediates being too short-lived for isolation.

5. Deprotonation

$\eta^4\text{-}C_7H_8Fe(CO)_3 \xrightarrow[-\ BuH]{n\text{-}BuLi} [\eta^3\text{-}C_7H_7Fe(CO)_3]^-$

In the case of larger cyclic ligands, application of the 18 VE rule does not generally allow unequivocal structural prediction because increasingly, only partial coordination of the π-perimeters is encountered. Partially coordinated seven-membered rings often show **fluxional behavior.**

STRUCTURE AND BONDING

The description of the η^7-C_7H_7-metal bond is analogous to that for η^5-C_5H_5 and η^6-C_6H_6. A common feature of all three carbocycles is the existence of π orbitals with symmetries a, e_1 and e_2 which, by interaction with the appropriate metal orbitals, generate bonds of σ-, π- und δ symmetry.

The crystal structure of $\mathbf{C_5H_5VC_7H_7}$ *reveals that the* V–C *distances to both rings are almost identical. Accordingly, the distances from the metal to the ring centers differ in that larger rings are approached more closely by the metal. This situation leads to relatively poor metal-ligand orbital overlap, so that centrosymmetric bonding of a first row transition metal to a large ring is less favorable than to a smaller ring (Rundle,* 1963*).*

Comparison of the structures of $\mathbf{C_7H_8Mo(CO)_3}$ *(left) and* $\mathbf{C_7H_7V(CO)_3}$ *(right): The triolefin complex shows considerable exo-bending of the non-coordinated* CH_2 *group as well as alternating* C–C *bond lengths; the tropylium complex, on the other hand, exhibits planar coordination of the* C_7H_7 *ring and equal* C–C *distances (Dunitz,* 1960; *Allegra,* 1961*).*

REACTIONS

(a) Redox Reactions

$$C_5H_5CrC_7H_7 + 1/2\,I_2 \longrightarrow [C_5H_5CrC_7H_7]^+ + I^-$$

The redox potential for the couple $C_5H_5CrC_7H_7^{0/+}$ (– 0.55 V *in* DME *versus* SCE) *is shifted anodically compared to the symmetrical isomer* $C_6H_6CrC_6H_6$ (– 0.69 V).

(b) Addition of Nucleophiles

$[C_7H_7Mo(CO)_3]^+ \xrightarrow{CH_3O^-} (CH_3O\text{-}C_7H_7)Mo(CO)_3$

(c) Ring Contraction

$[C_7H_7Cr(CO)_3]^+ \xrightarrow{NaC_5H_5} C_6H_6Cr(CO)_3$

Labeling experiments prove that the six-membered ring is formed from the tropylium ring rather than from the cyclopentadienyl ring, since reaction with $NaC_5H_4CH_3$ also gives (benzene)$Cr(CO)_3$ whereas treatment of $[C_7H_6CH_3Cr(CO)_3]^+$ with NaC_5H_5 generates (toluene)$Cr(CO)_3$ (Pauson, 1961).

(d) Formation of Dinuclear Complexes

$[C_7H_7Fe(CO)_3]^- + [Rh(CO)_2Cl]_2 \longrightarrow C_7H_7Fe(CO)_3Rh(CO)_2$

Partially coordinated seven-membered rings can form additional bonds to a second transition metal. This quite often results in the formation of a metal-metal bond as well as fluxional behavior of the bridging ligand (Takats, 1976).

15.4.6 C_8H_8 as a Ligand

In its versatility as a ligand, C_8H_8 (COT) resembles the acetylenes. The variety of COT complex chemistry is due to the following facts:

- C_8H_8 *can coordinate as a tub-shaped tetraolefin or as a planar aromatic anion* $C_8H_8^{2-}$
- *Potential coordination modes are* η^2, η^4, η^6 *and* η^8 *(in binuclear complexes also* η^3 *and* η^5*).*
- C_8H_8 *can function as a bridging ligand, supporting either syn- or anti-coordination.*
- *As an* $8e(C_8H_8)$ *or a* $10e(C_8H_8^{2-})$ *ligand cyclooctatetraene forms complexes with metals from the actinoid and lanthanoid series, in which, in varying degrees, metal f orbitals contribute to bonding.*

COMPLEXES WITH PLANAR C_8H_8 RINGS

$$
\begin{array}{lllll}
Ti(OC_4H_9)_4 & + C_8H_8 & + Al(C_2H_5)_3 & & \\
0.2 & : 0.4 & \quad 2 & \longrightarrow & Ti_2(C_8H_8)_3 \\
0.2 & : 2 & : 2 & \longrightarrow & Ti(C_8H_8)_2 \\
HfCl_4 & + 2\,Mg & + 2\,C_8H_8 & \longrightarrow & Hf(C_8H_8)_2 + 2\,MgCl_2 \\
Zr(allyl)_4 & + 2\,C_8H_8 & & \longrightarrow & Zr(C_8H_8)_2
\end{array}
$$

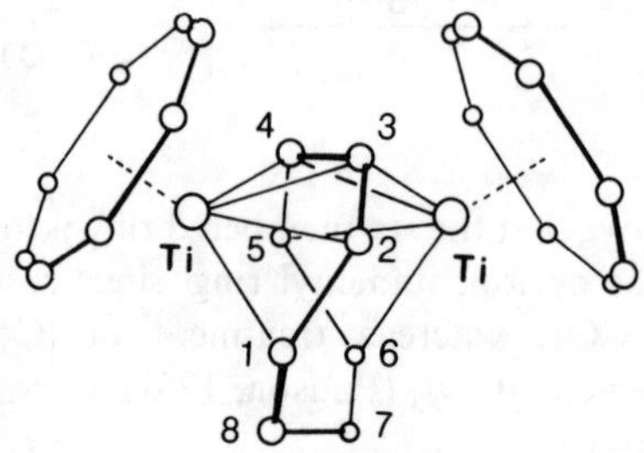

μ-[1–4η : 3–6η-C_8H_8] [(η^8-C_8H_8)Ti]$_2$: *yellow crystals, very air-sensitive, paramagnetic. One double bond of the bridging ligand remains uncoordinated; two carbon atoms are coordinated to* ***both*** *titanium atoms (Dietrich, 1966).*

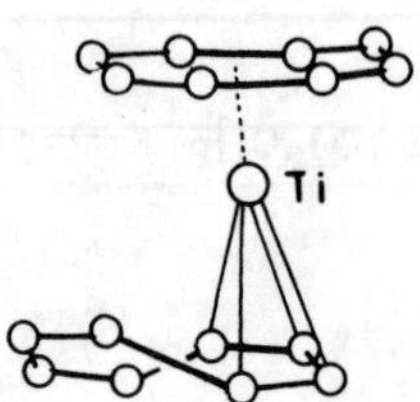

(η^8-C_8H_8) (η^4-C_8H_8)Ti: *Violet crystals, very air-sensitive. One* COT *ring is planar (all* d(C–C) = 141 pm), *while the other is puckered and bonded in a butadiene-like fashion. In solution,* ***fluxional behavior*** *is observed (Wilke, 1966).*

The **cyclooctatetraene complexes of the lanthanoids and actinoids** are of particular interest. A comparison of the two series reveals a marked difference in the extent to which

f orbitals participate in metal-ligand bonding. The $4f$ orbitals of the lanthanoids have no radial nodes and are deeply buried in the atomic core, so that their overlap with ligand orbitals is negligible. Therefore, the metal-ligand bond in lanthanoid-C_8H_8 complexes is predominantly ionic. In actinoid-C_8H_8 complexes, on the other hand, a considerable contribution from covalent interactions may be assumed since the $5f$ orbitals are subject to less shielding than $4f$ orbitals and therefore overlap more extensively with the ligand orbitals. Examples:

$$CeCl_3 + 2\ K_2C_8H_8 \xrightarrow{\text{diglyme}} [K(diglyme)][Ce(C_8H_8)_2]$$

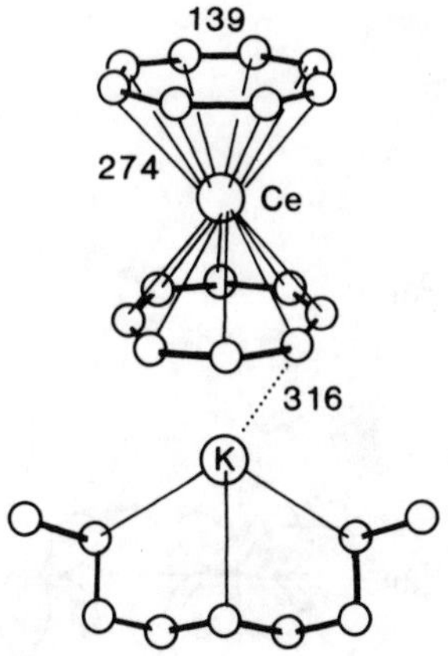

139

274 Ce

316

K

[K(diglyme)] [Ce(C_8H_8)$_2$]: *Pale-green crystals, paramagnetic, air- and* ***moisture-sensitive.*** *Both* COT *rings are planar and oriented in* ***staggered*** *conformation, symmetry* D_{8d}. *The solvated* K^+ *ion in this contact ion pair is attached centrosymmetrically to one of the* COT *rings. This type of structure is adopted by a number of lanthanoid* COT *complexes* (*Raymond*, 1972).

"Uranocene" (Streitwieser, Müller-Westerhoff, 1968):

$$UCl_4 + 2\ K_2C_8H_8 \xrightarrow{\text{THF}}$$

$$U(g) + 2\ C_8H_8 \xrightarrow[-196°]{\text{CC}}$$

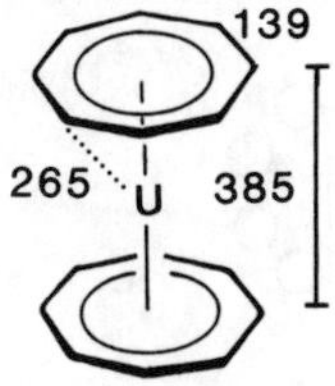

139

265 U 385

U(C_8H_8)$_2$: *Green crystals, paramagnetic, pyrophoric, but* ***stable to hydrolysis.***
Structure: *uniform, planar* COT *rings which are oriented in an* ***eclipsed*** *conformation, symmetry* D_{8h}, $d(C-C) = 139$ pm (Raymond, 1969).

ELECTRONIC STRUCTURE AND BONDING

(η^8-C_8H_8)$_2$)U is a 22 VE complex (8 + 8 + 6 or 10 + 10 + 2). Whereas for sandwich complexes of the d metals, this electron configuration would lead to the occupation of antibonding MO's (cf. the MO scheme of ferrocene p. 320), in the case of f metals further bonding MO's, capable of accepting additional electrons, can be formed through interactions of symmetries e_2 and e_3. In the following survey symmetry-adapted linear combinations (SALC) of the π MO's of two parallel $C_8H_8^{2-}$ ligands are matched with the appropriate metal orbitals:

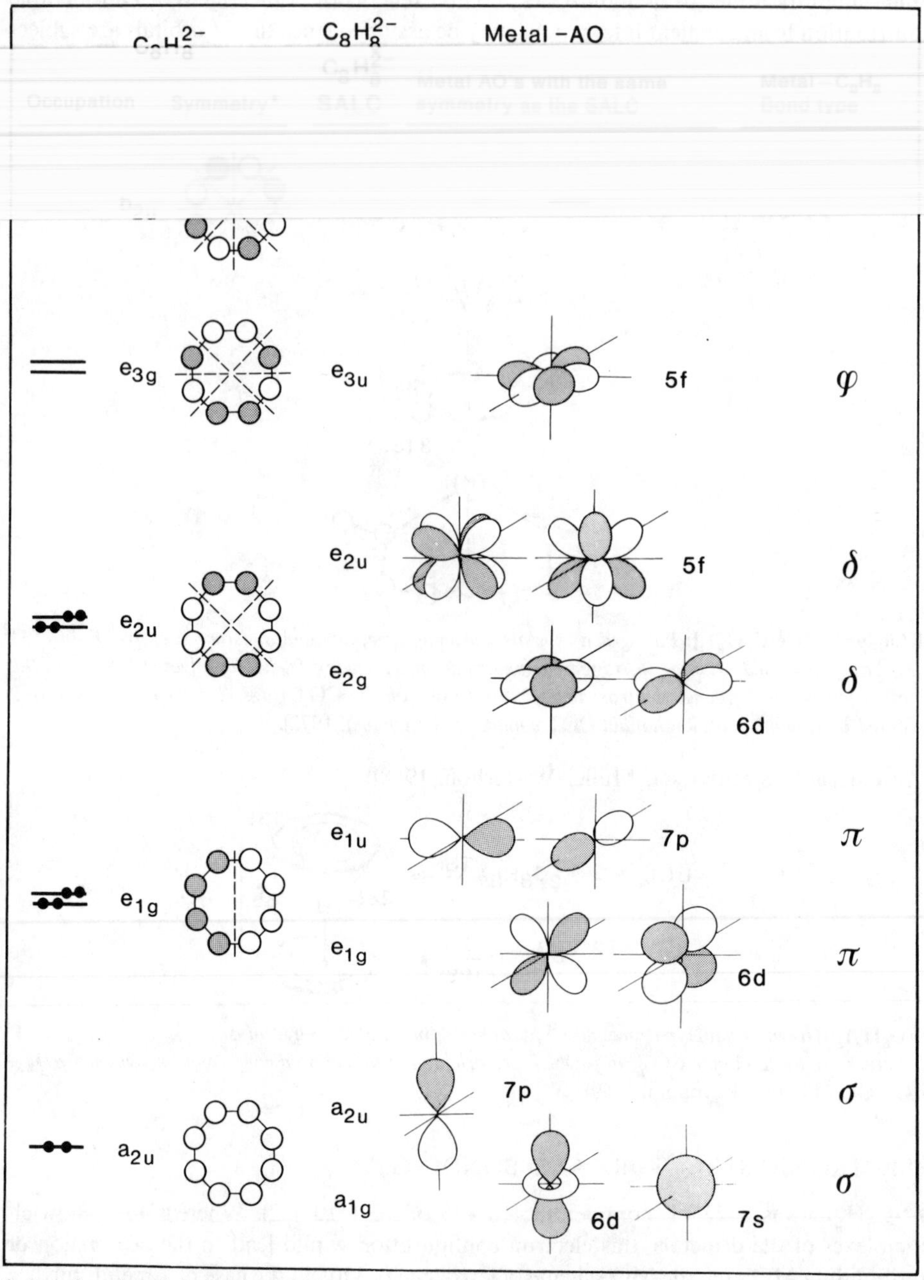

Occupation	$C_8H_8^{2-}$	$C_8H_8^{2-}$	Metal-AO		
	[illegible]				
	e_{3g}	e_{3u}	5f		φ
	e_{2u}	e_{2u}	5f		δ
		e_{2g}	6d		δ
	e_{1g}	e_{1u}	7p		π
		e_{1g}	6d		π
	a_{2u}	a_{2u}	7p		σ
		a_{1g}	6d	7s	σ

* *For the sake of clarity, ligand* MO's *are shown from above. Their phases invert in the plane of the paper (= nodal plane).*

The energetic sequence in the MO scheme for uranocene up to the level e_{2g} qualitatively resembles that for ferrocene (p. 320). However, instead of the essentially nonbonding level a_{1g} (Fe, $3d_{z^2}$) which follows in ferrocene, "uranocene" possesses molecular orbitals which are derived from $C_8H_8-U(5f)$ interactions, namely the frontier orbitals e_{2u} and e_{3u}. Thus, for the 22 VE complex uranocene, a triplet ground state ensues.

Frontier orbitals for "uranocene" and related complexes (adapted from Streitwieser, 1973):

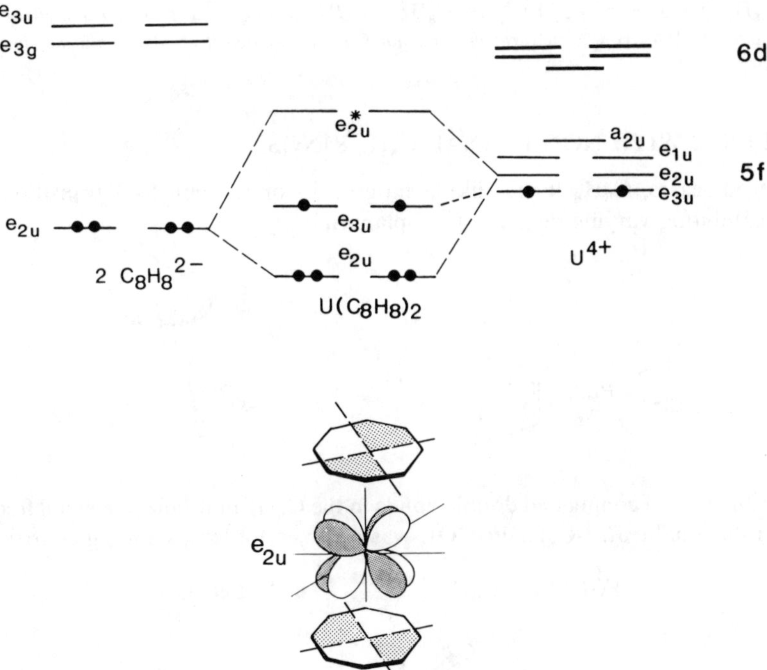

Interactions of symmetry e_{2u} cannot occur in sandwich complexes of the d metals. In the case of f metals, they allow the covalent bonding of large, electron-rich π perimeters. Contributions of U(5*f*) *orbitals to bonding in uranocene seem, however, to be dominated by those from* U(6*d*) *orbitals (Bursten, 1989).*

Apart from "uranocene", other binary actinoid COT complexes have also been prepared. The magnetic moments of these complexes can be rationalized on the basis of the above MO scheme.

$Th(C_8H_8)_2$ = 20 VE diamagnetic
$Pa(C_8H_8)_2$ = 21 VE paramagnetic
$U(C_8H_8)_2$ = 22 VE paramagnetic
$Np(C_8H_8)_2$ = 23 VE paramagnetic
$Pu(C_8H_8)_2$ = 24 VE diamagnetic

C_8H_8 also occurs in sandwich complexes of mixed ring size:

(*Kroon*, 1970).

COMPLEXES WITH NON-PLANAR C_8H_8 RINGS

In numerous cases, C_8H_8 ligates like a mono-, di- or triolefin to a transition metal, thereby exhibiting varying degrees of nonplanarity:

The coordination of conjugated double bonds in the C_8H_8 molecule to a metal frequently results in fluxionality. $(\eta^4\text{-}C_8H_8)Fe(CO)_3$ was a subject of long-standing controversy:

$$Fe(CO)_5 + C_8H_8 \xrightarrow[-2\,CO]{\text{Octane, }125\,^\circ C} C_8H_8Fe(CO)_3$$

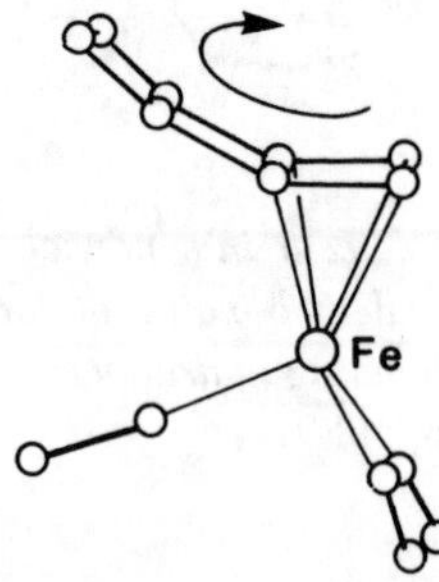

This complex cannot be catalytically hydrogenated although it appears to contain two free C=C *double bonds. In the* ^{1}H-*NMR spectrum, a single line is observed, even at low temperatures, the slow-exchange limit (cf. p. 303) not being reached even at* − 150 °C. *The* ^{13}C-*NMR spectrum, on the other hand, at* $T < -110\,^\circ C$ *displays four signals for the* η^4-C_8H_8 *ring which merge into a single line at* − 60 °C, $E_a = 34$ kJ/mol (*Cotton*, 1976). *The ring rotation apparently involves simultaneous ring distortion, so that the geometry of the complex is retained despite* ***successive* 1,2 *shifts.***
There are indications that fluxionality continues even in the solid state (*Fyfe*, 1972).

In cases where there is only partial coordination of a C_nH_n perimeter to a transition metal, isomerism can arise. $CpCoC_8H_8$ is an inseparable mixture of $(1,2\eta:5,6\eta\text{-}C_8H_8)CoCp$ and $(1\text{–}4\eta\text{-}C_8H_8)CoCp$, which slowly interconvert. The $1\text{–}4\eta$ complex, like $(\eta^4\text{-}C_8H_8)Fe(CO)_3$, is fluxional in solution (Geiger, 1979):

$$CpCo(CO)_2 + C_8H_8 \xrightarrow{h\nu} \underset{\substack{\eta^4\ (isolated)\\ rigid}}{(C_8H_8)CoCp} \rightleftharpoons \underset{\substack{\eta^4\ (conjugated)\\ fluxional}}{(C_8H_8)CoCp}$$

η^4 (isolated)
rigid

η^4 (conjugated)
fluxional

Two different bonding modes of cyclooctatetraene to the same central metal are present in $Fe(C_8H_8)_2$:

$$\left.\begin{array}{l} FeCl_3 + iso\text{-}C_3H_7MgCl + C_8H_8 \xrightarrow[-30\,°C]{Et_2O} \\ Fe(acac)_3 + Al(C_2H_5)_3 + C_8H_8 \xrightarrow[-10\,°C]{} \end{array}\right\} \longrightarrow Fe(C_8H_8)_2$$

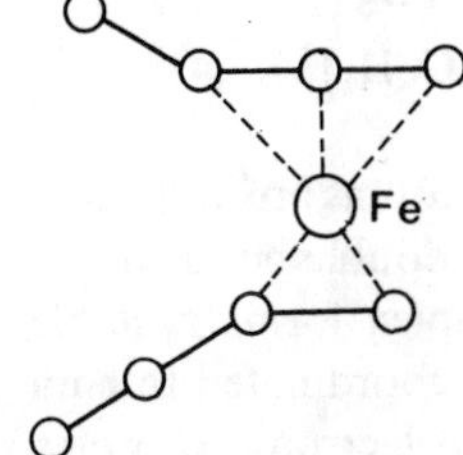

Black crystals, very air-sensitive, active as an oligomerization catalyst. Crystal structure of $\mathbf{Fe(C_8H_8)_2}$: *One* COT *ring coordinates with 3 double bonds* (η^6), *the other* COT *ring with 2 double bonds* (η^4). *The complex is fluxional in solution.*

In a number of binuclear complexes, cyclooctatetraene acts as a **bridging ligand,** e.g. in the binary complexes $M_2(C_8H_8)_3$ (M = Cr, Mo, W):

$$\left.\begin{array}{l} 2\,CrCl_3 + 3\,iso\text{-}C_3H_7MgBr + 3\,C_8H_8 \xrightarrow[Et_2O]{h\nu} \\ Cr(g) + 2\,C_8H_8(g) \xrightarrow[-196\,°C]{CC} \end{array}\right\} \longrightarrow Cr_2(C_8H_8)_3$$

$$WCl_4 + K_2C_8H_8 \xrightarrow{THF} W_2(C_8H_8)_3$$

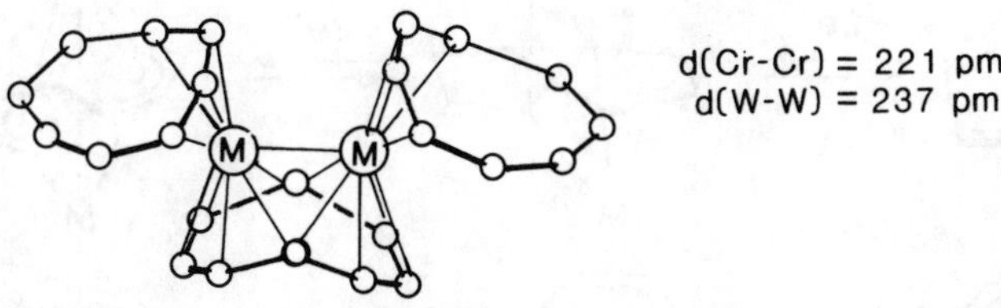

In both the $\mathbf{M_2(C_8H_8)_3}$ *complexes, the terminal rings are only* η^4*-coordinated* (*Krüger,* 1976). *Under the assumption of a metal-metal triple bond, the metals attain an* 18 VE *shell. Both complexes are fluxional in solution,* ^{1}H-*NMR: singlet,* δ 5.26 ppm. *Obviously, all three* COT *rings participate in the structural dynamics.*

The C_8H_8 ligands in $Ni_2(C_8H_8)_2$ are exclusively bridging:

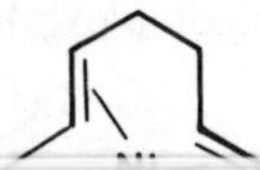

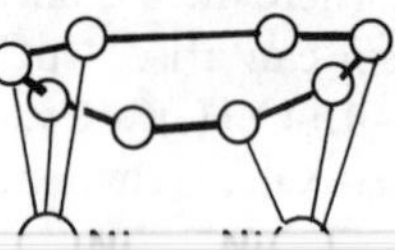

An interesting aspect of **cyclooctatetraene bridging** is the occurrence of **syn- and anti-variants**. The most thoroughly studied complexes in this respect are those of the composition $(C_5H_5M)_2C_8H_8$ (M = V, Cr, Fe, Ru, Co, Rh).
Preparative examples:

$$2\,CrCl_2 + 2\,NaC_5H_5 + K_2C_8H_8 \xrightarrow[2.\ 150\,°C]{1.\ 25\,°C} (C_5H_5Cr)_2C_8H_8$$

$$2\,[C_5H_5Ru(CH_3CN)_3]^+ + K_2C_8H_8 \xrightarrow[-40\,°C]{THF} (C_5H_5Ru)_2C_8H_8$$

$$[ClRhC_8H_8RhCl]_x + 2\,TlCp \longrightarrow (C_5H_5Rh)_2C_8H_8$$

$$(C_5H_5Rh)_2C_8H_8 + 2\,AgBF_4 \longrightarrow [(C_5H_5Rh)_2C_8H_8]^{2+}$$

In $(C_5H_5Co)_2C_8H_8$ and $(C_5H_5Rh)_2C_8H_8$ (36 VE), where both metal atoms are *anti*-coordinated, each metal is bonded to a separate pair of non-conjugated double bonds of the tub-shaped C_8H_8 ring, $(\eta^4:\eta^4)$. On oxidation, the rhodium complex forms a stable dication (34 VE), in which two carbon atoms of the C_8H_8 ring are coordinated to **both** metal atoms, $(\eta^5:\eta^5)$ (Geiger, Rheingold, 1984). The neutral isoelectronic complex $(C_5H_5Ru)_2C_8H_8$ displays the same "slipped-tripledecker" structure, which on oxidation converts into a "fly-over" structure (p. 277), in which the ring has opened and a metal-metal bond has formed (Geiger, Salzer, 1990).

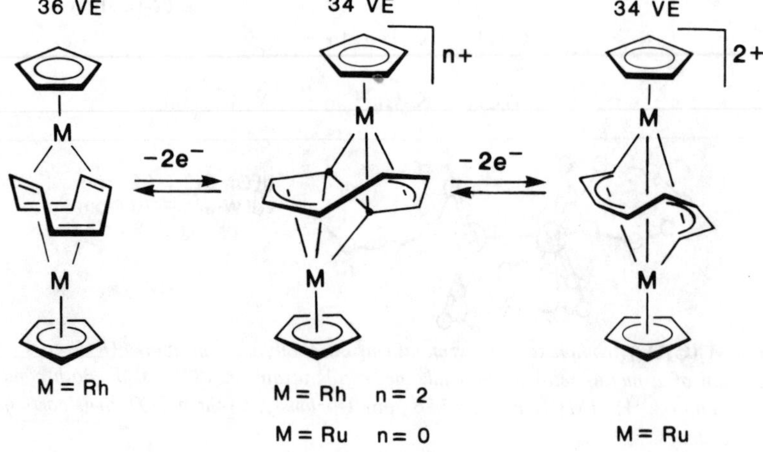

The "fly-over" chromium complex $(C_5H_5Cr)_2C_8H_8$ (32 VE), on the other hand, is thermally converted into its isomer (30 VE) with *syn*-bridging cyclooctatetraene ($\eta^5:\eta^5$) (Wilke, 1978).

32 VE
30 VE
Cr
ΔT
Cr

By way of summary it can be stated that for complexes of the type $(C_5H_5M)_2C_8H_8$ **decreasing number of valence electrons** leads to more **closed structures.** This trend is already noticeable from the structural change which accompanies the oxidation of $(C_5H_5Rh)_2C_8H_8$, (36 → 34 VE, see above). It continues with dramatic results in the oxidation of $(C_5H_5Ru)_2C_8H_8$ (34 VE → 34 VE) with an unprecedented ring-opening and concurrent formation of a metal-metal bond. A further increase in electron-deficiency, as in the chromium complexes $(C_5H_5Cr)_2C_8H_8$ (32 → 30 VE) completes the change from *anti-* to *syn*-coordination of the two C_5H_5M units, in this way also effecting a higher metal-metal bond order.

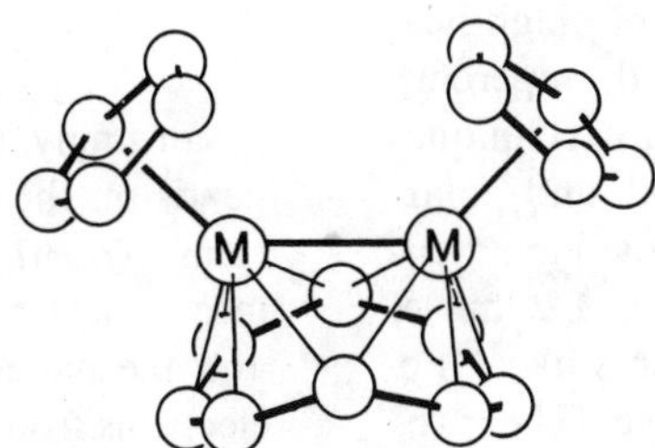

The complexes $\mathbf{(C_5H_5V)_2C_8H_8}$ *and* $\mathbf{(C_5H_5Cr)_2C_8H_8}$ *possess almost identical structures: in the chromium complex* (30 VE), *the bridging* C_8H_8 *ring consists of two virtually planar* C_5 *moieties which are inclined at a dihedral angle of* 134°, *and which have two carbon atoms in common. The* Cr–Cr *bond* (239 pm) *is formulated as a double bond. The vanadium complex* (28 VE, d(V–V) = 244 pm) *should contain a* V–V *triple bond if* 18 VE *shells are to be attained by the vanadium atoms.* $(C_5H_5)_2VC_8H_8$ *shows slight temperature-dependent paramagnetism, however, thereby implying that the "third* V–V *bond" is weak* (*Heck*, 1983).

Excursion:

Transition Metal NMR in Organometallic Chemistry

As in the chemistry of main-group organometallics (cf. p. [illegible]) for the transition elements as well, NMR spectroscopy of metal nuclei increasingly gains in importance. In the following, a few characteristics and applications will be outlined, using two representative examples, namely ^{57}Fe (2.2%, $I = 1/2$) and ^{59}Co (100%, $I = 7/2$). To underline the significance of transition metal NMR, it should be remembered that it was the detection of the magnetic resonance of ^{59}Co nuclei which led to the discovery of the chemical shift phenomenon (Proctor, Yu, 1951).

^{57}Fe-NMR

The relative receptivity of ^{57}Fe amounts to only 0.4% of that of ^{13}C and the nuclear spin $I = 1/2$ of ^{57}Fe, because of the absence of quadrupolar relaxation, may lead to saturation problems with concomitant low signal intensity. Nevertheless, the availability of high-field spectrometers equipped with superconducting magnets, special pulse techniques like the steady-state method, and polarization transfer experiments have rendered NMR studies on $I = 1/2$ nuclei with small magnetic moments like ^{57}Fe, ^{103}Rh or $^{107,109}Ag$ feasible. The pronounced increase in sensitivity achieved at high magnetic fields is a result of the more favorable Boltzmann distribution and the more efficient T_1 relaxation. For $I = 1/2$ nuclei in strong magnetic fields B_0, the relaxation rate is dominated by the anisotropy of the chemical shift, $1/T_1$ being proportional to B_0^2. In the case of Rh and Fe diolefin complexes in a magnetic field of strength B_0 = [illegible], nuclear spin lattice relaxation times T_1 are usually shorter than [illegible] (T_1 may even be as low as [illegible]). Transition metal NMR can then be detected even with conventional Fourrier transform (FT) techniques (Benn, 1986 R).

The **chemical shifts** $\delta^{57}Fe$ for diamagnetic compounds of iron encompass a range of about 12000 ppm. As an external standard, $Fe(CO)_5$ is usually employed. $Fe(CO)_5$ absorbs at the upper end of the field scale. In the survey (p. 371) a few inorganic coordination compounds of iron have been included for comparison. Since the paramagnetic shielding term σ_i^{para} dominates, the chemical shifts $\delta^{57}Fe$ of many iron complexes may be interpreted by means of Eq. (2) (p. 296) under the assumption of a mean excitation energy ΔE:

$$\delta_i \sim -\sigma_i^{para} \sim \langle r^{-3} \rangle_{nd} (Q_i + \sum_{i \neq j} Q_j)/\Delta E$$

Accordingly, the chemical shift $\delta^{57}Fe$ depends on the electron density at the nucleus (Q_i), the bond orders (ΣQ_j), the mean electronic excitation energy ΔE and the mean distance of the valence d electrons from the nucleus (for transition metals with partially filled d shells, the contribution from p electrons can be neglected). Therefore, **deshielding,** which is 1500 ppm stronger in ferrocene as compared to (butadiene)$Fe(CO)_3$, may be rationalized by the **smaller HOMO-LUMO gap** ($\delta \Delta E = 0.2$ eV) in the former as compared to the latter. An additional

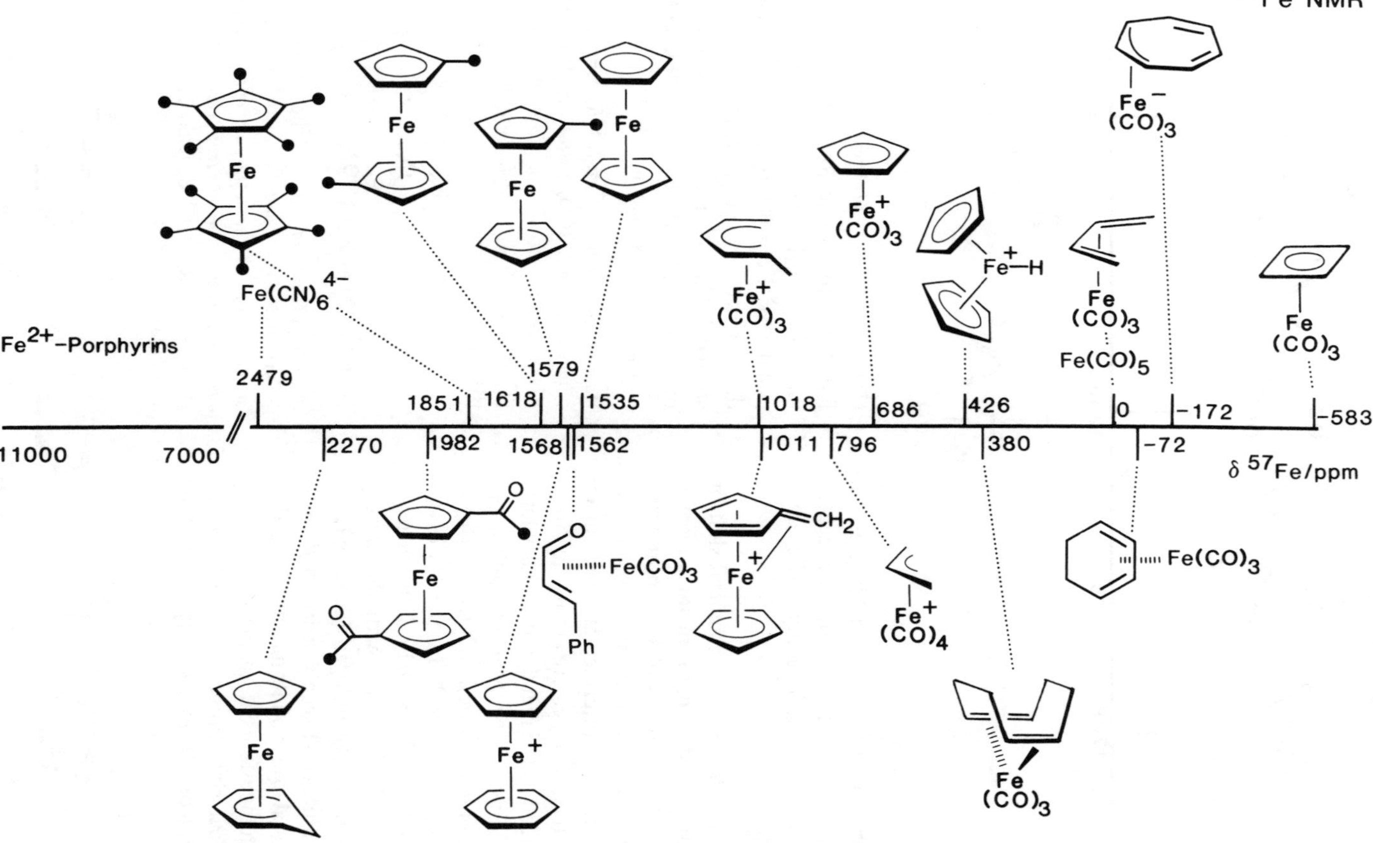

δ^{57}Fe *values for organometallics and inorganic coordination compounds of iron.*

Scheme I

contribution to the deshielding is the smaller charge density on the central metal in ferrocene, formally an Fe^{II} compound.

The dependence of the chemical shift on the **oxidation state of the metal** is a general trend, resonances of Fe^{II} compounds occurring at low fields and those of Fe^{0} compounds at intermediate and higher fields. If electronegative atoms are directly bonded to the metal, particularly large deshielding is observed [examples: (η^4-phenylbutene-3-one)$Fe(CO)_3$, Fe^{II}(porphyrine) complexes]. In the series (allyl)$Fe(CO)_3X$, X = Cl, Br, I, the chemical shifts $\delta^{57}Fe$ correlate with the electronegativity of the halogen (Scheme I).
The influence **charge density on the central metal** exerts on its magnetic shielding also manifests itself in the fact that the chemical shift range for neutral olefin- or diolefin iron carbonyl complexes ($550 > \delta^{57}Fe > -600$ ppm) clearly differs from the range for the complexes $[(allyl)Fe(CO)_4]^+$ and $[(dienyl)Fe(CO)_3]^+$ ($1500 > \delta^{57}Fe > 600$ ppm). This charge dependence is convincingly demonstrated by the $\delta^{57}Fe$ values for a series of structurally related but differently charged species (Scheme II).

As an application of ⁵⁷Fe-*NMR, consider the binuclear complex* $(CO)_3Fe(\mu\text{-}C_7H_7)Rh(C_7H_8)$ *which results from the reaction between* $[(C_7H_7)Fe(CO)_3]Li$ *and* [(*norbornadiene*)-$RhCl]_2$.

$\delta^{57}Fe$ − 223 ppm

A comparison of the $\delta^{57}Fe$ *values for* $[(C_7H_7)Fe(CO)_3]^-$ *and the binuclear product suggests that in the latter, no charge equilibration between the electron rich* Fe^0 *atom* (18 VE) *and* Rh^I (16 VE) *takes place* (*Salzer, v. Philipsborn*, 1982).

Scheme II

$\delta^{57}Fe$	1435	170	− 172 ppm

Contrary to expectation, the iron nucleus in protonated ferrocene and in ferrocenylcarbenium ions (p. 329) is strongly shielded as compared to ferrocene. This contradicts simple arguments which take only inductive effects into consideration; in fact, rehybridization of non-bonding Fe(3d) orbitals is generally deemed responsible for this unusual shielding (Koridze, 1983).

These examples clearly show that in order to rationalize chemical shifts of metal nuclei, all contributions to the paramagnetic shielding term must be assessed.

An interesting **stereoelectronic effect** is the dependence of the shifts δ^{57}Fe on olefin geometry and on the presence of alkyl substituents (v. Philipsborn, 1986 R):

Methylsubstitution at (butadiene)-$Fe(CO)_3$ causes deshielding of the ^{57}Fe resonance, a finding which cannot be explained by inductive effects. Instead, a correlation of δ^{57}Fe with changes in ligand geometry applies. Thus, strongest deshielding of the ^{57}Fe nuclei is displayed by those complexes which possess diene ligands with carbon atoms most extensively rehybridized towards a higher p orbital content (cf. 302). Similar deshielding effects are also encountered with methyl substituted $[(allyl)Fe(CO)_4]^+$ species (Scheme III).

These deshielding stereoelectronic effects are most likely caused by modifications of metal-ligand ($d_\pi - p_\pi$) overlap.

A correlation of a different kind is that between thermal and ^{57}Fe-NMR spectroscopic behavior: for (diene)$Fe(CO)_3$ – as well as for $[(allyl)Fe(CO)_4]^+$ higher thermal stability is parallelled by increased ^{57}Fe shielding.

δ^{57}Fe *Values for (diene)*$Fe(CO)_3$ *Complexes*

Diene	δ^{57}Fe	Diene	δ^{57}Fe
Cyclobutadiene	−583	Buta-1,3-diene	0
Cyclohexa-1,3-diene	− 72	Isoprene	+ 54
Cyclohepta-1,3-diene	+ 86	*trans*-Penta-1,3-diene	+ 31
Cycloocta-1,3-diene	+169	*cis*-Penta-1,3-diene	+119
Cyclopentadiene	+185	2-Methylpenta-2,4-diene	+179
Norborna-1,4-diene	+382	Hexa-2,4-dienal	+378
Cycloocta-1,5-diene	+380	Hexa-1,3-dien-5-on	+335

Scheme III

BF_4^- Fe(CO)4 Fe(CO)4 Fe(CO)4 Fe(CO)4

δ^{57}Fe 998 884 896 796 ppm

For the iron carbonyl complexes of cyclic diolefins, with increasing ring size, i.e. with increasing CCC bond angle in the s-**cis**-diene segments, deshielding of the ^{57}Fe nucleus increases. For 1,3-dienes, this effect is usually more pronounced than for 1,4- or 1,5-dienes. (Cyclopentadiene)$Fe(CO)_3$, whose ligand may be regarded either as a 1,3- or a 1,4-diene, assumes an intermediate value for δ^{57}Fe.

^{59}Co-NMR

In contrast to most transition metals, ^{59}Co nuclei boast a very high receptivity which is comparable to that of the proton. ^{59}Co ($I = 7/2$) also carries a high quadrupole moment, however (p. 24). Consequently, for organocobalt compounds of low symmetry and/or of large molecular dimension, due to the large field gradient and/or slow tumbling rates in solution, very large line widths may arise (5–30 kHz), rendering the detection of ^{59}Co resonances difficult. Furthermore, chemical shifts $\delta^{59}Co$ and the line widths $W_{1/2}$ in particular are temperature- and solvent dependent. Interpretations of these NMR parameters therefore call for identical conditions of measurement.

The influence that molecular symmetry exerts on the **line widths** $W_{1/2}$ is exemplified by the cobalamins (p. 201):

The **chemical shifts δ^{59}Co** are spread over a range of 18000 ppm; they are referred to $K_3[Co(CN)_6]$ as an external standard. Again, the dependence of the chemical shift on the oxidation state of the central metal is a general trend. However, because of the ambiguities which may be inherent in oxidation state assignments, there is considerable overlap between the different regions and Scheme IV should only be regarded as a crude guide line. There are also irritating exceptions: $[Cp_2Co^{III}]Cl$ absorbs at $\delta^{59}Co = -2400$ ppm, a spectral region typical for Co^{-I} complexes, and resonance for the standard $K_3[Co(CN)_6]$, a Co^{III} complex, occurs at 0 ppm.

	Conc. in mol L^{-1}	T in K	Solvent (pH)	$\delta^{59}Co$ in ppm	$W_{1/2}$ in kHz	Symmetry
K[Dicyanocobalamin]	0.06	313	D_2O (7)	4040	14.0	axial
Cyanocobalamin	0.01	313	D_2O (7)	4670	29.5	non-axial

In the axially symmetric dicyanocobalamin anion the nucleus ^{59}Co is shielded more strongly than in neutral cyanocobalamin. Moreover, the axially symmetric anion features smaller line widths (v. Philipsborn, 1986). ^{59}Co-NMR line widths therefore aid in the elucidation of molecular structure, an example being the differentiation between **cis/trans-** and **fac/mer** isomeric cobalt complexes (Yamasaki, 1968).

The significance of the $1/\Delta E$ factor in the paramagnetic shielding term σ_i^{para} emerges from a comparison of $\delta^{59}Co$ values for compounds of the type $CpCoL_2$. Contrary to expectations based on ligand electron affinities, ^{59}Co is shielded more strongly in $CpCo(CO)_2$ relative to $CpCo(C_2H_4)_2$. Apparently, the different magnitudes of the mean excitation energy ΔE dominate here. This conjecture is

Scheme IV

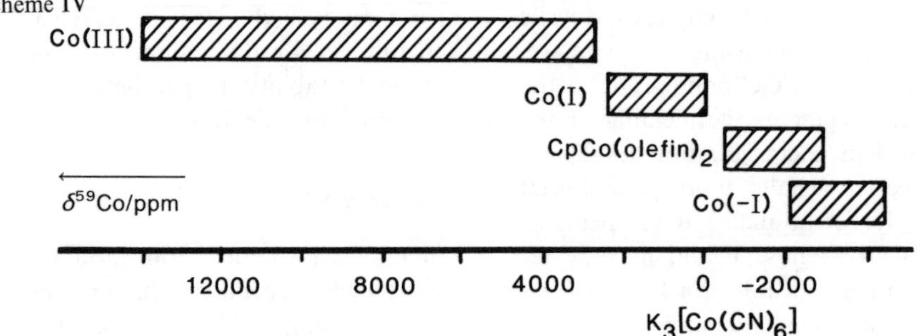

supported by the results of Extended Hückel calculations according to which the HOMO-LUMO gap in $CpCo(C_2H_4)_2$ is smaller by 0.3 eV as compared to $CpCo(CO)_2$ (Benn, 1985). The magnitude of the energy difference between the frontier orbitals HOMO and LUMO plays the dual role of influencing the magnetic shielding of ^{59}Co nuclei via the $1/\Delta E$ dependence, and of reflecting the strength of the metal-ligand bond. Therefore, besides correlations between ^{59}Co-NMR data and electronic spectra (Griffith, 1957) relationships between $\delta^{59}Co$ values and chemical behavior would be expected. Parallels of the latter type have in fact been observed. They can be illustrated with a recent example, the activity of organocobalt complexes in the homogeneously catalyzed synthesis of pyridine derivatives (Scheme V, Bönnemann, v. Philipsborn, 1984).

In this process, depending on the catalyst employed, singly and multiply substituted pyridine- as well as benzene derivatives are obtained from alkynes and nitriles (p. 277). For a series of complexes $(R\text{-}Cp)CoL_2$ and $(R\text{-indenyl})CoL_2$ as catalyst precursors, it has been demonstrated that their activity and selectivity correlates with $\delta^{59}Co$. The actual catalyst CpCo (14 VE) is formed via thermal ligand dissociation. **Precatalysts** of high **activity,** which operate even at low temperature, contain weakly shielded ^{59}Co nuclei [example: CpCo(COD), $\delta^{59}Co = -1176$ ppm]. The complex CpCo(cyclobutadiene), on the other hand, in which the central atom is strongly shielded ($\delta^{59}Co = -2888$ ppm), is catalytically ineffective. Obviously, cyclobutadiene is strongly bonded to cobalt (ΔE is large) and $CpCo(C_4H_4)$ is not capable of initiating the catalytic cycle by forming the coordinatively unsaturated species {CpCo}.

The **chemo- and regioselectivities** of these catalytic cyclizations are governed by the nature of the cyclopentadienyl ligand which remains attached to cobalt. Here as well, a correlation with NMR data was discovered in that the product ratio 1,3,5-trisalkyl-pyridine/1,2,5-trisalkyl-pyridine rises with increasing shielding of the ^{59}Co nucleus.

A quantitative interpretation of these findings is of course extremely difficult; correlations of the type discussed above are of high practical value, however, since they indicate a way to systematically search for catalytically active species by means of ^{59}Co-NMR.

Scheme V

Cp = Cyclopentadienyl- or Indenyl derivative

L_2 = Diolefin

15.5 Metal π-Complexes of Heterocycles

Unsaturated ring systems in which individual (or all) carbon atoms have been exchanged for heteroatoms can form transition-metal π-complexes under certain conditions. In some cases, this can also lead to the stabilization of heterocycles unknown in the free state.

15.5.1 Nitrogen-, Phosphorus- and Sulfur Heterocycles as Ligands

The ability of organic molecules containing heteroatoms like N, P, O or S to bind to transition metals is relevant in classical coordination chemistry as well as for the in vivo binding of metals. If the heteroatoms are part of rings such as pyridine, thiophene or the pyrrolyl anion, they usually retain their Lewis basicity and the rings tend to coordinate through their lone pairs rather than via the cyclic conjugated π system. The number of π complexes containing N-, P- and S-heterocyclic ligands is therefore relatively small compared to their carbocyclic counterparts.

FOUR-MEMBERED RING LIGANDS

The heterocycle **1,3-diphosphete** (= 1,3-diphosphacyclobutadiene), unknown in the free state, has been obtained through cyclodimerization of *t*-butylphosphaalkyne (p. 164) in the coordination sphere of cobalt (Nixon, Binger, 1986):

Co

2 t-BuC≡P

toluene, 25°

Co

P

P

t-Bu

t-Bu

d(C-P) = 181 pm

∢CPC = 81°

∢PCP = 99°

Like coordinated cyclobutadiene (p. 311), this 1,4-diphosphacyclobutadiene displays equal intra-ligand bond distances, although, due to unequal angles, the planar ring assumes a diamond shape.

FIVE-MEMBERED RING LIGANDS

$C_5H_5^-$(cyclopentadienyl), $C_4H_4N^-$(pyrrolyl), $C_4H_4P^-$(phospholyl) and $C_4H_4As^-$ (arsolyl) are *iso*-π-electronic. The synthesis of π complexes containing these heterocycles is therefore often analogous to that of corresponding cyclopentadienyl complexes. Azacyclopentadienyl complexes can be prepared in a redox reaction from pyrrole (Pauson, 1962):

$2\ C_4H_4NH + Mn_2(CO)_{10} \longrightarrow 2\ (C_4H_4N)Mn(CO)_3 + H_2 + 4\ CO$

or via a preformed pyrrolyl anion (Pauson, 1964):

$C_5H_5Fe(CO)_2I + K^+C_4H_4N^- \longrightarrow C_5H_5Fe(CO)_2(\sigma\text{-}NC_4H_4) \xrightarrow[-\,2CO]{40°} C_5H_5FeC_4H_4N$

Azaferrocene ($C_5H_5FeC_4H_4N$) forms orange-red, diamagnetic, easily sublimable crystals, which are isomorphous to ferrocene. Treatment with acids leads to protonation at the nitrogen atom. Compared to $C_5H_5^-$, the pyrrolyl anion $C_4H_4N^-$ is a weaker π-donor and/or a stronger π-acceptor. 1,1′-Diazaferrocene is as yet unknown.
For the elements phosphorus and arsenic, however, both single and multiple incorporation of the heteroatom into the ferrocene frame has been achieved:

$C_4H_4PPh + [CpFe(CO)_2]_2 \xrightarrow[xylene]{150°} CpFe(C_4H_4P)$ (Mathey, 1977)

$C_4H_4PPh \xrightarrow[THF,\ 20°]{\times 2,\ 4\ Li} 2\ C_4H_4PLi + 2\ PhLi \xrightarrow[-\,AlPh_3]{AlCl_3} 2\ C_4H_4PLi \xrightarrow[20°]{FeCl_2} Fe(C_4H_4P)_2$ (Mathey, 1986)

1,1′-Diarsaferrocene is accessible by the same route (Ashe, 1987).

1,1′-diphosphaferrocene via its phosphorus atoms is capable of σ-donor/π-acceptor interactions. It can act as a diphos-type chelating ligand and can become a building block for oligonuclear complexes. Example:

$Fe(C_4H_4P)_2 \xrightarrow{(CO)_4Fe(THF)} Fe[C_4H_4P\text{—}Fe(CO)_4]_2$

Pentaphosphaferrocene is formed in a surprisingly mundane fashion using P_4 as the source of phosphorus (Scherer, 1987):

$$[(C_5Me_5)Fe(CO)_2]_2 \xrightarrow[\text{xylene, } 150°, \text{ 15h}]{P_4}$$

The similarity of **thiophene** to benzene also extends to coordination chemistry. Like benzene, thiophene forms $M(CO)_3$ adducts, albeit of lower stability (Öfele, 1958):

$$C_4H_4S + Cr(CO)_6 \xrightarrow[-\,3CO]{85°}$$

d(Cr–C) 220 pm (thiophene)
177 pm (carbonyl)

(Dahl, 1965)

$C_4H_4SCr(CO)_3$ is obtained as orange-colored diamagnetic crystals which are isomorphous with $C_6H_6Cr(CO)_3$. Selenophene and tellurophene complexes are prepared in the same way (Öfele, 1966).

SIX-MEMBERED RING LIGANDS

For **pyridine,** σ coordination via its lone pair is highly preferred over bonding which utilizes the π-electron system. Therefore, the direct synthesis of η^6-pyridine metal complexes requires the N atom to be blocked by means of subsitution in the 2,6-positions (Lagowski, 1976):

$$2\ \ 2,6\text{-}Me_2C_6H_3N(g) + Cr(g) \xrightarrow{CC}$$

The parent complex **bis(η^6-pyridine)chromium** can be prepared by occupying the 2,6-positions with protecting groups, which can be removed after complex formation (Elschenbroich, 1988):

$$2\ \ 2,6\text{-}(Me_3Si)_2C_5H_3N\,(g) + Cr(g) \xrightarrow[\text{1. } -196°\text{ 2. } 30°]{CC}$$

1. O_2, 5 min, C_6H_6, H_2O
2. $Na_2S_2O_4$

Bis(η^6-pyridine)chromium differs from bis(benzene)chromium in its high thermal and solvolytic lability. π-complexes of the heavier group 15 heteroarenes **phosphabenzene** and **arsabenzene** like $(\eta^6\text{-}C_5H_5P)_2V$ and $(\eta^6\text{-}C_5H_5As)_2Cr$ are obtained directly from metal-atom ligand-vapor cocondensations, they are considerably more stable than the pyridine analogues (Elschenbroich, 1991, 1986).

The observation that for arsabenzene, η^1*-coordination (via the* As *atom) is disfavored as compared to* η^6*-coordination (via the* π*-electron system) concords with the Lewis basicities, which decrease in the order pyridine > phosphabenzene > arsabenzene. The* π*-donor character of the heteroarene is expected to follow an opposite trend, since the electropositive character increases in the order* N < P < As. *The disparate behaviour of the heteroarenes towards metal carbonyls conforms with this notion (Ashe,* 1977):

E–Mo(CO)$_5$ ← Mo(CO)$_5$THF — E-heteroarene — Mo(CO)$_3$Py$_3$ / Et$_2$O·BF$_3$ → (η^6-C$_5$H$_5$E)Mo(CO)$_3$

E = N,P,As E = As,Sb

Thus, as the atomic number of E increases, π coordination (η^6) dominates over σ coordination (η^1). In the particular case of arsabenzene, a σ → π rearrangement has been established:

As–Mo(CO)$_5$ —120°, −2CO→ (η^6-C$_5$H$_5$As)Mo(CO)$_3$

A spectacular result is the formation of **hexaphosphabenzene** P_6 as a bridging ligand in a tripledecker complex (Scherer, 1985):

$Me_5C_5(CO)_2Mo{\equiv}Mo(CO)_2C_5Me_5$ —P$_4$, xylene; 140°, 5h→

d(P-P) = 217 pm
d(Mo-P) = 254 pm

The formation of the P_6 ring can be regarded as a cyclotrimerization of three $|P{\equiv}P|$ units in analogy to the alkyne trimerization. The complex μ-(cyclo-P_6) $[Mo(C_5Me_5)]_2$ is obviously related to the cyclopentaarsyne complex μ-(cyclo-As_5) $[Mo(C_5H_5)]_2$ which has already been mentioned (p. 160). For substituted heterocycles (E = N, P, As, Sb), the respective products formed reflect the interplay of steric and electronic effects. For example, substituted pyridines, in which N-coordination is sterically hindered, react with $Cr(CO)_6$ to give $(\eta^6\text{-}R_nC_5H_{5-n}N)Cr(CO)_3$, while pyridine itself solely forms N-coordinated complexes $(C_5H_5N)_nCr(CO)_{6-n}$ ($n = 1-3$). (2,4,6-Triphenyl)phosphabenzene, depending on the metal carbonyl precursor, gives both types of complexes (Nöth, 1973):

(THF)Mo(CO)5 → Ph–C5H2P(Ph)2–P–Mo(CO)5

(mesitylene)Mo(CO)3 → (η⁶-2,4,6-Ph3C5H2P)Mo(CO)3

– 2CO | ΔT

15.5.2 Boron Heterocycles as Ligands

The first transition-metal complexes of boron-containing rings were synthesized as **carbaborane-metal complexes** by Hawthorne from 1967 onwards. While these compounds, consisting of metal cations and carbaboranyl anions, in many cases parallel the corresponding cyclopentadienyl compounds (p. 72), complex fragments built from metal atoms and simple boron heterocycles excel in their ability to form stacks **(multidecker complexes).**

Not all principally conceivable B, C heterocycles have yet been realized; some hitherto have only appeared as ligands in metal complexes.

The most important B, C heterocycles, serving as ligands in organometallic chemistry, are shown below:

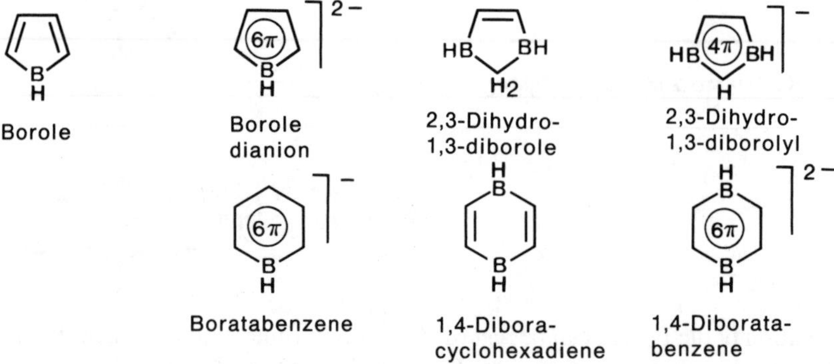

*The prefix **bora-** designates the replacement of a* CH *fragment by* B, ***borata-** the replacement of* CH *by* BH^-.

Neutral boron-carbon cycles like borole and diboracyclohexadiene have low-lying, unoccupied molecular orbitals. Consequently, these neutral ligands are strong electron acceptors, provided the substituents at boron are void of π-donor properties. The existence of anions such as boratabenzene, 1,4-diboratabenzene and the borol dianion are in keeping with this idea. In their coordination to metals, B, C heterocycles can be regarded as neutral ligands (the number of π-electrons being equal to the number of sp^2 carbon atoms) or as anions with 4 or 6 π electrons. Isoelectronic replacement of BH^- for CH relates the anionic species to carbocyclic aromatic systems (compare $C_4BH_5^{2-}$ and $C_5H_5^-$ or $C_5BH_6^-$ and C_6H_6).

HETEROCYCLES WITH ONE BORON ATOM

Borole Complexes

A versatile route to (η^5-borole)metal complexes is the dehydrogenative complexation of borolenes to metal carbonyls (Herberich, 1987 R):

+ $Co_2(CO)_8$

hexane 60°,48h
$-H_2$,-CO

T > 160°

30 VE

[(η^5-C_4H_4BMe)$Co(CO)_2$]$_2$ is isostructural with [(η^5-C_5H_5)$Fe(CO)_2$]$_2$ which is a consequence of the isoelectronic nature of the building units $C_4H_4BMe(4\pi) + Co(d^9)$ and $C_5H_5(5\pi) + Fe(d^8)$.

Complexes of Boratabenzene Derivatives

The first synthesis of a boratabenzene complex was accomplished by the treatment of cobaltocene with organoboron dihalides via an unusual ring expansion (Herberich, 1970):

2 Co —$C_6H_5BX_2$ / −[Cp_2Co]X→ ; X = Br, Cl

—Cp_2Co / −[Cp_2Co]X→ Co, B−C_6H_5

Today, boratabenzene complexes are known in a substantial number. Their preparation follows two general routes:

1. Independent Synthesis of the Boratabenzene Ligand (Ashe, 1975 R):

$$\text{R} = C_6H_5, CH_3, C(CH_3)_3$$
R=H (1979)

2. Bis(boratabenzene)cobalt as a Source for Boratabenzene (Herberich, 1976):

R=CH_3,C_6H_5

M'= V,Cr,Fe,Co Ru Os

*The structure of **bis(B-methyl-boratabenzene)-cobalt** reveals that the metal is not centrosymmetrically bonded, but has slipped away from boron. The differences in the* M–C *and* M–B *bond lengths are greater than would be expected from the sum of the covalent radii,* $d(\text{Co}-\text{C}_1) = 208$, $d(\text{Co}-\text{C}_2) = 217$, $d(\text{Co}-\text{C}_3) = 222$, $d(\text{Co}-\text{B}) = 228$ pm (*Huttner*, 1972).

HETEROCYCLES WITH TWO BORON ATOMS

2,3-Dihydro-1,3-diborolyl Complexes

The neutral ligand 2,3-dihydro-1,3-diborole can be regarded as a neutral $3e$ ligand (after elimination of a hydrogen atom) or as an anionic $4e$ ligand (after deprotonation):

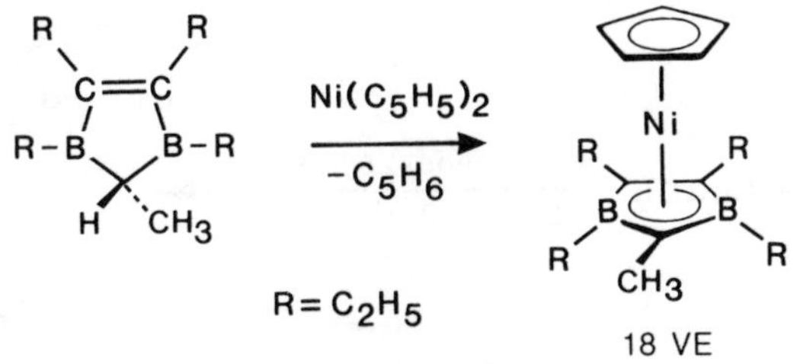

$Ni(C_5H_5)_2$, $-C_5H_6$

$R = C_2H_5$

18 VE

Siebert (1977)

The replacement of carbon by boron appears to enhance the ability of planar ring systems to engage in bifacial coordination to metal atoms. While a characteristic of the carbaboranes is the formation of three-dimensional clusters with incorporation of metal atoms, the five- and six-membered rings with more than one boron atom described here show a pronounced tendency to form **multidecker complexes.** As "stacking reagents", (cyclopentadienyl)metal carbonyls are of great utility (Siebert, 1977):

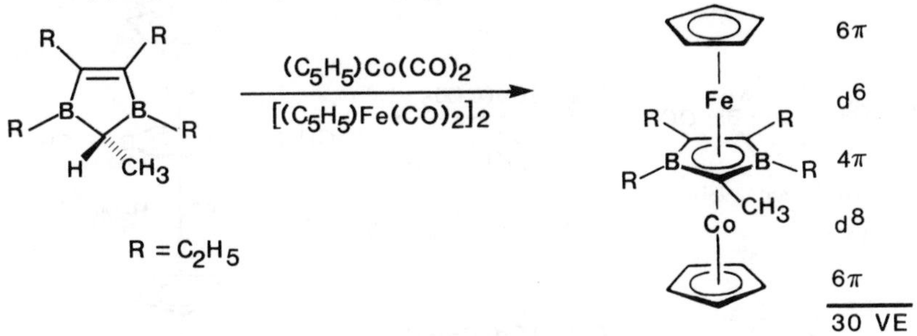

$(C_5H_5)Co(CO)_2$, $[(C_5H_5)Fe(CO)_2]_2$

$R = C_2H_5$

6π, d^6, 4π, d^8, 6π — 30 VE

This structural type is not limited to 30 VE complexes, a complete series with 29–34 VE having been realized: FeFe(29 VE), FeCo(30 VE), CoCo(31 VE), CoNi(32 VE), NiNi(33 VE), $[NiNi]^-$(34 VE).

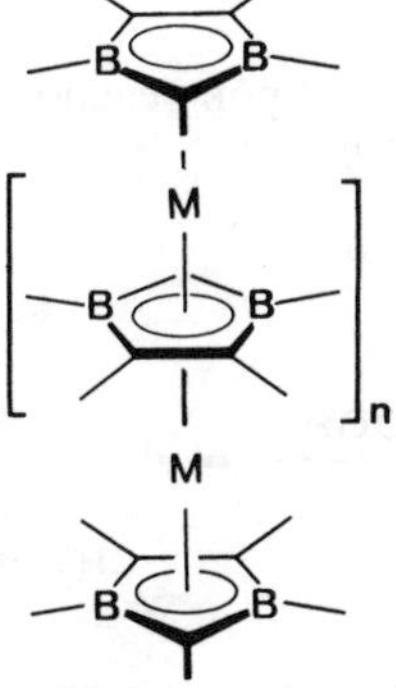

M = Ni, Rh

In addition to the ***tripledecker complexes,*** *which are now known in a large variety,* ***tetra-, penta-, and hexadecker complexes,*** *in which 1,3-diborolyl acts as the bridging ligand, have also been synthesized. The ultimate extension of this series has been reached by the preparation of* ***polydecker-sandwich complexes,*** *the nickel complex being a semiconductor and the* Rh *compound an insulator (Siebert, 1986).*

Complexes of 1,4-Diboracyclohexa-2,5-diene Derivatives

The unsubstituted heterocycle $C_4B_2H_6$ is as yet unknown. Its 1,4-difluoro-2,3,5,6-tetramethyl derivative, however, which is prepared from boron monofluoride and 2-butyne (p. 64), reacts with nickel tetracarbonyl to give a sandwich complex (Timms, 1975):

+ $Ni(CO)_4$ —— 1. toluene, 25°, 30 min; 2. $h\nu$, 12 h ——→ (FB/BF)Ni(FB/BF)

X-ray structural data point to bis-η^6-coordination. As expected from the isoelectronic nature of $\rangle C=\overline{O}$ and $\rangle B-\overline{\underline{F}}|$, this complex is isostructural with bis(duroquinone)-nickel (Schrauzer, 1964 R).
The synthesis of a B-alkylated derivative, coordinated to Co, proceeds as follows:

CH_3O-B ⟨ ⟩ $B-OCH_3$ —— $(C_5H_5)Co(CO)_2$, 80°, − 2 CO ——→ $(C_5H_5)Co(CH_3O-B\ C_4H_4\ B-OCH_3)$

(Herberich, 1980)

20° | CH_3MgI ↓

CH_3-B ⟨ ⟩ $B-CH_3$ —— $(C_5H_5)Co(C_2H_4)_2$, −2 C_2H_4, $-H_2$ ——→ $(C_5H_5)Co(CH_3-B\ C_4H_4\ B-CH_3)$ 18 VE

(Siebert, 1987)

The corresponding rhodium species, upon reaction with protic acids, converts into a tripledecker complex (Herberich, 1981):

$(C_5Me_5)Rh(CH_3-B\ C_4H_4\ B-CH_3)$ —— CF_3COOH ——→ $[(C_5Me_5)Rh(\mu\text{-}CH_3-B\ C_4H_4\ B-CH_3)Rh(C_5Me_5)]^{2+}$

6π
d⁶
6π
d⁶
6π
30 VE

15.5.3 Boron-Nitrogen Heterocycles

Cyclo-trisborazine $B_3N_3H_6$ is occasionally referred to as "inorganic benzene". Some of its alkyl derivatives actually form half-sandwich π-complexes which are, however, considerably less stable than the corresponding arene counterparts (Werner, 1969):

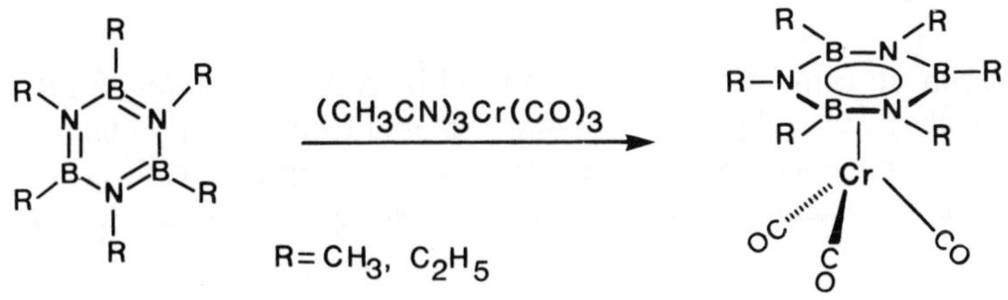

$(CH_3CN)_3Cr(CO)_3$

R = CH_3, C_2H_5

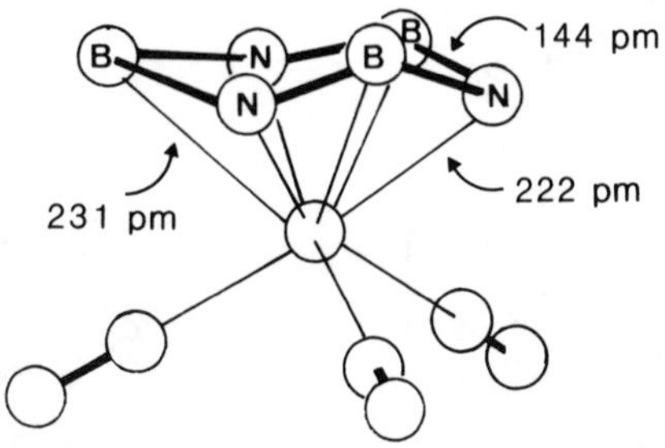

144 pm

231 pm

222 pm

Structure of **(hexaethyl-cyclo-trisborazine)Cr(CO)$_3$**: *In contrast to the free ligand, the ring is not exactly planar. The plane defined by the boron atoms deviates from the plane of the nitrogen atoms by 7 pm. The* Cr—B *and* Cr—N *distance differ to an extent which is expected from the different covalent radii of* B *and* N. *The* $Cr(CO)_3$ *unit is oriented such that the* CO *groups and the* N *atoms are trans-positioned* (*Huttner*, 1971).

The synthesis of sandwich complexes of cyclo-trisborazine has not yet been successful. Further examples involving boron-nitrogen heterocycles as ligands are the following:

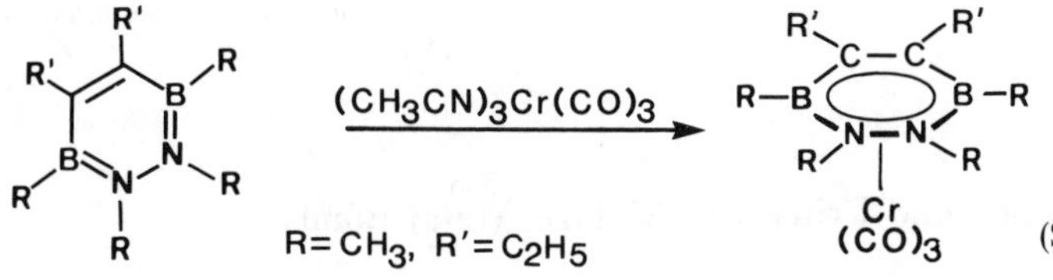

$(CH_3CN)_3Cr(CO)_3$

R = CH_3, R′ = C_2H_5 (Siebert, 1980)

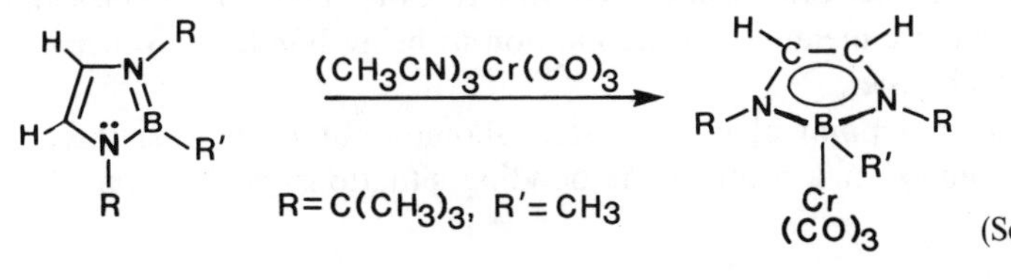

$(CH_3CN)_3Cr(CO)_3$

R = $C(CH_3)_3$, R′ = CH_3 (Schmid, 1981)

16 Metal – Metal Bonds and Transition Metal Atom Clusters

Oligonuclear metal complexes were first studied by A. Werner (1866 – 1919), pioneer in coordination chemistry. Oligonuclearity was achieved by means of bridging ligands and direct metal – metal bonds were absent. In modern complex chemistry metal – metal bonds play an important role; they are responsible for the formation of metal atom clusters.

Classical Werner Complexes
The properties of these oligonuclear complexes do not differ greatly from those of their mononuclear analogues.

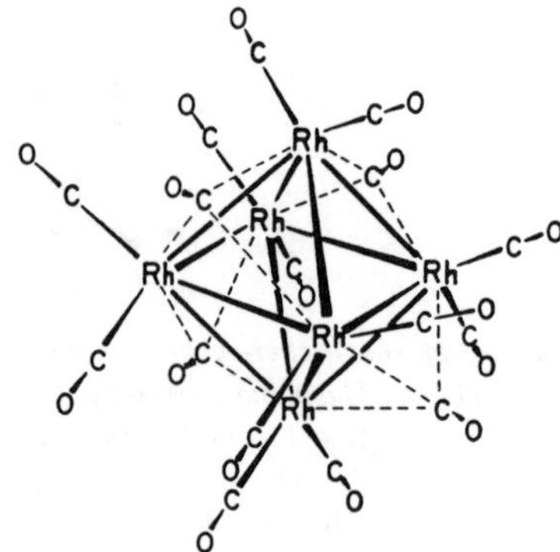

Metal Atom Clusters
Because of the presence of metal – metal bonds these compounds display novel properties. Frequently they cannot be described in terms of 2e 2c *bonds and through applications of the* 18 VE *rule.*

16.1 Formation and Criteria of Metal – Metal Bonds

A definition of **cluster compounds,** according to Cotton (1963), includes all molecules in which two or more metal atoms, in addition to being bonded to other non-metals, are bonded to each other.
From a structural point of view, clusters should contain at least three metal atoms. However, a unified treatment of the bonding situations profits from the inclusion of M_2 species.

Clusters of the main-group elements (i.e. Pb_9^{4-}, Bi_5^{3+}, Te_4^{2+}) are not treated here since they are void of metal – carbon bonds.

Apart from their fascinating structures, metal atom clusters possess properties which are highly interesting for potential practical applications:

- $[Rh_{12}(CO)_{34}]^{2-}$ *is effective in homogeneous catalysis of a Fischer-Tropsch synthesis:*

$$2\,CO + 3\,H_2 \xrightarrow[\text{pressure}]{250\,^\circ C} CH_2OHCH_2OH$$

(Pruett, Union Carbide)

- *Chevrel phases* MMo_6S_8 (*e.g.* $PbMo_6S_8$) *remain superconducting even in very strong magnetic fields. This is essential in the manufacture of electromagnets of extremely high field strength* (*Sienko*, 1983 *R*).
- *Certain clusters with* M≡M *quadruple bonds display photochemical properties with possible relevance for the conversion of solar energy* (*H. B. Gray*, 1981).
- *Cluster units* Fe_4S_4 *are encountered in the important redox enzyme ferredoxin* (*p.* 396).

Transition-metal atom clusters may be divided into three groups of which only group ③ will be treated in this text:

① **Naked clusters,** i.e. metal atoms in low states of aggregation. They are accessible in inert matrices and at low temperatures only and have to be studied by physical methods.
Example: Ag_6.

② **Metal halides and oxides, the metal being in a low oxidation state.**
Examples: $[Mo_6Cl_8]^{4+}$, $[Re_3Cl_{12}]^{3-}$.

③ **Oligonuclear metal carbonyls, nitrosyls and mixed-ligand complexes containing cyclopentadienyl groups.**
Example: $[CpFe(CO)]_4$.

Favorable conditions for the occurrence of **M – M bonds** include a **low formal oxidation state** and a **high atomic number** since under these circumstances widely protruding *d* orbitals with good overlap properties are encountered. M – M bonds are likely for elements which possess high enthalpies of atomization. In their cluster compounds these elements tend to maintain structural features of the metallic state.

Analogies between metal atom clusters and bulk metals:

- *Color* (*many closely spaced energy levels*)
- *Redox activity* (*relationship to metallic conductivity*)
- *Malleability* (*wide range of* M – M *bond distances*)

Diagnostic criteria for the presence of M – M bonds:

a. Thermochemical data. The chemical relevance of M – M bonds should be reflected in the thermochemical bond enthalpy *D*(M – M). Due to the difficulty in formulating correct equations for the combustion of organometallics (product analysis!), the number of experimentally determined bond enthalpies for M$\equiv$M multiple bonds is small.

Approximate value: $D(\mathrm{M{\equiv}M}) \approx 400-600$ kJ/mol
compare: $D(\mathrm{N{\equiv}N}) = 946$ kJ/mol (N_2 molecule)
$D(\mathrm{Mn-Mn}) = 160$ kJ/mol [$Mn_2(CO)_{10}$]

b. Bond distances $d(\mathrm{M-M})$ from X-ray diffraction. In the evaluation of a bond distance, the oxidation states of the metals involved as well as the nature of the ligands have to be taken into account. Assigning a bond order from structural data gets more difficult as the bond order increases, since with rising bond order the extent of bond shortening diminishes. Example:

	$d(\mathrm{Mo-Mo})$	$d(\mathrm{Mo{=}Mo})$	$d(\mathrm{Mo{\equiv}Mo})$	$d(\mathrm{Mo{\equiv}Mo})$
Cluster:	272 pm	242 pm	222 pm	210 pm
Mo(s):	278 pm			

c. Spectroscopic data. A measure of bond strength is also provided by the position of the metal – metal stretching mode in the Raman spectrum. Examples:

$\nu(\mathrm{Cr{\equiv}Cr})$	$\nu(\mathrm{Mo{\equiv}Mo})$	$\nu(\mathrm{Re{\equiv}Re})$
556 cm^{-1}	420 cm^{-1}	285 cm^{-1}

Considering the large masses, these are surprisingly high values.

d. Magnetic properties. Spin pairing is not synonymous with chemical bonding! Whereas for $Mn_2(CO)_{10}$ diamagnetism is a consequence of the metal – metal bond between the two paramagnetic units $Mn(CO)_5^{\bullet}$ (**direct exchange**), spin pairing can also arise without direct metal – metal interaction, in which case a ligand-mediated mechanism is responsible (**superexchange**).
An example for a superexchange interaction is provided by the magnetic properties of the ion $[Ru_2OCl_{10}]^{4-}$:

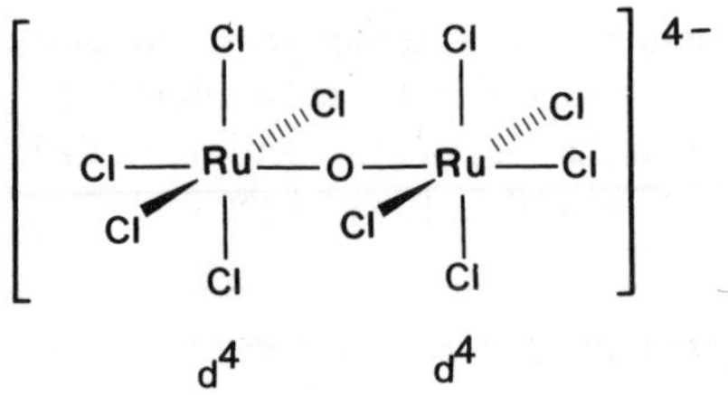

Ru^{IV} (d^4) *in an octahedral environment possesses the electronic configuration* $(xy)^2$ $(xz)^1$ $(yz)^1$. *Nevertheless, the dinuclear complex is diamagnetic.*

This diamagnetism may be rationalized by the idea that in the regions where the singly occupied metal orbitals overlap with the doubly occupied orbitals of the bridging ligands, antiparallel spin orientation accrues. The result is an antiparallel disposition of the electron spins on the two metal atoms, **a** (**antiferromagnetic coupling**). An alternative description proceeds in terms of $4e\,3c$ interactions which lead to a low-spin state for the dinuclear complex, **b**:

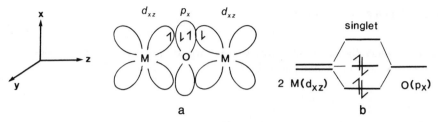

The combination $2\,M\,(d_{yz}) + O\,(p_y)$ is analogous.

If, on the other hand, two **orthogonal** singly occupied **metal orbitals** contribute to the interaction, Hund's rule calls for parallel spin orientation in the vicinity of the bridging atom leading to a preference for parallel spins on the metal atoms, **a (ferromagnetic coupling)**. The alternative description (two orthogonal $3e\,2c$ bonds) demands that in the dinuclear complex a degenerate pair of antibonding molecular orbitals is occupied by two electrons. Consequently, a high-spin state is encountered, **b**:

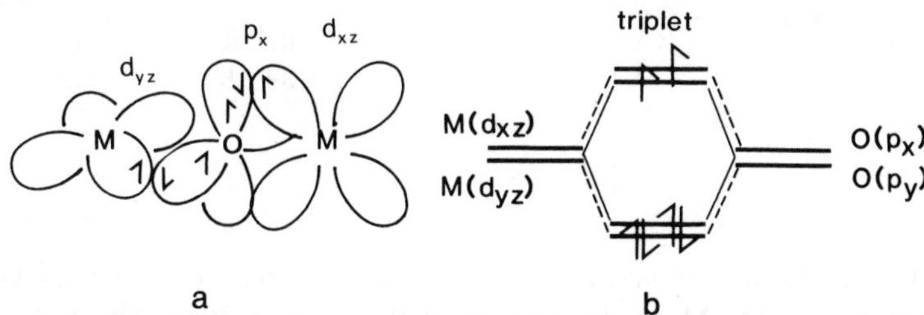

The terms ferromagnetic and antiferromagnetic originally served to characterize the magnetic behavior of extended solid state structures; increasingly they are used for descriptions on the molecular level as well. In the interpretation of a lowering or quenching of the magnetic moment of an oligonuclear complex which consists of units with unpaired electrons, the possible contribution of three mechanisms (direct, indirect antiferromagnetic and indirect ferromagnetic) must be invoked. Thus, a sizable proportion of work in the field of molecular magnetochemistry deals with the relative weight of these pathways. As a guideline it may be stated that under the provision of sufficient approach of the interacting centers, direct interactions are more effective than indirect interactions and that among the latter type, antiferromagnetic coupling is more effective than ferromagnetic coupling (see Gerloch, 1979 *R). Ambiguities may arise in the case of ligand-bridged oligonuclear complexes which possess* M – M *distances in the range* 300 – 500 pm; *here direct and indirect antiferromagnetic contributions can be of comparable magnitude and a decision for the dominance of one of the paths is tantamount to a decision for or against a prevailing chemical bond (example:* $[Cp_2TiCl]_2$, *p.* 341 *and R. L Martin,* 1965). *Therefore the study of magnetic properties eventually leads to statements of chemical relevance.*

In the following, a few metal atom clusters from the realm of organotransition-metal chemistry will be discussed. Oxo- and halometal clusters are part of inorganic coordination chemistry and will not be treated here. In the designation "*n*-nuclear", *n* implies the number of metal atoms in the cluster frame.

16.2 Dinuclear Clusters

The organometallic analogue of the well known ion $[Re_2Cl_8]^{2-}$ was first described by F. A. Cotton and G. Wilkinson in 1976:

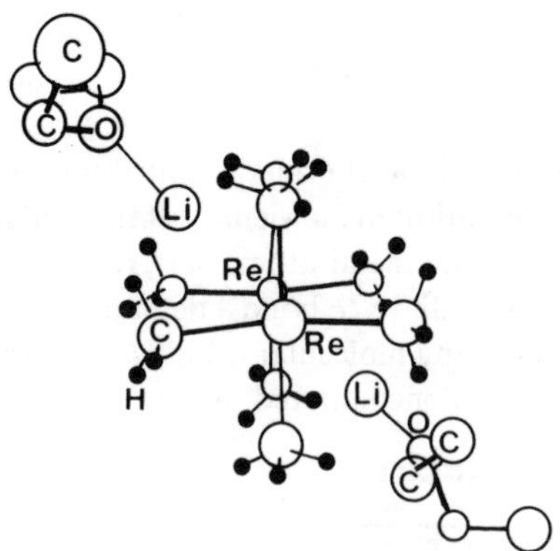

$$Re_2Cl_{10} \xrightarrow{LiCH_3,\ Et_2O} Li_2[Re_2(CH_3)_8] \cdot 2(C_2H_5)_2O$$

red, sensitive to air and moisture, stable at RT

$d(Re{\equiv}Re) = 218$ pm

compare $d(Re-Re) = 274$ pm (Re *metal*)

The short Re – Re bond distance and the eclipsed conformation (symmetry $\mathbf{D_{4h}}$) attest to the presence of an M≣M quadruple bond with a δ-component. The preference for a staggered conformation (D_{4d}) resulting from ligand-ligand repulsions should be even more pronounced for $L = CH_3$ as compared to $L = Cl$ considering the van der Waals radii of these two ligands. Notwithstanding, the ion $[Re_2(CH_3)_8]^{2-}$ also displays D_{4h} symmetry.

A qualitative MO scheme for $[Re_2(CH_3)_8]^{2-}$ may be set up in the following way (Cotton, 1978 R):

- Only Re(5*d*) AO's are considered as a first approximation
- Interactions between two Re atoms split the basis of the *d*-orbitals into the levels

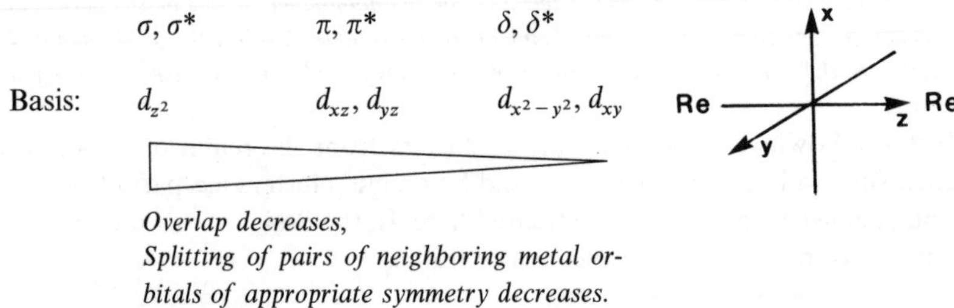

Overlap decreases,
Splitting of pairs of neighboring metal orbitals of appropriate symmetry decreases.

- One component each of the δ, δ^*-pairs interacts with a symmetry-adapted group orbital from eight CH_3 ligands; the other δ, δ^*-component is used for the $Re \overset{\delta}{-} Re$ bond.

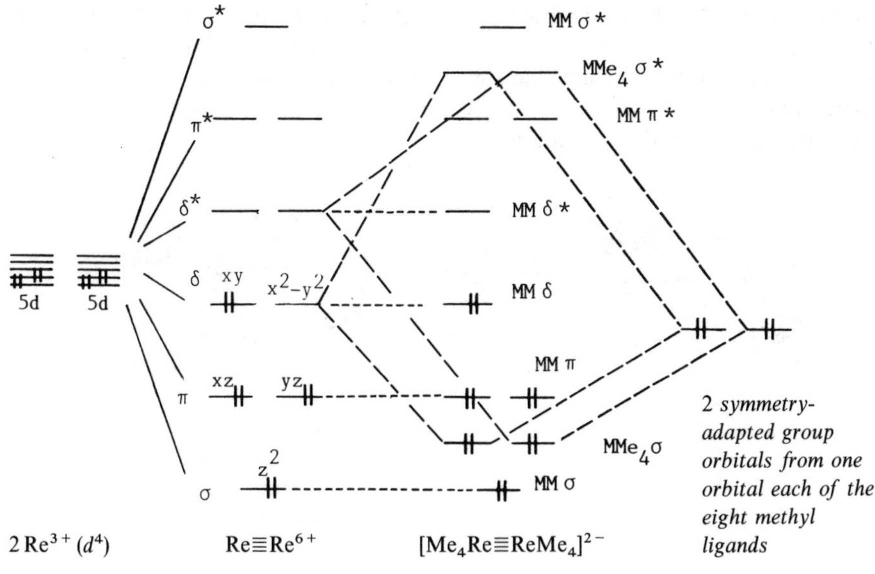

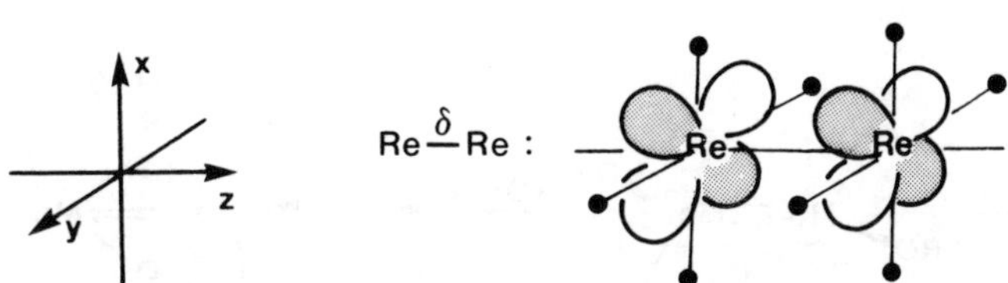

Additional contributions to Re–CH$_3$ bonding are provided by an admixture of Re(6s, p_x, p_y) AO's. Isoelectronic and isostructural with [Re$_2$(CH$_3$)$_8$]$^{2-}$ are the complex ions [M$_2$(CH$_3$)$_8$]$^{4-}$ (M = Cr, Mo). Occupation of the bonding MO MMδ is responsible for the realization of an eclipsed conformation of [Re$_2$(CH$_3$)$_8$]$^{2-}$ and its congeners. Twisting about the Re–Re bond axis would entail a decrease in overlap of the Re(d_{xy}) orbitals and thus a sacrifice in bond energy. Conversely, the representatives of another important class of dinuclear clusters, **M$_2$R$_6$**, assume a staggered conformation.

$$\text{Mo(NMe}_2)_4 + \text{Mo}_2(\text{NMe}_2)_6 \xleftarrow{\text{LiNMe}_2} \text{MoCl}_5 \xrightarrow[\text{Et}_2\text{O, 25}^\circ\text{C}]{\text{Me}_3\text{SiCH}_2\text{MgCl}} \text{Mo}_2(\text{CH}_2\text{SiMe}_3)_6$$

(Chisholm, 1974) (Wilkinson, 1972)

Structure of **Mo$_2$(CH$_2$SiMe$_3$)$_6$** *in the crystal, symmetry* **D$_{3d}$** *(idealized),* d(*Mo*≡Mo) = 217 pm, *compare*: d(Mo–Mo) = 278 pm (Mo metal). (*Skapski*, 1971)

R = CH$_2$SiMe$_3$

The Mo≡Mo triple bond linking two ions Mo^{3+} (d^3) can be described in terms of the electronic configuration $(\sigma_{z^2})^2$ $(\pi_{xz})^2$ $(\pi_{yz})^2$. In analogy to –C≡C–, this leads to an electron distribution which is cylindrically symmetrical along the z axis. As far as the $M \equiv M$ interaction is concerned, the bond energy is rotationally invariant and it is ligand-ligand repulsion that dictates the staggered conformation.

The **generation** of a metal–metal multiple bond from a single bond sometimes occurs spontaneously under ligand dissociation:

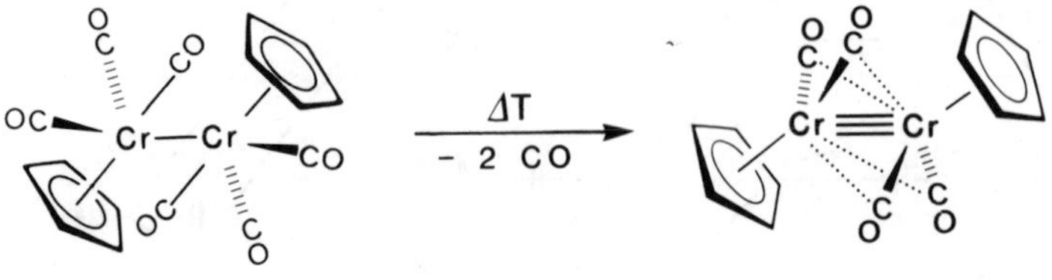

ΔT
– 2 CO

18 VE *complex*
congested

18 VE *complex*
relaxed
the CO ligands
are semibridging
(p. 225)

A **reduction** of metal–metal bond order is effected by oxidative addition:

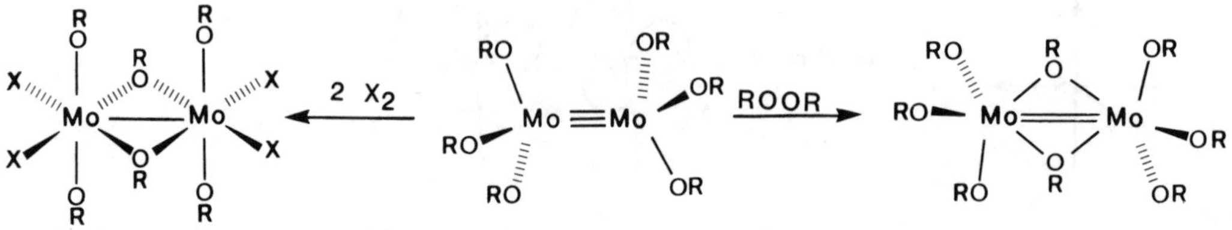

2 X_2 ← ROOR →

X = Cl, Br

16.3 Trinuclear Clusters

Examples of a **triangular arrangement M_3**:

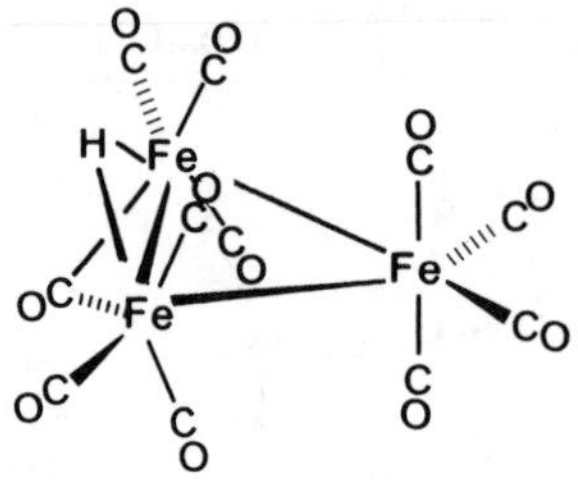

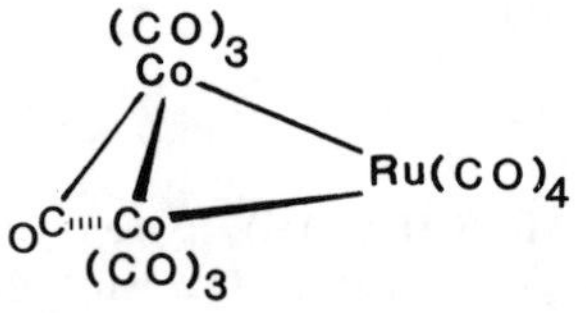

$[HFe_3(CO)_{11}]^-$
from $Fe_3(CO)_{12}$ *by means of a Hieber base reaction*

$RuCo_2(CO)_{11}$
from $Ru_3(CO)_{12}$
and $Co_2(CO)_8$

For triangular M_3 clusters the 18 VE rule demands the following total number of valence electrons Σ VE:

	M—M (3 single bonds)	M=M (one double bond)	M=M (two double bonds)	M=M (three double bonds)
"magic number" ΣVE:	**48**	46	44	42

$\Sigma VE = 3 \cdot 18 - m \cdot 2$ (m = number of metal–metal bonds)

Apparently, depending on the respective metal–metal bond orders, a considerable variety of valence-electron numbers may lead to triangular M_3 clusters, odd VE numbers also being encountered occasionally (L. Dahl, 1986 R):

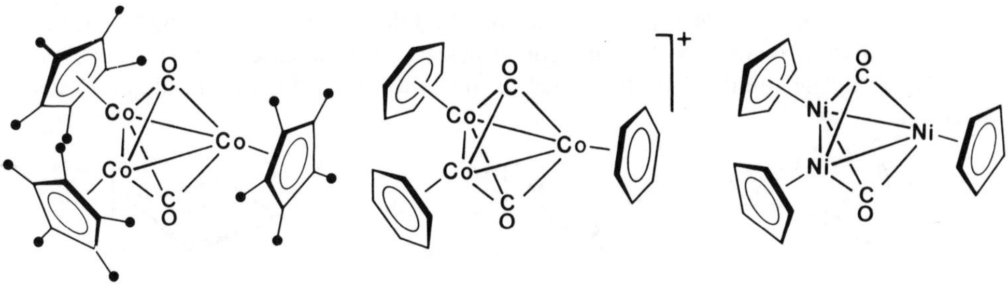

46 VE
from $C_5Me_5Co(CO)_2$
and $h\nu$

48 VE
from $Co_2(CO)_8$,
C_6H_6 *and* $AlBr_3$

49 VE
from $(C_5H_5)_2Ni$
and $Ni(CO)_4$

A compound with a **linear arrangement M_3** and an anticlinal conformation is formed in the following way (Herrmann, 1985):

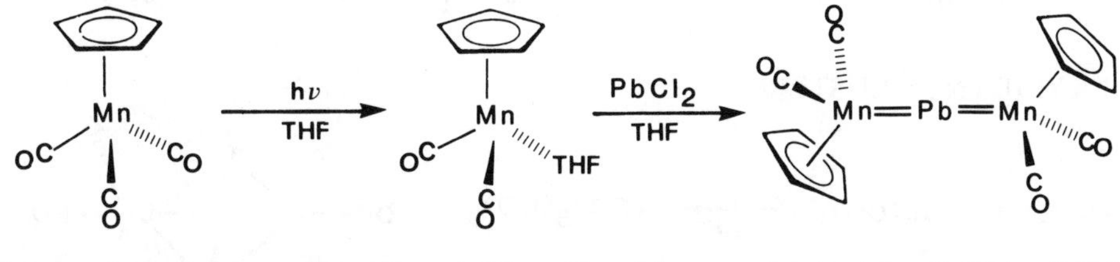

$h\nu$, THF; $PbCl_2$, THF

A related species featuring the unit Mn=Ge=Mn was obtained by E. Weiss (1981). The application of isolobal relationships (p. 396) reveals an analogy between $[(\eta\text{-}C_5H_5)Mn(CO)_2]_2Pb$ and $\mathbf{CO_2}$, since

$$Pb \leftrightarrow C \text{ and } (\eta\text{-}C_5H_5)Mn(CO)_2 \leftrightarrow Fe(CO)_4 \leftrightarrow CH_2 \leftrightarrow O$$

$d^6\text{-}ML_5$ $\qquad$ $d^8\text{-}ML_4$

16.4 Tetranuclear Clusters

BINARY METAL CARBONYLS

A widespread geometry for M_4 clusters is the **metallatetrahedrane** structure.

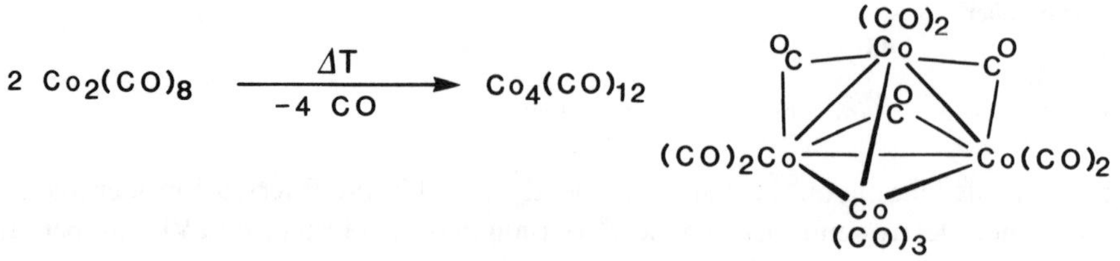

$$2\ Co_2(CO)_8 \xrightarrow[-4\ CO]{\Delta T} Co_4(CO)_{12}$$

As the "magic number" for the formation of a tetrahedral M_4 cluster one obtains $\boldsymbol{\Sigma}$ **VE** $= 4 \cdot 18 - 6 \cdot 2 =$ **60**. The presence $[Co_4(CO)_{12}]$ or absence $[Rh_4(CO)_{12}, Ir_4(CO)_{12}]$ of CO bridges does not affect the VE book-keeping. Like the generation of metal lattices, cluster formation is an aspect of electron deficiency. An increase of Σ VE above the respective "magic number" partially dispels this shortage and may result in **cluster opening:**

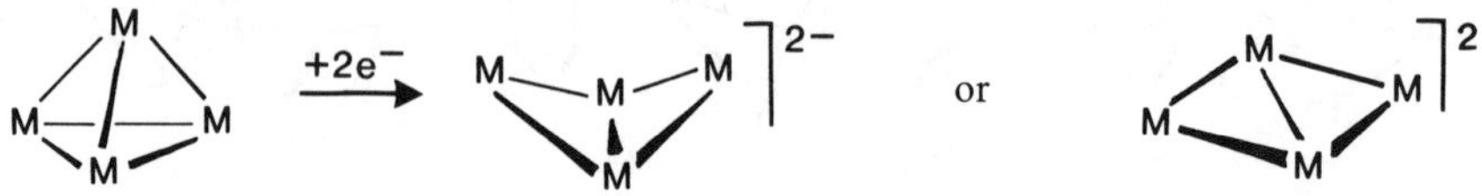

60 VE $\xrightarrow{+2e^-}$ 62 VE *Butterfly cluster* or 62 VE

An example is provided by the structure of the anion $[\mathbf{Re_4(CO)_{16}}]^{2-}$ (*Churchill*, 1967):

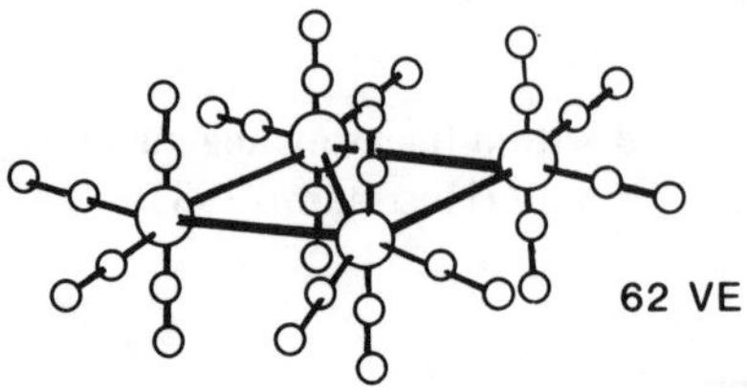

62 VE

ALKYLIDYNE CLUSTERS

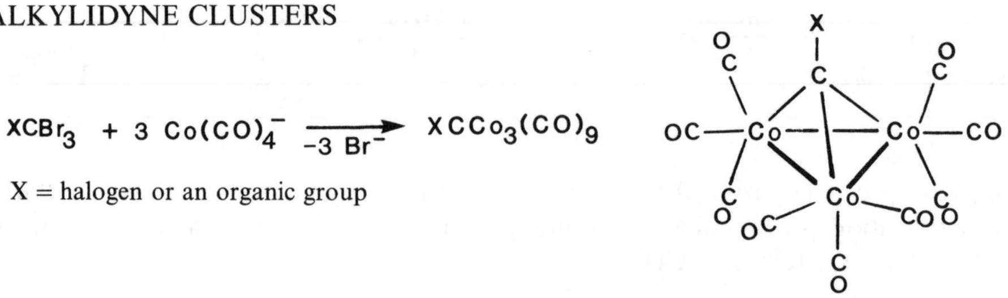

$$XCBr_3 + 3\ Co(CO)_4^- \xrightarrow[-3\ Br^-]{} XCCo_3(CO)_9$$

X = halogen or an organic group

$\boldsymbol{\Sigma}$ **VE** $=$ **50** $= 3 \cdot 18 + 1 \cdot 8 - 6 \cdot 2)$
TM MGE M–M bonds

Because of the participation of a main-group element which requires eight electrons only for a closed valence shell, the "magic number" for a tetrahedral frame is lower here.

CUBANE CLUSTERS $M_4(\eta\text{-}C_5H_5)_4(\mu_3\text{-}L)_4$

Four μ_3-ligands which cap the triangular faces of an M_4 tetrahedron generate a **hetero-cubane** structure M_4L_4. In transition-metal complexes the role of the μ_3-ligand is often assumed by CO:

12h reflux

14d reflux

$[(\eta\text{-}C_5H_5)Fe(CO)_2]_2$
purple red

$[Fe_4(\eta\text{-}C_5H_5)_4(\mu_3\text{-}CO)_4]$
greenish black

Examples with μ_3-L = S^{2-} are the **cubane clusters** $[M_4(\eta\text{-}C_5H_5)_4(\mu_3\text{-}S)_4]$. They are formed from the repective cyclopentadienylmetal carbonyls and sulfur or cyclohexene sulfide.

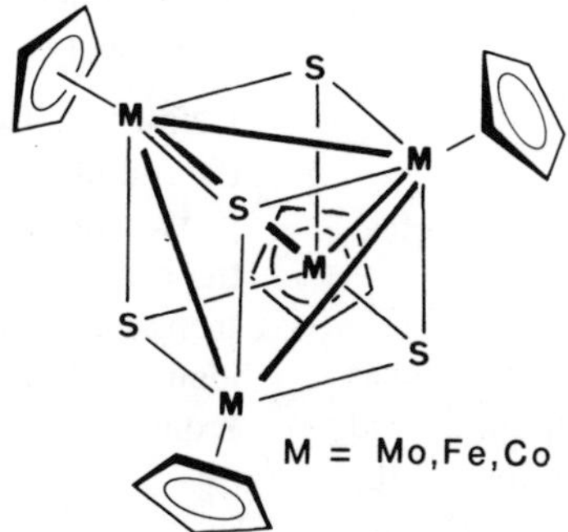

By means of **cyclovoltammetry** it has been demonstrated that the iron sulfur cubanes undergo four reversible electron transfer steps whereby the tetrahedral Fe_4 frame is preserved (Geiger, 1985 R):

$$Fe_4 \underset{-0.33\text{ V}}{\rightleftharpoons} Fe_4^{+} \underset{0.33\text{ V}}{\rightleftharpoons} Fe_4^{2+} \underset{0.88\text{ V}}{\rightleftharpoons} Fe_4^{3+} \underset{1.41\text{ V}}{\rightleftharpoons} Fe_4^{4+}$$

The oxidations are accompanied by increasing distortion of the cubane frame (Dahl, 1977). ^{57}Fe-Mössbauer spectra indicate, however, that the four iron atoms remain equivalent. This finding suggests delocalized bonding in the Fe_4 frame. The significance of these M_4S_4 clusters rests in their similarity to the active centers of iron-sulfur redox proteins (ferredoxin, rubredoxin) and possibly nitrogenase.

Excursion:

Structure and Bonding Conditions in Clusters, the Isolobal Analogy

A generalized treatment which embraces all types of clusters and metal–metal bonded species is not possible at present. Therefore, a variety of interpretations has been advanced (see B. F. G. Johnson, 1981 R):

	Cluster size	M_n
• 18 VE Rule (dating back to Sidgwick)	small	$n = 2-4$
• Skeletal Electron Theory (Wade, Mingos) • Isolobal Relations (R. Hoffmann)	medium	$n = 5-9$
• Closest Packing (metal lattices)	large	$n \geq 10$

The success in applying these approaches varies with the class of compounds and with cluster size. An evaluation of the power of these models requires comprehensive knowledge of descriptive organometallic chemistry which is out of the scope of the present text. In the following, the course of argumentation will be illustrated with a few selected examples. The 18 VE rule (p. 186 f) and Wade's rules (p. 68 f) were already introduced.

ISOLOBAL ANALOGIES

Applications of the skeletal electron theory (SET) necessitate an a priori division of the metal orbitals into skeleton-binding and ligand-binding functions; they depend on a preconception of the structure of the molecule to be treated. An alternative approach was devised by R. Hoffmann who calculated the **bonding characteristics of cluster fragments ML_n** in the frontier orbital region by means of the Extended-Hückel-method and compared them with the units CH_l and BH_m.

Definition: Two fragments are **isolobal** if the number, symmetry properties, approximate energy and shape of their frontier orbitals as well as the number of electrons occupying them are similar – not identical but similar.
Example:

$$\cdot CH_3 \overset{\text{isolobal}}{\longleftrightarrow} \cdot Mn(CO)_5$$

The symbol $\longleftrightarrow$ (with o) will be used not only for the fragments but also for the molecules they constitute.

Starting-point for the derivation of the lobal properties of organometallic fragments is the simplified MO *scheme of an octahedral complex. The removal of one ligand* L *converts a bonding* σ-MO *of the complex* ML_6 *into a nonbonding frontier orbital* ψ_{hy} *of the fragment* ML_5:

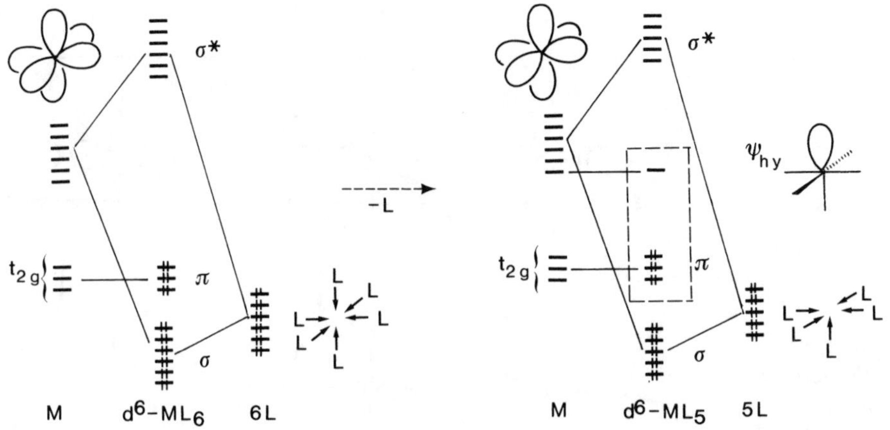

M d6-ML6 6L -L M d6-ML5 5L

The removal of 2 L *creates two, the removal of* 3 L *three new frontier orbitals* ψ_{hy}:

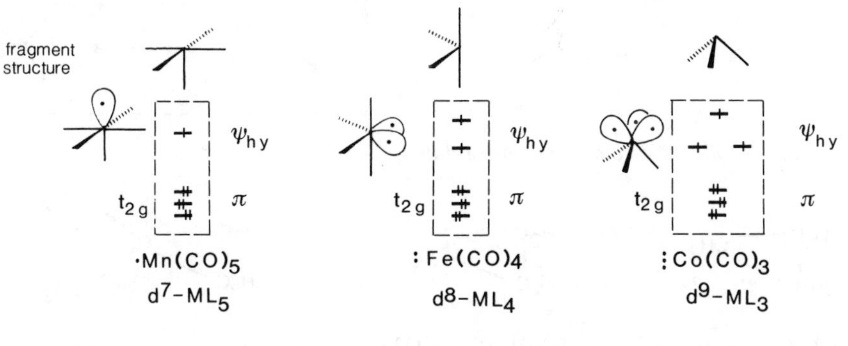

fragment structure

$\cdot Mn(CO)_5$ d^7-ML_5 $:Fe(CO)_4$ d^8-ML_4 $\vdots Co(CO)_3$ d^9-ML_3

The frontier orbitals t_{2g} *and* ψ_{hy} *are filled by n electrons of the central metal (configuration* d^n*). This provides* $Mn(CO)_5$ *with one,* $Fe(CO)_4$ *with two and* $Co(CO)_3$ *with three singly occupied orbitals with distinct spatial orientation. These organometallic fragments are complementary to organic analogues*:

d^7-$ML_5 \leftarrow\!\!\circ\!\!\rightarrow \cdot CH_3$ Methyl d^8-$ML_4 \leftarrow\!\!\circ\!\!\rightarrow :CH_2$ Methylene d^9-$ML_3 \leftarrow\!\!\circ\!\!\rightarrow :CH$ Methylidyne

The isolobal connection allows a **joint consideration of inorganic, organic and organometallic structures,** the relationships being based on the isolobal nature of the respective molecular fragments:

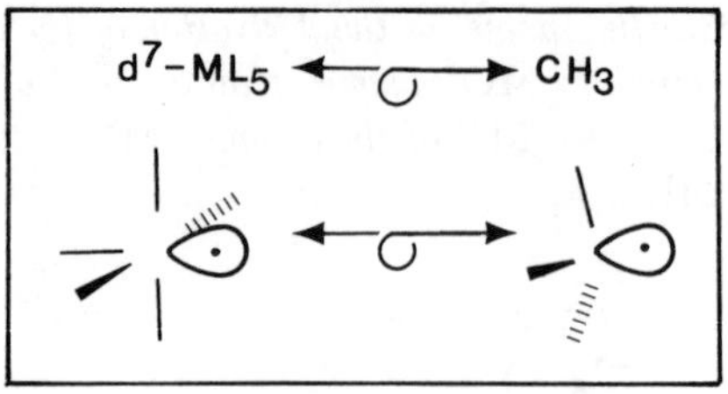

d^7-ML_5 ⟷ CH_3

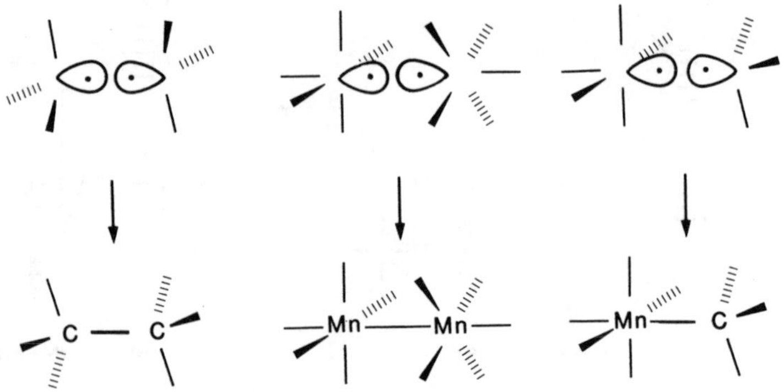

An analogy of the following compounds is implied:

C_2H_6 $Mn_2(CO)_{10}$ $(CO)_5MnCH_3$

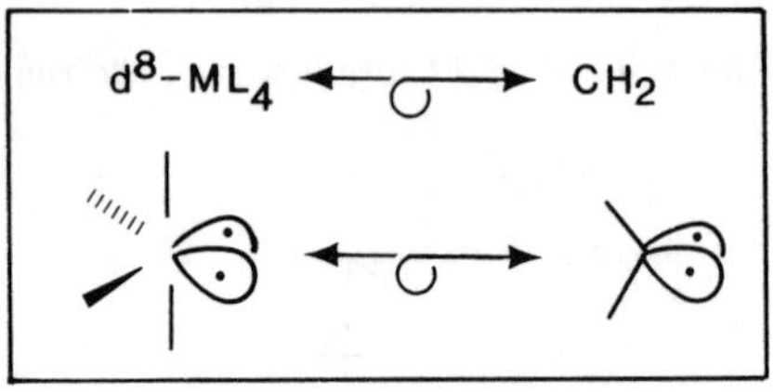

d^8-ML_4 ⟷ CH_2

Therefore, the following compounds are related:

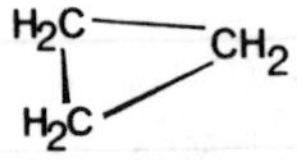

C_3H_6 (cyclopropane) 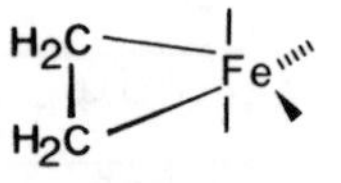 $(C_2H_4)Fe(CO)_4$ 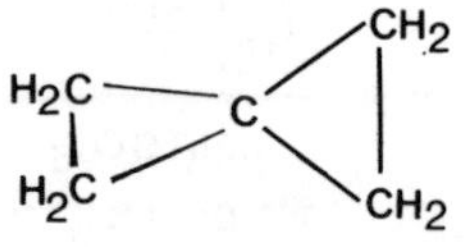 C_5H_8 (spiropentane)

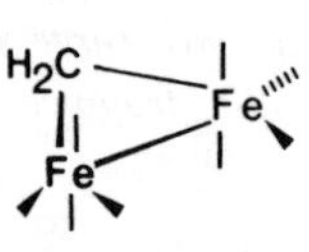

$(\mu$-$CH_2)Fe_2(CO)_8$ 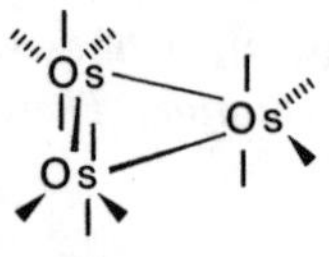 $Os_3(CO)_{12}$

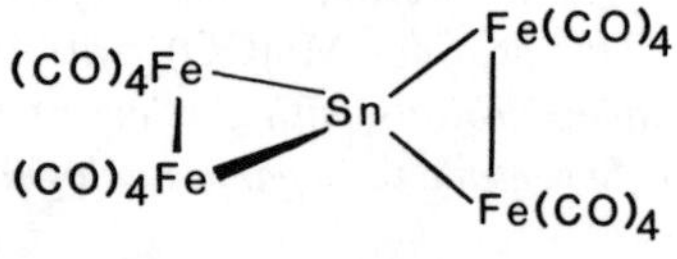

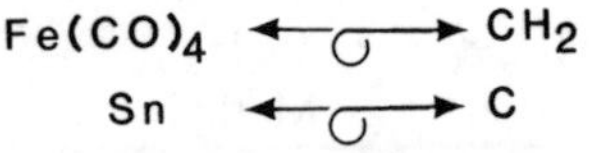

$Fe(CO)_4$ ⟷ CH_2

Sn ⟷ C

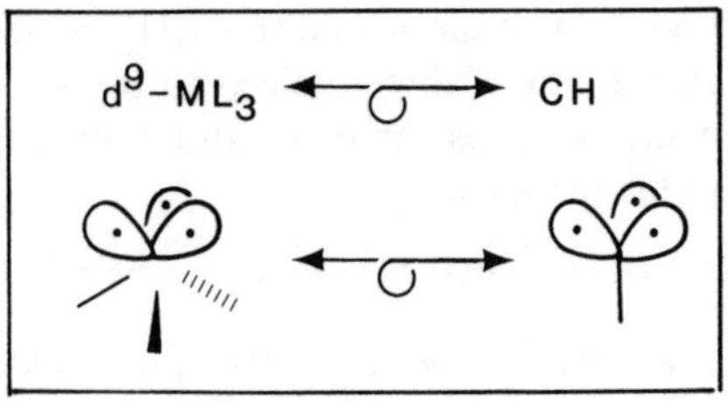

d^9-ML_3 ←σ→ CH

This leads to the following series of analogous clusters:

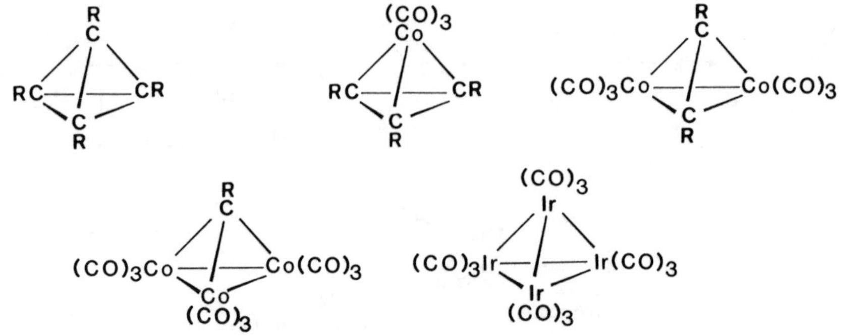

Even the molecules P_4, As_4 *and* Sb_4 *may be included into a larger family:*

d^9-ML_3 ←σ→ CH ←σ→ As,

hence the following tetrahedranes are analogous:

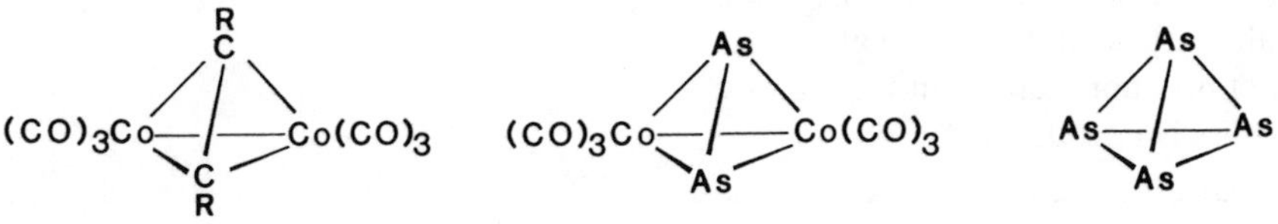

*Structural relationships between boranes and carbaboranes (**p. 70**) stem from the isolobal analogy*

BH^- ←σ→ CH

Inclusion of the ligand η^5-$C_5H_5^-$ which, as a donor of three π-electron pairs formally occupies three coordination sites, yields the analogies

$$Fe(CO)_4 \longleftrightarrow CpFe(CO)^- \longleftrightarrow CpRhCO \qquad (d^8\text{-}ML_4).$$

The following molecules then are related:

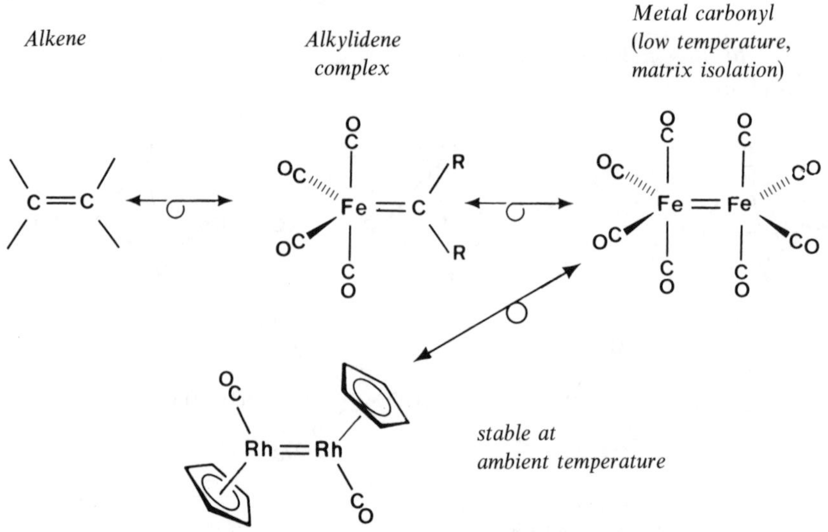

Alkene

Alkylidene complex

Metal carbonyl (low temperature, matrix isolation)

stable at ambient temperature

Another analogy is that between cyclopropane and μ-alkylidene complexes:

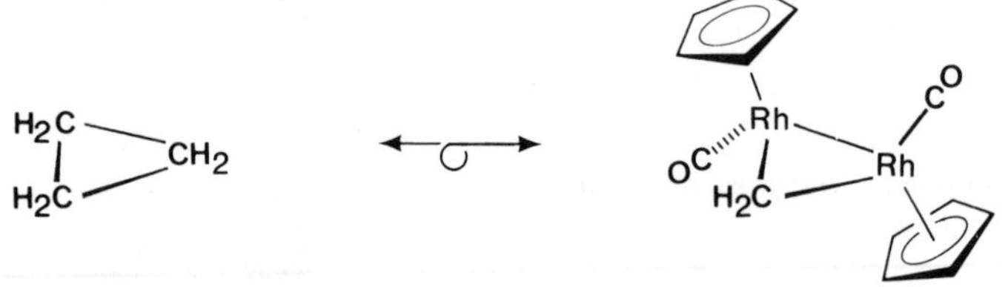

The realm of isolobal connections is considerably extended if the mutual replacement of σ-donor ligands and metal electron pairs is introduced:

Organic fragment	Coordination number of the metal on which the derivation of the isolobal fragments is based				
	9	8	7	6	5
CH_3	d^1-ML_8	d^3-ML_7	d^5-ML_6	d^7-ML_5	d^9-ML_4
CH_2	d^2-ML_7	d^4-ML_6	d^6-ML_5	d^8-ML_4	d^{10}-ML_3
CH	d^3-ML_6	d^5-ML_5	d^7-ML_4	d^9-ML_3	

Not for all of these general types have fragments yet been recognized. As a caveat it should also be stressed that applications of the isolobal principle do not necessarily furnish stable molecules: according to the relation $CH_2 \overset{\longleftrightarrow}{\scriptstyle o} Fe(CO)_4$ the existence of a molecule $(CO)_4Fe{=}Fe(CO)_4$ would be predicted. However, $Fe_2(CO)_8$ has only been observed in low temperature matrices; it readily accepts CO to yield the stable molecule $Fe_2(CO)_9$. Its analogue cyclopropanone, on the other hand, does not form spontaneously from $H_2C{=}CH_2$ and CO; it has to be prepared in a different way.

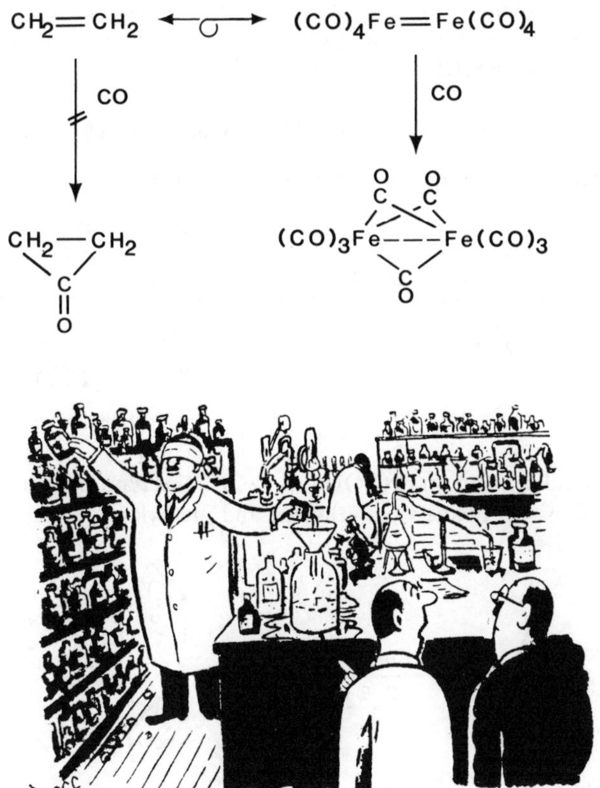

"That's Dr. Arnold Moore. He's conducting an experiment to test the theory that most great scientific discoveries were hit on by accident."
(Hoff, © 1957, *The New Yorker Magazine, Inc.)*

16.5 Approaches to Systematic Cluster Synthesis

As opposed to the formation of carbon—carbon bonds, the methods of cluster synthesis have not yet reached a stage where far-reaching generalizations would be warranted. Rather, many new cluster molecules owe their existence to serendipity, and writing balanced equations for their formation is fraught with difficulties. In an attempt to plan cluster synthesis, the conceptional assembly of a cluster from fragments with distinct frontier orbital characteristics is translated into a preparative reaction sequence. Since the structural chemistry of metal atom clusters is extremely varied, the role of organic chemistry as a supplier of models is limited. In the following paragraphs, a few promising approaches are sketched. They are arranged according to the reaction types build-up (I), expansion (II) and exchange (III).

I Cluster build-up starting from M⩵M and M⩵C multiple bonds

- **Addition of carbene-like fragments to M=M** (Stone, 1983)

Rh=Rh —[$Cr(CO)_5THF$]→ Rh—Rh—Cr$(CO)_5$

Trismetallacyclopropane

Because of the isolobal connection $Cr(CO)_5 \overset{}{\underset{O}{\longleftrightarrow}} CH_2$ this type of reaction corresponds to the formation of cyclopropane from ethylene and methylene.

Co=Ir —[$CpCo(C_2H_4)_2$, $-2\ C_2H_4$]→ Co_2Ir

The two CO ligands engage in unsymmetrical bridging (Herrmann, 1985).

- **Addition of carbene-like fragments to M≡C and to M=C** (Stone, 1982)
  Principle:

M≡CR —[M′]→ M—M′ (RC bridged) —[M″]→ M—M′—M″ (RC capped)

M′,M″ = coordinatively unsaturated fragment — *Dimetalla-cyclopropene* — *Trimetalla-tetrahedrane*

Example:

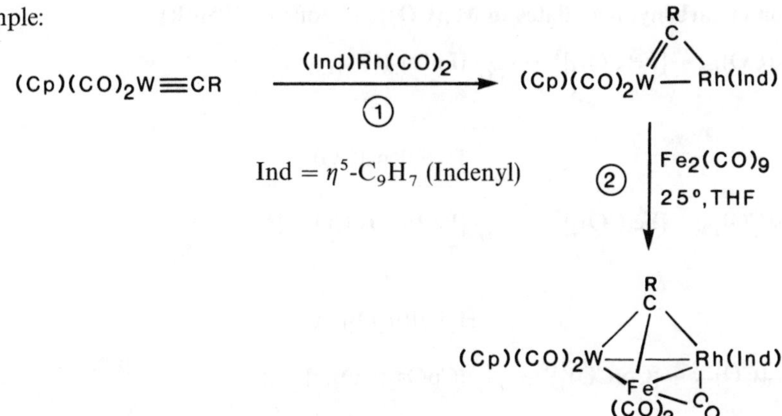

In reaction ① the fragment (Ind)Rh[d^8-ML_3] *behaves like a carbene although it is a species of the type d^8-ML_4 which was introduced as a carbene analogue (p. 398). The propensity of the late transition metals (d^8, d^{10}) to form stable* 16 VE *complexes calls for a* ***corollary in the classification of isolobal fragments.*** *Accordingly, the fragments d^8-ML_3 and d^{10}-ML_2 also are carbene-like and can add to multiple bonds.*

	Carbene analogues	*but also (for late TM)*
	d^8-ML_4	d^8-ML_3
	Cp(CO)Co, $(CO)_4Fe$	CpRh, (acac)(CO)Ir
and	d^6-ML_5	d^{10}-ML_2
	$Cp(CO)_2Re$, $(C_6H_6)(CO)_2Cr$	$(R_3P)_2Pt$, (1,5-COD)Pt

The cluster formed in reaction ② contains one element each of the three transition series and is chiral. Clusters of this kind show extremely high values of optical rotation (up to 10^4 degrees).

II Cluster expansion

- **Addition of the fragment Ph_3PAu** (Lauher, 1981)

The moieties H and Ph_3PAu can mutually replace each other in carbonyl metallates ($Ph_3PAu^+ \underset{\text{O}}{\longleftrightarrow} H^+$). Examples are the pairs $(CO)_4CoH$, $(CO)_4CoAuPPh_3$ and $(CO)_4FeH_2$, $(CO)_4Fe(AuPPh_3)_2$. This finding led to a general synthesis of transition-metal gold clusters from carbonyl metallates and Ph_3PAu^+:

$$[FeCo_3(CO)_{12}]^- \xrightarrow[-\ NO_3^-]{Ph_3PAuNO_3\,/\,Acetone} Ph_3PAuFeCo_3(CO)_{12}$$

This procedure has wide applicability.

- **Addition of carbonyl metallates to $M_3(CO)_{12}$** (Geoffroy, 1980 R)

$$Ru_3(CO)_{12} + [Fe(CO)_4]^{2-} \xrightarrow[-CO]{} [FeRu_3(CO)_{13}]^{2-} \xrightarrow{H^+} H_2FeRu_3(CO)_{13}$$

$$Ru_2Os(CO)_{12} + [Fe(CO)_4]^{2-} \xrightarrow[-CO]{} [FeRu_2Os(CO)_{13}]^{2-} \xrightarrow{H^+} H_2FeRu_2Os(CO)_{13}$$

$$Os_3(CO)_{12} + [Co(CO)_4]^{-} \xrightarrow[-CO]{} [CoOs_3(CO)_{13}]^{-}$$

In this way triangular 48 VE clusters may be transformed into tetrahedral 60 VE clusters which contain various combinations of the elements Fe, Ru, Os, and Co in their frame. These reactions rarely proceed in a uniform fashion and product composition is very sensitive to reaction temperature and duration.

- **Cluster expansion by main-group elements** (Vahrenkamp, 1983)

$(CO)_4Ru$, $Co(CO)_3$, $Co(CO)_3$, CO

Ph_2Se_2 → Se; $(CO)_3Ru$, $Co(CO)_3$, Co $(CO)_3$

$RAsH_2$ → R–As; $(CO)_3Ru$, $Co(CO)_3$, Co $(CO)_3$

$$Se^+ \longleftrightarrow RAs^+ \longleftrightarrow Co(CO)_3$$

$$Ru(CO)_3^- \longleftrightarrow Co(CO)_3$$

These isolobal analogies relate the Co_2Ru(arsinidyne) *cluster to the binary metal carbonyl* $Co_4(CO)_{12}$.

III Metal exchange using isolobal fragments

$$Os_3(CO)_{12} \xrightarrow[-\ CO]{[Re(CO)_5]^-} [ReOs_2(CO)_{12}]^-$$

$(CO)_4$Os, Os$(CO)_4$, Os$(CO)_4$ → $(CO)_4$Os, Os$(CO)_4$, Re$(CO)_4$

$$d^8\text{-}ML_4\text{:}\quad Re(CO)_4^- \longleftrightarrow Os(CO)_4$$

(Mays, 1972)

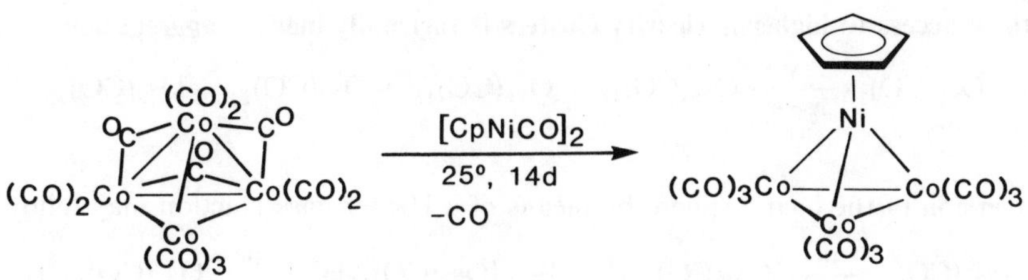

d^9-ML_3: CpNi $\longleftrightarrow$ $Co(CO)_3$ (Vahrenkamp, 1982)

In contrast to the pyrolytic procedures used in the infancy of cluster synthesis, these reactions proceed under mild conditions and consequently are more specific. In the examples stated above, the high nucleophilicity of the carbonyl metallate $[Re(CO)_5]^-$ *and the reactivity of the* 17 VE *species* CpNi(CO), *which exists in equilibrium with its dimer, are decisive. The mechanistic study of this type of metal exchange reaction is still at an early stage – among the three conceivable sequences, cluster fragmentation/reconstitution, elimination/addition and addition/elimination, the third variant is currently best supported by experimental evidence* (*Vahrenkamp*, 1985 *R*).

16.6 Pentanuclear and Higher Nuclearity Clusters

The structures of medium-sized clusters are rationalized in two different ways:

- The simplest approach is to relate the number of valence electrons of a particular cluster to the "magic number" for the equiapical polyhedron. The magic numbers are generated according to the prescription:

$$\textbf{magic number} = 18 \cdot \begin{array}{c}\text{number}\\ \text{of vertices}\\ \text{(TM atoms)}\end{array} - 2 \cdot \begin{array}{c}\text{number}\\ \text{of edges}\\ \text{(M–M bonds)}\end{array}$$

  Obviously, in this way the inert gas rule is fulfilled for the constituent metal atoms. The following magic numbers thus arise: **48** (triangle), **60** (tetrahedron), **72** (trigonal bipyramid), **74** (square pyramid), **86** (octahedron), **90** (trigonal prism), **120** (cube).
- Alternatively, the d^x-ML_y fragments of the framework are "translated" into the corresponding $BH^{+,0,-}$ fragments, using isolobal analogies, and **Wade's rules** are employed thereafter. The plethora of cluster geometries calls for an extension of the relations between the number n of apical units, the number of skeletal bonding electron pairs and the corresponding structural type:

Skeletal bonding electron pairs	Wade type	Cluster geometry
$n-1$		(n – 2)polyhedron, 2 faces capped
n		(n – 1)polyhedron, 1 face capped
$n+1$	**closo**	(n)polyhedron
$n+2$	**nido**	(n + 1)polyhedron, 1 vertex unoccupied
$n+3$	**arachno**	(n + 2)polyhedron, 2 vertices unoccupied
$n+4$	**hypho**	(n + 3)polyhedron, 3 vertices unoccupied

A path of access to **higher nuclearity clusters** is **thermally induced aggregation**:

$$Os_3(CO)_{12} \xrightarrow[\text{pressure tube}]{\Delta T} Os_5(CO)_{16} + Os_6(CO)_{18} + Os_7(CO)_{21} + Os_8(CO)_{23}$$

A conversion of the ligand sphere by means of a Hieber base reaction may follow:

$$Os_5(CO)_{16} \xrightarrow[-CO_2]{OH^-} [Os_5(CO)_{15}]^{2-} \xrightarrow{H^+} [Os_5(CO)_{15}H]^- \xrightarrow{H^+} Os_5(CO)_{15}H_2$$

A rationalization of the structures of the products from $Os_3(CO)_{12}$ pyrolysis proceeds as follows:

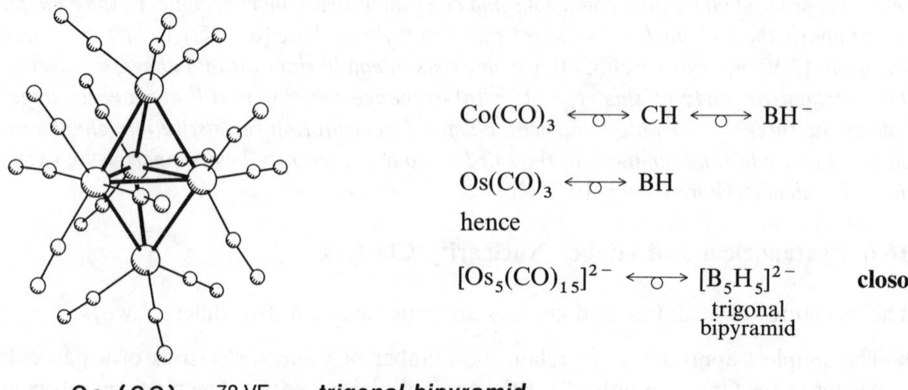

$Co(CO)_3 \leftrightarrow CH \leftrightarrow BH^-$ (isolobal)

$Os(CO)_3 \leftrightarrow BH$ (isolobal)

hence

$[Os_5(CO)_{15}]^{2-} \leftrightarrow [B_5H_5]^{2-}$ (isolobal) **closo**

trigonal bipyramid

$Os_5(CO)_{16}$ 72 VE ***trigonal bipyramid***

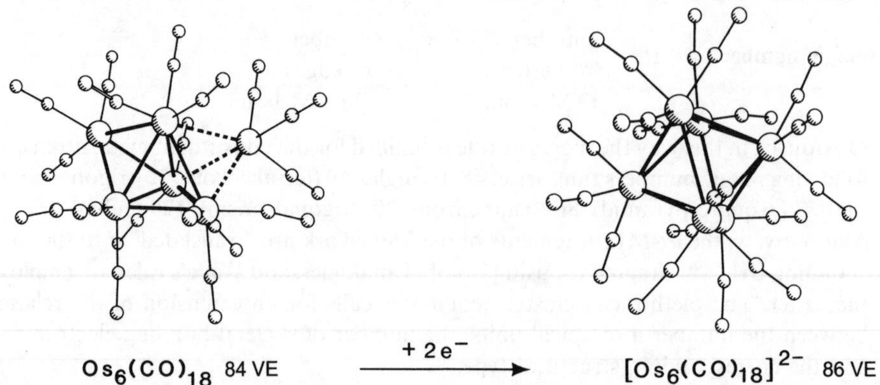

$Os_6(CO)_{18}$ 84 VE $\xrightarrow{+2e^-}$ $[Os_6(CO)_{18}]^{2-}$ 86 VE

For an octahedral structure $Os_6(CO)_{18}$ is "short of one electron pair". Therefore, a structure with a higher degree of M–M connectivity is encountered: ***monocapped trigonal bipyramid.***

$[Os_6(CO)_{18}]^{2-}$ is a ***regular octahedron*** *since $n = 6$ apical units hold $(n + 1) = 7$ skeletal electron pairs (**closo structure**).*

Accordingly: $Os_7(CO)_{21}$ monocapped octahedron, 6 + 1 Os atoms,
$Os_8(CO)_{23}$ bicapped octahedron, 6 + 2 Os atoms,
as in the iso(valence)electronic anion $[Os_8(CO)_{22}]^{2-}$.

Another method of synthesis is the **reductive aggregation** of mononuclear precursors:

$$\mathrm{Ni(CO)_4} \xrightarrow{\text{Na/THF}} \underset{\text{yellow}}{[\mathrm{Ni_5(CO)_{12}}]^{2-}} \xrightarrow{\mathrm{Ni(CO)_4}} \underset{\text{red}}{[\mathrm{Ni_6(CO)_{12}}]^{2-}} + 4\,\mathrm{CO}$$

(Chini, 1976)

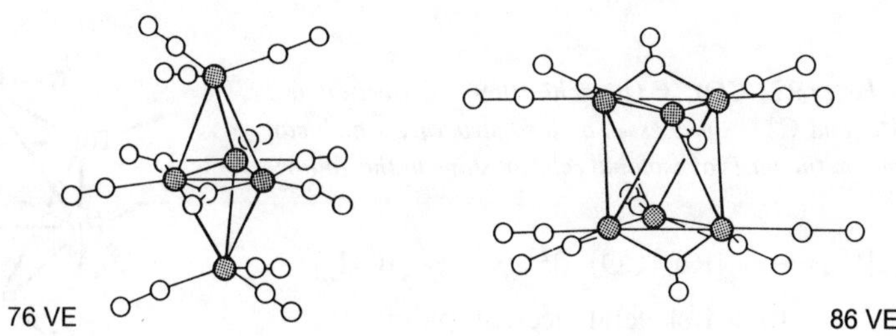

$$(\eta^5\text{-}\mathrm{C_5H_5})_2\mathrm{Ni} \xrightarrow[\text{THF}]{\mathrm{Na^+C_{10}H_8^-}} \underset{d^9\text{-}\mathrm{ML_3}}{[(\eta^5\text{-}\mathrm{C_5H_5})\mathrm{Ni}]_6} + \ldots$$

(Dahl, 1980)

The cavity in the center of medium-sized clusters has the appropriate dimension to allow the incorporation of first-row main group elements like B, C, and N. These **interstitial atoms** may add to cluster stability. Note the unusual coordination numbers of C, N and H in these situations!

(CO)3Fe
(CO)3Fe Fe(CO)3
C
(CO)3Fe Fe(CO)3

$Fe_5(CO)_{15}C$ 74 VE square pyramid

*The **carbido-cluster*** $Fe_5(CO)_{15}C$ *is formed in low yield if* $Fe_2(CO)_9$ *or* $Fe_3(CO)_{12}$ *are heated in hydrocarbon solvents in the presence of alkynes like* 1-*pentyne (Hübel, 1962). According to IR spectroscopy, bridging* CO *ligands are absent. The five-coordinate* C *atom is located slightly below the* Fe_4 *plane. Formal dissection into* $[Fe_5(CO)_{15}]^{4-}$ *and* C^{4+} *followed by an application of the isolobal principle and Wade's rules leads to a **nido** structure for the* Fe_5 *frame:*

$$\mathrm{Fe(CO)_3} \longleftrightarrow \mathrm{BH}$$

$$[\mathrm{Fe_5(CO)_{15}}]^{4-} \longleftrightarrow [\mathrm{B_5H_5}]^{4-}$$

nido ($n + 2$ skeletal electron pairs)

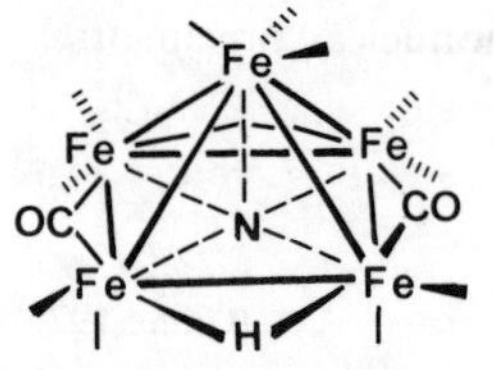

The ***nitrido-cluster*** $[Fe_5(CO)_{14}N]^-$ *and its protonated form* $HFe_5(CO)_{14}N$ *also display* ***nido*** *structures. They are prepared from* $Na_2Fe(CO)_4$, $Fe(CO)_5$ *and* $NOBF_4$ *at* 145 °C (*Muetterties*, 1980).

The ***carbido-cluster*** $Ru_6(CO)_{17}C$ (*Bianchi*, 1969) – *dissection into* $[Ru_6(CO)_{17}]^{4-}$ *and* C^{4+} – *processes a closo structure which can be rationalized on the basis of isolobal relationships in the following way*:

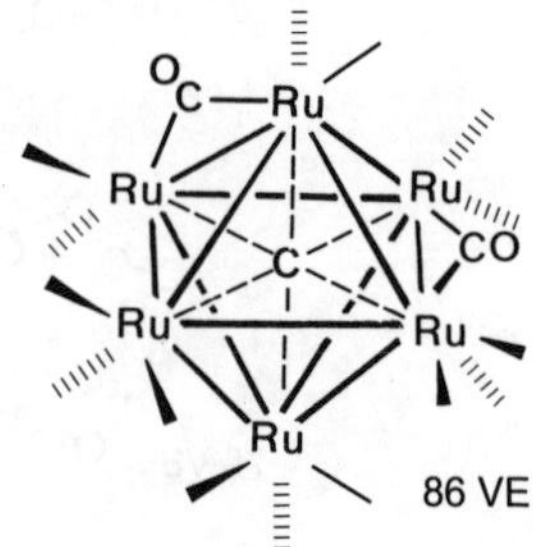

$$[Ru_6(CO)_{17}]^{4-} \longleftrightarrow [Ru_6(CO)_{18}]^{2-} \longleftrightarrow [B_6H_6]^{2-}$$

($n + 1$ skeletal electron pairs)

closo

In the context of the bonding mode of hydrogen in interstitial metal hydrides, the preparation and structural characterization of the **hydrido cluster** $[HCo_6(CO)_{15}]^-$ is relevant (Longoni, Bau, 1979):

$$Co_2(CO)_8 \xrightarrow[\text{EtOH}]{\Delta T} [Co_6(CO)_{15}]^{2-} \underset{H_2O}{\overset{\text{conc. HCl}}{\rightleftharpoons}} [HCo_6(CO)_{15}]^-$$

d(Co–Co) 251 pm; d(Co–Co) 258 pm

86 VE

Structure (neutron diffraction) of the cluster anion $[\mathbf{HCo_6(CO)_{15}}]^-$ *in the crystal: the* Co_6 *octahedron holds ten terminal* CO *ligands, four unsymmetrical* CO *bridges and one symmetrical* CO *bridge. Upon raising the* pH, *the* H *atom readily leaves the center of the* Co_6 *octahedron. As opposed to conventional carbonyl metal hydrides, the* H *atom in the cluster center is strongly deshielded,* $\delta^1H = 23$ ppm (*see p.* 306).

The octahedral structure (closo) of the ion $[Co_6(CO)_{15}]^{2-}$ and of its protonated form $[HCo_6(CO)_{15}]^-$ is concordant with Wade's rules because of the analogies

$$[Co_6(CO)_{15}]^{2-} \longleftrightarrow [Ru_6(CO)_{18}]^{2-} \longleftrightarrow [B_6H_6]^{2-} \qquad \textbf{(closo)}$$

The reliability of predictions which are based on the analogy of TM carbonyl clusters to equiapical borane polyhedra is limited. Whereas the structural chemistry of **higher boranes** is dominated by **icosahedral geometry,** which is also found in elemental boron, **higher nuclearity TM clusters** tend to form **closest packed structures** mimicking the metallic state. Furthermore, magic number considerations become somewhat nebulous for the late transition metals which incline towards 16 and 14 VE configurations (p. 189). The magic numbers rest on the presumption of localized 2e 2c M–M single bonds. To the extent that partial multiple bond character or delocalized antibonding electrons cannot be excluded, the predictive value of the valence electron count suffers.
It is already with $[Ni_5(CO)_{12}]^{2-}$ that a comparison with the corresponding borane leads to inconsistencies. This cluster anion was found to have a trigonal bipyramidal structure (p. 407) which according to

$$Ni(CO)_2\ (d^{10}\text{-}ML_2) \overset{}{\underset{\circ}{\longleftrightarrow}} Os(CO)_3\ (d^8\text{-}ML_3) \underset{\circ}{\longleftrightarrow} BH$$

$$[Ni_5(CO)_{10}]^{2-} \underset{\circ}{\longleftrightarrow} [B_5H_5]^{2-} \quad \textbf{(closo)}$$

would be predicted for the – hitherto unknown – anion $[Ni_5(CO)_{10}]^{2-}$.
The octahedral cluster $[CpNi)_6$ (p. 407) in the spirit of Wade's rules as well as in terms of the magic number criterion is "in excess of four electrons".

$$(\eta^5\text{-}C_5H_5)Ni\ (d^9\text{-}ML_3) \underset{\circ}{\longleftrightarrow} Co(CO)_3 \underset{\circ}{\longleftrightarrow} CH \underset{\circ}{\longleftrightarrow} BH^-$$

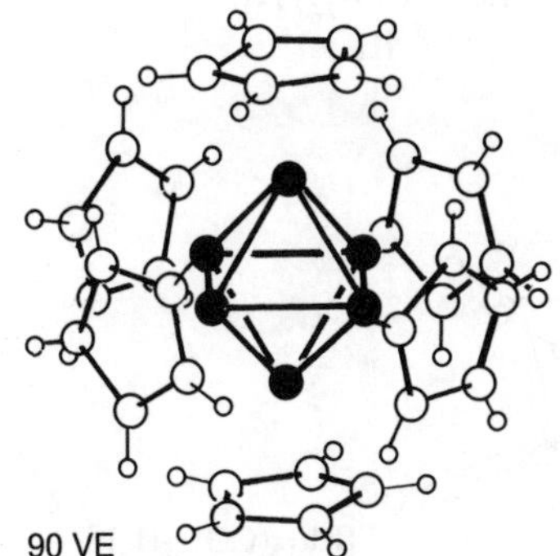

An analogy between **(CpNi)$_6$** *and* $[B_6H_6]^{6-}$ *– would suggest for this* Ni_6 *cluster an arachno structure derived from a square antiprism. Apparently, in this particular case the steric requirement of the ligands* C_5H_5 *must be considered which favors the highly symmetrical closo structure.*

A trigonal prismatic arrangement of six metal atoms is found in the carbido cluster anion $[Rh_6(CO)_{15}C]^{2-}$ (Albano, 1973):

$$Rh_4(CO)_{12} \xrightarrow[\substack{2.\ CHCl_3 \\ 3.\ [Me_3NCH_2Ph]Cl}]{1.\ NaOH/MeOH,\ CO} \underset{70\%}{[Me_3NCH_2Ph]_2[Rh_6(CO)_{15}C]} \quad (90\ VE)$$

The carbon atom is situated in the center of a trigonal Rh_6 prism with bridging and terminal CO ligands at the edges and corners, respectively. Contrarily, in the binary metal carbonyl $Rh_6(CO)_{16}$ the Rh atoms span an octahedron (=trigonal antiprism). It would appear that the central carbido C atom triggers the unusual disposition of the metal atoms in $[\mathbf{Rh_6(CO)_{15}C}]^{2-}$. The following stereoview of this cluster anion also serves to demonstrate an alternative way of looking at the structure of oligonuclear metal carbonyls (B. F. G. Johnson, 1981 R):

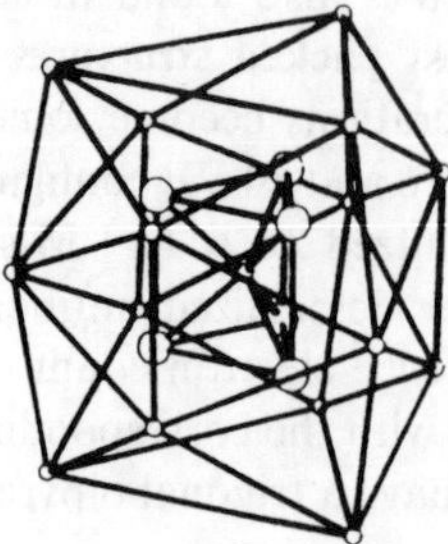

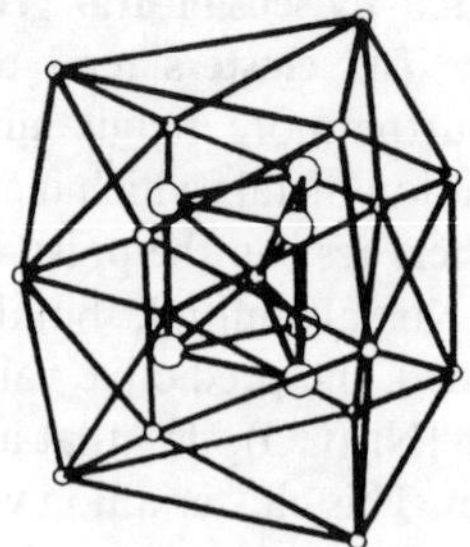

One recognizes a concentric arrangement of one C atom, an Rh_6 prism and a $(CO)_{15}$ deltahedron (= polyhedron with triangular faces). The positions of the terminal and the bridging CO groups are easily spotted. In this structural approach, it is assumed that the geometry of the ligand**sphere** is governed by a sterically optimal distribution of the external CO molecules (effective radius ≈ 300 pm) on the surface of a sphere as well as by the steric demands of the internal M_n cluster.

One of the most interesting aspects of multiatomic clusters is the **gradual transition from the molecular to the sub-microcrystalline metallic state.** By way of conclusion this is exemplified for another higher cluster of rhodium (Chini, 1975):

$$[Rh_{12}(CO)_{30}]^{2-} \xrightarrow[50\,°C]{H_2,\ 1\ bar} 2[Rh_6(CO)_{15}H]^- \xrightarrow[10\ h]{80\,°C} [Rh_{13}(CO)_{24}H_3]^{2-}$$

(i.a.)

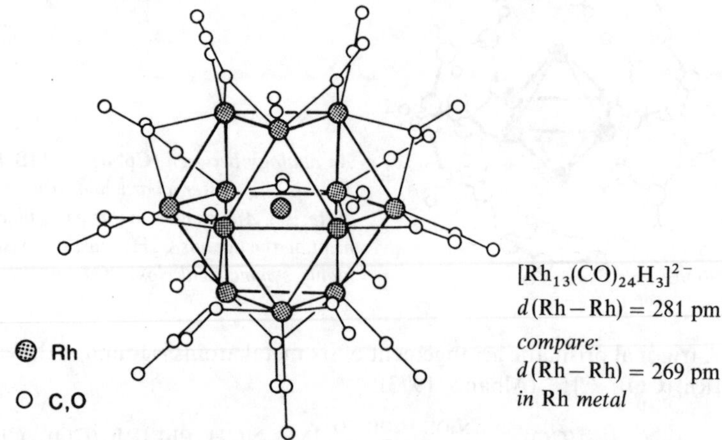

$[Rh_{13}(CO)_{24}H_3]^{2-}$
$d(Rh-Rh) = 281$ pm
compare:
$d(Rh-Rh) = 269$ pm
in Rh *metal*

The 13 Rh atoms lie in three parallel planes; they form a cluster of symmetry D_{3h} and display **hexagonal closest packing.** The central Rh atom is connected to additional Rh atoms only; one half of the CO groups is in terminal, the other half in bridging positions. This cluster ion may be regarded as the smallest conceivable metal crystal with carbon monoxide **chemisorbed** at its surface. The connection to heterogeneous catalysis is obvious.

17 Organometallic Catalysis

pro memoria:

A catalyst accelerates the rate of a thermodynamically feasible reaction by opening a lower activation energy pathway. If alternative routes exist, a catalyst can enhance product selectivity by accelerating just one of the competing reaction sequences.

MODE OF ACTION OF TRANSITION-METAL COMPLEXES IN CATALYSIS

- *The coordination of the reaction partners to a transition metal brings them into* ***close proximity****, thus promoting the reaction (Example: cyclooligomerization of alkynes,* *p.* 276*).*
- *Through coordination to a transition metal, a reaction partner can become* ***activated*** *for subsequent reactions (Example: hydrogenation of alkenes,* *p.* 427*).*
- *The coordination of an organic substrate to a transition metal can* ***facilitate nucleophilic attack*** *(Example:* $PdCl_2$*-catalyzed oxidation of ethylene to acetaldehyde,* *p.* 425*).*

Catalytically active systems therefore must possess **vacant coordination sites** or be able to generate them in a primary dissociation step.

HETEROGENEOUS CATALYSIS

The vacant coordination site is located at a phase boundary (solid/liquid, solid/gas), i.e., only the surface atoms are catalytically active. A principal advantage is the easy recovery of the catalyst. Disadvantages include low specificity, relatively high reaction temperatures and difficulties in the mechanistic study.

HOMOGENEOUS CATALYSIS

The catalyst can be tailor-made by ligand variation and is obtained reproducibly; high specificity is thus achieved and the catalysis can often be carried out at low temperatures. Ideally, the catalyst complex is stable in more than one coordination number and, through fine-tuning of chemical bond strength (variation of the ligands), capable of holding a substrate molecule **selectively but not too tightly**.
The readiness of a particular metal to exist in a coordinatively unsaturated state is pronounced at the end of the transition series (M d^8, d^{10}, p. 189). Thus, the most important complexes which are active in homogeneous catalysis are those which contain the "late transition metals" Ru, Co, Rh, Ni, Pd and Pt.
The mechanisms of catalysis are studied more easily in the homogeneous than in the heterogeneous case; catalyst recovery is, however, fraught with difficulties. A possible

remedy is to attach a homogeneously active complex to **a polymeric support** (Pittman, 1974):

$EtOCH_2Cl$, $SnCl_4$; $LiPPh_2$; $RhCl(PPh_3)_3$

CH_2Cl CH_2Cl

CH_2PPh_2 CH_2PPh_2

CH_2 CH_2

$(PPh_3)_2RhClPPh_2$ $Ph_2PRhCl(PPh_3)_2$

Supported Wilkinson's catalyst (cf. p. 427)

Another variant, already in industrial usage, is the **catalysis in a liquid/liquid two-phase system**. In this approach, the ligand tris(metasulfonato)triphenylphosphane is employed, which renders the catalyst water-soluble. The catalyst can be recovered quantitatively from the organic products (Application: hydroformylation of propene to give *n*-butyraldehyde, the Ruhrchemie/Rhone-Poulenc process, cf. p. 437).

17.1 Catalytic Reactions and the 16/18 VE Rule

Discussions of organometallic reaction sequences may be systematized by the definition of five general reaction types and their reverse reactions (Tolman, 1972). Characteristic variables pertaining to these reaction types are:

ΔVE *Change in the number of valence electrons at the central metal*
ΔOS *Change in the oxidation state of the metal (usual convention: -hydride, -alkyl, -allyl and -cyclopentadienyl are regarded as anions)*
ΔC.N. *Change in the coordination number*

For olefinic ligands, the classifications Lewis-base ligand association and oxidative addition are limiting cases for a continuum of intermediate types. Which term most adequately describes the bonding situation has to be decided either on the basis of the electronic properties of the incoming alkene or through inspection of structural details of the product as determined by X-ray analysis (Example, p. 413).

Elementary organometallic reaction steps

Reaction	ΔVE	ΔOS	ΔC.N.	Example
→ Lewis-acid ligand dissociation	0	0	−1	$CpRh(C_2H_4)_2SO_2 \rightleftarrows CpRh(C_2H_4)_2 + SO_2$
← Lewis-acid ligand association	0	0	+1	
→ Lewis-base ligand dissociation	−2	0	−1	$NiL_4 \rightleftarrows NiL_3 + L$
← Lewis-base ligand association	+2	0	+1	
→ Reductive elimination	−2	−2	−2	$H_2Ir^{III}Cl(CO)L_2 \rightleftarrows H_2 + Ir^{I}Cl(CO)L_2$
← Oxidative addition	+2	+2	+2	
→ Insertion (migratory)	−2	0	−1	$MeMn(CO)_5 \rightleftarrows MeCOMn(CO)_4$
← Extrusion	+2	0	+1	
→ Oxidative coupling	−2	+2	0	$(C_2F_4)_2\,Fe^{0}(CO)_3 \rightleftarrows (CF_2)_4Fe^{II}(CO)_3$
← Reductive decoupling	+2	−2	0	

Example:

$(Ph_3P)_4Pt$

$-2\ PPh_3$

$(Ph_3P)_2Pt$

C_2H_4 $C_2(CF_3)_4$

$(Ph_3P)_2Pt\text{---}\|(CH_2)_2$ $(Ph_3P)_2Pt[C(CF_3)_2]_2$

$Pt^{0}(d^{10})$ *π-complex*
Lewis-base ligand association

$Pt^{II}(d^{8})$ *σ-complex*
(metallacyclopropane)
oxidative addition

The formulation of a plausible reaction mechanism for an organometallic catalysis is assisted by the 16/18 VE rule (Tolman, 1972):

① *Diamagnetic organometallic complexes of the transition metals exist in significant amounts at normal temperatures only if the central metal atom has* 18 *or* 16 *valence electrons. A concentration can be regarded as "significant" if it is detectable by spectroscopy or in kinetic studies.*

② *Organometallic reaction sequences proceed by elementary steps which involve intermediates having* 18 *or* 16 *valence electrons.*

The conditions ① and ② considerably limit the number of possible reactions which a particular complex is likely to undergo:

Elementary step	expected for complexes with the configuration:
Lewis-acid ligand dissociation Lewis-acid ligand association	16, 18 VE
Lewis-base ligand dissociation Reductive elimination Insertion Oxidative coupling	18 VE
Lewis-base ligand association Oxidative addition Extrusion Reductive decoupling	16 VE

It should be stressed, however, that the 16/18 VE criterion is increasingly regarded as an undue restraint (Kochi, 1986 R). Thus, associatively activated CO substitution at $Mn_2(CO)_{10}$ proceeds via the 17 VE intermediate $\cdot Mn(CO)_5$ (p. 247), and **electron-transfer(ET) catalysis** (p. 233) of substitution reactions at 18 VE species is explained by intermediates with 17 VE (oxidatively initiated) or 19 VE (reductively initiated). The ET-catalyzed substitution of CO by phosphanes in (methylidyne)cobaltcarbonyl clusters is a good example (P. H. Rieger, 1981). The reaction is initiated by a mere 0.01 equivalents of electrons, supplied either by electrochemical reduction or through the addition of minute amounts of benzophenone ketyl:

$$PhC[Co(CO)_3]_3 + PPh_3 \xrightarrow[- CO]{[+e^-]} PhC[Co(CO)_3]_2[Co(CO)_2PPh_3]$$

Since in organometallic reaction mixtures traces of O_2 can generate 17 VE radical cations or, likewise, reducing components 19 VE radical anions, a radical pathway cannot be excluded a priori, even in those cases which currently are rationalized on the basis of the 16/18 VE criterion.

Homogeneous catalysis proceeds via a series of organometallic reaction steps, coupled to each other in such a way that they form a loop. This will be illustrated in the following sections for several reactions, some of which are of vast industrial importance. The discussion will be organized according to increasing number of components involved in the catalyzed reaction. These cycles have not always have been worked out in detail; often they take on the character of plausible assumptions, of working hypotheses or even caricatures.

When referring to a "catalyst", it should be borne in mind that the actual catalyst of a reaction is not always known. The term catalyst in this book therefore often refers to the "precatalyst", usually a well-known substance like $(PPh_3)_3Rh(CO)H$, which releases PPh_3 to generate the active catalyst $(PPh_3)_2Rh(CO)H$. In industrial processes, the catalyst sometimes forms directly in the reactor from simple precursors like a metal salt and the appropriate ligands, e.g. phosphanes, or from the bulk chemicals employed in the process.

17.2 Transition-Metal Assisted Valence Isomerizations

These reactions entail the rearrangements of σ frameworks in small hydrocarbon rings **(skeletal rearrangements)** with relief of ring strain. They are usually initiated by the oxidative addition of a C–C bond to a transition metal. Examples:

$(Ph_3P)_3RhCl$, 50°; $- L_3RhCl$; Ag^+

Cubane; $(Ph_3P)_3RhCl$

Ag^+; Ag; +; $-Ag^+$; ≡; Cuneane

hν; Rh(I), Pd(II) complexes

Quadricyclane; Norbornadiene

As concerted processes, the transformations quadricyclane → norbornadiene and cubane → cyclooctatetraene would be symmetry-forbidden (Woodward-Hoffmann rules). The presence of transition-metal ions may, however, open up low-energy multistep routes, rendering these rearrangements possible. In some cases, organometallic intermediates can be isolated. Example:

$[Rh^{I}(CO)_2Cl]_2$; $Rh^{III}-Cl$ $(CO)_2$; Ph_3P, $-[Ph_3PRh^{I}(CO)Cl]_2$; C=O

The product of the initial oxidative addition to Rh^{I} can release Rh^{I} either directly or after intervening *β*-hydride elimination, C=C double bonds being formed (K. C. Bishop, 1976 R):

Valence isomerizations are also catalyzed by iron carbonyl moeties in a variety of ways. Occasionally, organic ligands which are unstable in free form are thereby stabilized:

The free ligand bicyclo[4.2.0]octatriene would rapidly rearrange to its valence isomer cyclooctatetraene.
Metal-catalyzed cyclobutene ring opening affords a dimer of benzene coordinated to $Fe(CO)_3$:

After oxidative decomplexation, in a *retro*-Diels-Alder reaction monomeric benzene is immediately formed (Grimme, 1977).

Excursion:
Interaction of Transition-Metal Fragments with Cyclopropanes

The opening of a strained ring is preceded by the coordination of a $C \overset{\sigma}{-} C$ bond to the transition metal. In the case of cyclopropane, it is instructive to examine the relationship between the "bent bond" in the C_3H_6 skeleton and the conventional $(\sigma + \pi)$ double bond in alkenes.

Cyclopropane

C_3 bonding, occupied

C_3 antibonding, unoccupied

The small bond angle in the three-membered ring induces a higher p character for the σ bond relative to sp^3 hybridization and an overlap region lying outside the C–C axis (so-called ***Walsh orbitals****).*

Cyclopropane complex

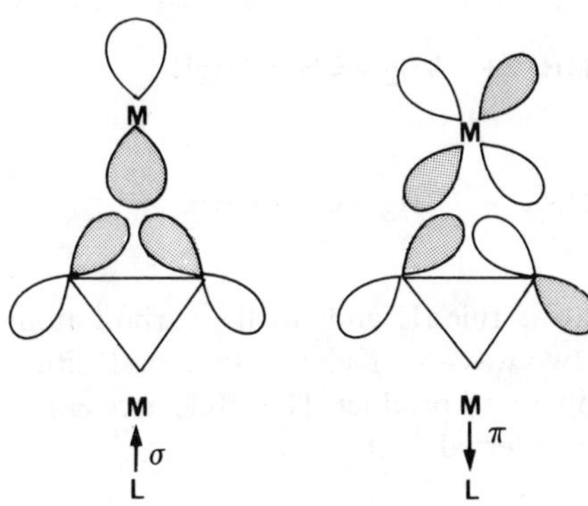

Olefin complex

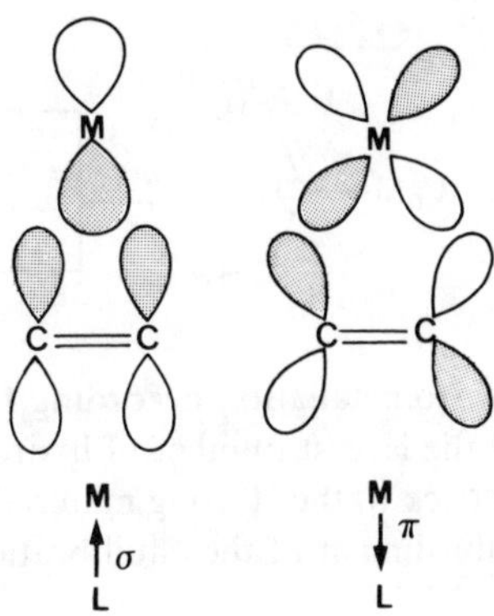

While in the case of metal-olefin coordination, the $C \overset{\sigma}{-} C$ bond of the ligand is retained, it is virtually lost on metal-cyclopropane coordination: structural data for transition-metal cyclopropane complexes point to metallacyclobutane character.

The following reaction can therefore be classified as an oxidative addition to Pt(II) or as an insertion of Pt(II) into cyclopropane, respectively, ring strain being diminished. Besides their role in valence isomerizations, metallacyclobutanes also qualify as intermediates in the olefin metathesis mechanism (p. 419).

$[(C_2H_4)PtCl_2]_2$ + △ $\xrightarrow[-C_2H_4]{py}$ (platinacyclobutane: ring carbons 1, 2, 3; Pt bearing Cl, Cl, py, py) $\xrightarrow{CN^-}$ △

∢ $C_1C_2C_3/C_3PtC_1 = 168°$,
$C_1–C_3 = 255$ pm, $Pt–C_2 = 269$ pm

17.3 Isomerization of Unsaturated Molecules

Catalysts of composition L_nMH or combinations of the type $(R_3P)_2NiCl_2 + H_2$ can accelerate a rearrangement to the thermodynamically most stable olefin, i.e. they can catalyze the transformations terminal $\rightleftharpoons$ internal and isolated $\rightleftharpoons$ conjugated. Essential steps are:

$$L_nMH + CH_2{=}CH{-}CH_2R \rightleftharpoons L_nM(H)\cdots(CH_2{=}CH{-}CH_2R) \rightleftharpoons L_nM{-}CH(CH_3)(CH_2R)$$

$$L_nM{-}CH(CH_2{-}H)(CHR{-}H) \longrightarrow \begin{cases} L_nMH + CH_2{=}CH{-}CH_2R \\ L_nMH + CH_3{-}CH{=}CHR \end{cases}$$

After hydrometallation according to Markovnikov's rule (H adds to the carbon atom bearing the largest number of hydrogen atoms), two ways for β-elimination exist, either leading back to the starting material or to the isomerized product. This cycle accelerates the establishment of the equilibration terminal $\rightleftharpoons$ internal.

$HCo(CO)_4 \rightarrow HCo(CO)_3 + CO$

$CH_2{=}CH{-}CH_2OH$ + $HCo(CO)_3$ ① → $CH_2{=}CH{-}CH_2OH \cdots H{-}Co(CO)_3$ ② → $CH_3{-}CH(Co(CO)_3){-}CH(H)OH$ ③ → $HCo(CO)_3$ + $[CH_3{-}CH{=}CHOH]$ → CH_3CH_2CHO

Olefin isomerization

Since $HCo(CO)_3$ is the actual catalyst here, $HCo(CO)_4$ is an example of a "precatalyst" (p. 415).

If $DCo(CO)_4$ is introduced into the cycle, the only deuterated product found is CH_2DCH_2CHO. This proves the Markovnikov direction of the olefin insertion ②.

17.4 Arylation/Vinylation of Olefins (Heck Reaction)

In this reaction, a vinylic hydrogen atom is replaced by a vinyl-, benzyl- or aryl group. Example:

The catalyst is a Pd^0 complex, formed **in situ** from $Pd(OAc)_2$, Et_3N and PPh_3.

With regard to the mechanism of the Heck reaction (1982 R), the following cycle has been proposed: oxidative addition ① of R′X to the Pd^0 species first gives *trans*-$R'PdL_2X$. In order to prevent rapid decomposition of this intermediate by β-elimination, R′X can only be an aryl-, benzyl- or vinyl halide. After olefin insertion ② into the Pd–C bond, the β-elimination ③ proceeds, the substituted olefin being released. The catalyst PdL_2 is regenerated in step ④ by reaction with Et_3N, which is consumed in stoichiometric amounts:

Heck reaction

17.5 Alkene Metathesis

Principle:

$$\begin{matrix} CH_2{=}CHR \\ + \\ CH_2{=}CHR \end{matrix} \xrightleftharpoons{\text{cat.}} \begin{matrix} CH_2 & & CHR \\ \| & + & \| \\ CH_2 & & CHR \end{matrix}$$

A possible application of olefin metathesis is the catalytic conversion of propene into ethylene (for ensuing polymerization) and pure 1-butene (to be used in the production of

C_5 aldehyde via an oxo process). This reaction type is of less technical importance than, for instance, hydroformylation or olefin oxidation (Wacker process). Alkene metathesis is, however, intriguing from a mechanistic point of view, since it is the first homogeneous catalysis for which **carbene complex intermediates** have been established. Typical catalysts are $Mo(CO)_6$ on Al_2O_3 (heterogeneous) and $WCl_6/EtOH/2\,Et_2AlCl$ (homogeneous). The mechanism of alkene metathesis has been intensively studied in recent years. Currently, the following sequence is favored **(Chauvin mechanism)**:

$$\mathrm{abC{=}M}\ (\mathbf{1}) + \mathrm{cdC{=}Cef} \rightleftharpoons \mathrm{abC{=}M(cdC{=}Cef)}\ (\mathbf{2}) \rightleftharpoons \mathbf{3} \rightleftharpoons \mathrm{(abC{=}Ccd)M{=}Cef} \rightleftharpoons \mathrm{abC{=}Ccd} + \mathrm{M{=}Cef}$$

The catalytically active species is the carbene complex **1**. The metallacyclobutane intermediate **3** has two possibilities for ring opening. If the ring opens in a way which differs from the reverse reaction of its formation, the result is a metathesis (redistribution of alkylidene units). In support for the Chauvin mechanism, the following stoichiometric reaction may be cited (Casey, 1974):

$$\mathrm{(CO)_5W{=}CPh_2 + CH_2{=}CMe_2 \longrightarrow (CO)_5W{=}CMe_2 + CH_2{=}CPh_2}$$

$$\mathrm{(CO)_4W({=}CPh_2)(Me_2C{=}CH_2) \rightleftharpoons [(CO)_4W{-}CPh_2{-}CH_2{-}CMe_2] \rightleftharpoons (CO)_4W({=}CMe_2)(CH_2{=}CPh_2)}$$

(each first and last species interconverting with the corresponding pentacarbonyl via $+CO$ / $-CO$)

The tungstacyclobutane intermediate in this mechanism has not yet been isolated. The likely role of a metallacycle in the alkylidene redistribution is, however, supported by the following result (Puddephatt, 1976):

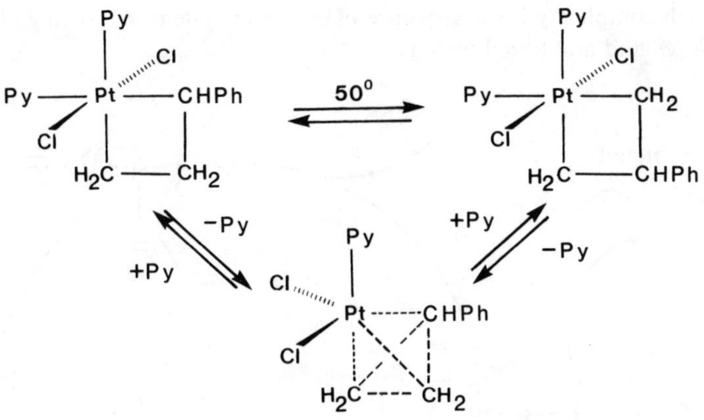

The isomerization of the platinacyclobutane is blocked in the presence of excess pyridine. The *formation of a catalytically active carbene complex* from WCl_6 and the cocatalyst Me_2AlCl could occur as follows:

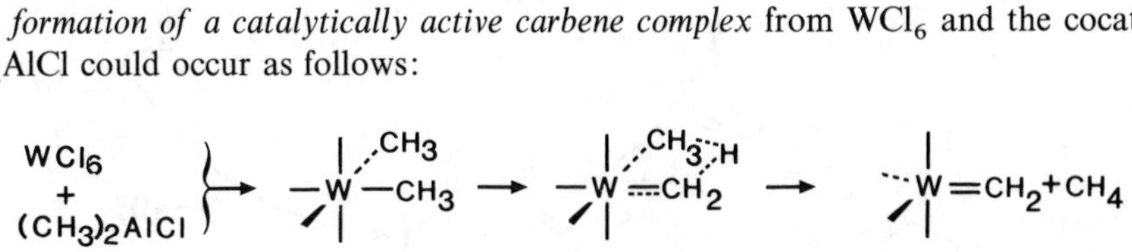

This proposal is based on observations by Muetterties: when WCl_6 is activated by $(CH_3)_2Zn$, CH_4 is formed; if the reaction is performed in deuterated solvents, the methane produced contains no deuterium.

17.6 Oligomerization and Polymerization

CYCLIZATIONS

The catalytic **cyclization of alkynes** to give cyclooctatetraene, discovered by Reppe (BASF), has already been mentioned (p. 278):

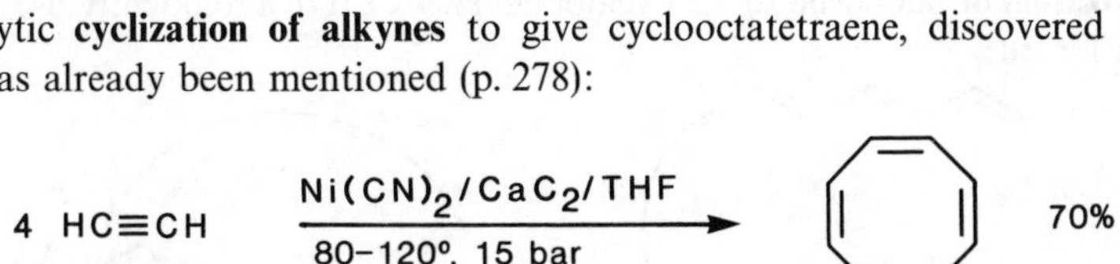

$$4\ HC{\equiv}CH \xrightarrow[80-120^\circ,\ 15\ \text{bar}]{Ni(CN)_2/CaC_2/THF} C_8H_8 \quad 70\%$$

Monosubstituted alkynes yield tetrasubstituted cyclooctatetraene derivatives whereas disubstituted alkynes are unreactive. In the presence of PPh_3 (1 : 1 relative to Ni), one coordination site at Ni is blocked and benzene forms instead of cyclooctatetraene. Despite extensive investigations (cf. Colborn, 1986), the mechanism of this classical organometallic reaction has not been fully elucidated. Possibly, the dinuclear complex $Ni_2(C_8H_8)_2$ (p. 368) acts as the catalytically active species (Wilke, 1988).

A topic of high complexity is the sequence of events in systems consisting of butadiene, Ni(0) complexes and additional ligands:

Diallyl

+2

Ni 1

2

>-40°

3 Ni

>-40°

"Naked nickel"

trans,trans,trans-
1,5,9,- Cyclododecatriene

Butadiene trimerization

This **cyclotrimerization of butadiene** to cyclododecatriene, initiated by bis(η^3-allyl)nickel, was one of the first organometallic reactions subjected to careful mechanistic study (Wilke, from 1960 onwards). Compound **3** can be synthesized independently and channeled into the catalytic cycle. Note the change in coordination number (3, 4), oxidation state (Ni^0, Ni^{II}) and valence electrons (18, 16) during the course of this cycle.

Complexes of type **2**, bearing a phosphane ligand L instead of η^2-C_4H_6, effect the **cyclodimerization of butadiene** to 1,5-cyclooctadiene (COD), a frequently used chelating cycloolefin ligand:

Ni +2 + 2L

L=PR3

1,5-COD

An alternative catalytic cyclotrimerization of butadiene leads to *cis,trans,trans*-1,5,9-cyclododecatriene (10,000 t/a, Chem. Werke Hüls), which is subsequently converted into dodecane-1,12-dicarboxylic acid, a polyamide building block:

$$\xrightarrow[R_3Al_2Cl_3]{TiCl_4} \quad \xrightarrow{H_2} \quad \xrightarrow[2.\ HNO_3]{1.\ O_2} \begin{array}{c} COOH \\ | \\ (CH_2)_{10} \\ | \\ COOH \end{array}$$

OLEFIN POLYMERIZATION

The "growth reaction" (Ziegler)

$$R_2AlC_2H_5 + (n-1)\ CH_2{=}CH_2 \xrightarrow[100\ \text{bar}]{90-120\,°C} R_2Al(CH_2CH_2)_nH$$

competes with the dehydroalumination

$$R_2AlCH_2CH_2R' \longrightarrow R_2AlH + CH_2{=}CHR'$$

and is limited to the formation of linear aliphatics with a maximum chain length of about C_{200} (cf. p. 77). During a series of systematic studies aimed at producing higher-chain polymers by means of the growth reaction, one particular batch gave the opposite result, namely the quantitative formation of 1-butene from ethylene. A trace of a nickel compound, adventitiously present in the autoclave, was discovered as the cause. In the ensuing search for other transition-metal complexes which would also lead to premature termination of chain growth, a highly polymeric product was unexpectedly obtained when using zirconium acetylacetonate and triethylaluminum.

It was soon realized that the most effective catalyst resulted from the combination $TiCl_4/AlEt_3$ and that ethylene polymerization could be carried our at a pressure as low as 1 bar (**Mülheim normal-pressure polyethylene process,** K. Ziegler, 1955 R):

$$CH_2{=}CH_2 \xrightarrow[25\,°C,\ 1\ \text{bar}]{TiCl_4/AlEt_3} \text{polyethylene}$$

molecular weight 10^4-10^5

In contrast to high-pressure polyethylene (ICI, 1000–3000 bar, catalyst: max. 0.01 % O_2), low-pressure polyethylene is strictly linear.

The application of Ziegler-type catalysts to the polymerization of propylene and the establishment of a link between the stereochemical structure and the bulk properties of the polymer was the contribution of G. Natta (1955):

$$CH_2{=}CH{-}CH_3 \xrightarrow[25\,°C,\ 1\ \text{bar}]{TiCl_4/AlEt_3} \text{polypropylene}$$

molecular weight 10^5-10^6

Owing to its stereoregular **isotactic structure,** low-pressure polypropylene has excellent material properties such as high density, hardness and tensile strength (Natta).

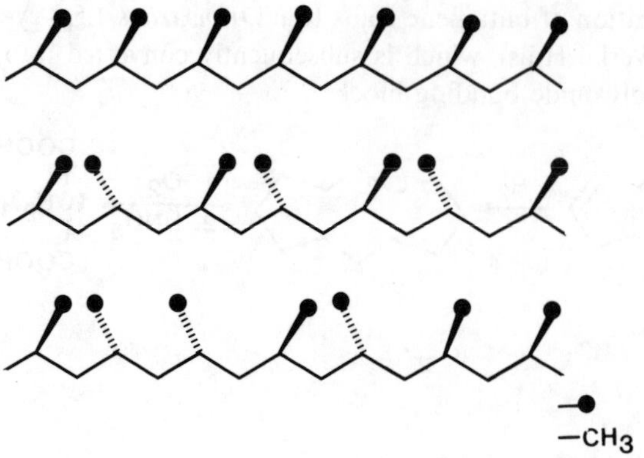

isotactic
all C atoms have the same configuration

syndiotactic
regular change of configuration

atactic
irregular change of configuration

$-CH_3$

Although plausible conceptions about the mechanism of the **Ziegler-Natta polymerization** and, in particular, the regiospecificity exist, there is little experimental evidence. The reaction is a heterogeneous catalysis in which fibrous $TiCl_3$, alkylated on its surface, is regarded as the catalytically active species (Cossee, 1964 R).

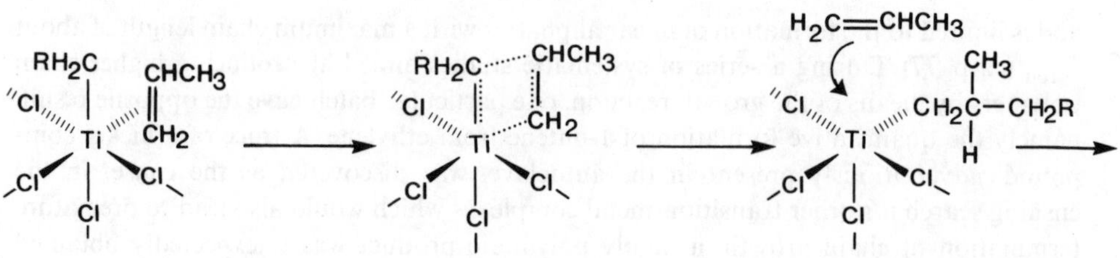

Chain propagation

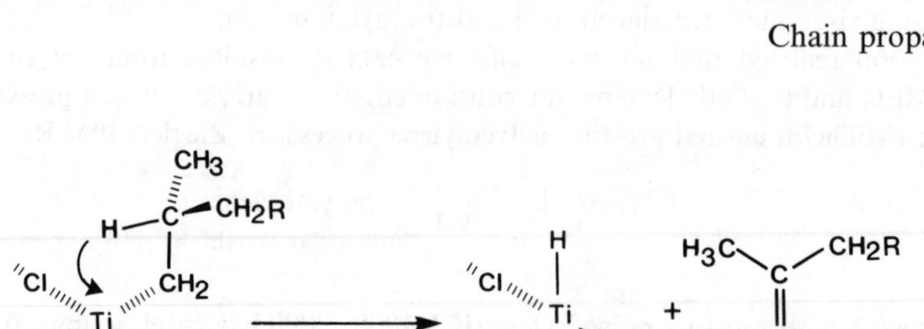

Chain termination

In that the insertion of CO into TM−C and of >C=C< into TM−H is frequently encountered, it is surprising that the central step of the **Cossee mechanism**, namely the insertion of >C=C< into a TM−C bond, is almost without precedent. One of the few examples is provided by the interconversion $CpCo(CD_3)_2(\eta^2\text{-}C_2H_4) \rightarrow CpCo(CD_3)CH_2CH_2CD_3)$ reported by Bergmann (1979).

Recently, an observation by Natta (1957) that species formed from $Cp_2TiCl_2/AlCl_3$ can act as homogeneous catalysts in ethylene polymerization, has been reexamined. Of particular significance is the discovery of homogeneous isotactic polymerization of α-olefins, brought about by a catalyst generated from a chiral zirconocene halide and methylalumoxane (Brintzinger, Kaminsky, 1985).

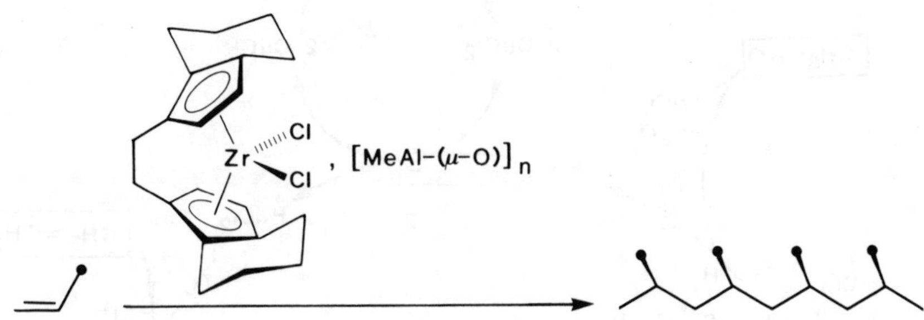

The actual catalyst here is thought to be the coordinatively unsaturated cation Cp_2ZrMe^+ (14 VE) which mimics the catalytic sites of surface-alkylated $TiCl_3$ of the classical Ziegler-Natta system by causing isotactic polymerisation in a sequence of π-coordination and insertion steps (Jordan, 1988 R).

17.7 Olefin Oxidation (Wacker Process)

The commercially important organic intermediates acetaldehyde, vinyl chloride and vinyl acetate can be produced from acetylene. However, for economic reasons, the switch to ethylene as a chemical feedstock was highly desirable.

The fact that ethylene-chloro complexes of palladium readily decompose in water to acetaldehyde and palladium metal has been known for a century. The reoxidation of Pd^0 to Pd^{II} by O_2, mediated by the couple Cu^+/Cu^{2+}, finally led to the design of a catalytic cycle (**Wacker process,** Jira, 1962 R):

(1) $C_2H_4 + PdCl_2 + H_2O \longrightarrow CH_3CHO + Pd + 2\,HCl$

(2) $Pd + 2\,CuCl_2 \longrightarrow PdCl_2 + 2\,CuCl$

(3) $2\,CuCl + 2\,HCl + 1/2\,O_2 \longrightarrow 2\,CuCl_2 + H_2O$

Total: $C_2H_4 + 1/2\,O_2 \longrightarrow CH_3CHO$

Alkenes of the type $RCH{=}CHR'$ and $RCH{=}CH_2$ are converted to ketones (Example: propylene → acetone).

The following rate equation was derived for the Wacker process (Baeckvall, 1979):

$$\frac{d[CH_3CHO]}{dt} = k \cdot \frac{[C_2H_4]\,[PdCl_4^{2-}]}{[H^+]\,[Cl^-]^2}$$

Based on detailed kinetic studies and various labelling experiments, a catalytic cycle of the Wacker process which conforms with this rate law has been proposed by Henry (1982):

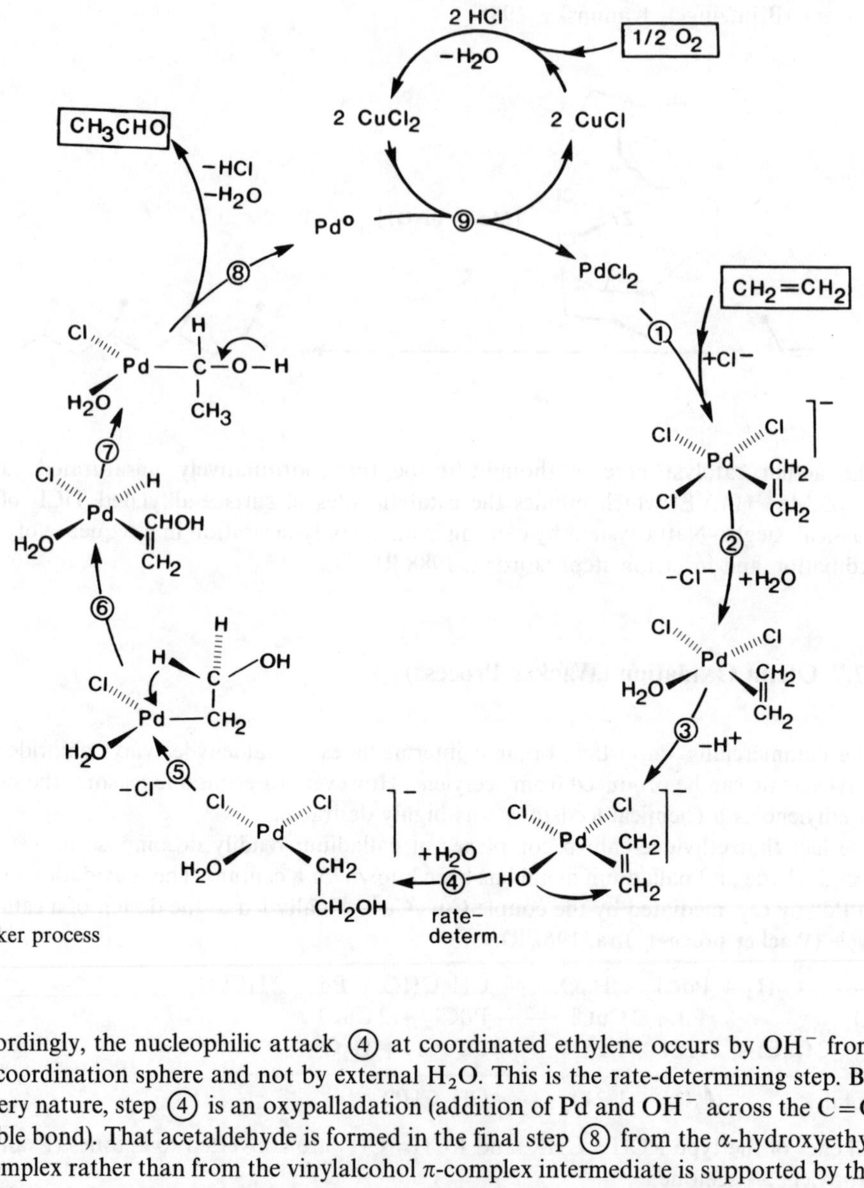

Wacker process

Accordingly, the nucleophilic attack ④ at coordinated ethylene occurs by OH^- from the coordination sphere and not by external H_2O. This is the rate-determining step. By its very nature, step ④ is an oxypalladation (addition of Pd and OH^- across the C=C double bond). That acetaldehyde is formed in the final step ⑧ from the α-hydroxyethyl σ-complex rather than from the vinylalcohol π-complex intermediate is supported by the fact that no deuterium is found in the product if the catalysis is performed in D_2O. The formation of the α-hydroxyethyl palladium complex from the β-hydroxyethyl precursor probably occurs via β-hydride elimination ⑥ and migratory insertion ⑦.
The nature of the products of olefin oxidation depends on the reaction medium:

Solvent			**Product**
water	⟶	$[CH_2{=}C(H)(OH)] \longrightarrow CH_3{-}C(=O)H$	(acetaldehyde)
acetic acid	⟶	$CH_2{=}C(H)(OCOCH_3)$	(vinyl acetate)
alcohol	⟶	$CH_2{=}C(H)(OCH_3)$	(vinyl ether)
inert, non-aqueous	⟶	$CH_2{=}C(H)(Cl)$	(vinyl chloride)

17.8 Hydrogenation of Alkenes

The activation of molecular hydrogen by transition-metal complexes in solution is long known. Intermediate hydride complexes could, however, rarely be isolated. The following three types of reaction present possible ways in which dihydrogen can be activated for hydrogenation in homogeneous solution:

- *Oxidative addition*: $(Ph_3P)_3RhCl + H_2 \longrightarrow (Ph_3P)_3Rh(H)_2Cl$
- *Heterolytic cleavage*: $[RuCl_6]^{3-} + H_2 \longrightarrow [RuCl_5H]^{3-} + H^+ + Cl^-$
- *Homolytic cleavage*: $2\,[Co(CN)_5]^{3-} + H_2 \longrightarrow 2\,[Co(CN)_5H]^{3-}$

The first complex to serve as a hydrogenation catalyst in homogeneous solution was the compound $(Ph_3P)_3RhCl$ (**Wilkinson's catalyst**, 1965). This species catalyzes the hydrogenation of alkenes, alkynes and other unsaturated molecules at 25 °C and a pressure of 1 bar.

A thorough mechanistic study of this system was undertaken by Halpern (1976):

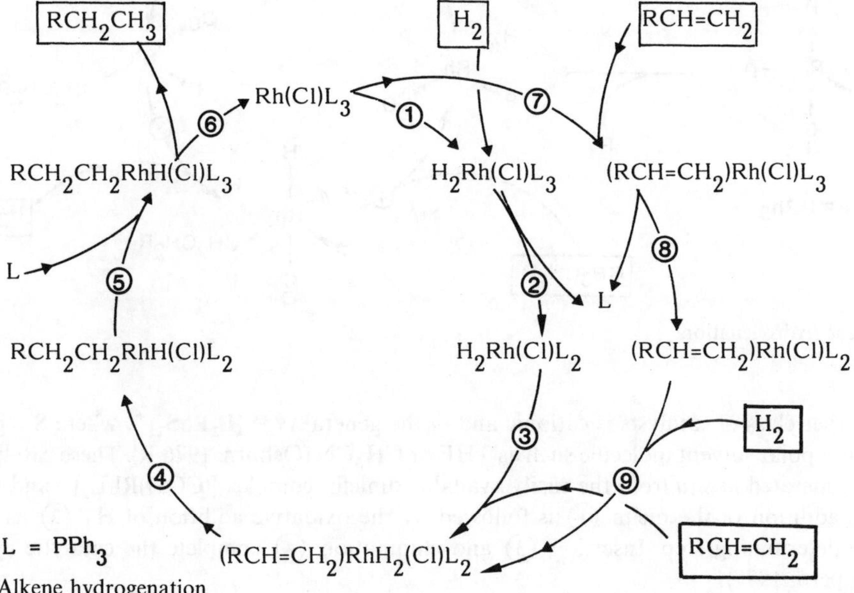

Alkene hydrogenation

The simplified reaction scheme on p. 427 shows two routes which differ in the order in which H_2 and the olefin are taken up. As step ⑨ is slow, the sequence ① → ② → ③ should take preference over ⑦ → ⑧ → ⑨. The decisive step is the **insertion** ④:

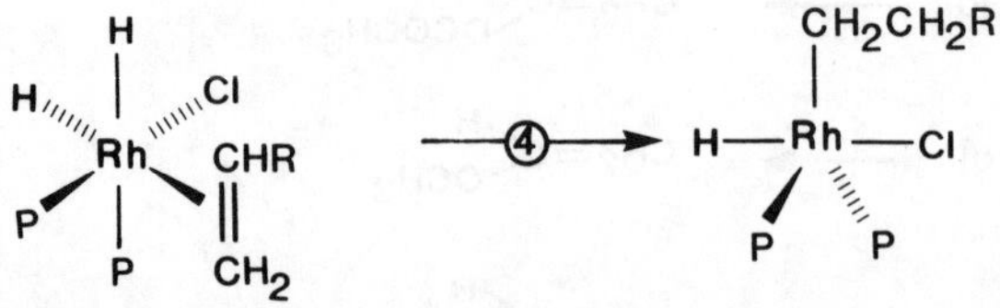

Cis-coordination of the phosphane ligands in these non-isolable intermediates is suggested by special NMR techniques (J. M. Brown, 1984). The rate determining nature of the intramolecular migratory insertion ④ also follows from an ab initio MO *study of the full catalytic cycle (Morokuma, 1988).*

While $(Ph_3P)_3RhCl$ catalyzes the hydrogenation of terminal **and** internal olefins, **$(PPh_3)_2Rh(CO)H$ is highly selective,** catalyzing hydrogenation of **terminal** olefins only (Wilkinson, 1968). A drawback of this catalyst is the concomitant isomerization of terminal to internal olefins, (cf. p. 418), the latter being hydrogenated very slowly (Strohmeier, 1973).

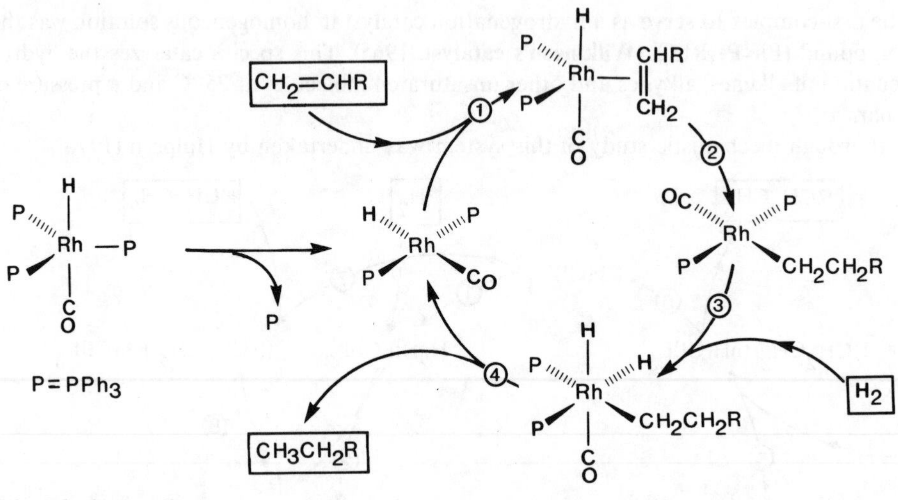

Alkene hydrogenation

Another class of catalysts is cationic and of the general type **$[L_2RhS_2]^+$**, where S represents a polar solvent molecule such as THF or CH_3CN (Osborn, 1976 R). These catalysts are generated *in situ* from the easily available diolefin complex $[(COD)RhL_2]^+$ and H_2. The addition of the olefin ① is followed by the oxidative addition of H_2 ② as the rate-determining step. Insertion ③ and elimination ④ complete the catalytic cycle (Halpern, 1977):

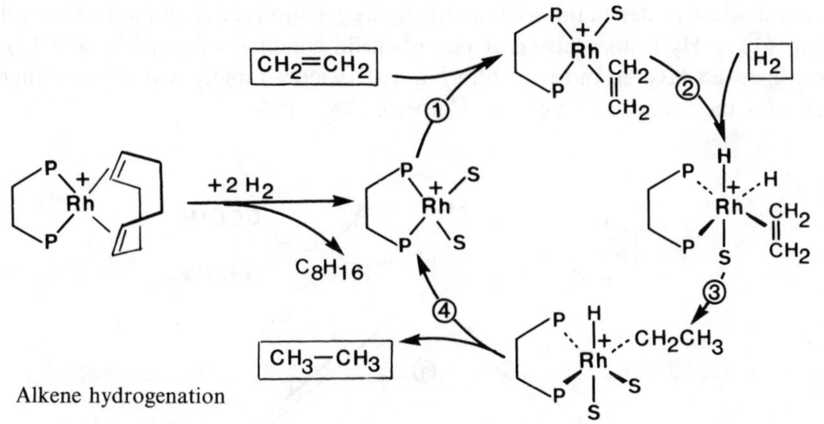

Alkene hydrogenation

Asymmetric Hydrogenation

If the catalyst of the variety $(L_2RhS_2)^+$ bears an optically active diphosphane, prochiral unsaturated molecules can be hydrogenated to chiral products (asymmetric induction). In many cases, high optical purity has been achieved. Examples of chiral chelating diphosphane ligands are the following:

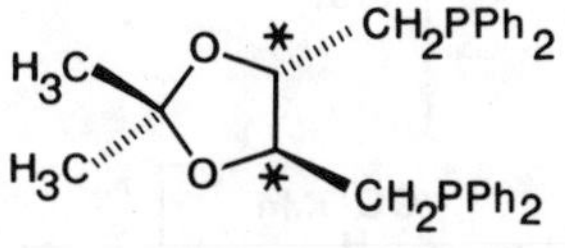

DIOP (Kagan, 1972)

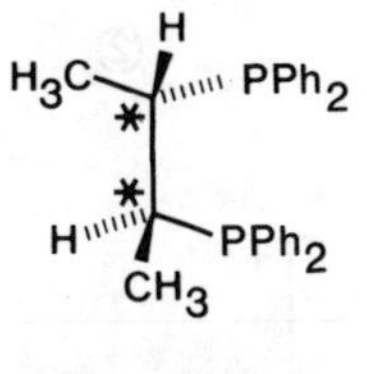

CHIRAPHOS (Bosnich, 1977)

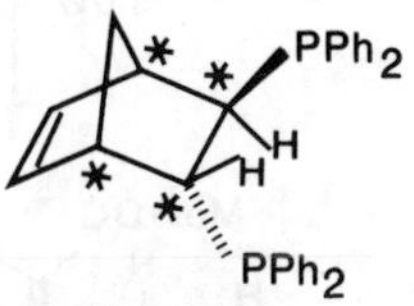

NORPHOS (Brunner, 1979)

The chirality of the diphosphane ligand can be centered either at a carbon atom of the ligand framework or at phosphorus. A commercial application of asymmetric hydrogenation was the manufacture of a derivative of the chiral amino acid L-Dopa, which is effective in the treatment of Parkinson's disease:

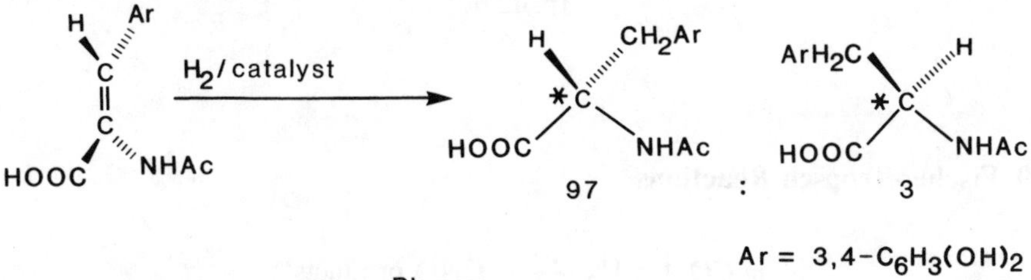

H_2/catalyst

97 : 3

Ar = 3,4-$C_6H_3(OH)_2$

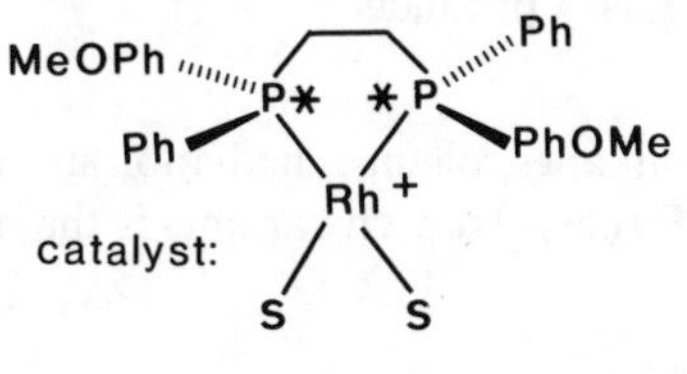

catalyst:

MONSANTO (1977)

(Knowles, 1986 R)

The enantioselective step in this asymmetric hydrogenation is thought to be the oxidative addition ② of H_2 to the diastereomeric η^2-olefin complexes formed in step ①. The diastereomer less favored in the equilibrium reacts more rapidly with H_2 and therefore accounts for the enantiomeric excess (Halpern, 1983 R):

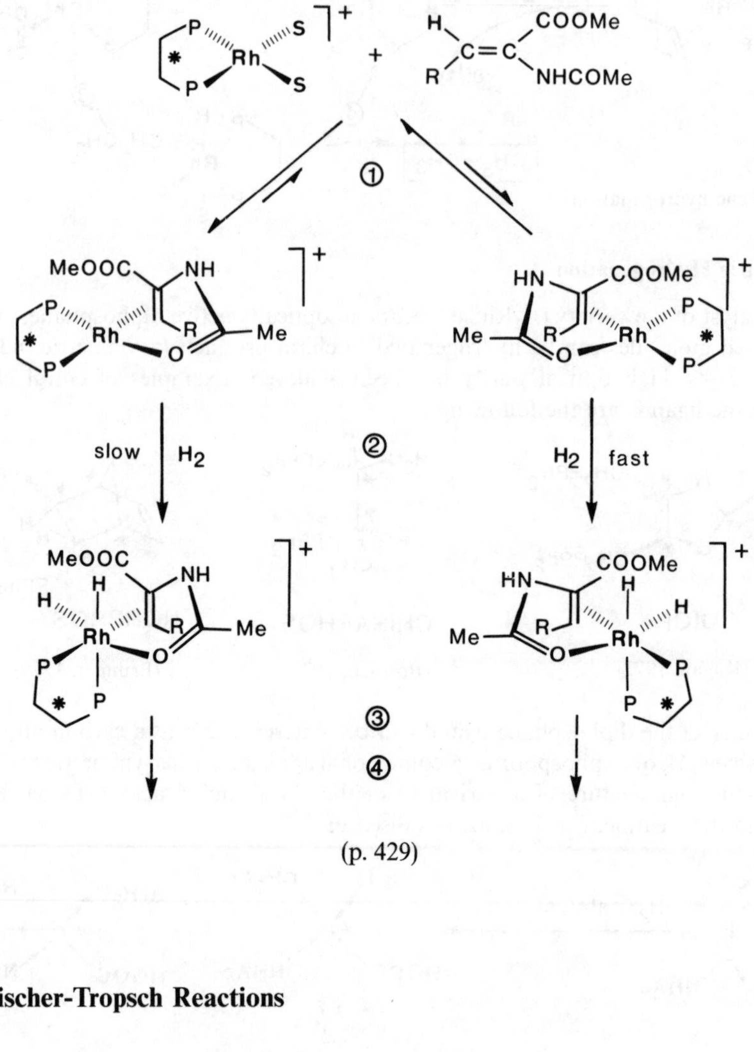

slow H_2

H_2 fast

(p. 429)

17.9 Fischer-Tropsch Reactions

$$\underset{\text{synthesis gas}}{m\,CO + n\,H_2} \xrightarrow{\text{cat.}} \text{“CHO products”}$$

Depending on the reaction conditions, liquid alkanes, olefins, methanol and/or higher alcohols are formed. A common feature of all Fischer-Tropsch variants is the use of CO as the source of carbon (**“C_1 chemistry”**).

Generation of synthesis gas:

- *Controlled combustion of petroleum*
- *"Reforming" natural gas (which consists mainly of* CH_4*)*

$$CH_4 + H_2O \rightleftharpoons CO + 3\,H_2 \quad (\Delta H = +\,205\ \text{kJ/mol})$$

$$CH_4 + \tfrac{1}{2}O \rightleftharpoons CO + 2\,H_2 \quad (\Delta H = -\,35\ \text{kJ/mol})$$

At present, this is the main source of syngas.

- *Coal gasification*

$$C + H_2O \rightleftharpoons CO + H_2 \quad (\Delta H = +\,131\ \text{kJ/mol})$$

The production of syngas from coal in the (distant) future may well replace the process based on natural gas. This would, however, require that a catalysis of the water gas shift reaction be developed to enhance the H_2 content in syngas (cf. p. 432).

Classical Fischer-Tropsch processes are heterogeneously catalyzed. One problem is low product selectivity. Examples:

$$\textit{Synthesis gas} \xrightarrow[170-200\,°C,\ 1\ bar]{Co/ThO_2/MgO/silica\ gel} \underset{50\%}{\textit{gasoline}} + \underset{25\%}{\textit{diesel oil}} + \underset{25\%}{\textit{waxes}}$$

(*Ruhrchemie A.G., from* 1936 *on, discontinued in Germany after World War II*)

$$\underset{3.5\ :\ 1}{H_2 + CO} \xrightarrow[320-340\,°C,\ 25\ bar]{Fe\ oxide\ cat.} \underset{70\%}{\textit{gasoline}} + \textit{diesel oil, waxes etc.}$$

The availability of cheap coal as a feedstock for syngas enables South Africa to satisfy about half of its gasoline demand in this manner (Sasol plant).

The introduction of homogeneous processes for the production of synthetic fuels from coal seems unlikely owing to the large plant size these processes would necessitate. With regard to the selective preparation of **other** Fischer-Tropsch products (short-chain olefins, glycols etc.), homogeneously catalyzed processes are potentially interesting, however ("New era of C_1 chemistry"). Example:

$$3\,H_2 + 2\,CO \xrightarrow[3000\,bar/250°]{[Rh_{12}(CO)_{34}]^{2-}} HOCH_2{-}CH_2OH$$

(Pruett, Union Carbide, 1974)

At lower pressures, methanol is formed rather than glycol. This homogeneously catalyzed reaction has not yet been implemented, largely due to the drastic conditions required. There remains, however, a heavy industrial demand for glycol (anti-freeze, building block for polyesters etc.).

Mechanistic models of the Fischer-Tropsch reaction are based on the grossly simplified reaction sequences sketched on p. 432. **M** can be an active site on the catalyst surface (heterogeneous catalysis) or the vacant coordination site of a transition-metal complex in solution (homogeneous catalysis). In order to elucidate the detailed mechanisms, potential organometallic intermediates are synthesized and their reactivity patterns studied.

$M \xrightarrow{CO} M{-}CO \xrightarrow{H_2} M{-}C(=O){-}H \xrightarrow{H_2} M{-}CH_3 \xrightarrow{CO} M{-}COCH_3$

η^1-formyl complex $\xrightarrow{H_2}$ η^2-formaldehyde complex ($M(O)CH_2$) $\xrightarrow{H_2}$ $M{-}OCH_3$ $\xrightarrow{H_2}$ CH_3OH + $M{-}H$

$M{-}CH_3 \xrightarrow{H_2} H{-}M{-}CH_3 \rightarrow M + CH_4$

$M{-}COCH_3 \xrightarrow{H_2} M{-}CH_2{-}CH_3 \xrightarrow{CO} M{-}COCH_2CH_3 \rightarrow$

The first experimental proof for the reduction of coordinated CO by H_2 was supplied by Bercaw (1976):

$$(C_5Me_5)_2Zr(CO)_2 \xrightarrow[1.5\ bar/25^0]{H_2/h\nu} (C_5Me_5)_2Zr(H)(OCH_3) \xrightarrow{2\ HCl} (C_5Me_5)_2ZrCl_2 + CH_3OH + H_2$$

17.10 The Water-Gas Shift Reaction

$$H_2O + CO \underset{T > 200\,°C}{\overset{\text{Fe/Cu cat. (heterogeneous)}}{\rightleftharpoons}} H_2 + CO_2 \qquad (\Delta H = -\ 42\ \text{kJ/mol})$$

The importance of this equilibrium, also known as "conversion", lies in

- H_2 *enrichment of synthesis gas for Fischer-Tropsch reactions*
- *Exploitation of the reducing power of* CO *to produce* H_2 *from* H_2O *under mild conditions*
- *Removal of* CO *from "mixed gas"* (50% N_2, 30% CO, 15% H_2, 5% CO_2) *for the Haber-Bosch process*

The relatively high reaction temperature, unfavorable for the position of the equilibrium, is dictated by the non-optimal properties of the currently available catalysts. The development of a homogeneous catalysis for the water gas shift reaction has assumed high priority. The following cycle illustrates a possible approach (King, 1978):

$HFe(CO)_4^-$ $\xrightarrow{①,\ +CO}$ $Fe(CO)_5 + H^-$; $H^- + H_2O \rightarrow OH^- + H_2$; $Fe(CO)_5 + OH^- \xrightarrow{②}$ $(CO)_4Fe{-}C(=O){-}OH$ $\xrightarrow{③}$ $HFe(CO)_4^- + CO_2$

Watergas shift reaction

17.11 Monsanto Acetic Acid Process

This involves the homogeneously catalyzed **carbonylation of methanol** to yield acetic acid ("C_1 chemistry"):

$$CH_3OH + CO \xrightarrow[180\,^\circ C/30\,bar]{Rh\ cat./I^-} \underset{99\,\%}{CH_3COOH} \qquad \text{Capacity} \approx 10^6\ t/a$$

The catalysis most likely proceeds by the following mechanism (D. Forster, 1979 R):

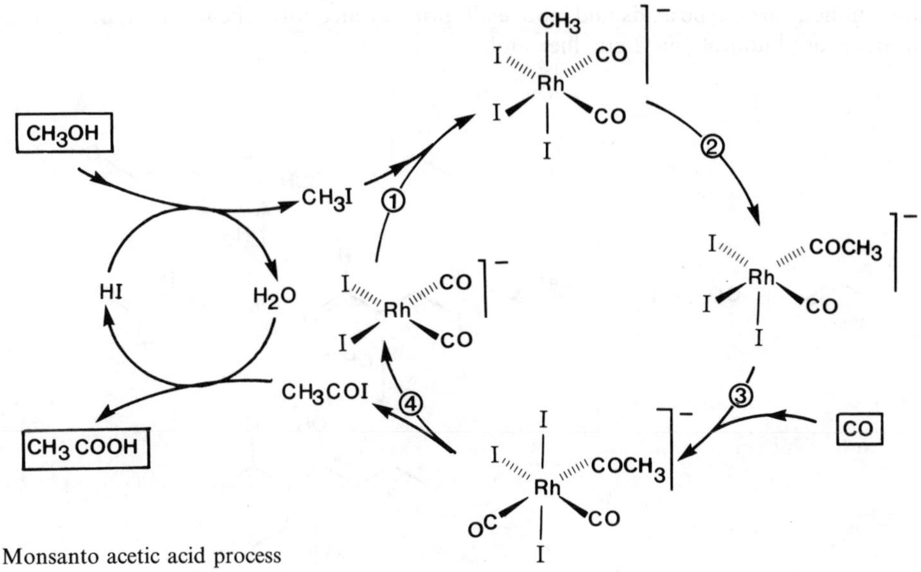

Monsanto acetic acid process

Since a large portion of the industrially produced acetic acid is converted into acetic anhydride to be used as an acetylation reagent, the direct formation of **acetic anhydride** by means of homogeneous catalysis is highly desirable. This target has been reached in the **Tennessee Eastman/Halcon SD process,** a carbonylation of methyl acetate in use since 1983:

$$CO + 2\,H_2 \longrightarrow CH_3OH \xrightarrow{CH_3COOH} CH_3CO_2CH_3 \xrightarrow{CO} (CH_3CO)_2O$$

An outstanding feature of this process is the fact that its C_1 building blocks are derived exclusively from coal. The acetic acid required is obtained as a byproduct in the acetylation of cellulose. As in the Monsanto acetic acid process, *cis*-$[(CO)_2RhI_2]^-$ serves as the catalyst and the overall cycle seems to share a certain similarity with that of the acetic acid process (Polichnowski, 1986).

17.12 Hydroformylation (Oxo Reaction)

The hydroformylation of olefins is the largest-scale industrial process that is homogeneously catalyzed, with a capacity of about 5×10^6 t/a in oxo compounds and their follow-up products. The reaction was discovered in 1938 by O. Roelen (Ruhrchemie A.G.), who prepared propionaldehyde from ethylene and synthesis gas by means of a cobalt catalyst:

$$CH_2{=}CH_2 + CO + H_2 \xrightarrow[\substack{90-150\,°C \\ 100-400\text{ bar}}]{Co_2(CO)_8} CH_3CH_2C\begin{smallmatrix}{=}O\\ \diagdown H\end{smallmatrix}$$

Today, C_3-C_{15} aldehydes are produced by the oxo process and subsequently converted into amines, carboxylic acids and, above all, primary alcohols. The most important oxo products are butanol and 2-ethylhexanol:

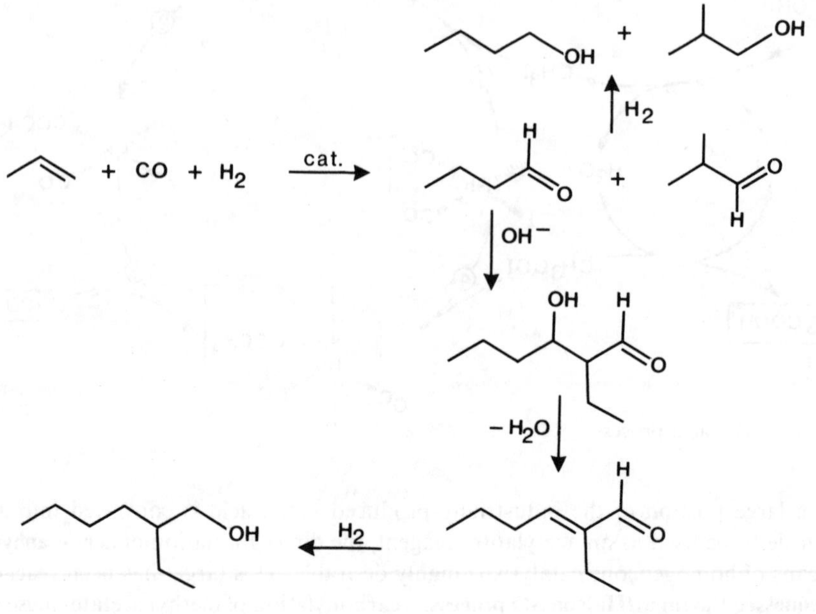

+ CO + H_2 cat. → ... + ... ; H_2 ; OH ; OH ; OH^- ; OH H ; $-H_2O$; H ; OH ← H_2

In the oxo reaction, **H and HCO are formally added** across a double bond; it is therefore referred to as **"hydroformylation"**.

A **mechanism** for the hydroformylation process was suggested by R.F. Heck and D.S. Breslow in 1961 (see p. 435 above). Experimental verification of this mechanism on the basis of kinetic measurements will be exceedingly difficult due to the large number of variables. The rate-determining step is possibly ⑥, the oxidative addition of H_2. Alternative mechanisms have been proposed by E. Oltay (1976, study under actual process conditions, assumption of a 20 VE intermediate) and by T. L. Brown (1980, radical mechanism).

$HCo(CO)_4$
① − CO
$HCo(CO)_3$
② + $CH_2{=}CHR$
$HCo(CO)_3(CH_2{=}CHR)$
③
$RCH_2CH_2Co(CO)_3$
④ + CO
$RCH_2CH_2Co(CO)_4$
⑤
$RCH_2CH_2COCo(CO)_3$
⑥ + H_2
$RCH_2CH_2COCo(H_2)(CO)_3$
⑦ → RCH_2CH_2CHO + $HCo(CO)_3$

Hydroformylation

Disadvantages of this process are:

- *Catalyst losses, since* $HCo(CO)_4$ *is labile and highly volatile*
- *Loss of alkene through hydrogenation in a reaction competing with hydroformylation*
- *Inherent difficulties in mechanistic studies.*

Mechanistic details of the oxo reaction are often deduced from results obtained on other, similar complexes like $(Ph_3P)_3Rh(CO)H$.
Typical mechanistic questions are the following:

$$\underset{\textbf{alkyl}}{RCH_2CH_2Co(CO)_3} + CO \xrightarrow{④\,⑤} \underset{\textbf{acyl}}{RCH_2CH_2COCo(CO)_3}$$

Is the CO *insertion an inter- or an intramolecular process? In the latter case, is it the alkyl group or the* CO *ligand that migrates?*
These questions have been unequivocally answered by IR spectroscopy on a model system labeled with ^{13}CO:

$CH_3Mn(CO)_5$ $\xrightarrow[+\ ^*CO]{Et_2O,\ 25°}$ cis-$CH_3C(O)Mn(CO)_4(^*CO)$ $\xrightarrow[-CO,\ ^*CO]{\text{heptane},\ 100°}$ $CH_3Mn(CO)_4{}^*CO$ (cis 50%, trans 25%) + $CH_3Mn(CO)_5$ 25%

*cis-product only, no *CO in the acetyl group*

$CH_3{}^*C(O)Mn(CO)_5$ $\xrightarrow[-\ CO]{\text{heptane},\ 100°}$ cis-$CH_3Mn(CO)_4(^*CO)$

cis-product only

Conclusion: The CO *group of the acetyl group originates in the coordination sphere of* $CH_3Mn(CO)_5$; *i.e.,* ***CO insertion*** *proceeds* ***intramolecularly.*** *The stereochemistry observed implies that the* ***alkyl group migrates*** *to the* ***cis****-positioned* CO *group and that the vacant coordination site, thus created, is occupied by a new ligand like* CO *or phosphane* (*Calderazzo*, 1977 *R*).

A more recent outgrowth, the **Union Carbide hydroformylation process,** utilizes $(Ph_3P)_3Rh(CO)H$ instead of $HCo(CO)_4$ as a precatalyst. This procedure circumvents some of the disadvantages mentioned for $HCo(CO)_4$:

$H(CO)Rh(PPh_3)_3$
$-PPh_3$
RCH_2CH_2CHO
$CH_2{=}CHR$
H_2
CO
① ② ③ ④ ⑤ ⑥

Union Carbide hydroformylation process

A goal of great commercial interest is the development of a catalytic oxo process which selectively yields **linear products**. In fact, replacement of cobalt by rhodium and the introduction of bulky phosphane ligands markedly increases the linear/branched product ratio. The reason for this could lie in the higher stability of the square-planar intermediate formed in step ② with its *trans*-oriented phosphane ligands and a terminal alkyl group.

Compared to the cobalt-catalyzed oxo reactions, the rhodium-catalyzed process operates under rather mild conditions (100 °C, 10–20 bar).
An important further development of the Union Carbide oxo process is that of a two-phase liquid/liquid reaction system, made possible by the introduction of water soluble (phosphane)rhodium catalysts, mentioned on p. 412 (Rhone-Poulenc/Ruhrchemie, 1984).

The classical oxo reaction represents the hydroformylation of olefins. In an attempt to entirely switch to coal as a chemical feedstock, the **hydroformylation of the C_1 building block formaldehyde** is currently being explored. It is hoped that the drastic reaction conditions ($p = 3000$ bar) which have so far prevented the commercial application of the Union Carbide ethylene glycol process can be attenuated. The starting material formaldehyde itself is easily accessible from the Fischer-Tropsch product methanol. The following scheme traces the formation of **ethylene glycol from synthesis gas** (Monsanto, 1983):

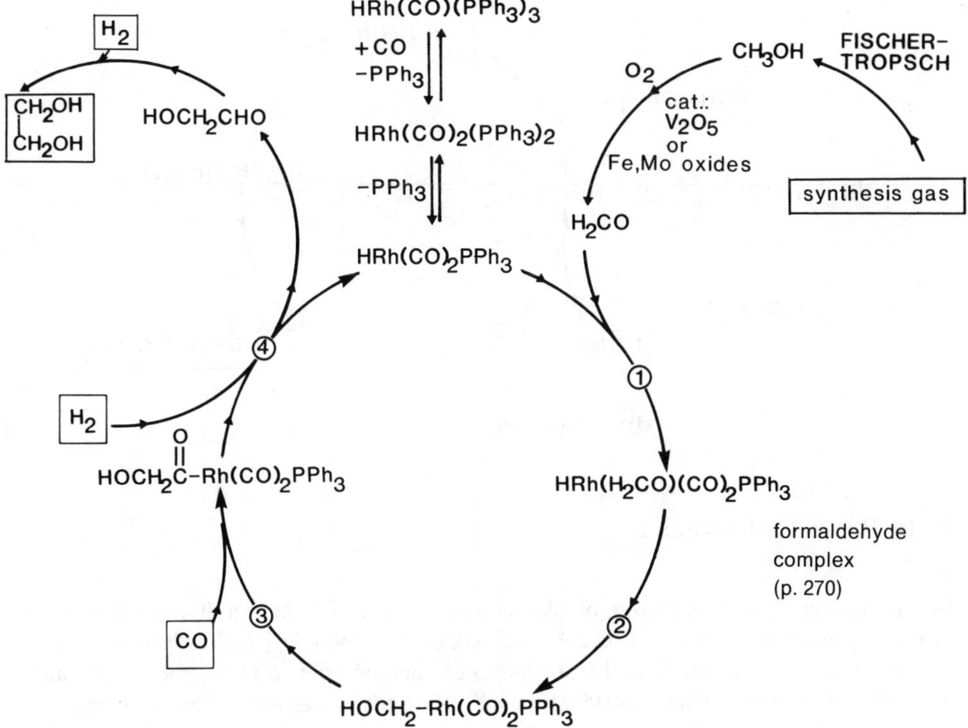

Ethylen glycol from synthesis gas (Monsanto, 1983)

This process is still at an experimental stage.

17.13 Hydrocyanation

Certain complexes, mainly of nickel and palladium, catalyze the addition of HCN to olefins (Tolman, 1986 R). An important commercial application of this homogeneous catalysis is the production of adiponitrile $NC(CH_2)_4CN$ from butadiene and HCN (Du Pont):

+ HCN → CN → CN

HCN

NC CN

$H_2N(CH_2)_6NH_2$ ← H_2

Polyamide building block

The active catalyst $NiH(CN)L_2$ is formed from NiL_4 by ligand dissociation and oxidative addition of HCN:

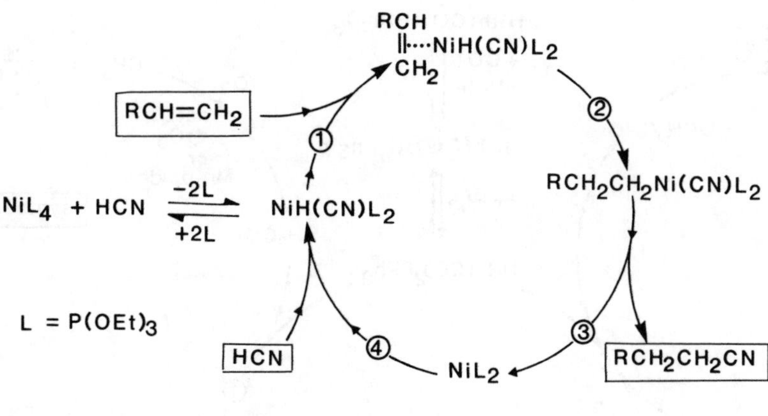

Hydrocyanation

17.14 Reppe Carbonylation

In this process, the carbonylation of alkenes and alkynes with the simultaneous addition of HX (molecule with active H) is achieved. $HCo(CO)_4$, $Ni(CO)_4$ or $Fe(CO)_5$ can serve as a precatalyst. An example is the production of acrylic ester from acetylene, CO, and ROH (BASF, Röhm & Haas, 140,000 t/a each). A disadvantage are the hazards encountered in handling acetylene.

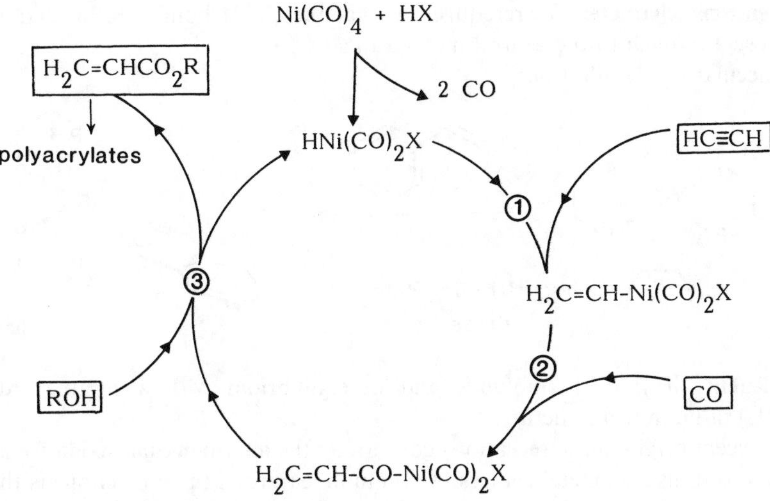

Reppe carbonylation

17.15 Activation of C – H Bonds in Alkanes – a Topical Problem in Organometallic Research

One of the most exacting tasks facing homogeneous catalysis is the **intermolecular** activation of saturated hydrocarbons. The decisive step, the oxidative addition of C – H to the metal, formally resembles the addition of H_2, but in most cases is thermodynamically unfavorable due to the weaker nature of M – C as compared to M – H bonds (Halpern 1985 R):

$$L_nM + \begin{matrix} H \\ | \\ H \end{matrix} \longrightarrow L_nM\begin{matrix} \diagup H \\ \diagdown H \end{matrix} \qquad \Delta H < 0$$

$$L_nM + \begin{matrix} C \\ | \\ H \end{matrix} \longrightarrow L_nM\begin{matrix} \diagup C \\ \diagdown H \end{matrix} \qquad \Delta H > 0$$

Thus, (hydrido)σ-alkyl transition-metal complexes are known in small number only; they tend to decompose by reductive elimination of alkanes (cf. the mechanism of homogeneously catalyzed hydrogenation of alkenes, Section 17.8). Oxidative addition of **arene C – H bonds** to transition metals has been known for some time, however.
Intramolecular C – H addition:

$(Ph_3P)_3Ir^{I}Cl \longrightarrow (Ph_3P)_2(H)Ir^{III}Cl(C_6H_4PPh_2)$

This reaction type is called orthometallation or, more generally, **cyclometallation** (M. I. Bruce, 1977 R). The widespread occurrence of such cyclometallations reflects their

neutral entropic character. A prerequisite is that the C–H bonds are located so as to allow close approach to the central metal atom.
Intermolecular C–H addition:

$P\frown P$ = $Me_2PCH_2CH_2PMe_2$
dmpe

(Chatt, 1965)

$(dmpe)_2RuH(C_{10}H_7)$ is thermolabile and in equilibrium with a complex $(dmpe)_2$-$Ru(C_{10}H_8)$ of unknown structure.
Of more recent origin are observations concerning the **intermolecular oxidative addition of alkanes** to transition metal centers. A common feature of these reactions is the prior generation of a complex of low coordination number and high electron density at the central metal. These coordinatively unsaturated species react readily even with non-activated aliphatic C–H bonds. What is particularly remarkable about this reaction is its high selectivity: the intermolecular reaction of the organometallic intermediate with saturated hydrocarbons is clearly favored over the reaction with the C–H bonds of its own ligands!

$h\nu$, $C(CH_3)_4$, $-H_2$

C_6F_{12}, $h\nu$, CH_4, $-CO$, 10 bar

45°

cy = cyclohexyl

(R. G. Bergman, 1982) (W. A. G. Graham, 1983) (G. M. Whitesides, 1986)

In the case of $[cy_2P(CH_2)_2Pcy_2]PtH(C_5H_{11})$, a coordinatively unsaturated Pt^0 complex is formed via reductive elimination of neopentane; this complex is especially suitable for illustrating important aspects of C–H activation:

- *Intramolecular cyclometallation is suppressed here since the incorporation of the organophosphanes into a chelating ligand prevents the* C–H *bonds from approaching the platinum atom.*
- *The* Pt–C *bond from the intermolecular reaction should be relatively strong as inter-ligand repulsion, which often weakens an* M–C *bond, is minimal here.*
- *The isolobal relationship likens the reaction of the fragment* L_2Pt *with* C–H *bonds to the insertion of singlet carbenes into* C–H *bonds.*

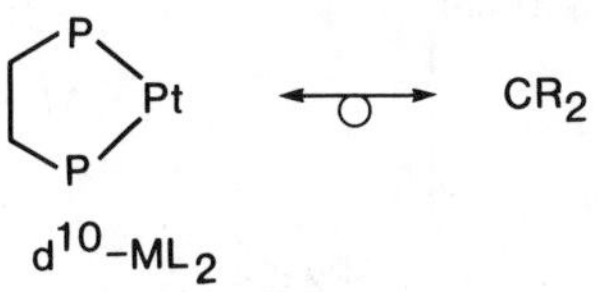

P–Pt–P (chelate) ⟷ CR_2

d^{10}-ML_2

- *The exposure of the* Pt *atom in the fragment* L_2Pt *is reminiscent of that of the vertex atoms at heterogeneously active metal surfaces.*

C–M activation bears resemblance to TM-catalyzed valence isomerisation and to the activation of H_2 since in all three cases a **side-on coordinated σ-bond** may be involved:

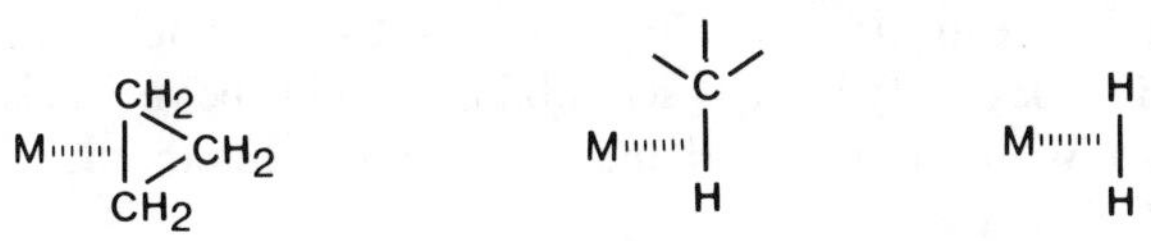

M⋯(CH₂–CH₂ of cyclopropane) (p. 416) M⋯(C–H) (Brookhart, 1988 R) M⋯(H–H) (Kubas, 1988 R)

Although no η^2(R–H)σ complex has yet been isolated, evidence for its intermediacy in reductive eliminations and, as required by the principle of microscopic reversibility, in the oxidative additions of R–H, is accumulating. Recent studies by Norton (1989) have demonstrated that:

- reductive elimination of methane from $Cp_2W(H)CH_3$ is an intramolecular process (no cross products are obtained):

$$\begin{bmatrix} Cp_2W(D)CH_3 \\ + \\ Cp_2W(H)CD_3 \end{bmatrix} \xrightarrow{\Delta T \text{ or } h\nu} CH_3D + CHD_3 + [Cp_2W]$$

- this elimination is governed by an inverse isotope effect in that the formation of CD_4 from $Cp_2W(D)CD_3$ occurs more rapidly than the formation of CH_4 from $Cp_2W(H)CH_3$.
- hydrogen scrambling between the methyl and hydride ligands

$$Cp_2W(D)CH_3 \rightleftharpoons Cp_2W(H)CH_2D$$

is faster than methane elimination.

From these findings, an energy profile can be drawn which implies that the reductive elimination of CH_4 from $Cp_2W(H)CH_3$ as well as the yet unobserved reverse process, oxidative addition of CH_4 to Cp_2W, would proceed through the reversible formation of a methane σ complex.

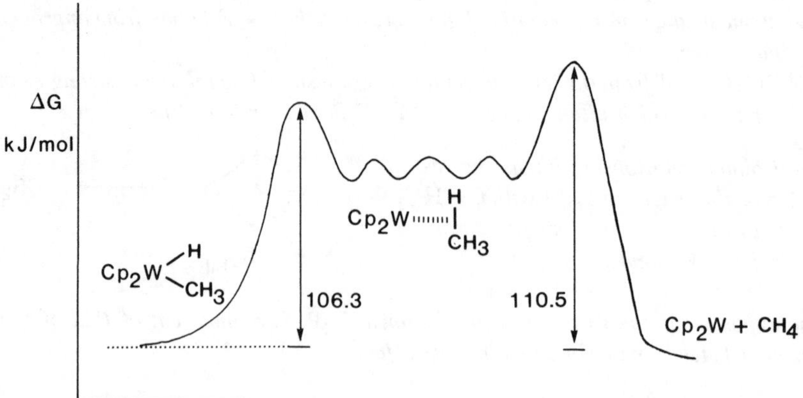

The inverse isotope effect arises because in the intermediate σ complex hydrogen is bonded to carbon (strong bond) rather than to tungsten (weak bond). Therefore, the pre-equilibrium concentration of Cp_2W-η^2-$(D-CD_3)$ should slightly exceed that of Cp_2W-η^2-$(H-CH_3)$. Hydrogen scrambling would be a consequence of rapid migration of Cp_2W from one methane C–H bond to another in the σ complex Cp_2W-η^2-$(H-CH_3)$.

C–H activation, albeit of a somewhat different nature, is also effected by certain **organolanthanoids**, a class of compounds currently receiving considerable attention. In order to appreciate the mechanistic difference compared with transition metal induced C–H activation, a few remarks on organolanthanoids are called for.
Since, ultimately, chemical behavior is an outcome of the nature of the chemical bond, the most pertinent gradation to note is the degree of **covalency of the metal carbon bond** which decreases in the sequence **transition metal-R > actinoid-R > lanthanoid-R**, organolanthanoids being predominantly ionic. This trend is supported by numerous chemical, spectroscopic and theoretical results (Bursten, 1989 R). The main factor responsible for this gradation involves the nature of the valence orbitals, which changes from 3,4,5*d* (transition metals) to 4*f* (lanthanoids), the actinoids assuming an intermediate position, in that 5*f* as well as 6*d* orbitals engage in bonding. In the case of the lanthanoid elements, the highly contracted nature of the 4*f* orbitals results in efficient shielding by filled 5*s* and 5*p* orbitals rendering overlap of Ln(4*f*) – with ligand orbitals vanishingly small. Thus, $Ln^{2+,3+}$ ions are expected to resemble the main-group metal ions $M^{2+,3+}$, because electrostatic factors and steric considerations, rather than metal-ligand orbital interactions, govern structure and chemical reactivity. This general statement may be illustrated by the following observations:

- Organolanthanoid compounds are extremely **air- and moisture sensitive**, a reflection of the highly carbanionic character of the organic ligand and the oxophilicity of the

$Ln^{2+,3+}$ ions. Consequently, $LnCp_3$ complexes, like $MgCp_2$ act as Cp-transfer agents, a property not generally shared by the cyclopentadienyl compounds of the actinoids and the transition metals.

– The lack of extensive metal(4*f*)-ligand orbital overlap prevents the lanthanoid cations from acting as π-bases. Therefore, **π-backbonding effects** should be virtually **absent**. To date, no species containing Ln–CO bonds that are stable at ambient temperature has been prepared, whereas $(Me_3Si{-}\eta^5{-}C_5H_4)_3U$ reversibly absorbs carbon monoxide (Andersen, 1986). The product $(Me_3Si{-}\eta^5{-}C_5H_4)_3UCO$ features the stretching frequency $\nu_{CO} = 1976\ cm^{-1}$ ($\Delta\nu_{CO} = -167\ cm^{-1}$) which signifies $U \overset{\pi}{-} CO$ backbonding. In contrast to CO, the more polar ligand RNC (p. 240) *does* coordinate to Cp_3Ln. However, as judged from the IR stretching frequency in $Cp_3Yb(CN{-}c{-}C_6H_{11})$, $\nu_{CN} = 2203\ cm^{-1}$ ($\Delta\nu_{CN} = +67\ cm^{-1}$), cyclohexyl isocyanide essentially acts as a σ-donor ligand to the trivalent lanthanoid ion (E. O. Fischer, 1966).

– The dominance of ionic character in the lanthanoid-ligand bond is also demonstrated by the variable hapticity, that the ligand rings display in the binary lanthanoid cyclopentadienyls Cp_3Ln. To cite an example, in $(MeC_5H_4)_3Nd$ each Nd atom is η^5 bonded to three cyclopentadienide rings and η^1 bonded to another ring of an adjacent $(MeC_5H_4)_3Nd$ unit (Burns, 1974). In this way, the central metal maximizes the number of cation-anion contacts compatible with its size. The same phenomenon is observed in the ionic cyclopentadienyl complexes of the group 1 and 2 metals (e.g., Cp_2Be, p. 40). Another structural feature common to alkaline earths- and lanthanoid cyclopentadienyls is the **bent structure** of the metallocenes $(Me_5C_5)_2M$ (M = Ca, Sr, Ba and Eu, Sm, Yb, respectively).

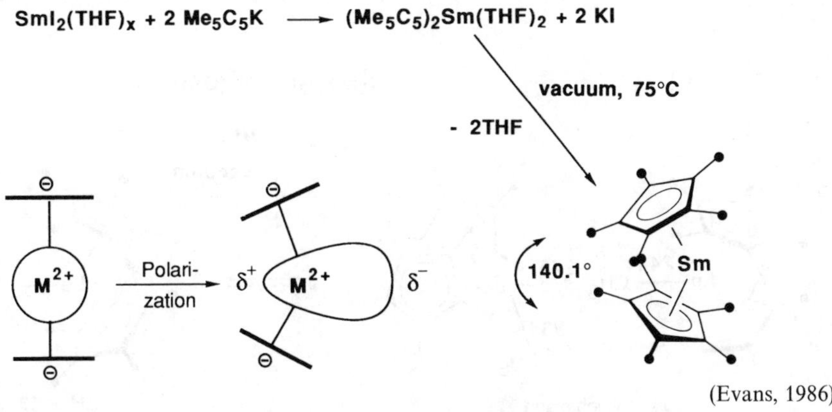

(Evans, 1986)

The seemingly anomalous bent geometry of these metallocenes resists simple explanations based on steric, traditional electrostatic or covalency principles. Therefore, as mentioned on p. 47, strongly differing approaches have been put forward in this context. A qualitative explanation is based on the *polarizable ion model*, first introduced by Rittner

(1951). Accordingly, total electrostatic bonding is maximized in the bent structure because then the dipole moment induced on the central cation interacts favorably with the two adjacent anions. Since a certain extent of anion-anion repulsion has to be offset for this effect to operate, it comes as no surprise that only the ionic metallocenes of relatively large M^{2+} ions are bent, whereas Cp_2Mg is axially symmetric (p. 42).
Besides the binary compounds Cp_3Ln, which are known for all of the lanthanoids, and the complexes Cp^*_2Ln which have been obtained only for the lanthanoids accessible in the divalent state (Ln = Eu, Sm, Yb), most of organolanthanoid chemistry deals with ternary compounds Cp_2LnX (X = halogen, H, alkyl, aryl) and their solvates. While it is true that the chemistry of these molecules changes rather little upon traversing the 4*f* element series, an intriguing aspect is the possibility of fine-tuning chemical reactivity by *gradually varying the ionic radius* from $r(La^{3+}) = 103$ to $r(Lu^{3+}) = 86$ pm, keeping the molecular composition unaltered. Furthermore, because of their comparatively large size, lanthanoids in their compounds respond sensitively to the *degree of steric saturation* imposed by the ligands. Once again, C_5Me_5 has proved to be the ligand of choice, and compounds of the type $(C_5Me_5)_2LnR$ have furnished remarkable results. The pattern

$(C_5Me_5)_2$ Ln ◀ R ($R = CH_3$, H)

↑	↑
Highly electropositive, electrophilic center	Highly ionic nucleophilic alkyl or hydride

(Evans, 1985)

suggests that $(C_5Me_5)_2LnR$ bears resemblance to related compounds of trivalent titanium or of the main group element aluminum. In fact, it even surpasses the reactivity of its more conventional congeners as will be seen. The key compound, solvent-free coordinatively unsaturated $(C_5Me_5)_2LuCH_3$, exists in a monomer ⇄ dimer equilibrium (Watson, 1983):

$(C_5Me_5)_2LuCH_3(OEt_2)$

NEt_3
vacuum

Lu —244— CH_3 ·····276····· Lu —234— CH_3 ⇌ 2 Lu—CH_3

25°C, cyclohexane
solution: 85%
solid state: 100%

15%

ΔH = 53 kJ/mol

$(C_5Me_5)_2LuCH_3$ serves admirably as a soluble model for Ziegler-Natta catalysis (p. 423), mimicking the initiation ①, propagation and termination steps of the industrially used heterogeneous catalysts (Parshall, 1985 R):

$Cp^*_2Lu–CH_3$

① $CH_2{=}CHMe$ / INS → $Cp^*_2Lu\text{-}CH_2CHMe_2$ → ($CH_2{=}CHMe$ / INS) → propagation (growth reaction)

② $CH_2{=}CMe_2$ / $-CH_4$ → Cp^*_2Lu–CH=CMe$_2$

③ $SiMe_4$ / $-CH_4$ → $Cp^*_2Lu\text{-}CH_2SiMe_3$ — metallation (C-H activation)

④ $^{13}CH_4$ / $-CH_4$ → $Cp^*_2Lu–^{13}CH_3$

A more astounding property pertaining to the topic of this section is the ability of $Cp^*_2LuCH_3$ to react with C–H bonds of extremely low acidity. Whereas vinylic metallation ② and metallation of tetramethylsilane ③ are precedential, C–H *activation of the ultimately unreactive alkane* CH_4 ④ by a compound stable at room temperature is unique (Watson, 1983). A mechanism involving the conventional steps of oxidative addition and reductive elimination may safely be excluded since the requisite changes in oxidation state, $Lu^{III}(f^{14})/Lu^{V}(f^{12})$ and $Lu^{III}(f^{14})/Lu^{I}(f^{14}d^2)$, are unreasonable. A clue to the possible path effecting C–H activation is provided by the dimeric structure of $[Cp^*_2LuCH_3]_2$. Perhaps the type of force responsible for pre-coordination of CH_4 to the monomer $Cp^*_2LuCH_3$ is the same as that accounting for the generation of a $Lu–CH_3 \cdots Lu$ bridge in the dimer $[Cp^*_2LuCH_3]_2$ (or that causing intercluster bonding in solid $[LiCH_3]_4$, p. 20). The transition state could then undergo an electrocyclic rearrangement with concomitant proton transfer, the final result being methyl group exchange:

Lu—CH_3 ⇌ Lu(CH_3)···CH_4 ⇌ [Lu(CH_3)(H)(CH_3)]‡

C–H activation processes of the kind described in this final section have not yet found practical applications, as they have resisted incorporation into catalytic cycles. They demonstrate, however, that under appropriate conditions the reaction of non-activated C–H bonds with metal centers is thermodynamically and kinetically feasible and may therefore pave the way for further advances.

18 Literature

For a more detailed treatment of special topics in organometallic chemistry, students are referred to "Comprehensive Organometallic Chemistry" by Wilkinson, Stone and Abel. In keeping with the nature of the present text as a 'concise introduction', only a limited number of monographs and specialized review articles are included in the following list of references.

History

Thayer, J. S., Organometallic Chemistry, A Historical Perspective *Adv. Organomet. Chem.,* 1975, **13,** 1.

Parshall, G. W., Trends and Opportunities for Organometallic Chemistry in Industry, *Organometallics,* 1987, **6,** 687.

Detailed Accounts and Reference Books

Wilkinson, G., Stone, F. G. A., Abel, E. W., eds., "Comprehensive Organometallic Chemistry", Pergamon Press: Oxford, 1982, 9 volumes.

Wilkinson, G., Gillard, R. D., McCleverty, J. E., "Comprehensive Coordination Chemistry", Pergamon Press: Oxford, 1987, 7 volumes.

Barton, D., Ollis, W. D., "Comprehensive Organic Chemistry", Pergamon Press: Oxford, 1979, Vol. 3.

Gmelin, "Handbook of Inorganic Chemistry", Springer Verlag: Berlin (a detailed treatise of organometallic compounds arranged by element).

Römpps Chemie Lexikon, 8. Ed., Franckh'sche Verlagshandlung: Stuttgart (1979–) (inter alia technical applications and references to review articles).

Buckingham, J., ed., "Dictionary of Organometallic Chemistry", Chapman and Hall: London, 1984, 3 volumes and 5 supplements.

Emsley, J., "The Elements", Clarendon Press: Oxford, 1989 (a collection of chemical, nuclear and electron shell properties)

Textbooks **MAIN-GROUP AND TRANSITION-METAL ORGANOMETALLICS**

Coates, G. E., Green, M. L. H., Wade, K., "Organometallic Compounds", 3rd Ed., Methuen: London, 1967, 1968, 2 volumes.

Powell, P., "Principles of Organometallic Chemistry", 2nd Ed., Chapman and Hall: London, 1988.

Haiduc, I., Zuckerman, J. J., "Basic Organometallic Chemistry", Walter de Gruyter: Berlin, 1985. (excellent bibliography, only in the hard-cover version).

Parkins, A. W., Poller, R. C., "An Introduction to Organometallic Chemistry', Macmillan: London, 1986.

TRANSITION-METAL ORGANOMETALLICS

Collman, P., Hegedus, L. S., Norton, J. R., Finke, R. G., "Principles and Applications of Organotransition Metal Chemistry", 2nd Ed. University Science Books: Mill Valley, 1987.

Lukehart, C. M., "Fundamental Transition Metal Organometallic Chemistry", Brooks/Cole: Monterey, 1985.

Yamamoto, A. "Organotransition Metal Chemistry", Wiley: New York, 1986.

Pearson, A. J., "Metallo-Organic Chemistry", Wiley: New York, 1985.

Crabtree, R. E., "The Organometallic Chemistry of the Transition Metals", Wiley: New York, 1988.

Hartley, F. R., Patai, S., eds., "The Chemistry of the Metal-Carbon Bond", Wiley: New York, 1982, 1985, 1986.
Vol. 1: The Structure, Preparation, Thermochemistry and Characterization of Organometallic Compounds (predominantly transition-metal compounds). Vol. 2: The Nature and Cleavage of Metal-Carbon Bonds. Vol. 3: Carbon-Carbon Bond Formation using Organometallic Compounds. Vol. 4: The Use of Organometallic Compounds in Organic Synthesis.

Marks, T. J., Fischer, R. D., eds., "Organometallic Chemistry of the *f*-Elements", Reidel: Dordrecht, 1979.

Kegley, S. E., Pinchas, A. R., "Problems and Solutions in Organometallic Chemistry", University Science Books: Mill Valley, 1986.

Wilkins, R. G., "Kinetics and Mechanism of Reactions of Transition Metal Complexes", 2nd Ed., VCH: Weinheim 1991.

INORGANIC CHEMISTRY

Cotton, F. A., Wilkinson, G., "Advanced Inorganic Chemistry", 5th Ed., Wiley: New York, 1988.

Greenwood, N. N., Earnshaw, A., "Chemistry of the Elements", Pergamon Press: Oxford, 1984.

Huheey, J. E., "Inorganic Chemistry – Principles of Structure and Reactivity", 3rd Ed., Harper and Row: New York, 1983.

Purcell, K. F., Kotz, J. C., "Inorganic Chemistry", W. B. Saunders: Philadelphia, 1977.

Review Series

Specialist Periodical Report, The Royal Society of Chemistry, London: Organometallic Chemistry (annual reviews, from 1971).

Advances in Organometallic Chemistry.

Advances in Inorganic Chemistry and Radiochemistry.

Progress in Inorganic Chemistry.

Coordination Chemistry Reviews.

Mechanisms of Inorganic and Organometallic Reactions.

Angewandte Chemie.

Accounts of Chemical Research.

Chemical Reviews.

Comments on Inorganic Chemistry.

Specialized Journals

Journal of Organometallic Chemistry.
Organometallics.
Synthesis and Reactivity in Inorganic and Organometallic Chemistry.
Applied Organometallic Chemistry.
Advanced Materials
Chemistry of Materials
All Journals of Inorganic and Organic Chemistry.
General chemical journals.

Synthetic Methods in Organometallic Chemistry

Brauer, G., "Handbuch der Präparativen Anorganischen Chemie", F. Enke: Stuttgart, 1975, 1978, 1981, 3 volumes (includes an introduction to organometallic laboratory techniques).

Eisch, J. J., King, R. B., eds. "Organometallic Syntheses", Academic Press: New York, Vol. 1, 1965, Vol. 2, 1981, Elsevier: Amsterdam, Vol. 3, 1986, Vol. 4, 1988.

Shriver, D. F., Drezdon, M. A., "The Manipulation of Air-sensitive Compounds", 2nd Ed., Wiley: New York, 1986.

Wayda, A. L., Darensbourg, M. Y., eds., "Experimental Organometallic Chemistry – A Practicum in Synthesis and Characterization", ACS Symposium Series No. 357: Washington D.C., 1987.

Brandsma, L., Verkruijsse, H., "Preparative Polar Organometallic Chemistry", Springer: Berlin 1988.

"Preparative Inorganic Reactions", Vol. 1, 1964–..., Wiley: New York.

"Inorganic Syntheses", Vol. I, 1939–..., Wiley: New York.

Houben-Weyl, "Methoden der Organischen Chemie", G. Thieme: Stuttgart (covers almost all classes of organyls and π complexes).

J. J. Zuckerman, ed., The Formation of Bonds to C, Si, Ge, Sn, Pb, in "Inorganic Reactions and Methods", Vol. 10–12, VCH: Weinheim, 1988–....

Timms, P. L., Turney, T. W., Metal Atom Synthesis of Organometallic Compounds, *Adv. Organomet. Chem.* 1977, **15,** 53.

Blackborrow, J. R., Young, D., "Metal Vapor Synthesis in Organometallic Chemistry", Springer: Berlin, 1979.

McGlinchey, M. J., Metal Atoms in Organometallic Synthesis, in: Hartley, F. R., Patai, S., eds., "The Chemistry of the Metal-Carbon Bond", Vol. 1, Wiley: New York, 1982.

Connelly, N. G., Synthetic Applications of Organotransition-metal Redox Reactions, *Chem. Soc. Rev.,* 1989, **18**, 153.

Applications to Organic Synthesis (Transition Metals)

Alper, H., "Transition Metal Organometallics in Organic Synthesis", Academic Press: New York, 1976, 1978, 2 volumes.

Davies, S. G., "Organotransition Metal Chemistry: Applications to Organic Synthesis", Pergamon Press: Oxford, 1982.

de Meijere, A., tom Dieck, H., eds., "Organic Synthesis via Organometallics" (First Symposium, Hamburg 1966), Springer: Berlin, 1987.

Werner, M., Erker, G., eds., "Organic Synthesis via Organometallics" (Second Symposium, Würzburg, 1988), Springer: Berlin, 1989.

Dötz, K. H., Hoffmann, R. W., eds., "Organic Synthesis via Organometallics" (Third Symposium, Marburg 1990), Vieweg: Braunschweig, 1991.

see also: Literature to Ch. **17**

Spectroscopic Methods **GENERAL**

Drago, R. S., "Physical Methods in Chemistry", W. B. Saunders: Philadelphia, 1977.
Ebsworth, E. A. V., Rankin, D. W. H., Cradock, S., "Structural Methods in Inorganic Chemistry", Blackwell: Oxford, 1987.
Specialist Periodical Report, Spectroscopic Properties of Inorganic and Organometallic Compounds, The Royal Society of Chemistry, London, (annual reviews, from 1967).

IR/RAMAN

Nakamoto, K., "Infrared and Raman Spectra of Inorganic and Coordination Compounds", 3rd Ed., Wiley: New York, 1978.
Braterman, P. S., "Metal-Carbonyl Spectra", Academic Press: New York, 1975.
Weidlein, J., Müller, U., Dehnicke, K., "Schwingungsfrequenzen I", MGE, 1981, "Schwingungsfrequenzen II", TM, 1986, G. Thieme: Stuttgart.
Darensbourg, M. Y. and D. J., Infrared Determination of Stereochemistry in Metal Complexes, *J. Chem. Ed.* 1970, **47,** 33; 1974, **51,** 787.
Kettle, S. F., The Vibrational Spectra of Metal Carbonyls, *Top. Curr. Chem.,* 1977, **71,** 111.

NMR

Günther, H., "NMR-Spektroskopie", 2nd Ed., (^{1}H-NMR), G. Thieme: Stuttgart, 1983.
Kalinowski, H. O., Berger, S., Braun, S., "^{13}C-NMR-Spektroskopie", G. Thieme: Stuttgart, 1984.
Lamar, G. N., Horrocks, W. D., Holm, R. H., "NMR of Paramagnetic Molecules", Academic Press: New York, 1973.
Jackman, L. M., Cotton, F. A., "Dynamic NMR Spectroscopy", Academic Press: New York, 1975.
Harris, R. K., Mann, B. E., eds., "NMR and the Periodic Table", Academic Press: New York, 1978 [see also: *Chem. Soc. Rev.,* 1976, **5,** 1].
Mann, B. E., Taylor, B. F., "^{13}C-NMR Data for Organometallic Compounds", Academic Press: New York, 1981.
Dechter, J. J., NMR of Metal Nuclides I: The Main Group Metals, *Prog. Inorg. Chem.* 1982, **29,** 285; NMR of Metal Nuclides II: The Transition Metals, *Prog. Inorg. Chem.*1985, **33,** 393.
Mason, J., Patterns of Nuclear Magnetic Shielding of Transition Metal Nuclei, *Chem. Rev.,* 1987, **87,** 1299.
Benn, R., Rufinska, A., High Resolution Metal NMR Spectroscopy of Organometallic Compounds, *Angew. Chem. Int. Ed. Engl.,* 1986, **25,** 861.
von Philipsborn, W., Transition Metal NMR Spectroscopy – a Probe into Organometallic Structure and Catalysis, *Pure Appl. Chem.,* 1986, **58,** 513.
Rehder, D., Applications of Transition Metal NMR Spectroscopy in Coordination Chemistry, *Chimia,* 1986, **40,** 186.
Günther, H., Moskau, D., Bast, P., Schmalz, D., Modern NMR Spectroscopy of Organolithium Compounds, *Angew. Chem. Int. Ed. Engl.,* 1987, **26,** 1212.
Jolly, P. W., Mynott, R., Applications of ^{13}C-NMR to Organotransition Metal Complexes, *Adv. Organomet. Chem.,* 1981, **19,** 257.
Mann, B. E., Recent Developments in NMR Spectroscopy of Organometallic Compounds, *Adv. Organometal. Chem.,* 1988, **28,** 397.

ESR

Wertz, J. E., Bolton, J. R., "Electron Spin Resonance", McGraw-Hill: New York, 1972.

Kirmse, R., Stach, J., "ESR-Spektroskopie, Anwendungen in der Chemie", Akademie Verlag: Berlin, 1985.

Goodman, B. A., Raynor, J. B., Electron Spin Resonance of Transition Metal Complexes, *Adv. Inorg. Chem. Radiochem.*, 1970, **13,** 136.

MÖSSBAUER

Herber, R. H., "Chemical Mössbauer Spectroscopy", Plenum Press: New York, 1984.

Herber, R. H., Chemical Aspects of Mössbauer Spectroscopy, *Progr. Inorg. Chem.*, 1967, **8,** 1.

Parish, R. V., Mössbauer Spectroscopy, *Chem. in Britain,* 1985, **21,** 546, 740.

Bancroft, G. M., Platt, R. M., Mössbauer Spectra of Inorganic Compounds: Bonding and Structure, *Adv. Inorg. Chem. Radiochem.*, 1972, **15,** 59.

Gibb, T. C., "Principles of Mössbauer Spectroscopy", Chapman and Hall: London, 1980.

Long, G. J., ed., "Mössbauer Spectroscopy Applied to Inorganic Chemistry, Plenum: New York, 1984.

MS

Müller, J., Decomposition of Organometallic Compounds in the Mass Spectrometer (IE-MS), *Angew. Chem. Int. Ed. Engl.*, 1972, **11,** 653.

Litzow, M. R., Spalding, T. R., "Mass Spectrometry of Inorganic and Organometallic Compounds, Elsevier: Amsterdam, 1973.

Beckey, H. D., Schulten, H. R., Field Desorption Mass Spectrometry, *Angew. Chem. Int. Ed. Engl.*, 1975, **14,** 403.

Miller, J. M., Fast Atom Bombardment Mass Spectrometry and Related Techniques (FAB-MS), *Adv. Inorg. Chem. Radiochem.*, 1984, **28,** 1.

UV-VIS

Wrighton, M. S., The Photochemistry of Metal Carbonyls, *Chem. Rev.*, 1974, **74,** 401.

Adamson, A. W., Fleischauer, P. D., eds., "Concepts of Inorganic Photochemistry", Wiley: New York, 1974.

Geoffroy, G. L., Wrighton, M. S., "Organometallic Photochemistry", Academic Press: New York, 1975.

Wrighton, M. S., ed., Inorganic and Organometallic Photochemistry, *Adv. Chem. Ser. ACS* 1978, 168.

Geoffroy, G. L., Wrighton, M. S., Organometallic Photochemistry, *J. Chem. Ed.* 1983, **60**, 861.

PES

Cauletti, C., Furlani, C., Gas Phase UV Photoelectron Spectroscopy as a Tool for the Investigation of Electronic Structures of Coordination Compounds, *Comments Inorg. Chem.*, 1985, **5,** 29.

Green, J. C., Gas Phase Photoelectron Spectra of *d*- and *f*-Block Organometallic Compounds, *Struct. Bonding,* 1981, **43,** 37.

Van Dam, H., Oskam, A., UV-Photoelectron Spectroscopy of Transitionmetal Complexes, *Transition Met. Chem., A Series of Advances,* 1985, **9,** 125.

Lichtenberger, D. L., Kellog, G. E., Experimental Quantum Chemistry: Photoelectron Spectroscopy of Organotransition-metal Complexes, *Acc. Chem. Res.* 1987, **20**, 379.

STRUCTURE

Haaland, A., Organometallic Compounds Studied by Gas Phase Electron Diffraction, *Top. Curr. Chem.,* 1975, **53**, 1.

Orben, A. G., Brammer, L., Allen, F. H., Kennard, O., Watson, D. G., Taylor, R., Tables of Bond Lenghts determined by X-Ray and Neutron Diffraction, Part 2. Organometallic Compounds and Coordination Complexes of the d- and f-Block Metals, *J. Chem. Soc. Dalton* 1989, S 1–S 83.

Bernal, I., ed., "Stereochemistry of Organometallic and Inorganic Compounds", 1. 2, 3 ... Elsevier: Amsterdam, 1986–....

Angermund, K., Claus, K. H., Goddard, R., Krüger, C., High-Resolution X-ray Crystallography – An Experimental Method for the Description of Chemical Bonds, *Angew. Chem. Int. Ed. Engl.*, 1985, **24**, 237

Special Topics

Matteson, D. S., "Organometallic Reaction Mechanisms of the Nontransition Elements", Academic Press: New York, 1974.

Kochi, J. K., "Organometallic Mechanisms and Catalysis", Academic Press: New York, 1978.

Atwood, J. D., "Mechanismus of Inorganic and Organometallic Reactions", Brooks/Cole: Monterey, 1985.

Zuckerman, J. J., ed., Electron-Transfer and Electrochemical Reactions; Photochemical and other Energized Reactions, *Inorg. React. Meth.*, 1986, **15**.

Connelly, N. G., Geiger, W. E., The Electron-Transfer Reactions of Organotransition Metal Complexes, *Adv. Organomet. Chem.* 1984, **23,** 1 and 1985, **24,** 87.

Suslick, K. S., ed., "High-Energy Processes in Organometallic Chemistry", ACS Symposium Series **333,** 1987.

Cole-Hamilton, D. J., Williams, J. O., eds., "Mechanisms of Reactions of Organometallic Compounds with Surface", Plenum Press: New York, 1989.

Lappert, M. F., Lednor, W. E., Free Radicals in Organometallic Chemistry, *Adv. Organomet. Chem.*, 1976, **14,** 345.

Trogler, W. C., ed., "Organometallic Radical Processes", J. Organomet. Chem. Library, **22,** Elsevier: Amsterdam, 1990.

Jones, P. R., Organometallic Radical Anions, *Adv. Organomet. Chem.,* 1977, **15,** 273.

Klabunde, K. J., "Chemistry of Free Atoms and Particles", Academic Press: New York, 1980.

Cotton, F. A., Fluxional Organometallic Molecules, *Acc. Chem. Res.,* 1968, **1**, 257.

Jutzi, P., Fluxional η^1-Cyclopentadienyl Compounds of Main-Group Elements, *Chem. Rev.,* 1986, **86,** 983.

Thayer, J. S., "Organometallic Compounds and Living Organisms", Academic Press: New York, 1984.

Craig, P. J., ed., "Organometallic Compounds in the Environment, Principles and Reactions", Longman: Harlow, 1986.

Ryabov, A. D., The Biochemical Reactions of Organometallics with Enzymes and Proteins, *Angew. Chem. Int. Ed. Engl.* 1991, **30**, 931.

Selected Reviews for each Chapter

MAIN-GROUP ORGANOMETALLICS

O'Neill, M. E., Wade K., Structural and Bonding Relationships among Main Group Organometallic Compounds, in: "Comprehensive Organometallic Chemistry", 1982, Vol. 1, 1.

Rheingold, A. L., ed., "Homoatomic Rings, Chains and Macromolecules of the Main Group Elements". Elsevier: Amsterdam, 1977.

Matteson, D. S., "Organometallic Reaction Mechanisms of the Nontransition Elements", Academic Press: New York, 1974.

Szwarc, M. S., ed., "Ions and Ion Pairs in Organic Reactions", Wiley: New York, 1972, 1974, 2 volumes.

Jutzi, P., π-Bonding to Main-Group Elements, *Adv. Organomet. Chem.*, 1986, **26,** 217.

5

Setzer, W. N., Schleyer, P. v. R., X-Ray Structural Analyses of Organolithium Compounds, *Adv. Organomet. Chem.*, 1985, **24,** 353.

Schade, C., Schleyer, P. v. R., Sodium, Potassium, Rubidium, Cesium: X-ray Structural Analysis of their Organic Compounds, *Adv. Organomet. Chem.* 1987, **27,** 169.

Kaufmann, E., Raghavachari, K., Reed, A. E., Schleyer, P. v. R., Methyllithium and its Oligomers. Structural and Energetic Relationships, *Organometallics,* 1988, **7,** 1597.

Fraenkel, G., Hsu, H., Su, B. M., The Structure and Dynamic Behaviour of Organolithium Compounds in Solution, ^{13}C-, ^{6}Li- and ^{7}Li NMR, in Bach, R., ed., "Lithium Current Applications in Science, Medicine and Technology", Wiley: New York, 1985.

Schlosser, M., "Struktur and Reaktivität polarer Organometalle. Eine Einführung in die Chemie organischer Alkali- und Erdalkalimetallverbindungen", Springer: Berlin, 1973.

Ebel, H. F., "Die Acidität der CH-Säuren", G. Thieme: Stuttgart, 1969.

Reutov, O. A., Beletskaya, I. P., Butin, K. P., "CH-Acids", Pergamon Press: Oxford, 1978.

Wakefield, B. J., "The Chemistry of Organolithium Compounds", Pergamon Press: Oxford, 1974.

Wakefield, B. J., "Organolithium Methods", Academic Press: London, 1988.

Reetz, M. T., Organotitanium Reagents in Organic Syntheses. A Simple Means to Adjust Reactivity and Selectivity of Carbanions, *Topics Curr. Chem.,* 1982, **106,** 1.

Reetz, M. T., "Organotitanium Reagents in Organic Synthesis", Springer: Heidelberg, 1986.

6.1

Walborsky, H. M., Mechanism of Grignard Reagent Formation. The Surface Nature of the Reaction, *Acc. Chem. Res.,* 1990, **23**, 286.

Whitesides, G. M., Lawrence, L. M., Trapping of Free Alkyl Radical Intermediates in the Reaction of Alkyl Bromides with Magnesium, *J. Am. Chem. Soc.* 1980, **102,** 2493.

Bogdanovic, B., Mg-Anthracene Systems and their Applications in Synthesis and Catalysis, *Acc. Chem. Res.*, 1988, **21,** 261.

Rieke, R. D., Preparation of Highly Reactive Metal Powders and their Use in Organic and Organometallic Synthesis, *Acc. Chem. Res.,* 1977, **10,** 301.

Faegri, K., Almlöf, J., Lüthi, H. P., The Geometry and Bonding of Magnesocene. An ab-initio MO-LCAO Investigation, *J. Organomet. Chem.,* 1983, **249,** 303.

Ashby, E. C., Laemmle, J. T., Stereochemistry of Organometallic Compound Addition to Ketones, *Chem. Rev.,* 1975, **75**, 521.

Holm, T., Electron Transfer from Alkylmagnesium Compounds to Organic Substrates. *Acta Chem. Scand.* 1983, **B37,** 567.

Jutzi, P., The Versatility of the Pentamethylcyclopentadienyl Ligand in Main-Group Chemistry, *Comments Inorg. Chem.*, 1987, **6**, 123.

6.2

Jones, P. R., Desio, P. J., The Less Familiar Reactions of Organocadmium Reagents, *Chem. Rev.* 1978, **78**, 491.
Larock, R. C., Organomercury Compounds in Organic Synthesis, *Angew. Chem. Int. Ed. Engl.*, 1978, **17**, 27.
Rabenstein, D. L., The Aqueous Solution Chemistry of Methylmercury and its Complexes, *Acc. Chem. Res.* 1978, **11**, 100.
Jensen, S., Jernelöv, A., Biological Methylation of Mercury in Aquatic Organisms, *Nature*, 1969, **223**, 753.
Rabenstein, D. L., The Chemistry of Methylmercury Toxicology, *J. Chem. Ed.*, 1978, **55**, 292.

7.1.

Rundle, R. E., Electron Deficient Compounds, *J. Am. Chem. Soc.*, 1947, **69**, 1327.
Wade, K., Boranes: Rule-Breakers become Pattern-Makers, *New Scientist*, 1974, 615.
Wade, K., Structural and Bonding Patterns in Cluster Chemistry, *Adv. Inorg. Chem. Radiochem.* 1976, **18**, 1.
Rudolph, R. W., Boranes and Heteroboranes: A Paradigm for the Electron Requirements of Clusters, *Acc. Chem. Res.*, 1976, **9**, 446.
Muetterties, E. L., ed., "Boron Hydride Chemistry", Academic Press: New York, 1975.
Onak, T. P., "Organoborane Chemistry", Academic Press: New York, 1975.
Brown, H. C., Negishi, Ei-Ichi, Boraheterocycles via Cyclic Hydroboration, *Tetrahedron*, 1977, **33** 2331.
Kölle, P., Nöth, H., The Chemistry of Borinium- and Borenium-Ions, *Chem. Rev.* 1985, **85**, 399.
Nöth, H., Wrackmeyer, B., NMR of Boron Compounds in: *NMR: Basic Principles and Progress*, 1978, **14**, 1–461.
Hawthorne, M. F., The Chemistry of Polyhedral Species derived from Transition Metals and Carboranes, *Acc. Chem. Res.*, 1968, **1**, 281.
Grimes, R. N., Carbon-rich Carboranes and their Metal Derivatives, *Adv. Inorg. Chem. Radiochem.*, 1983, **26**, 55.

7.2

Ziegler, K., Organoaluminum Compounds, in Zeiss, H., "Organometallic Chemistry", Reinhold: New York, 1960.
Eisch, J. J., Karl Ziegler – Master Advocate for the Unity of Pure and Applied Research, *J. Chem. Ed.*, 1983, **60**, 1009.
Zietz, J. R. Organoaluminum Compounds, in: "Ullmann's Encyclopedia of Industrial Chemistry" 5th Ed., A1, 1985, 543.
Mole, T., Jeffery, E. A., "Organoaluminum Compounds", Elsevier: Amsterdam, 1972.
Schmidbaur, H., Arene Complexes of Monovalent Gallium, Indium and Thallium, *Angew. Chem. Int. Ed. Engl.*, 1985, **24**, 893.

8.1

Aylett, B. J., "Organometallic Compounds", 4th Ed., Vol. 1, Part 2: Groups IV and V, Chapman and Hall: London, 1979.

Walsh, R., Bond Dissociation Energy Values in Silicon-containing Compounds and some of their Implications, *Acc. Chem. Res.*, 1981, **14**, 246.

Eaborn, C., Cleavage of Aryl-Silicon and Related Bonds by Electrophiles, *J. Organomet. Chem.*, 1975, **100,** 43.

Colvin, E. W., "Silicon in Organic Synthesis", Butterworths: London, 1981.

Reich, H., ed., Recent Developments in the Use of Silicon in Organic Synthesis, *Tetrahedron,* 1983, **39,** 839–1009.

Sakurai, H., Reactions of Allylsilanes and Applications to Organic Synthesis, *Pure Appl. Chem.*, 1982, **54,** 1.

Sommer, L. H., "Stereochemistry, Mechanism and Silicon", McGraw Hill: New York, 1965.

Corriu, R. J. P., Guerin, C., Moreau, J. J. E., Stereochemistry at Silicon, *Top. Sterochem.*, 1984, **15,** 43.

Speier, J. L., Homogeneous Catalysis of Hydrosilation by Transition Metals, *Adv. Organomet. Chem.,* 1979, **17,** 407.

Fritz, G., Carbosilanes, *Angew. Chem. Int. Ed. Engl.*, 1987, **26,** 1111.

West, R., The Polysilane High Polymers, *J. Organomet. Chem.*, 1986, **300,** 327.

McMahon, R. J., Organometallic π-Complexes of Silacycles, *Coord. Chem. Rev.*, 1982, **47,** 1.

Barton, T. J., Reactive Intermediates from Organosilacycles, *Pure Appl. Chem.*, 1980, **52,** 615.

Seyferth, D., The Elusive Silacyclopropanes, *J. Organomet. Chem.*, 1975, **100,** 237.

Harrison, J. F., The Structure of Methylene, *Acc. Chem. Res.*, 1974, **7,** 378.

Nefedov, O. M., Kolesnikov, S. P., Ioffe, A. I., Group IVb Carbene Analogs – Structure and Reactivity, *J. Organomet. Chem. Library,* 1977, **5,** 181.

Gusel'nikov, L. E., Nametkin, N. S., Formation and Properties of Unstable Intermediates Containing Multiple $p_\pi - p_\pi$ Bonded Group IVb Metals, *Chem. Rev.*, 1979, **79,** 529.

Schaefer III, H. F., The Silicon – Carbon Double Bond: A Healthy Rivalry between Theory and Experiment, *Acc. Chem. Res.*, 1982, **15,** 283.

West, R., Chemistry of the Silicon-Silicon Double Bond, *Angew. Chem. Int. Ed. Engl.*, 1987, **26,** 1201.

Masamune, S., Strained-Ring and Double-Bond Systems Consisting of the Group 14 Elements Si, Ge and Sn, *Angew. Chem. Int. Ed. Engl.* **30**, 1991, 920.

Miller, R. D., Michl, J., Polysilane High Polymers, Chem. Rev., 1989, **89,** 1359.

Corriu, R. J. P., Henner, M., The Siliconium Ion Question, *J. Organomet. Chem.*, 1974, **74,** 1.

Aylett, B. J., Some Aspects of Silicon-Transition-Metal Chemistry, *Adv. Inorg. Chem. Radiochem.*, 1982, **25,** 1.

8.2

Dräger, M., Organometallic Compounds with Homonuclear Group IVB – Group IVB Bonds, *Comments Inorg. Chem.* 1986, **5,** 201.

Satgé, J. Reactive Intermediates in Organogermanium Chemistry, *Pure Appl. Chem.*, 1984, **56,** 137.

Molloy, K. C., Zuckerman, J. J., Structural Organogermanium Chemistry, *Adv. Inorg. Chem. Radiochem.*, 1983, **27,** 113.

Satgé, J., Multiply Bonded Germanium Species, *Adv. Organomet. Chem.* 1982, **21,** 241.

Neumann, W. P., Germylenes and Stannylenes, *Chem. Rev.*, 1991, **91**, 311.

Neumann, W. P., Die organischen Verbindungen und Komplexe von Germanium, Zinn und Blei, *Naturwissenschaften,* 1981, **68,** 354.

Colomer, E., Corriu, R. J. P., Chemical and Stereochemical Properties of Compounds with Silicon- or Germanium-Transition Metal Bonds, *Top. Curr. Chem.*, 1981, **96,** 79.

8.3

Pereyre, M., Quintard, J. P., Organotin Chemistry for Synthetic Applications, *Pure Appl. Chem.*, 1981, **53,** 2401.

Pereyre, M., Quintard, J. P., Rahm, A., "Tin in Organic Synthesis", Butterworths: London, 1987.

Veith, M., Recktenwald, O., Structure and Reactivity of Monomeric, Molecular Tin(II) Compounds, *Top. Curr. Chem.*, 1982, **104,** 1

Connolly, J. W., Hoff, C., Organic Compounds of Divalent Tin and Lead, *Adv. Organomet. Chem.* 1981, **19,** 123.

Evans, C. J., Karpel, S., eds., "Organotin Compounds in Modern Technology", J. Organomet. Chem. Library, **16,** Elsevier: Amsterdam, 1985.

9

Aylett, B. J., "Organometallic Compounds". 4th Ed., Vol. 1, Part 2: Groups IV and V, Chapman and Hall: London, 1979.

Schmidbaur, H., Pentaalkyls and Alkylidene Trialkyls of the Group V Elements, *Adv. Organomet. Chem.*, 1976, **14** 205.

Hellwinkel, D., Penta- and Hexaorganyl Derivatives of the Main-Group V Elements, *Top. Curr. Chem.*, 1983, **109,** 1.

Okawara, R., Matsumura, Y., Recent Advances in Organoantimony Chemistry, *Adv. Organomet. Chem.*, 1976, **14,** 187.

McAuliffe, C. A., Levason, W., "Phosphine, Arsine and Stibine Complexes of the Transition Elements", Elsevier: Amsterdam, 1979.

Ashe III, A. J., The Group V Heterobenzenes: Arsabenzene, Stibabenzene and Bismabenzene, *Top. Curr. Chem.*, 1982, **105,** 125.

Ashe III, A. J., Thermochromic Distibines and Dibismuthines, Adv. Organomet. Chem., 1990, **30,** 77.

Sun, H., Hrovat, D. A., Borden, W. T., Why are π-Bonds to Phosphorus More Stable towards Addition Reactions than π-Bonds to Silicon?, *J. Am. Chem. Soc.*, 1987, **109,** 5275.

Cowley, A. H., Stable Compounds with Double Bonding between the Heavier Main-Group Elements, *Acc. Chem. Res.*, 1984, **17,** 386.

Scherer, O. J., Phosphorus, Arsenic, Antimony, and Bismuth Multiply Bonded Systems with Low Coordination Number – Their Role as Complex Ligands, *Angew. Chem. Int. Ed. Engl.*, 1985, **24,** 924.

10

Magnus, P. D., Organic Selenium and Tellurium Compounds, in: "Comprehensive Organic Chemistry", 1979, Vol. 3, 491.

Reich, H. J., Functional Group Manipulation Using Organoselenium Reagents, *Acc. Chem. Res.*, 1979, **12,** 22.

Paulmier, C., "Selenium Reagents and Intermediates in Organic Synthesis", Pergamon Press: Oxford, 1986.

Reich, H., Organoselenium Oxidation in: Trahanowsky, W. S., ed., "Oxidation in Organic Chemistry", Academic Press: New York, Part C, 1978, 1.

Bechgaard K., et al., Superconductivity in an Organic Solid: Bis(tetramethyltetraselenafulvalenium) perchlorate $(TMTSF)_2ClO_4$, *J. Am. Chem. Soc.*, 1981, **103,** 2440.

11

Posner, G. H., "An Introduction to Synthesis using Organocopper Reagents", Wiley: New York, 1980.

Normant, J. F., Stoichiometric versus Catalytic Use of Copper(I)-Salts in the Synthetic Use of Main-Group Organometallics, *Pure Appl. Chem.*, 1978, **50,** 709.

Normant J. F., et al., Organocopper Reagents for the Synthesis of Saturated and α,β-Ethylenic Aldehydes and Ketones. *Pure Appl. Chem.*, 1984, **56,** 91.

Fanta, P. E., Ullmann Biaryl Synthesis, *Synthesis*, 1974, 9.

Schmidbaur, H., Inorganic Chemistry with Ylids, *Acc. Chem. Res.*, 1975, **8,** 62.

Schmidbaur, H. Phosphorus Ylides in the Coordination Sphere of Transition Metals: An Inventory, *Angew. Chem. Int. Ed. Engl.* 1983, **22,** 907.

Anderson, G. K., The Organic Chemistry of Gold, *Adv. Organomet. Chem.*, 1982, **20,** 39.

TRANSITION-METAL ORGANOMETALLICS

12

Mitchell, P. R., Parish, R. V., The Eighteen Electron Rule, *J. Chem. Ed.*, 1969, **46,** 811.

Pilar, F. L., 4s is Always Above 3d, *J. Chem. Ed.*, 1978, **55,** 2.

13

Wilkinson, G., The Long Search for Stable Transition Metal Alkyls, *Science*, 1974, **185,** 109.

Schrock, R. R., Parshall, G. W., σ-Alkyl and σ-Aryl Complexes of the Group 4–7 Metals, *Chem. Rev.*, 1976, **76,** 243.

Bruce, M. I., Cyclometallation Reactions, *Angew. Chem. Int. Ed. Engl.*, 1977, **16,** 73.

Chappell, J. D., Cole-Hamilton, D. J., The Preparation and Properties of Metallacyclic Compounds of the Transition Elements, *Polyhedron*, 1982, **1,** 739.

Halpern, J., Determination and Significance of Transition-Metal-Alkyl Bond Dissociation Energies, *Acc. Chem. Res.*, 1982, **15,** 238.

Connor, J. A., Thermochemical Studies of Organo-Transition Metal Carbonyls, *Top. Curr. Chem.*, 1977, **71,** 71.

Martinho Simões, J. A.:, Beauchamp, J. L., Transition Metal Hydrogen and Metal-Carbon Bond Strengths: The Keys to Catalysis, Chem. Rev., 1990, **90,** 629.

Marks, T. J., "Bonding Energetics in Organometallic Compounds", ACS Symp. Ser., 1990, 428.

Marks T. J., Sonnenberger, D. C., Morss, L. R., Organo-*f*-Element Thermochemistry, *Organometallics*, 1985, **4,** 352.

Shriver, D. S., Transition Metal Basicity, Acc. Chem. Res., 1970, **3,** 231.

Abeles, R. H., Dolphin, D., The Vitamin B_{12} Coenzyme, *Acc. Chem. Res.*, 1976, **9,** 114.

Guengerich, F. P., Macdonald, T. L., Chemical Mechanisms of Catalysis by Cytochromes P-450. A Unified View, *Acc. Chem. Res.*, 1984, **17,** 9.

Ortiz de Montellano, P. R., ed., "Cytochrome P 450", Plenum Press: New York, 1986.

14.2

Nast, R., Coordination Chemistry of Metal Alkyl Compounds, *Coord. Chem. Rev.*, 1982, **47,** 89.

14.3

Schrock, R. R., Alkylidene Complexes of Niobium and Tantalum, *Acc. Chem. Res.,* 1979, **12,** 98.
Fischer, H., Kreissl, F. R., Schubert, U., Hofmann, P., Dötz, K. H., Weiss, K., "Transition Metal Carbene Complexes", VCH Publishers: Weinheim, 1984.
Nugent, W. A., Mayer, J. M., "Metal-Ligand Multiple Bonds. The Chemistry of Transition Metal Complexes Containing Oxo, Nitrido, Imido, Alkylidene, or Alkylidyne Ligands", Wiley: New York, 1988.
Taylor, T. E., Hall, M. B., Theoretical Comparison between Nucleophilic and Electrophilic Transitionmetal Carbenes, Using Generalized MO and CI Methods, *J. Am. Chem. Soc.,* 1984, **106,** 1576.
Dötz, K. H., Carbene Complexes in Organic Synthesis, *Angew. Chem. Int. Ed. Engl.,* 1984, **23,** 587.
Brookhart, M., Studabaker, W. B., Cyclopropanes from Reactions of Transition-Metal-Carbene Complexes with Olefins, *Chem. Rev.,* 1987, **87,** 411.

14.4

Schrock, R., High-Oxidation-State Molybdenum and Tungsten Alkylidyne Complexes, *Acc. Chem. Res.,* 1986, **19,** 342.
Roper, W. R., Platinum Group Metals in the Formation of Metal-Carbon Multiple Bonds, *J. Organomet. Chem.* 1986, **300,** 167.
Fischer H., Kreissl, F. R., Schrock, R. R., Schubert, U., Hofmann, P., Weiss, K., "Carbyne Complexes", VCH Publishers: Weinheim, 1988.

14.5

Abel, E. W., Stone, F. G. A., The Chemistry of Transition-Metal Carbonyls: Structural Considerations, *Quart. Rev.,* 1969, **23,** 325; Synthesis and Reactivity, *Quart. Rev.,* 1970, **24,** 498.
Johnson, J. B., Klemperer, W. G., A Molecular Orbital Analysis of Electronic Structure and Bonding in Chromium Hexacarbonyl, *J. Am. Chem. Soc.* 1977, **99,** 7132.
Wender, I., Pino, P., "Organic Syntheses via Metal Carbonyls", Wiley: New York, 1968, 1977, 2 volumes.
Cotton, F. A., Metal Carbonyls: Some New Observations in an Old Field, *Prog. Inorg. Chem.,* 1976, **21,** 1.
Dobson, G. R., Trends in Reactivity for Ligand-Exchange Reactions of Octahedral Metal Carbonyls, *Acc. Chem. Res.,* 1976, **9,** 300.
Basolo, F., Polyhedron, Kinetics and Mechanisms of CO Substitution of Metal Carbonyls, Polyhedron, 1990, **9**, 1503.
Werner, H., Complexes of Carbon Monoxide and Its Relatives: An Organometallic Family Celebrates its Birthday, *Angew. Chem. Int. Ed. Engl.*, 1990, **29**, 1077.
Hershberger, J. W., Klingler, R. J., Kochi, J. K., Electron-Transfer Catalysis, *J. Am. Chem. Soc.* 1982, **104,** 3034; 1983, **105,** 61.
Stiegman, A. E., Tyler, D. R., Reactivity of Seventeen- and Nineteen-Valence Electron Complexes in Organometallic Chemistry, *Comments Inorg. Chem.,* 1986, **5,** 215.
Bau R., et al., Structures of Transition-Metal Hydride Complexes, *Acc. Chem. Res.,* 1979, **12,** 176.
Pearson, R. G., The Transition-Metal Hydrogen Bond, *Chem. Rev.,* 1985, **85,** 41.
Collman, J. P., Disodium Tetracarbonylferrate – a Transition-Metal Analog of a Grignard Reagent, *Acc. Chem. Res.* 1975, **8,** 342.

14.6

Broadhurst, P. V., Transition Metal Thiocarbonyl Complexes, *Polyhedron,* 1985, **4,** 1801.

14.7

Singleton, E., Oosthuizen, H. E., Metal Isocyanide Complexes, *Adv. Organomet. Chem.,* 1983, **22,** 209.

Malatesta, L., Bonati, F., "Isocyanide Complexes of Metals", Academic Press: New York, 1969.

15.1

Herberhold, M., "Metal π-Complexes", Vol. 2, Elsevier: Amsterdam, 1974.

Mingos, D. M. P., Bonding of Unsaturated Organic Molecules to Transition Metals, in: "Comprehensive Organometallic Chemistry", 1982, Vol. 3, 1.

Akermark, B., Roos, B., et al., Chemical Reactivity and Bonding of Ni-ethene Complexes. An ab initio MO-SCF Study, *J. Am. Chem. Soc.,* 1977, **99,** 4617.

Deganello, G., "Transition-Metal Complexes of Cyclic Polyolefins", Academic Press: New York, 1979.

Körner von Gustorf, E. A., Grevels, F. W., Fischler, I., "The Organic Chemistry of Iron", Academic Press: New York, 1978, 1981, 2 volumes.

Stone, F. G. A., "Ligand-Free" Platinum Compounds, *Acc. Chem. Res.,* 1981, **14,** 318.

Albright, T. A., Rotational Barriers and Conformations in Transition-Metal Complexes, *Acc. Chem. Res.,* 1982, **15,** 149.

Brookhart, M., Green, M. L. H., Carbon-Hydrogen-Transition Metal Bonds, *J. Organomet. Chem.,* 1983, **250,** 395.

Crabtree, R. H., Hamilton, D. G., H – H, C – H, and Related sigma-Bonded Groups as Ligands, *Adv. Organomet. Chem.,* 1988, **28,** 299.

Bäckvall, J. E., Palladium in Some Selective Oxidation Reactions, *Acc. Chem. Res.* 1983, **16,** 335.

Darensbourg, D. J., Kudarosky, R. A., The Activation of Carbon Dioxide by Metal Complexes, *Adv. Organomet. Chem.,* 1983, **22,** 132.

15.2

Vollhardt, K. P. C., Cobalt-Mediated [2 + 2 + 2]-Cycloadditions: A Maturing Synthetic Strategy, *Angew. Chem. Int. Ed. Engl.* 1984, **23,** 539.

Bönnemann, H., Organocobalt Compounds in the Synthesis of Pyridines – An Example of Structure-Effectivity Relationships in Homogeneous Catalysis, *Angew. Chem. Int. Ed. Engl.,* 1985, **24,** 248.

Nicholas, K. M., Chemistry and Synthetic Utility of Cobalt-Complexed Propargyl-Cations, *Acc. Chem. Res.,* 1987, **20,** 207.

Pauson, P. L., The Khand-Reaction, Tetrahedron, 1985, **41,** 5855.

Schore, N. E., Transition-Metal-Mediated Cycloaddition Reactions of Alkynes in Organic Synthesis, *Chem. Rev.,* 1988, **88,** 1081.

15.3

Wilke G., et al., π-Allyl Metal Complexes, *Angew. Chem. Int. Ed. Engl.,* 1966, **5,** 151.

Davies, S. G., Green, M. L. H., Mingos, D. M. P., Nucleophilic Addition to Organotransition Metal Cations Containing Unsaturated Hydrocarbon Ligands – A Survey and Interpretation, *Tetrahedron,* 1978, **34,** 3047.

Ernst, R. D., Metal-Pentadienyl Chemistry, *Acc. Chem. Res.* **18,** 56.
Pearson, A. J., Tricarbonyl(diene)iron Complexes: Synthetically Useful Properties, *Acc. Chem. Res.,* 1980, **13,** 463.
Tsuji, J., 25 Years in the Organic Chemistry of Palladium, *J. Organomet. Chem.,* 1986, **300,** 281.

15.4.2

Efraty, A., Cyclobutadiene Metal Complexes, *Chem. Rev.,* 1977, **77,** 691.

15.4.3

Wilkinson, G., The Iron Sandwich. A Recollection of the First Four Months, *J. Organomet. Chem.,* 1975, **100,** 273.
Pauson, P. L., Aromatic Transition-Metal Complexes – the First 25 Years, *Pure Appl. Chem.* 1977, **49,** 839.
Warren, K. D., Ligand Field Theory of Metal Sandwich Complexes, *Struct. Bonding,* 1976, **27,** 45.
Raymond, K. N., Eigenbrot, Jr., C. W., Structural Criteria for the Mode of Bonding of Organoactinides and -lanthanides and Related Compounds, *Acc. Chem. Res.,* 1980, **13,** 276.
Elian, M., Chen, M. M. L., Mingos, D. M. P., Hoffmann, R., Comparative Bonding Study of Conical Fragments, *Inorg. Chem.,* 1976, **15,** 1148.
Lauher, J. W., Elian, M., Summerville, R. H., Hoffmann, R., Triple-Decker Sandwiches, *J. Am. Chem. Soc.* 1976, **98,** 3219.
Werner, H., New Varieties of Sandwich Complexes, *Angew. Chem. Int. Ed. Engl.,* 1977, **16,** 1.
Werner, H., Electron-Rich Half-Sandwich Complexes – Metal Bases *par excellence*, *Angew. Chem. Int. Ed. Engl.,* 1983, **22,** 927.
Poli, R., Monocyclopentadienyl Halide Complexes of the d- and f-block Elements, *Chem. Rev.,* 1991, **91**, 509.
Jonas, K., Reactive Organometallic Compounds form Metallocenes, *Pure Appl. Chem.* 1984, **56,** 63; Reactive Organometallic Compounds Obtained from Metallocenes and Related Compounds and Their Synthetic Applications, *Angew. Chem. Int. Ed. Engl.,* 1985, **24,** 295.
Schwartz, J., Organozirconium Compounds in Organic Synthesis: Cleavage Reactions of Carbon-Zirconium Bonds, *Pure Appl. Chem.,* 1980, **52,** 733.
Watts, W. E., Ferrocenylcarbocations and Related Species, *J. Organomet. Chem. Libr.,* 1979, **7,** 399.
Dombrowski, K. E., Baldwin, W., Sheats, J. E., Metallocenes in Biochemistry, Microbiology and Medicine, *J. Organomet. Chem.* 1986, **302**, 281.
Lauher, J. W., Hoffmann, R., Structure and Chemistry of Bis(cyclopentadienyl)-ML_n Complexes, *J. Am. Chem. Soc.,* 1976, **98,** 1729.
Bursten, B. E., Strittmatter R. J., Cyclopentadienyl – Actinide Complexes: Bonding and Electronic Structure, *Angew. Chem. Int. Ed. Engl.*, 1991, **30**, 1069.

15.4.4

Silverthorn, W. E., Arene Transition Metal Chemistry, *Adv. Organomet. Chem.,* 1975, **13,** 47.
Gastinger, R. G., Klabunde, K. J., π-Arene Complexes of the Group VIII Transition Metals, *Transition Met. Chem.* 1979, **4,** 1.
Green, M. L. H., The Use of Atoms of the Group IV, V, VI Transition Metals for the Synthesis of Zerovalent Arene Compounds, *J. Organometal. Chem.* 1980, **200,** 119.
Clack, D., Warren, K. D., Metal-Ligand Bonding in 3d Sandwich Complexes, Struct. Bonding 1980, **39,** 1.

Muetterties, E. L., Bleeke, J. R., Wucherer, E. J., Albright, T. A., Structural, Stereochemical and Electronic Features of Arene-Metal Complexes, *Chem. Rev.*, 1982, **82,** 499.

Albright, T. A., Hofmann, P., Hoffmann, R., Lillya, C. P., Dobosh, P. A., Haptotropic Rearrangements of Polyene-ML_n Complexes, *J. Am. Chem. Soc.*, 1983, **105,** 3396.

15.4.6

Deganello, G., "Transition-Metal Complexes of Cyclic Polyolefins", Academic Press: New York, 1979.

Marks, T. J., Chemistry and Spectroscopy of *f*-Element Organometallics, *Prog. Inorg. Chem.*, 1979, **25,** 223.

Raymond, K. N., Eigenbrot, Jr., C. W., Structural Criteria for the Mode of Bonding of Organoactinides and -lanthanides and Related Compounds, *Acc. Chem. Res.*, 1980, **13,** 276.

Bursten, B. E., Burns, C. J., Covalency in f-Element Organometallic Complexes: Theory and Experiment, *Comments Inorg. Chem.*, 1989, **2**, 61.

15.5

Pannell, K. H., Kalsotra, B. L., Parkanyi, C., Heterocyclic π-Complexes of the Transition Metals, *J. Heterocyclic Chem.*, 1978, **15,** 1057.

Mathey, F. H., Fischer, J., Nelson, J. H., Complexing Modes of the Phosphole Moiety, *Struct. Bonding,* 1983, **55,** 153.

Grimes, R. N., Metal Sandwich Complexes of Cyclic Planar and Pyramidal Ligands Containing Boron, *Coord. Chem. Rev.*, 1979, **28,** 47.

Herberich, G. E., Boron Ring Systems as Ligands to Metals, in: "Comprehensive Organometallic Chemistry", 1982, Vol. 1, 381.

Siebert, W., Boron Heterocycles as Ligands in Transition-Metal Chemistry, *Adv. Organomet, Chem.*, 1980, **18,** 301.

Siebert, W., 2,3-Dihydro-1,3-diborole-Metal Complexes with Activated C–H Bonds: Building Blocks for Multilayered Sandwich Compounds, *Angew. Chem. Int. Ed. Engl.*, 1985, **24,** 943.

Herberich G. E., et al., (η^5-Borol)metall-Komplexe, *J. Organomet. Chem.*, 1987, **319,** 9.

16

Cotton, F. A., Quadruple Bonds and Other Metal to Metal Bonds, *Chem. Soc. Rev.*, 1975, **4,** 27.

Chisholm, M. H., ed., "Reactivity of Metal-Metal Bonds", ACS Symposium Series **155**: Washington, 1981.

Cotton, F. A., Walton, R. A., "Multiple Bonds Between Metal Atoms", John Wiley: New York, 1982.

Cotton, F. A., Multiple Metal-Metal Bonds, *J. Chem. Ed.*, 1983, **60,** 713.

Cotton, F. A., Chisholm, M. H., Bonds between Metal Atoms – A New Mode of Transition-Metal Chemistry, *Chem. Eng. News,* 1982, June 28, 40.

Gerloch, M., A Local View in Magnetochemistry, *Prog. Inorg. Chem.*, 1979, **26,** 1.

Muetterties, E. L., Molecular Metal Clusters, *Science,* 1977, **196,** 839.

Mingos, D. M. P., "Introduction to Cluster Chemistry", Prentice Hall: Englewood Cliffs, 1990.

Mingos, D. M. P., Bonding Models for Ligated and Bare Clusters, *Chem. Rev.,* 1990, **90**, 383.

Shriver, D. F., Kaesz, H. D., Adams, R. D., eds., "The Chemistry of Metal Cluster Complexes", VCH Publishers, Weinheim, 1990.

Chini, P., Large Metal Carbonyl Clusters, *J. Organomet. Chem.*, 1980, **200,** 37.

Johnson, B. F. G., Lewis, J., Transition-Metal Molecular Clusters, *Adv. Inorg. Chem. Radiochem.* 1981, **24,** 225.

Johnson, B. F. G., ed., Recent Advances in the Structure and Bonding of Cluster Compounds, *Polyhedron,* 1984, **3,** 1277.

Johnson, B. F. G., The Structure of Simple Binary Carbonyls, *JCS Chem. Commun.,* 1976, 211.

Johnson, B. F. G., Benfield, R. E., Stereochemistry of Transition-Metal Carbonyl Clusters, *Topics in Stereochem.,* 1981, **12,** 253.

Hoffmann, R., Building Bridges Between Inorganic and Organic Chemistry (Nobel Lecture), *Angew. Chem. Int. Ed. Engl.,* 1982, **21,** 711.

Gray, H. B., Maverick, A. W., Solar Chemistry of Metal Complexes, *Science,* 1981, **214,** 1201.

Stone, F. G. A., Metal-Carbon and Metal-Metal Multiple Bonds as Ligands in Transition-Metal Chemistry: The Isolobal Connection, *Angew. Chem. Int. Ed. Engl.,* 1984, **23,** 89.

Vahrenkamp, H., Basic Metal Cluster Reactions, *Adv., Organomet. Chem.,* 1983, **22,** 169.

Geiger, W. E., Connelly, N. H., The Electron-Transfer Reactions of Polynuclear Organotransition-Metal Complexes, *Adv. Organomet. Chem.,* 1985, **24,** 87.

Bradley, J. C., The Chemistry of Carbido-Carbonyl Clusters, *Adv. Organomet. Chem.,* 1983, **22,** 1.

Kaesz, H. D., An Account of Studies into Hydrido-Metal Complexes and Cluster Compounds, *J. Organomet. Chem.,* 1980, **200,** 145.

Gladfelter, W. L. Organometallic Metal Clusters Containing Nitrosyl and Nitrido Ligands, *Adv. Organomet. Chem.,* 1985, **24,** 41.

17

Tolman, C. A., The 16 and 18 Electron Rule in Organometallic Chemistry and Homogeneous Catalysis. *Chem. Soc. Rev.,* 1972, **1,** 337.

Hartley, F. R., "Supported Metal Complexes. A New Generation of Catalysts", Reidel: Dordrecht, 1985.

Kochi, J. K., "Organometallic Mechanisms and Catalysis", Academic Press: New York, 1978.

Kochi, J. K., Electron Transfer and Transient Radicals in Organometallic Chemistry, *J. Organomet. Chem.,* 1986, **300,** 139.

Bishop III, K. C., Transition Metal Catalysed Rearrangements of Small Ring Organic Molecules, *Chem. Rev.,* 1976, **76,** 461.

Astruc, D., Electron-Transfer Chain Catalysis in Organotransition Metal Chemistry, *Angew. Chem. Int. Ed. Engl.* 1988, **27,** 643.

Parshall, G. W., Industrial Applications of Homogeneous Catalysis. A Review, *J. Mol. Catal,* 1978, **4,** 243.

Masters, C., "Homogeneous Transition-Metal Catalysis – a Gentle Art", Chapman and Hall: London, 1981.

Stone, F. G. A., West, R., eds., Catalysis and Organic Synthesis, *Adv. Organomet. Chem.,* 1979, **17,** 1–492.

Waller, J. F., Recent Achievements, Trends and Prospects in Homogeneous Catalysis. *J. Mol. Catal.,* 1985, **31,** 123.

Casey, C. P., ed., Symposium: Industrial Applications of Organometallic Chemistry and Catalysis, *J. Chem. Ed.,* 1986, **63,** 188.

Heck, R. F., Palladium-Catalysed Reactions of Organic Halides with Olefins, *Acc. Chem. Res.,* 1979, **12,** 146.

Herrison, J.-P., Chauvin, Y., Catalyse de transformation des Oléfines par les complexes de tungstène, *Makromol. Chem.* 1970, **141,** 161.

Grubbs, R. H., Alkene and Alkyne Metathesis Reactions, in: "Comprehensive Organometallic Chemistry", 1982, Vol. 8, 499.

Schrock, R. R., On the Trail of Metathesis Catalysts, *J. Organomet. Chem.* 1986, **300,** 249.

Jolly, P. W., Wilke, G., "The Organic Chemistry of Nickel", Academic Press: New York, 1974, 1975, 2 volumes.

Sinn, H., Kaminsky, W., Ziegler-Natta Catalysis, *Adv. Organomet. Chem.,* 1980. **18,** 99.

Baeckvall, J. E., Akermark, B., Ljunggren, S. O., Stereochemistry and Mechanism for the Palladium(II)-Catalysed Oxidation of Ethene in Water (the Wacker-Process), *J. Amer. Chem. Soc.* 1979, **101,** 2411.

Knowles, W. S., Asymmetric Hydrogenation, *Acc. Chem. Res.* 1983, **16,** 106.

Bosnich, B. S., Asymmetric Synthesis Mediated by Transition-Metal Complexes, *Topics Stereochem.,* 1981, **12,** 119.

Bosnich, B., Asymmetric Catalysis, *Chem. in Britain,* 1984, **20,** 808.

Brunner, H., Enantioselective Catalysis with Transition-Metal Complexes, *J. Organomet. Chem.,* 1986, **300,** 39.

Nogradi, M., "Stereoselective Synthesis", VCH Publishers: Weinheim, 1987.

Muetterties, E. L., Stein, J., Mechanistic Features of Catalytic Carbon Monoxide Hydrogenation Reactions, *Chem. Rev.,* 1979, **79,** 479.

Ford, P. C., Rokocki, A., Nucleophilic Activation of Carbon Monoxide: Applications to Homogeneous Catalysis by Metal Carbonyls of the Water Gas Shift and Related Reactions, *Adv. Organomet. Chem.,* 1988, **28,** 139.

Behr, A., Carbon Dioxide as an Alternative C_1 Synthetic Unit: Activation by Transition-Metal Complexes, *Angew. Chem. Int. Ed. Engl.* 1988, **27,** 661.

Braunstein, P., Reactions of Carbon Dioxide with Carbon-Carbon Bond Formation Catalyzed by Transition-Metal Complexes, *Chem. Rev.* 1988, **88,** 747.

Henderson, S., Henderson, R. A., The Nucleophilicity of Metal-Complexes Toward Organic Molecules, *Adv. Phys. Org. Chem.,* 1987, **23,** 1.

Tolman, C. A., Hydrocyanation, *J. Chem. Ed.,* 1986, **63,** 199.

Constable, E. C., Cyclometallated Complexes Incorporating a Heterocyclic Donor Atom, *Polyhedron,* 1984, **3,** 1037.

Bergman, R. G., Activation of Alkanes with Organotransition-Metal Complexes, *Science,* 1984, **223,** 902.

Halpern, J., Activation of C–H Bonds by Metal Complexes: Mechanistic, Kinetic and Thermodynamic Considerations, *Inorg. Chim. Acta,* 1985, **100,** 41.

Crabtree, R. H., Organometallic Chemistry of Alkanes, *Chem. Rev.,* 1985, **85,** 245.

Watson, P. L., Parshall, G. W., Organolanthanoids in Catalysis, Acc. Chem. Res. 1985, **18,** 51.

Nomenclature

Leigh, G. J., ed., "Nomenclature of Inorganic Chemistry: Recommendations 1990 (IUPAC)", Blackwell Scientific Publications: Oxford, 1990

Appendix
Symbols and Abbreviations

a	isotropic hyperfine coupling constant (ESR)
ab	antibonding
Ac	acetyl
acac	acetylacetonate
AIBN	azoisobutyronitrile
AN	acetonitrile
AO	atomic orbital
Ar_F	perfluoroaryl
Ar	aryl
b	bonding
bipy	2,2′-bipyridyl
B.M.	Bohr magneton
bp	boiling point (°C)
Bu	1-butyl
Bz	benzyl
CC	cocondensation
c-Hex	cyclohexyl
C.N.	coordination number
COD	1,5-cyclooctadiene
COMC	Comprehensive Organometallic Chemistry
COT	cyclooctatetraene
Cp	cyclopentadienyl
Cp*	pentamethylcyclopentadienyl
D	bond dissociation energy (kJ/mol)
d	bond length (pm)
dmpe	1,2-bis(dimethylphosphino)-ethane
dppe	1,2-bis(diphenylphosphino)-ethane
δ	chemical shift (NMR)
Δ_0	splitting t_{2g}/e_g (octahedral field)
DBPO	di-*t*-butylperoxide
DME	1,2-dimethoxyethane
DMF	dimethylformamide
DMSO	dimethylsulfoxide
E	element
E	energy
E^0	standard reduction potential for $Ox + e^- \rightarrow Red$
EH	Extended Hückel
El	electrophile
$\bar{E}$	mean bond enthalpy (kJ/mol)
EN	electronegativity
en	ethylenediamine
ET	electron transfer
EPR	electron spin resonance (ESR)
$2e3c$	two electron three center bond
EXT	extrusion (= deinsertion)
(g)	gas
g	g-value (EPR)
γ	magnetogyric ratio
HMPA	hexamethylphosphoric tris-amide
HOMO	highest occupied molecular orbital
INS	insertion
J	scalar nuclear spin-spin coupling constant (NMR)
(l)	liquid
LUMO	lowest unoccupied molecular orbital
M	metal
Me	methyl
MGE	main-group element
MO	molecular orbital (method)
MOCVD	metal organic chemical vapor desposition

mp	melting point (°C)
n	normal
Nu	nucleophile
ν	stretching frequency (cm^{-1})
nb	nonbonding
OA	oxidative addition
Ox	oxinate
OS	oxidation state
Ph	phenyl
Pr	1-propyl
PE	petrol ether
py	pyridine
r	radius (pm)
RE	reductive elimination
R_F	perfluoroalkyl
RT	room temperature
σ	shielding (NMR)
(s)	solid
SCE	saturated calomel electrode
t	tertiary, ton
T	temperature
T	Tesla
$t_{1/2}$	half life
THF	tetrahydrofuran
TM	transition metal
TMEDA	*N*,*N*,*N'*,*N'*-tetramethylethylenediamine
Vi	vinyl
VSEPR	valence shell electron pair repulsion (Gillespie-Nyholm-model)
VE	valence electron
VB	valence bond (method)

Nomenclature of Organometallic Compounds

Organometallic compounds containing only $M \overset{\sigma}{-} C$ bonds are named according to the IUPAC Rules D-3. Ligands are listed in alphabetical order, followed by the name of the metal (cations taking precedence over anions):

$(CH_3)_2(C_2H_5)_2Sn$	Diethyldimethyltin

Group 14 and 15 organyls can also be named by considering them to be derivatives of their binary hydrides (ending -ane, analogues to alkanes):

$(CH_3)_3P$	Trimethylphosphane

Anionic complexes are given the ending -ate, together with the charge on the anion or the oxidation number of the central metal atom:

$Na[(C_2H_5)_3Sn]$	Sodium triethylstannate (1−) or (II)

Cations are also characterized by a charge or oxidation number:

$[(C_6H_{11})_3Sn]^+$	Tricyclohexyltin (1+) or (IV)

Complexes containing unsaturated hydrocarbons coordinated to a metal require a special nomenclature; this is also found in the IUPAC rules (volume 2, group 1, Rule 7.4). In π-complexes, at least two carbons of the ligand are bonded to the metal. However, use of the expression "π-coordinated" alone is too imprecise, since the exact nature of the bonding (σ, π) is often uncertain. Therefore, it seems wise to indicate the atoms bonded to the metal atom in a manner completely independent of theoretical implications (F. A. Cotton).

The precision with which a compound is named is largely determined by the complexity of its structure.

1. Designation of Stoichiometric Composition Only

$K[PtCl_3(C_2H_4)]$	Potassium trichloro(ethylene)platinate (1−)
$[Fe(C_5H_5)_2]$	Di (cyclopentadienyl)iron
$[Fe(CO)_3C_8H_8]$	Tricarbonyl(cyclooctatetraene)iron

2. Designation of Structure

If all the unsaturated C atoms are coordinated to the metal, the name of the ligand is preceeded by η- (pronounced eta). The number of coordinated C atoms can be additionally indicated by a numerical right superscript and η then reads as -hapto (e.g., η^3 = trihapto, η^4 = tetrahapto, η^5 = pentahapto etc.).

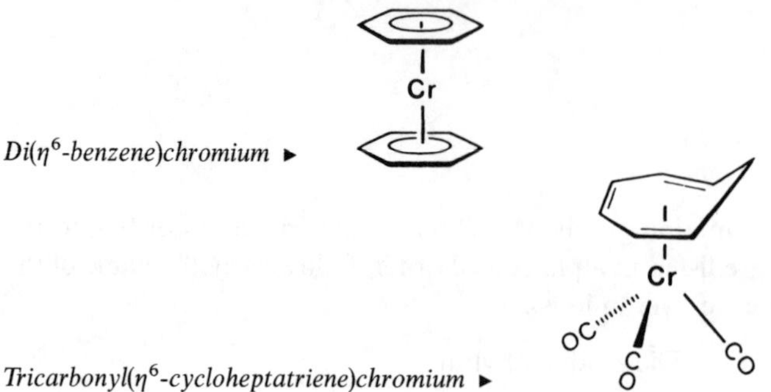

Di(η^6-benzene)chromium ▸

Tricarbonyl(η^6-cycloheptatriene)chromium ▸

3. Exact Designation of the Coordinated Atoms

If not all the unsaturated C atoms of a ligand are involved in bonding, or if a ligand possesses several bonding modes, the locations of the ligating atoms appear in numerical sequence before the hapto symbol η. If the coordinated carbon atoms are in a consecutive sequence, only the first and last positions are indicated.

◂ (1,2 : 5,6-*η-Cyclooctatetraene*) (*η^5-cyclopentadienyl*)*cobalt*

Tricarbonyl(4-7-*η-octa*-2,4,6-*trienal*)*iron* ▸

When it is desired to stress that a ligand is bonded to a single atom, the prefix σ- may be used:

Dicarbonyl(η^5-cyclopentadienyl)-(σ-cyclopentadienyl)iron ▸

Ligands which coordinate simultaneously to two or more metals (bridging ligands) are given the prefix μ (mu). A metal-metal bond is indicated by italicized atomic symbols enclosed in parentheses at the end of the name:

Di-μ-carbonyl-bis(tricarbonyl)-cobalt (Co—Co) ▸

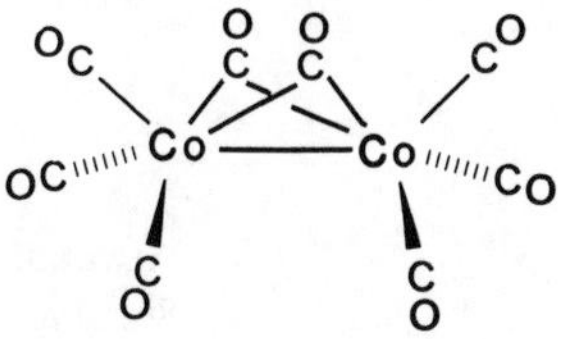

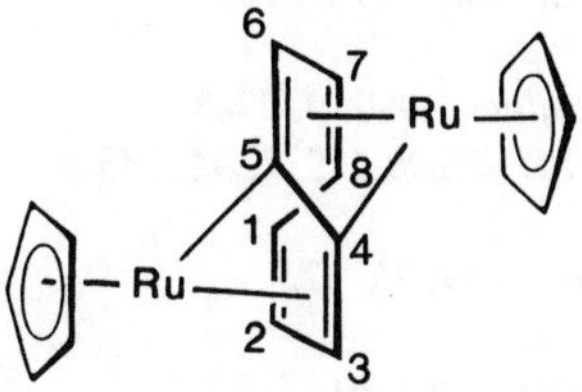

◂ *trans-μ*[1-5 : 4-8-*η-Cyclooctatetraene*]-*bis* [(*cyclopentadienyl*)*ruthenium*]

For further details relating to stereochemical nomenclature and notation see T. E. Sloan in “Comprehensive Coordination Chemistry”, Vol. 1, 1987, 109.

Author Index

Ahlrichs, R.
22: *Chem. Phys. Lett.*, 1986, **2**, 172
Albano, V. G.
409: *J. Chem. Soc. (Dalton)* 1973, 651
Allegra, G.
360: *Ric. Sci. Rend.*, 1961, **1**, (IIA) 362
Allred, A. L., 8
Alt, H.
250: *Angew. Chem., I. E.*, 1984, **23**, 766
Amma, E. L.
136: *Inorg. Chem.*, 1979, **18**, 751
146: *Inorg. Chem.*, 1974, **13**, 2429
Ammeter, J. H.
325: *J. Am. Chem. Soc.*, 1974, **96**, 7833
Andersen, R. A.,
443: *J. Am. Chem. Soc.*, 1986, **108**, 335
Andrews, L.
22: *J. Chem. Phys.* 1967, **47**, 4834
Andrews, M. A.
240: *Inorg. Chem.* 1977, **16**, 496
Angelici, R. J.
238: *J. Am. Chem. Soc.*, 1968, **90**, 3282
238: *J. Am. Chem. Soc.*, 1973, **95**, 7516
239: *Inorg. Chem.*, 1987, **26**, 452
239: *Inorg. Chem.*, 1977, **16**, 1173
Arnett, E. M.
277: *J. Am. Chem. Soc.*, 1964, **86**, 4729
Ashby, E. C.
42: *Pure Appl. Chem.*, 1980, **52**, 545
Ashe, A. J.
64: *J. Am. Chem. Soc.*, 1971, **93**, 1804
99: *J. Am. Chem. Soc.*, 1970, **92**, 1233
158: *Organometallics*, 1984, **3**, 337
161: *J. Am. Chem. Soc.* 1971, **93**, 3293
377: *Angew. Chem. I. E.*, 1987, **26**, 229
379: *J. Am. Chem. Soc.*, 1977, **99**, 8099
382: *J. Am. Chem. Soc.*, 1975, **97**, 6865
Astruc, D.
356: *Tetrahedron*, 1983, **39**, 4027
Atherton, N. M.
36: *Trans. Faraday Soc.*, 1966, **62**, 1702
Atwood, J. L.
86: *Angew. Chem. I. E.*, 1987, **26**, 485

Baeckvall, J. E.
425: *J. Am. Chem. Soc.*, 1979, **101**, 2411
Baird, M. C.
333: *Organometallics*, 1990, **9**, 2248
356: *J. Organomet. Chem.*, 1972, **44**, 383
Barbier, P.
2: *J. Chem. Soc.*, 1899, **76**, 323
Bartell, L. S.
89: *J. Chem. Phys.*, 1964, **41**, 717
Barton, T. J.
108: *J. Organomet. Chem.*, 1972, **42**, C21
110: *J. Organomet. Chem.*, 1979, **179**, C17
120: *J. Am. Chem. Soc.*, 1973, **95**, 3078
Barton, D. H. R.
169: *J. Chem. Soc. Chem. Comm.*, 1975, 539
Bau, R.
113: *J. Am. Chem. Soc.*, 1987, **109**, 5123
258: *Inorg. Chem.*, 1975, **14**, 2653
406: *Angew. Chem. I. E.*, 1979, **18**, 80
Baudler, M.
159: *Angew. Chem. I. E.*, 1985, **24**, 991
Beauchamp, J. L.
198: *Chem. Rev.*, 1990, **90**, 629
Bechgaard, K.
167: *J. Am. Chem. Soc.*, 1981, **103**, 2440
Beachley, Jr., O. T.
90: *J. Am. Chem. Soc.*, 1986, **108**, 4666
Becker, G.
164: *J. Mol. Struct.*, 1981, **75**, 283
Behrens, U.
326: *Angew. Chem. I. E.*, 1987, **26**, 147
329: *J. Organomet. Chem.*, 1979, **182**, 89
Benn, R.
43: *Chem. Ber.*, 1986, **119**, 1054
84: *J. Organomet. Chem.*, 1987, **333**, 169
370: *Angew. Chem. I. E.*, 1986, **25**, 861
375: *Organometallics*, 1985, **4**, 2214

Bennett, M. A.
275: *Angew. Chem. I. E.*, 1988, **27**, 941
355: *J. Organomet. Chem.*, 1979, **175**, 87
Bent, H. A.
9: *J. Chem. Ed.*, 1960, **37**, 616
260: *J. Chem. Ed.*, 1960, **37**, 616
Bercaw, J. E.
323: *J. Am. Chem. Soc.*, 1978, **100**, 3078
432: *J. Am. Chem. Soc.*, 1976, **98**, 6733
Bergman, R. G.
4: *J. Am. Chem. Soc.*, 1983, **105**, 3929
424: *J. Am. Chem. Soc.*, 1979, **101**, 3973
440: *J. Am. Chem. Soc.*, 1982, **104**, 352
Berndt, A.
63: *Angew. Chem. I. E.*, 1985, **24**, 788
Berke, H.
217: *Z. Naturforsch.*, 1980, **35b**, 86
Berry, D. H.
111: *J. Am. Chem. Soc.*, 1990, **112**, 452
Bertheim, A., 158
Bianchi, M.
406: *J. Chem. Soc. Chem. Comm.*, 1969, 596
Bickelhaupt, F.
46: *Heterocycles*, 1977, **7**, 237
164: *Angew. Chem. I. E.*, 1979, **18**, 395
Binger, P.
376: *Angew. Chem. I. E.*, 1986, **25**, 644
Birch, A. J.
253: *J. Chem. Soc. (A)*, 1968, 332
Birnbaum, K.
252: *Ann. Chem.*, 1868, **145**, 67
Bishop, K. C.
416: *Chem. Rev.*, 1976, **76**, 461
Blom, R.
47: *Acta. Chem. Scand.*, 1987, **A41**, 24;
J. Chem. Soc. Chem. Comm., 1987, 768
Boche, G.
23: *Angew. Chem. I. E.*, 1986, **25**, 104
Bock, H.
107: *J. Am. Chem. Soc.*, 1979, **101**, 7667
109: *J. Am. Chem. Soc.*, 1980, **102**, 429
Boeckelheide, V.
355: *Tetrahedron Lett.* 1980, 4405
Bogdanovic, B.
41: *Angew. Chem. I. E.*, 1980, **19**, 818
41: *Angew. Chem. I. E.*, 1990, **29**, 223
Bonati, F.
181: *J. Organomet. Chem.*, 1973, **59**, 403
Borden, W. T.
162: *J. Am. Chem. Soc.*, 1987, **109**, 5275
Bosnich, B.
429: *J. Am. Chem. Soc.*, 1977, **99**, 6262
Boudjouk, P.
110: *Chem. Abstr.*, 1984, **105**, 6549n
Boyd, T. A.
3: *Ind. Engl. Chem.*, 1922, **14**, 894
Bönnemann, H.
277: *Angew. Chem. I. E.*, 1985, **24**, 248
277: *Synthesis*, 1974, 575
375: *J. Organomet. Chem.*, 1984, **272**, 231
Breslow, D. S.
434: *J. Am. Chem. Soc.*, 1961, **83**, 4023
Breslow, R.
310: *J. Am. Chem. Soc.*, 1957, **79**, 5318
Breunig, H. J.
159: *Z. Anorg. Allg. Chem.*, 1983, **502**, 175
Braunstein, P.
272: *Chem. Rev.*, 1988, **88**, 747
Brintzinger, H. H.
245: *J. Organomet. Chem.*, 1977, **127**, 87
425: *Angew. Chem. I. E.*, 1985, **24**, 507
Brockway, L. O.
226: *Angew. Chem.*, 1964, **76**, 553
Brook, A.
109: *J. Chem. Soc. Chem. Comm.*, 1981, 191
Brookhart, M. S.
215: *Chem. Rev.*, 1987, **87**, 411
268: *Prog. Inorg. Chem.*, 1988, **36**, 1
441: *Prog. Inorg. Chem.*, 1988, **36**, 1
Brown, D. S.
52: *Acta Cryst. B*, 1978, **34**, 1695
Brown, H. C.
3: *J. Am. Chem. Soc.*, 1956, **78**, 5694
4: *Angew. Chem.*, 1980, **92**, 675
60: *J. Am. Chem. Soc.*, 1956, **78**, 5694
61: *J. Am. Chem. Soc.*, 1961, **81**, 486
Brown, J. M.
428: *J. Chem. Soc. Chem. Comm.*, 1984, 914
Brown, T. L.
21: *Pure Appl. Chem.*, 1970, **23**, 447
247: *J. Am. Chem. Soc.*, 1985, **107**, 5700
249: *J. Am. Chem. Soc.*, 1977, **99**, 2527
434: *J. Am. Chem. Soc.*, 1980, **102**, 2494
Bruce, M. I.
439: *Angew. Chem. I. E.*, 1977, **16**, 73
Brunner, H.
338: *Angew. Chem. I. E.*, 1971, **10**, 249
429: *Angew. Chem. I. E.*, 1979, **18**, 620
Buckingham, A. D.
307: *J. Chem. Soc.*, 1964, 2747

Reihlen, H.
253: *Ann.*, 1930, **482**, 161
Reppe, W.
3: *Ann*, 1948, **560**, 1
278: *Ann*, 1948, **560**, 1
421: *Ann*, 1948, **560**, 1
Reutov, O. A.
16: *Russ. Chem. Rev.*, 1976, **45**, 330
Rheingold, A. L.
159: *Inorg. Chem.*, 1973, **12**, 2845
160: *J. Am. Chem. Soc.*, 1982, **104**, 4727
368: *J. Am. Chem. Soc.*, 1984, **106**, 3052
Rieger, P. H.
414: *J. Chem. Soc. Chem. Comm.*, 1981, 265
Rieke, R. D.
41: *Acc. Chem. Res.*, 1977, **10**, 301
Rittner, E. S.
443: *J. Chem. Phys.* 1951, **19**, 1030
Rochow, E. G.
3: *J. Chem. Ed.*, 1965, **42**, 41
96: *J. Chem. Ed.*, 1965, **42**, 41
100: *J. Chem. Ed.*, 1965, **42**, 41
101: *J. Chem. Ed.*, 1965, **42**, 41
Roelen, O.
3: *Angew. Chem.*, 1948, **60**, 62
434: *German Patent*, 1938, 849 548
Roos, B.
256: *J. Am. Chem. Soc.*, 1977, **99**, 4617
Roper, W. D.
211: *J. Am. Chem. Soc.*, 1983, **105**, 5939
218: *J. Organomet. Chem.*, 1986, **300**, 167
219: *J. Am. Chem. Soc.*, 1980, **102**, 1206
239: *J. Am. Chem. Soc.*, 1980, **102**, 1206
270: *J. Am. Chem. Soc.*, 1979, **101**, 503
271: *J. Organomet. Chem.*, 1978, **159**, 73
Rosenblum, M.
267: *Acc. Chem. Res.*, 1974, **7**, 122
328: *J. Am. Chem. Soc.*, 1966, **88**, 4178
328: *J. Am. Chem. Soc.*, 1960, **82**, 5249
Rotruck, J. T.
171: *Science*, 1973, **179**, 588
Rudolph, R. W.
69: *Inorg. Chem.*, 1972, **11**, 1974
Rundle, R. E.
39: *J. Chem. Phys.*, 1950, **18**, 1125
360: *J. Am. Chem. Soc.*, 1963, **85**, 481
Rybinskaya, M. I.
332: *J. Organomet. Chem.*, 1987, **336**, 187
Sabatier, P., 2
Sacconi, L.
311: *Angew. Chem. I. E.*, 1980, **19**, 931
Sakurai, H.
64: *Chem. Lett.*, 1987, 1451
98: *Tetrahedron Lett.*, 1978, 3043
116: *Angew. Chem. I. E.*, 1989, **28**, 55
120: *J. Chem. Soc. Chem. Comm.*, 1971, 1581
Salzer, A.
267: *Helv. Chim. Acta*, 1987, **70**, 1487
304: *Helv. Chim. Acta*, 1987, **70**, 1487
368: *J. Am. Chem. Soc.*, 1990, **112**, 7113
372: *Helv. Chim. Acta*, 1982, **65**, 1145
Satge, J.
118: *J. Organomet. Chem.*, 1973, **56**, 1
Saveant, J. M.
203: *J. Am. Chem. Soc.* 1978, **100**, 3221
Schafarik, A.
2: *Ann.*, 1859, **109**, 206
Scherer, O. J.
378: *Angew. Chem. I. E.*, 1987, **26**, 59
379: Angew. Chem. *I. E.*, 1985, 24, 351
Schlemper, E. O.
125: *Inorg. Chem*, 1966, **5**, 507, 511
Schlenk, W.
2: *Chem Ber.* 1917, **50**, 262
43: *Chem. Ber.*, 1929, **62**, 920
44: *Chem. Ber.*, 1929, **62**, 920
Schleyer, P.v. R.
22: *Organometallics*, 1988, **7**, 1597
63: *J. Am. Chem. Soc.*, 1981, **103**, 2589
Schmid, G.
385: *Chem. Ber.*, 1981, **114**, 495
Schmidbaur, H.
91: *Angew. Chem. I. E.*, 1985, **24**, 893
92: *Angew. Chem. I. E.*, 1987, **26**, 338
173: *Angw. Chem. I. E.*, 1973, **12**, 415
179: *Acc. Chem. Res.*, 1975, **8**, 62
Schneider, J. J.
332: *Angew. Chem. I. E.*, 1991, **30**,
Schnöckel, H.
86: *Angew. Chem. I. E.*, 1991, **30**, 564
Schoeller, W. W.
107: *Inorg. Chem.*, 1987, **26**, 1081
Schrauzer, G. N.
202: *Angew. Chem. I. E.*, 1976, **15**, 417
203: *Chem. Ber.*, 1964, **97**, 3056
384: *Adv. Organomet. Chem.*, 1964, **2**, 1
Schrock, R. R.
212: *J. Am. Chem. Soc.*, 1975, **97**, 6577

213: *J. Am. Chem. Soc.*, 1975, **97**, 6578
219: *J. Am. Chem. Soc.*, 1980, **102**, 6608
274: *J. Am. Chem. Soc.*, 1979, **101**, 263
Schützenberger, M. P.
2: *Ann.*, 1868, **15**, 100
Schwartz, J.
341: *Angew. Chem. I. E.*, 1976, **15**, 333
Schwarz, K.
171: *J. Am. Chem. Soc.*, 1957, **79**, 3292
Schweig, A.
313: *Chem. Phys. Lett.*, 1986, **124**, 140
Schweiger, A.
349: *Mol. Phys.*,1984, **53**, 585
Schweizer, M. E., 1
Seebach, D.
32: *Helv. Chim. Acta*, 1980, **63**, 2451
Semmelhack, M. F.
268: *J. Am. Chem. Soc.*, 1984, **106**, 2715
354: *J. Organomet. Chem. Library*, 1976, **1**, 361
355: *Pure Appl. Chem.*, 1981, **53**, 2379
Seyferth, D.
54: *J. Am. Chem. Soc.*, 1964, **86**, 2730
164: *Organometallics*, 1982, **1**, 859
Sharpless, K. B.
167: *J. Am. Chem. Soc.*, 1973, **95**, 6137
168: *J. Am. Chem. Soc.*, 1973, **95**, 6137
169: *J. Am. Chem. Soc.*, 1973, **95**, 6137
Sheldrick, G. M.
88: *J. Chem. Soc (A)*, 1970, 28
Shore, S. G., 74, see p. 452: Muetterties
Shriver, D. F.
196: *Acc. Chem. Res.*, 1970, **3**, 231
Siebert, W.
63: *Angew. Chem. I. E.*, 1985, **24**, 759
383: *Angew. Chem. I. E.*, 1977, **16**, 468, 857
383: *Angew. Chem. I. E.*, 1986, **25**, 105
384: *Z. Naturforsch.*, 1987, **42b**, 186
385: *J. Organomet. Chem.*, 1980, **191**, 15
Sienko, M. J.
387: *Inorg. Chem.* 1983, **22**, 3773
Sidgwick, N. V., 186, 396
Skapski, A. C.
391: *J. Chem. Soc. Chem. Comm.*, 1971, 1079
Skell, P. S.
255: *J. Am. Chem. Soc.*, 1974, **96**, 626
344: *J. Am. Chem. Soc.*, 1974, **96**, 1945
344: *J. Am. Chem. Soc.*, 1973, **95**, 3337
Smart, J. C.
330: *J. Am. Chem. Soc.*, 1979, **101**, 3853
Smidt, J.
3: *Angew. Chem.*, 1959, **71**, 284
280: *Angew. Chem.*, 1959, **71**, 284
425: *Angew. Chem.*, 1959, **71**, 176
Sloan, T. E., 460
Sommer, L. H.
114: see p. 453
Sondheimer, F.
176: *J. Am. Chem. Soc.*, 1962, **84**, 260
Sowerby, D. B.
156: *J. Chem. Soc. (Dalton)*, 1979, 1430
Speier, J. L.
96: *Adv. Organomet. Chem.*, 1979, **17**, 407
Stadtman, T. C.
171: *Proc. Natl. Acad. Sci. USA*, 1984, **81**, 57
Steudel, R.
170: *Angew. Chem. I. E.*, 1967, **6**, 653
Stock, A.
66: *J. Chem. Ed.*, 1965, **42**, 41
70: *Chem. Ber.*, 1912, **45**, 3544
Stone, F. G. A.
254: *J. Chem. Soc. (Dalton)*, 1977, 271
402: *J. Chem. Soc.*, Chem. Comm., 1983, 759
402: *J. Chem. Soc. (Dalton)*, 1982, 2475
Strauss, S. H.
181: *J. Amer. Chem. Soc.*, 1991, **113**, 6277
Streitwieser, A.
4: *J. Am. Chem. Soc.*, 1968, **90**, 7364
22: *J. AM. Chem. Soc.*, 1976, **98**, 4778
363: *J. Am. Chem. Soc.*, 1968, **90**, 7364
365: *J. Am. Chem. Soc.*, 1973, **95**, 8644
Strohmeier, W.
145: *Z: Elektrochem.*, 1962, **66**, 823
428: *J. Organomet. Chem.*, 1973, **47**, C37
Stucky, G. D.
46: *J. Am. Chem. Soc.*, 1969, **91**, 2538
47: *J. Organomet. Chem.* 1974, **80**, 7
80: *J. Am. Chem. Soc.*, 1974, **96**, 1941
268: *J. Am. Chem. Soc.*, 1980, **102**, 981
Sutin, N.242
243: *J. Am. Chem. Soc.*, 1980, **102**, 1309

Takats, J.
264: *Inorg. Chem.*, 1976, **15**, 3140
361: *J. Am. Chem. Soc.*, 1976, **98**, 4810
Taube, H.
323: *Inorg. Chem.* 1987, **26**, 1309
Tebboth, J. A.
315: *J. Chem. Soc.*, 1952, 632

Ziegler, K.
3: *Adv. Organomet. Chem.*, 1968, **6**, 1
4: *Angew. Chem.*, 1964, **76**, 545
75: *Adv. Organomet. Chem.*, 1968, **6**, 1
77: *Adv. Organomet. Chem.*, 1968, **6**, 1
78: *Angew. Chem.*, 1964, **76**, 545
423: *Angew. Chem.*, 1955, **67**, 541
Zink, J. I.
246: *J. Am. Chem. Soc.*, 1981, **103**, 2635
Zuckerman, J. J.
133: *Inorg. Chem.*, 1973, **12**, 2522
135: *J. Organomet. Chem.*, 1985, **282**, C1

Subject Index